K. R. Lang

Astrophysical Formulae

Volume II: Space, Time, Matter and Cosmology

Third Enlarged and Revised Edition

With 41 Figures and 22 Tables

Springer

Kenneth R. Lang
Tufts University
Department of Physics and Astronomy
Robinson Hall
Medford, MA 02155, USA

Cover picture: Hot stars with powerful winds shape this glowing gas into a comet-like apparition. It is called a cometary globule, or CG for short, and designated CG 4. This cometary globule is surrounded by a diffuse red glow of hydrogen liberated from it by the radiation of nearby stars. These same stars are probably responsible for creating the green, dusty tail that absorbs some of the light from background reflection nebulosity. Courtesy of David Malin, Anglo-Australian Telescope Board.

Library of Congress Cataloging-in-Publication Data

Lang, Kenneth R. Astrophysical formulae: a compendium for the astronomer, astrophysicist and physicist / Kenneth R. Lang. – 3rd enl. and rev. ed. p. cm. – (Astronomy and astrophysics library, ISSN 0941-7834) Includes bibliographical references and index. Contents: v. 1. Radiation, gas processes, and high energy astrophysics – v. 2. Space, time, matter and cosmology. ISBN 3-540-61267-X (v. 1: alk. paper). – ISBN 3-540-64664-7 (v. 2: alk. paper)
1. Astrophysics–Formulae. I. Title. II. Series. QB461.L36 1998 523.01'02'12–dc21 89-27747 CIP

The 2nd Edition was published as a single volume. This 3rd Edition has 2 volumes.

ISSN 0941-7834
ISBN 3-540-64664-7 3rd Edition (Vol. II) Springer-Verlag Berlin Heidelberg New York
ISBN 3-540-55040-2 2nd Edition Springer-Verlag Berlin Heidelberg New York

Printed in Germany

Typesetting: SPS Madras, India
Cover design: *design & production* GmbH, Heidelberg

10684474 55/3120-5 4 3 2 1 0 - Printed on acid-free paper

ASTRONOMY AND ASTROPHYSICS LIBRARY

Springer
Berlin
Heidelberg
New York
Barcelona
Hong Kong
London
Milan
Paris
Singapore
Tokyo

ASTRONOMY AND ASTROPHYSICS LIBRARY

Theory of Orbits (2 volumes)
Volume 1: Integrable Systems and Non-perturbative Methods
Volume 2: Perturbative and Geometrical Methods
By D. Boccaletti and G. Pucacco

Galaxies and Cosmology
By F. Combes, P. Boissé, A. Mazure and A. Blanchard

The Solar System 2nd Edition By T. Encrenaz and J.-P. Bibring

Nuclear and Particle Physics of Compact Stars
By N. K. Glendenning

The Physics and Dynamics of Planetary Nebulae By G. A. Gurzadyan

Astrophysical Concepts 2nd Edition By M. Harwit

Stellar Structure and Evolution By R. Kippenhahn and A. Weigert

Modern Astrometry By J. Kovalevsky

Observational Astrophysics 2nd Edition
By P. Léna, F. Lebrun and F. Mignard

Supernovae Editor: A. Petschek

General Relativity, Astrophysics, and Cosmology
By A. K. Raychaudhuri, S. Banerji and A. Banerjee

Tools of Radio Astronomy 2nd Edition
By K. Rohlfs and T. L. Wilson

Atoms in Strong Magnetic Fields
Quantum Mechanical Treatment and Applications
in Astrophysics and Quantum Chaos
By H. Ruder, G. Wunner, H. Herold and F. Geyer

The Stars By E. L. Schatzman and F. Praderie

Gravitational Lenses By P. Schneider, J. Ehlers and E. E. Falco

Relativity in Astrometry, Celestial Mechanics and Geodesy By M. H. Soffel

The Sun An Introduction By M. Stix

Galactic and Extragalactic Radio Astronomy 2nd Edition
Editors: G. L. Verschuur and K. I. Kellermann

Reflecting Telescope Optics (2 volumes)
Volume I: Basic Design Theory and its Historical Development
Volume II: Manufacture, Testing, Alignment, Modern Techniques
By R. N. Wilson

Galaxy Formation By M. S. Longair

To Marcella

Ebbene, forse voi credete
che l'arco senza fondo della volta spaziale
sia un vuoto vertiginoso di silenzi.
Vi posso dire allora che verso
questa terra sospettabile appena
l'universo giá dilaga di pensieri.

(Mario Socrate, *Favole paraboliche*)

Well, maybe you think
That the endless arch of the space vault
Is a giddy, silent hollowness.
But I can tell you that,
Overflowing with thought,
the universe is approaching
this hardly guessable earth.

(Mario Socrate, *Parabolic Fables*)

Preface to the Third Edition

When the first edition of **Astrophysical Formulae** was published in 1974, it was designed as a fundamental reference for the student and researcher in the field of astrophysics. Here the reader could find that long-forgotten formula or reference to the more comprehensive article. Anyone interested in learning about a new area of astrophysics could first turn to **Astrophysical Formulae** to obtain an introductory background and guidance to the relevant literature. In all these ways, the book has succeeded far beyond my initial expectations, becoming a widely–used, standard reference.

I am grateful to the many astronomers and astrophysicists who have told me how useful this book has been to them. Such comments have helped sustain my writing and led to this revised, updated version. It could not have been written without the caring, steadfast support of my wife, Marcella. This third edition remains dedicated to her; we courted and were married during the writing and publication of the first edition.

It is more disappointing than astonishing that today's astronomers and astrophysicists so quickly forget the exciting moments of yesterday's science, and often inadequately acknowledge the important work of their predecessors. This can be partly attributed to the accelerated pace of our celestial science; it is difficult to stay abreast of rapid developments in one's own field, let alone others. By providing a comprehensive reference for fundamental formulae and original articles, **Astrophysical Formulae** provides a foundation for overcoming these difficulties. It can also help researchers in all areas to enjoy the development and exciting new discoveries of astronomy and astrophysics.

This completely revised, third edition of **Astrophysical Formulae** more than doubles the number of formulae and references found in the previous versions. There are now more than 4 000 formulae. The references are now gathered together alphabetically for the entire book, rather than by chapter as in previous versions. The authors, titles and journal information have now been provided for more than 5 000 journal articles!

Astrophysical Formulae has now been divided into two volumes. The first **Volume I: Radiation, Gas Processes and High Energy Astrophysics,** contains the classical physics that underlies astrophysics, updated with recent work and references in their applications to the cosmos. **Volume II: Space, Time, Matter and Cosmology** includes many of the topics that are under active scrutiny by contemporary astronomers and astrophysicists.

The first **Volume I: Continuum Radiation, Monochromatic (Line) Radiation, Gas Processes, and High Energy Astrophysics** retains the classical material of previous versions, including the suggestions of the Russian translators of the first edition. Some new sections have now been provided, such as those on helioseismology, solar neutrinos, neutrino oscillations, neutrino emission from stellar collapse and supernovae, and energetic particles and radiation from solar flares. These chapters have additionally been updated with references to important review articles and books. Many of the tables have been removed from these chapters, since more comprehensive tabular information can now be found in Kenneth R. Lang's **Astrophysical Data: Planets and Stars** (New York: Springer-Verlag 1992).

New formulae and supporting data in **Volume I** include, in order of presentation, those dealing with: synchrotron radiation, gyrosynchrotron radiation, synchrotron self-absorption, bremsstrahlung, dispersion measures and time delays of pulsar emission, longitudinal waves, Čerenkov radiation, inverse Compton radiation, interstellar dust, interstellar extinction, interstellar polarization, interstellar scintillation, transition probabilities, emission lines from planetary nebulae and H II regions, recombination lines, LS coupling, 21-cm line of interstellar hydrogen, Fraunhofer lines, solar emission lines, Zeeman effect and cosmic magnetic fields, gravitational redshifts, impact broadening, polytropes, Strömgren radius and its expansion, white dwarf stars and neutron stars, equations of state for degenerate gases, limiting masses for compact objects, rotation of fluid masses, Roche limit, Roche lobes, solar and stellar coronas, solar and stellar winds, heating of the Sun's chromosphere and corona, sound waves and solar oscillations, depth of the solar convection zone, rotation and gas currents inside the Sun, shock waves, gravitational collapse in the presence of a magnetic field, star formation in molecular clouds, magnetohydrodynamic waves, accretion, variable stars, nonradial stellar oscillations, fluid instabilities, pinch instability, magnetic reconnection, fundamental particles, nucleosynthesis, atomic mass excesses, thermonuclear reaction rates, electron screening, neutrinos, neutron star cooling, origin and abundance of the elements, spallation reactions of cosmic ray particles with interstellar matter, formation of lithium, beryllium, and boron, supernova explosions, supernova remnants, supernova SN 1987A, the solar neutrino problem, neutrino astrophysics, gamma-ray lines from solar flares, nonthermal hard X-ray bremsstrahlung of solar flares, particle acceleration in solar flares, observed cosmic rays, ultra-high-energy cosmic rays, shock wave acceleration of cosmic rays, and gamma ray bursts.

New advances, both observational and theoretical, have come at an ever increasing rate, including the discovery of the cosmic microwave background radiation with its anisotropy; improved tests of the special and general theories of relativity; new measurements of the extragalactic distance scale and age of the Universe; the discovery of, and search for, invisible dark matter; the discovery of the binary pulsar and its implications for gravitational radiation; the mature development of radio and X-ray astronomy; the discovery of gravitational lenses; the development of theories for accretion onto compact objects; a new knowledge of white dwarfs, neutron stars and pulsars; the discovery of candidate black holes at the centers of galaxies and the establishment of a comprehensive theory for black holes; a new inflationary model for the earliest stages of the Universe; the discovery of superluminal motions in quasars and nearby micro-quasars; the development of theories for active galactic nuclei; improved computations of big bang nucleosynthesis; new observations of deuterium and helium throughout the cosmos; and dramatic improvements in our knowledge of the shape, structure, content and formation of the Universe. All of these comparatively recent developments are included in a completely new **Volume II: Space, Time, Matter and Cosmology**. I am indebted to my colleagues for the opportunity to learn about so many captivating new ideas and discoveries.

Over one hundred of the seminal contributions to twentieth century astronomy and astrophysics, through to the year 1975, are now reproduced in, **A Source Book in Astronomy and Astrophysics 1900–1975** by Kenneth R. Lang and Owen Gingerich (Cambridge, Mass.: Harvard University Press 1979). It contains a dozen new English translations of articles originally in German, French or Dutch, and includes a historical introduction to each. It is heartening to realize that every article in the **Source Book** was previously referenced in the first edition of **Astrophysical Formulae**, and that several of these articles include the work of subsequent winners of the Nobel Prize in Physics – including Subramanyan Chandrasekar, William A. Fowler, Antony Hewish, Arno A. Penzias, Sir Martin Ryle, and Robert F. Wilson. When the **Source Book** is updated to the turn of the millennium in the year 2000, it will surely consist of articles referenced in this third edition of **Astrophysical Formulae**.

The library at the Harvard-Smithsonian Center for Astrophysics has been an invaluable resource, and special thanks go to Seth Redfield for helping to locate material there.

I have profited from the advice of many experts who have read individual sections of this version. Those who have read portions of the new material and have supplied critical comment include: Neil Ashby, John N. Bahcall, Roger D. Blandford, Margaret Geller, Jack Harvey, Mark Haugan, Bruce Partridge, Martin Rees, David Schramm, and Kenneth Seidelmann. Persons who have provided suggestions for new formulae and references include: E. H. Avrett, K. Borkowski, R. Catchpole, K. P. Dere, T. Forbes, J. Franco, V. G. Gurzadyan, H. J. Haubold, M. Hernanz, R. W. John, K. I. Kellermann, K. Krisciunas, W. Kundt, M. A. Lee, J. Linsley, G. P. Malik, W. C. Martin, J.-M. Perrin, V. A. Razin, D. O. Richstone, N. N. Shefov, B. V. Somov, J. Terrell, R. Woo, S. Woosley, and J. Yang.

Medford
October, 1998

Kenneth R. Lang

Preface

This book is meant to be a reference source for the fundamental formulae of astrophysics. Wherever possible, the original source of the material being presented is referenced, together with references to more recent modifications and applications. More accessible reprints and translations of the early papers are also referenced. In this way the reader is provided with the often ignored historical context together with an orientation to the more recent literature. Any omission of a reference is, of course, not meant to reflect on the quality of its contents. In order to present a wide variety of concepts in one volume, a concise style is used and derivations are presented for only the simpler formulae. Extensive derivations and explanatory comments may be found in the original references or in the books listed in the selected bibliography which follows. Following the convention in astrophysics, the c.g.s (centimeter-gram-second) system of units is used unless otherwise noted. To conserve space, the fundamental constants are not always defined, and unless otherwise noted they have the meaning and value given in the tables of physical constants and astrophysical constants provided at the beginning and end of this book.

A substantial fraction of this book was completed during two summers as a visiting fellow at the Institute of Theoretical Astronomy, Cambridge, and I am especially grateful for the hospitality and courtesy which the members of the Institute have shown me. I am also indebted to the California Institute of Technology for the freedom to complete this book. The staff of the scientific periodicals library of the Cambridge Philosophical Society and the library of the Hale observatories are especially thanked for their aid in supplying and checking references. Those who have read portions of this book and have supplied critical comment and advice include Drs. L. H. Aller, H. Arp, R. Blandford, R. N. Bracewell, A. G. W. Cameron, E. Churchwell, D. Clayton, W. A. Fowler, J. Greenstein, H. Griem, J. R. Jokipii, B. Kuchowicz, M. G. Lang, A. G. Michalitsanos, R. L. Moore, E. N. Parker, W. H. Press, M. J. Rees, J. A. Roberts, J. R. Roy, W. L. W. Sargent, W. C. Saslaw, M. Schmidt, I. Shapiro, P. Solomon, E. Spiegel, and S. E. Woosley.

Medford
October, 1974

Kenneth R. Lang

Contents

Contents for Volume I

5. Space, Time, Matter and Cosmology

"There is not, in strictness speaking, one fixed star in the heavens, ... there can hardly remain a doubt of the general motion of all the starry systems, and consequently of the solar one amongst the rest."

Sir F.W. Herschel, 1753

"A luminous star, of the same density as the Earth, and whose diameter should be two hundred and fifty times larger than that of the Sun, would not, in consequence of its attraction, allow any of its rays to arrive at us. It is therefore possible that the largest luminous bodies in the Universe may, through this cause, be invisible."

P.S. Laplace, 1796

"The laws of physical phenomena must be the same for a fixed observer as for an observer who has a uniform motion of translation relative to him.... There must arise an entirely new kind of dynamics which will be characterized above all by the rule that no velocity can exceed the velocity of light."

H. Poincaré, 1904

5.1 Position

5.1.1 Location on the Earth's Surface

The geodetic, or geographic, surface position is defined by a grid of great circles on a spherical Earth. A great circle divides the sphere in half, and the name comes from the fact that no greater circles can be drawn on a sphere. A great circle halfway between the North and South Poles is called the equator, because it is equally distant between both poles.

Circles of longitude are great circles that pass around the Earth from pole to pole perpendicular to the equator (Fig. 5.1). Each circle of longitude intersects the equator in two points that are 180 degrees apart. Geographers decided, in 1884, that the half-circle corresponding to 0 degrees longitude corresponds to the circle of longitude passing through the old Royal Observatory in Greenwich, England; it was designated as the "Prime Meridian", the starting point for counting longitudes. The longitude of any point on the Earth's surface is the angle, λ, measured westward from the intersection of the Prime Meridian with the equator to the equatorial intersection of the circle of longitude that passes through the point.

The latitude is the angle, ϕ, measured northward (positive) or southward (negative) along the circle of longitude from the equator to the point. To complete the geodetic, or geographic, description of a point on the Earth's

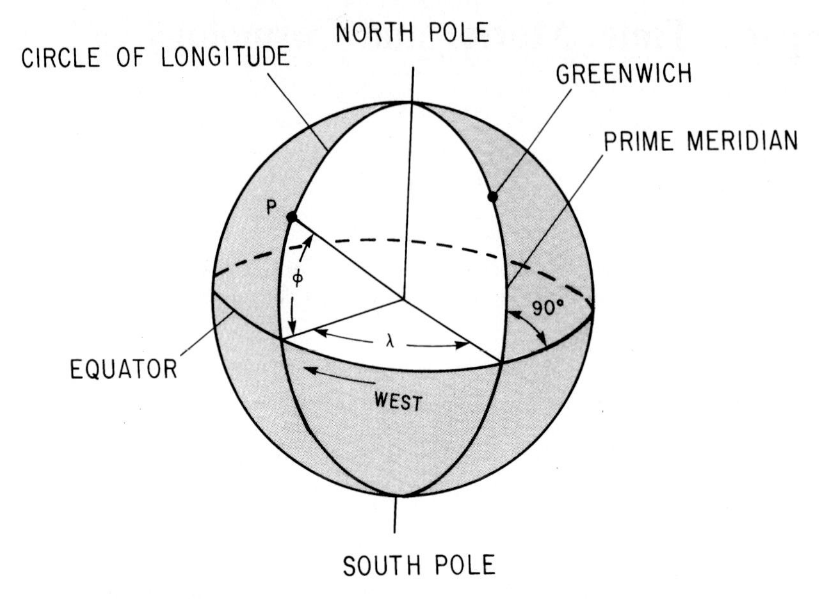

Fig. 5.1. Great circles through the poles create circles of longitude. They are perpendicular to the equator where they intersect it. The circle of longitude that passes through Greenwich England is called the Prime Meridian. The longitude of any point, P, is the angle lamda, λ, measured westward along the equator from the intersection of the Prime Meridian with the equator to the equatorial intersection of the circle of longitude that passes through the point. The latitude is the angle phi, ϕ, measured northward (positive) or southward (negative) along the circle of longitude from the equator to the point. In this example, the point P corresponds to San Francisco

surface, one specifies the height, h, above sea level. So the geodetic coordinates are λ, ϕ, and h. The geodetic longitude, latitude and height of all observatories currently engaged in professional programs of astronomical observations are given in *The Astronomical Almanac*, where they are listed in alphabetical order by geographical location and indexed according to the observatory name. In this tabulation, east longitudes and north latitudes are considered to be positive.

Portable radio receivers are used to determine precise locations anywhere on land or sea, or in the air. They detect signals from a satellite network of orbiting atomic clocks, each constantly beaming radio announcements of the exact time. An example is the 24-satellite Air Force Global Positioning System. The satellite's signals take a few milliseconds to arrive, and the delay is converted into distance by multiplication with the velocity of light. Then, the time differences from several satellites are translated by triangulation into an exact position accurate to about one hundred meters.

At geodetic latitude ϕ:

$$1 \text{ degree of latitude} = (110.575 + 1.110 \sin^2 \phi) \times 10^5 \text{ cm}$$

$$1 \text{ degree of longitude} = [(111.320 + 0.373 \sin^2 \phi) \cos \phi] \times 10^5 \text{ cm} \ . \quad (5.1)$$

The geodetic coordinates are referred to a reference ellipsoid (an ellipse of revolution), which is normally geocentric, and is defined by its equatorial radius, a_e, and flattening, f. That is, for geodetic reference purposes, the Earth's surface at sea level may be represented by revolving an ellipse of eccentricity, e, and major axis, a_e, about the polar axis. The flattening factor, f, is related to the eccentricity, e, by the equation

$$e^2 = 2f - f^2 \ . \tag{5.2}$$

The flattening factor is alternatively called the ellipticity or the oblateness factor. The major axis, a_e, is the equatorial radius, which is related to the polar radius, a_p, by the equation

$$a_p = a_e(1 - f) = a_e(1 - e^2)^{\frac{1}{2}} \ , \tag{5.3}$$

and the mean radius, $\langle a \rangle$, is given by

$$\langle a \rangle = (a_e^2 a_p)^{\frac{1}{3}} \ . \tag{5.4}$$

The International Astronomical Union (IAU) 1976 system of astronomical constants includes (Lang, 1992)

$$\begin{aligned} a_e &= 6.378140 \times 10^8 \text{ cm} \\ a_p &= 6.356755 \times 10^8 \text{ cm} \\ f &= 0.00335281 = 1/298.257 \ . \end{aligned} \tag{5.5}$$

Values of the equatorial radius and flattening of other geodetic reference spheroids are given by Seidelmann (1992) and in *The Astronomical Almanac.*

The radius, r, of the surface of the Earth geoid is specified by the equation

$$r = a_e(1 - f \sin^2 \phi) \ . \tag{5.6}$$

The Earth's surface may be represented by a geoid of constant potential, U, given by

$$U = V - \frac{1}{2}\omega^2 r^2 \cos^2 \phi \tag{5.7}$$

where r and ϕ are, respectively, the radius and geodetic latitude of a point on the surface, the angular velocity, or rate of rotation, of the Earth is

$$\omega = 7.292115 \times 10^{-5} \text{ rad s}^{-1} \ , \tag{5.8}$$

and the gravitational potential, V, satisfies Laplace's equation in polar coordinates (Laplace, 1782, 1817)

$$\nabla^2 V = \frac{1}{r^2}\frac{\partial}{\partial r}\left(r^2 \frac{\partial V}{\partial r}\right) + \frac{1}{r^2 \sin\theta}\frac{\partial}{\partial\theta}\left(\sin\theta \frac{\partial V}{\partial\theta}\right) + \frac{1}{r^2 \sin^2\theta}\frac{\partial^2 V}{\partial\lambda^2} = 0 \ , \tag{5.9}$$

where $\theta = \pi/2 - \phi$, and the term involving the longitude, λ, may be discounted by assuming rotational symmetry about the z axis. Laplace's equation may then be solved to give (Legendre, 1789, 1817)

$$V = -\frac{GM}{r}\sum_{n=0}^{\infty} J_n \left(\frac{a_e}{r}\right)^n P_n(\theta) \ , \tag{5.10}$$

where the zonal harmonics, J_n, are dimensionless constants, the $P_n(\theta)$ are Legendre polynomials, the Newtonian gravitational constant

$$G = 6.672 \times 10^{-8}\,\text{dyn}\,\text{cm}^{-2}\,\text{g}^{-2} \ , \tag{5.11}$$

and the mass of the Earth is (Lang, 1992)

$$M = 5.9742 \times 10^{27}\ \text{g} \ . \tag{5.12}$$

Assuming that the origin of the coordinate system is the center of mass, $J_1 = 0$, and taking $J_0 = 1$, the solution to Laplace's equation for the Earth takes the approximate form

$$V \approx -\frac{GM}{r} + \frac{GMa_e^2 J_2}{2r^3}(3\sin^2\phi - 1) \ . \tag{5.13}$$

Under the assumption that U is everywhere a constant, it follows that

$$f = \frac{a_e - a_p}{a_e} = \frac{3}{2}J_2 + \frac{1}{2}m \ , \tag{5.14}$$

where the ratio of centrifugal acceleration at the equator to the gravitational acceleration at the equator is given by

$$m = \frac{\omega^2 a_e^3}{GM} \ . \tag{5.15}$$

For the Earth, the dynamical form factor, J_2, is (Lang, 1992)

$$J_2 = 0.001082626 \ , \tag{5.16}$$

and other harmonic coefficients for the Earth are:

$$J_3 = -0.254 \times 10^{-5} \ , \tag{5.17}$$

and

$$J_4 = -0.161 \times 10^{-5} \ . \tag{5.18}$$

Detailed formulae for measuring the higher coefficients are given by (Lecar, Sorenson, and Eckels, 1959; O'Keefe, Eckels, and Squires, 1959; King-Hele, 1961; Kozai, 1959, 1961, 1966), and in the book by Kaula (1966).

The surface gravitational acceleration, g, is given by

$$g = -\nabla U \ . \tag{5.19}$$

It follows from the previous equations that

$$g = g_e\left[1 + \left(\frac{5m}{2} - f\right)\sin^2\phi\right] \ , \tag{5.20}$$

where the surface equatorial gravity is given by

$$g_e = \frac{-GM}{a_e^2}\left(1 + \frac{3}{2}J_2 - m\right) \ . \tag{5.21}$$

This equation is known as Clairaut's theorem after Clairaut's (1743) derivation of the result by assuming that the rotating body is a homogeneous fluid ellipsoid. By retaining higher terms, the following result is obtained.

$$g = g_e\left[1 + \left(\frac{5}{2}m - f - \frac{17}{14}mf\right)\sin^2\phi + \left(\frac{f^2}{8} - \frac{5}{8}mf\right)\sin^2 2\phi\right] \ . \tag{5.22}$$

Using the values of f, m, J_2, and a_e given in the previous equations for the Earth, we obtain,

$$g_e = 978.0327 \text{ cm s}^{-2} \ , \tag{5.23}$$

and the surface gravity at sea level and geodetic (geographic) latitude ϕ is (Lang, 1992)

$$g = g_e(1 + 0.0053024 \sin^2 \phi - 0.0000058 \sin^2 2\phi) \ . \tag{5.24}$$

The position of a point on the Earth can also be given in relationship to the Earth's center by specifying the geocentric longitude, latitude, and distance, λ, ϕ', ρ, or the geocentric equatorial rectangular coordinates, x, y, z.

The difference between the geodetic latitude, ϕ, and the geocentric latitude, ϕ', is given by

$$\phi - \phi' = (692.74 \sin 2\phi - 1.16 \sin 4\phi) \quad \text{seconds of arc} \ . \tag{5.25}$$

Conversion from geodetic (geographic) to geocentric coordinates is straight forward using the following relationships (Seidelmann, 1992):

$$\begin{aligned} x &= \rho \cos \phi' \cos \lambda = (a_e C + h) \cos \phi \cos \lambda \ , \\ y &= \rho \cos \phi' \sin \lambda = (a_e C + h) \cos \phi \sin \lambda \ , \\ z &= \rho \sin \phi' = (a_e S + h) \sin \phi \end{aligned} \tag{5.26}$$

where a_e is the equatorial radius of the spheroid, and C and S are auxiliary functions that depend on the geodetic latitude and the flattening f of the reference spheroid. It follows from the properties of the ellipse that

$$C = (\cos^2 \phi + (1 - f)^2 \sin^2 \phi)^{-\frac{1}{2}}, \qquad S = (1 - f)^2 C \ . \tag{5.27}$$

The geocentric distance, or radius, is often given in units of the equatorial radius of the reference spheroid, obtained by dividing the value given here by a_e.

Series expansions, which contain terms up to f^3, for S, C, ρ, and $\phi - \phi'$ are

$$\begin{aligned} S &= 1 - \frac{3}{2} f + \frac{5}{16} f^2 + \frac{3}{32} f^3 - \left(\frac{1}{2} f - \frac{1}{2} f^2 - \frac{5}{64} f^3 \right) \cos 2\phi \\ &\quad + \left(\frac{3}{16} f^2 - \frac{3}{32} f^3 \right) \cos 4\phi - \frac{5}{64} f^3 \cos 6\phi \ , \\ C &= 1 + \frac{1}{2} f + \frac{5}{16} f^2 + \frac{7}{32} f^3 - \left(\frac{1}{2} f + \frac{1}{2} f^2 + \frac{27}{64} f^3 \right) \cos 2\phi \\ &\quad + \left(\frac{3}{16} f^2 + \frac{9}{32} f^3 \right) \cos 4\phi - \frac{5}{64} f^3 \cos 6\phi \ , \\ \rho &= 1 - \frac{1}{2} f + \frac{5}{16} f^2 + \frac{5}{32} f^3 + \left(\frac{1}{2} f - \frac{13}{64} f^3 \right) \cos 2\phi \\ &\quad - \left(\frac{5}{16} f^2 + \frac{5}{32} f^3 \right) \cos 4\phi + \frac{13}{64} f^3 \cos 6\phi \ , \\ \phi - \phi' &= \left(f + \frac{1}{2} f^2 \right) \sin 2\phi - \left(\frac{1}{2} f^2 + \frac{1}{2} f^3 \right) \sin 4\phi + \frac{1}{3} f^3 \sin 6\phi \ . \end{aligned} \tag{5.28}$$

The expressions for ρ and $\phi - \phi'$ are for points on the spheroid ($h = 0$), and the latter quantity is sometimes known as the "reduction in latitude" or "the angle of the vertical," and it is of order 10′ in midlatitudes. To a first approximation when h is small, the geocentric radius is increased by h/a and the angle of the vertical is unchanged. The height h refers to a height above the reference spheroid and differs from the height above mean sea level (i.e., above the geoid) by the "undulation of the geoid" at the point.

Transformation of geocentric to geodetic coordinates is not a simple matter. Borkowski (1987) provides an exact solution to this problem, algorithms for the transformation are given by Borkowski (1989), and the exact solution is reproduced in Seidelmann (1992). It is based on using an expression for the reduced latitude in a solvable fourth-degree polynomial.

Several intermediate values are computed in order to perform a solution:

$$
\begin{aligned}
r &= \sqrt{x^2 + y^2} = (a_e C + h)\cos\phi \\
E &= \frac{\left[a_p z - \left(a_e^2 - a_p^2\right)\right]}{a_e r} \ , \\
F &= \frac{\left[a_p z + \left(a_e^2 - a_p^2\right)\right]}{a_e r} \ , \\
P &= \tfrac{4}{3}(EF + 1) \ , \\
Q &= 2(E^2 - F^2) \ , \\
D &= P^3 + Q^2 \ , \\
v &= \left(D^{\frac{1}{2}} - Q\right)^{\frac{1}{3}} - \left(D^{\frac{1}{2}} + Q\right)^{\frac{1}{3}} \ , \\
G &= \frac{1}{2}\left[\sqrt{E^2 + v} + E\right] \ , \\
t &= \sqrt{\left[G^2 + \frac{F - vG}{2G - E}\right]} - G \ .
\end{aligned}
\tag{5.29}
$$

Finally, the latitude and ellipsoidal height are computed from:

$$
\begin{aligned}
\phi &= \arctan\left[a_e \frac{1 - t^2}{2 a_p t}\right] \\
h &= (r - a_e t)\cos\phi + (z - a_p)\sin\phi \ .
\end{aligned}
\tag{5.30}
$$

If $D < 0$, e.g., if less than about 45 km from the Earth's center, the following equation should be used for v in order to avoid the use of complex numbers:

$$
v = 2\sqrt{-P}\cos\left\{\frac{1}{3}\arccos\left[\frac{Q}{P}(-P)^{-\frac{1}{2}}\right]\right\} \ . \tag{5.31}
$$

To obtain the proper sign (and solution of the fourth-degree polynomial since up to four solutions actually exist), the sign of a_p should be set to that of z before beginning. Borkowski also notes that, of course, this solution is singular for points at the z-axis ($r = 0$) or on the xy-plane ($z = 0$). Additionally for points close to those conditions, some roundoff error may be avoided and the accuracy improved slightly by replacing the value of v with

$$-\frac{v^3 + 2Q}{3P} \ . \tag{5.32}$$

Finally,

$$\lambda = \arctan\frac{y}{x} \ . \tag{5.33}$$

An iterative procedure for calculating the geodetic coordinates from x, y and z is also given in *The Astronomical Almanac*.

5.1.2 The Celestial Sphere and Astronomical Coordinates

Astronomical coordinates are measured on the celestial sphere, an imaginary sphere concentric with the Earth, on which the stars and other astronomical objects are placed (Fig. 5.2). The celestial poles are the pivotal points of the sky's apparent daily rotation, and the celestial equator is the Earth's equator projected onto the sky. The point where the Sun crosses the celestial equator going northward in spring is called the Vernal Equinox. The Vernal Equinox is sometimes called the first point of Aries, and is given the symbol γ.

On the equinox the Sun lies in the Earth's equatorial plane, so the twilight zone that separates night and day then cuts the Earth in equal parts and the Sun rises due east at all points on the Earth's surface. Thus, on the Vernal, or Spring, Equinox (March 21) the day and night are equally long, except for a small difference due to the diameter of the Sun, as they are on the Autumnal Equinox (September 23). The solstices mark the dates when the Sun is farthest from the equator. On June 21, the Sun is farthest north and the days in the northern hemisphere are longest; on December 22, the Sun is farthest south and the days in the northern hemisphere are shortest.

Positions on the celestial sphere are defined by angles along great circles. By analogy with terrestrial longitude, right ascension, α, is a star's celestial longitude, but it is measured eastward along the celestial equator from the Vernal Equinox. The right ascension is expressed in hours and minutes of time, with twenty-four hours in the complete circle of 360 degrees. For conversion between time and angle:

$$\begin{aligned}
&1 \text{ hour of time or } 1\text{ h} = 15 \text{ degrees or } 15^\circ \\
&1 \text{ minute of time or } 1\text{ m} = 15 \text{ minutes of arc or } 15' \\
&1 \text{ second of time or } 1\text{ s} = 15 \text{ seconds of arc or } 15'' \\
&1 \text{ degree or } 1^\circ = 4 \text{ minutes of time or } 4\text{ m} \\
&1 \text{ minute of arc or } 1' = 4 \text{ seconds of time or } 4\text{s} \ .
\end{aligned} \tag{5.34}$$

Just as latitude is a measure of an object's distance from the equator of the Earth, declination, δ, is a star's angular distance from the celestial equator. It is positive in the north and negative in the south. Right ascension, α, and declination, δ, specify a position on the celestial sphere in equatorial coordinates.

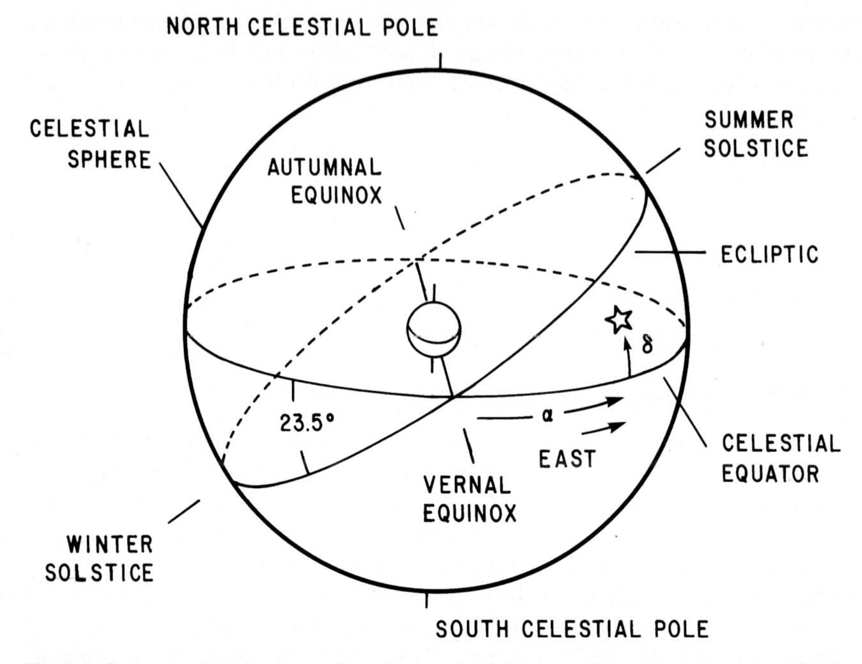

Fig. 5.2. Stars are placed upon an imaginary celestial sphere. The celestial equator divides the sky into northern and southern halves, and the ecliptic is the apparent annual path of the Sun on the celestial sphere. The celestial equator crosses the ecliptic at the equinoxes. Every celestial object has two star coordinates, called equatorial coordinates; they are right ascension and declination. Right ascension is the angle alpha, α, measured eastward along the celestial equator from the Vernal Equinox to the foot of the great circle that passes through the object. Declination is the angular distance delta, δ, from the celestial equator to the object along the great circle that passes through the object (positive to the north and negative to the south)

At a location having the latitude ϕ, a star of declination δ reaches a maximum altitude, $h_{\max}$, at its upper culmination and a minimum altitude, $h_{\min}$, at its lower culmination given by

$$\begin{aligned} h_{\max} &= 90^\circ - |\phi - \delta| \\ h_{\min} &= -90^\circ + |\phi + \delta| \ . \end{aligned} \tag{5.35}$$

Stars with $\delta > 90^\circ - \phi$ always remain above the horizon (circumpolar stars); those with $\delta < -(90^\circ - \phi)$ never rise above the horizon.

The convention of measuring right ascension eastward was chosen because it makes the celestial sphere into the face of a star clock. The hand of the clock is the local meridian, the north-south line passing through the observer's zenith. When the Vernal Equinox is on the local meridian, the Local Sidereal Time, or star time, is said to be 0 hours. At that moment, the right ascension of stars on the meridian is also 0 hours. As each succeeding hour of Sidereal Time passes, the right ascension of the stars on the meridian increases one hour.

At any moment, the Local Sidereal Time equals the right ascension of stars on the local meridian. And if we were to travel instantaneously east, the right ascension of stars overhead would increase at the rate of 1 hour of right ascension for every 15 degrees of terrestrial longitude.

In other words, the Local Sidereal Time is the hour angle of the Vernal Equinox, and

$$\text{Hour Angle} = \text{Local Sidereal Time} - \text{Right Ascension}$$

or

$$\mathrm{HA} = \mathrm{LST} - \alpha \ . \tag{5.36}$$

As the Earth rotates, the stars slide by and the celestial sphere appears to rotate once each day around the celestial axis (passing through the celestial North and South Poles). However, the local terrestrial clock is based on the rising and setting Sun, or Universal Time. It ticks ahead 24 hours of Sidereal Time in 23 hours 56 minutes 04 seconds of Universal Time (also see Sect. 5.3.7).

The Vernal Equinox will be on the local meridian on midnight of the day of Autumnal Equinox (approximately September 23), the Local Sidereal Time, or Star Time, will be exactly 0 hours, and the terrestrial clock (Universal Time or Sun Time) will also read zero. The next night at midnight, the local star time will be 0 hours 3 minutes 56 seconds; and midnight on the celestial clock will increase 2 hours for each succeeding month.

There are four sets of astronomical coordinates, called the ecliptic, equatorial, galactic and horizontal coordinate systems, that refer to different points and great circles on the celestial sphere.

The ecliptic coordinate system: The celestial latitude, β, of the celestial object is the angle in degrees from the ecliptic to the object as measured along the circle of celestial latitude which passes through the object. The celestial latitude is positive or negative, respectively, when measured north or south of the ecliptic. The celestial longitude, λ, is the angle in degrees measured eastward along the ecliptic from the Vernal Equinox to the foot of the circle of celestial latitude which passes through the object.

The equatorial coordinate system: The declination, δ, of the celestial object is the angle in degrees from the celestial equator to the object as measured along the hour circle which passes through the object. The declination is positive or negative, respectively, when measured north or south of the celestial equator. The right ascension, α, is the angle in degrees (or hours) measured from the Vernal Equinox along the celestial equator toward the east to the foot of the hour circle which passes through the object. The hour angle is the angular distance in degrees (or hours) measured along the celestial equator from the foot of the meridian toward the west to the foot of the hour circle which passes through the object.

The galactic coordinate system: The galactic latitude, b, of the celestial object is the angle in degrees from the galactic equator to the object as measured along the circle of galactic latitude which passes through the object. The galactic latitude is positive or negative, respectively, when measured north or south of the galactic equator. The galactic longitude, l, is the angle in degrees measured eastward along the galactic equator from the galactic center to the foot of the circle of galactic latitude which passes through the object.

The distribution of open star clusters in galactic coordinates is shown in Fig. 5.3. It lies in the galactic plane. The globular star-cluster distribution is illustrated in Fig. 5.4. It is centered about the galactic center.

In 1958 the International Astronomical Union adopted the equatorial coordinates for the epoch 1950.0 (Blauuw, Gum, Pawsey, Westerhout, 1960)

Galactic Center

$$\alpha\,(1950.0) = 17\,\mathrm{h}\,42.4\,\mathrm{min}$$
$$\delta\,(1950.0) = -28.92 \text{ degrees} \tag{5.37}$$

Galactic North Pole

$$\alpha\,(1950.0) = 12\,\mathrm{h}\,49\,\mathrm{min}$$
$$\delta\,(1950.0) = +27.40 \text{ degrees} \tag{5.38}$$

The position angle, θ, of the galactic center from this pole is $\theta = 123°$. Murray (1989) transforms this position for the pole to the epoch J2000.0:

Galactic North Pole

$$\alpha\,(2000.0) = 12\,\mathrm{h}\ 51\,\mathrm{min}\ 26.2755\,\text{seconds} \tag{5.39}$$
$$\delta\,(2000.0) = 27°07'\,41.704''$$

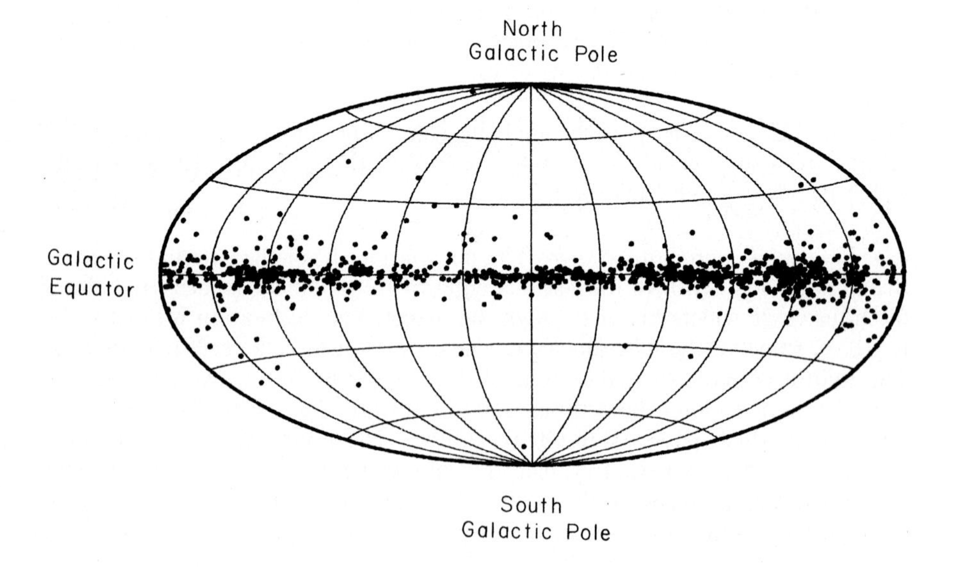

Fig. 5.3. Distribution of open clusters in galactic coordinates. These relatively young clusters of stars lie along the galactic plane, the site of current star formation

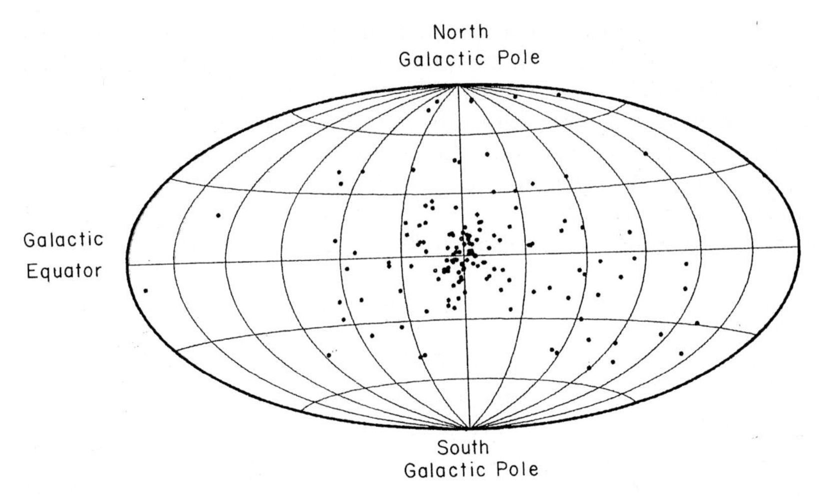

Fig. 5.4. Distribution of globular clusters in galactic coordinates. These old clusters of stars date back to the origin of our Galaxy. They are distributed within a spherical galactic halo that is centered at the galactic center, about 8.5 kiloparsecs from the Sun

There is a slight asymmetry between star counts at the north and south Galactic poles, with more stars visible toward the south pole of the Galaxy than toward the north. This means that the Sun lies north of the galactic plane. Cohen (1995) combined infrared and ultraviolet star counts, from the InfraRed Astronomical Satellite, IRAS, and the FAr Ultraviolet Space Telescope, FAUST, to obtain a solar distance of

$$z_0 = 15.0 \pm 0.5 \,\text{parsecs north} \tag{5.40}$$

above the galactic plane. Hammersley et al. (1995) used data from the Cosmic Microwave Background Explorer, COBE, the Two-Micron Galactic Survey, TMGS, and IRAS to find

$$z_0 = 15.5 \pm 3 \,\text{parsecs north} \ , \tag{5.41}$$

or about 50.5 ± 10 light-years above the plane of our Galaxy. Humphreys and Larsen (1995) find that the Sun lies 67 ± 11 light-years north of that plane.

The local supercluster: De Vaucouleurs (1953, 1958, 1959, 1960) has presented evidence for a Local Supergalaxy, or Local Supercluster. The majority of bright galaxies lie within ± 30 degrees from the equator of this system. The north supergalactic pole is at $l = 15$ degrees, $b = +5$ degrees, and the origin of the supergalactic longitude is at $l = 105$ degrees and $b = 0$ degrees. The plane of the Local Supercluster contains several of the nearest clusters of galaxies, including the Virgo, Hydra, Centaurus, Ursa Major, Perseus and Pavo clusters (Peebles, 1993); this plane forms a great circle on the sky, that runs approximately perpendicular to the Milky Way.

The horizontal or horizon coordinate system: The altitude, a, of the celestial object is the angle in degrees from the horizon to the object as measured along the vertical circle which passes through the object. The altitude is positive or negative, respectively, when measured above or below the horizon. The zenith distance, z, is $90^\circ - a$. The azimuth, A, is the angle in degrees measured from the north point through east along the horizon to the foot of the vertical circle which passes through the object.

The reference points and great circles defining the four astronomical coordinate systems are:

Celestial Equator: The great circle defined by the intersection of the celestial sphere and the projection of the plane of the Earth's equator.

Circle of Celestial Latitude: Any great circle which passes through the poles of the ecliptic, and is therefore perpendicular to the ecliptic.

Circle of Galactic Latitude: Any great circle which passes through the galactic poles and is therefore perpendicular to the galactic equator.

Ecliptic: The mean great-circle path of the Sun on the celestial sphere during the course of a year. The plane of the ecliptic is inclined at a mean angle of $23^\circ 26' 21.448''$, for epoch J2000.0, to the plane of the celestial equator. More exact values of this inclination are tabulated as the obliquity of the ecliptic in the *The Astronomical Almanac*.

Galactic Center: The center of our Galaxy. The equatorial coordinates of the galactic center are given in the preceding section on the galactic coordinate system.

Horizon: The great circle defined by the intersection of the celestial sphere and that plane which is perpendicular to the observer's plumb line at the position of the observer.

Hour Circle: Any great circle which passes through the north and south celestial poles and is therefore perpendicular to the celestial equator.

Hour Angle: Local Sidereal Time – Right Ascension.

Meridian: The great circle which is perpendicular to the horizon and passes through the zenith and the north celestial pole.

North and South Celestial Poles: The respective points of intersection of the celestial sphere with the northward and southward prolongation of the Earth's axis of rotation.

North Galactic Pole: The point of intersection of the celestial sphere with the northward prolongation of the rotation axis of the Galaxy. Its equatorial coordinates are given in the preceding section on the galactic coordinate system.

North Point: The northward point of intersection of the meridian and the horizon.

Vernal Equinox: That point of intersection of the ecliptic and the celestial equator which occurs when the Sun is going from south to north. The Vernal Equinox is sometimes called the first point of Aries. The hour angle and transit time of the Vernal Equinox at the Greenwich observatory are tabulated in *The Astronomical Almanac*.

Zenith and Nadir: The respective points of intersection of the celestial sphere with the upward and downward prolongation of the observer's plumb line.

5.1.3 Transformation of Astronomical Coordinates

The formulae connecting the galactic longitude, l, and galactic latitude, b, to the right ascension, α, and declination, δ, are given below. They have been used by the Lund Observatory (1961) to provide tables for the conversion of coordinates in the epoch 1950.0 using the galactic center and galactic north pole given in the preceding section on the galactic coordinate system. A graphical conversion chart between equatorial (1950.0) and galactic coordinates is shown in Fig. 5.5.

Equatorial to Galactic Coordinate Conversion

Right ascension, α Galactic latitude, b

Declination, δ Galactic longitude, l

$$\begin{aligned}
&\cos b\cos(l-33^\circ) = \cos\delta\cos(\alpha-282.25^\circ),\\
&\cos b\sin(l-33^\circ) = \cos\delta\sin(\alpha-282.25^\circ)\cos 62.6^\circ + \sin\delta\sin 62.6^\circ,\\
&\sin b = \sin\delta\cos 62.6^\circ - \cos\delta\sin(\alpha-282.25^\circ)\sin 62.6^\circ,\\
&b = \sin^{-1}\{\cos\delta\cos 27.4^\circ\cos(\alpha-192.25^\circ) + \sin\delta\sin 27.4^\circ\},\\
&l = \tan^{-1}\{\sin\delta - \sin b\sin 27.4^\circ/[\cos\delta\sin(\alpha-192.25^\circ)\cos 27.4^\circ]\} + 33^\circ
\end{aligned} \tag{5.42}$$

Galactic to Equatorial Coordinate Conversion

Galactic latitude, b Right ascension, α

Galactic longitude, l Declination, δ

$$\begin{aligned}
&\cos\delta\sin(\alpha-282.25^\circ) = \cos b\sin(l-33^\circ)\cos 62.6^\circ - \sin b\sin 62.6^\circ,\\
&\sin\delta = \cos b\sin(l-33^\circ)\sin 62.6^\circ + \sin b\cos 62.6^\circ,\\
&\delta = \sin^{-1}\{\cos b\cos 27.4^\circ\sin(l-33^\circ) + \sin b\sin 27.4^\circ\},\\
&\alpha = \tan^{-1}\{\cos b\cos(l-33^\circ)/[\sin b\cos 27.4^\circ - \cos b\sin 27.4^\circ\\
&\qquad - \sin(l-33^\circ)]\} + 192.25^\circ
\end{aligned} \tag{5.43}$$

Ecliptic to Equatorial Coordinate Conversion

The formulae connecting the longitude, λ, and latitude, β, to the right ascension, α, and declination, δ, are

$$\begin{aligned}
\cos\delta\cos\alpha &= \cos\beta\cos\lambda\\
\cos\delta\sin\alpha &= \cos\beta\sin\lambda\cos\varepsilon - \sin\beta\sin\varepsilon\\
\sin\delta &= \cos\beta\sin\lambda\sin\varepsilon + \sin\beta\cos\varepsilon\\
\cos\beta\sin\lambda &= \cos\delta\sin\alpha\cos\varepsilon + \sin\delta\sin\varepsilon\\
\sin\beta &= \sin\delta\cos\varepsilon - \cos\delta\sin\alpha\sin\varepsilon\ ,
\end{aligned} \tag{5.44}$$

where ε is the obliquity of the ecliptic which is tabulated in *The Astronomical Almanac* and has the mean value of $\langle\varepsilon\rangle = 23^\circ 26' 21.448''$ at epoch J(2000.0).

Horizontal, or Horizon, to Equatorial Coordinate Conversion

The formulae connecting the azimuth, A, and altitude, a, to the hour angle, h, and declination, δ, are

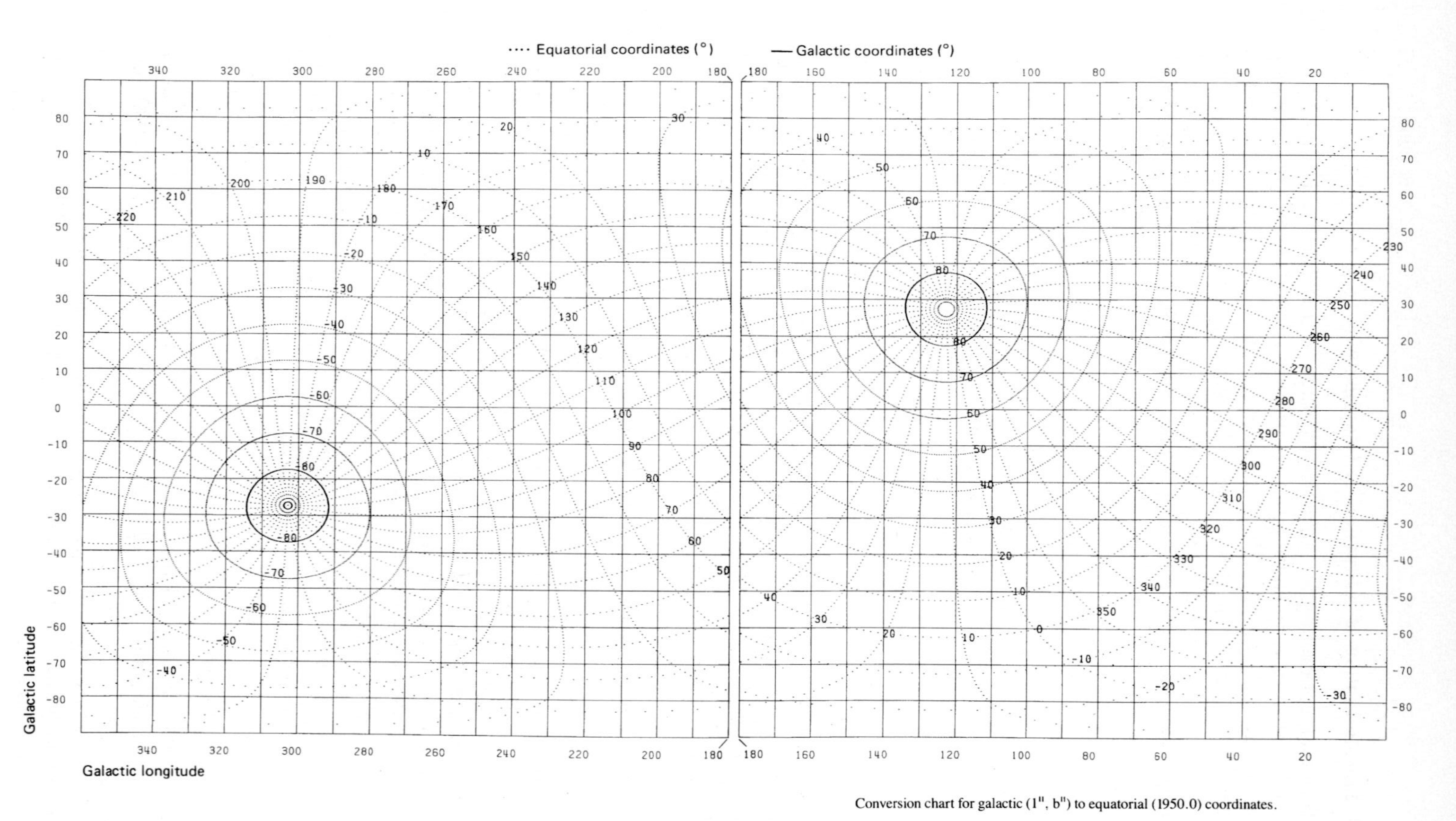

Fig. 5.5. Chart for the conversion of equatorial α, δ (1950.0) coordinates to galactic coordinates (l, b) or vice versa

$$\begin{aligned}
\cos a \sin A &= -\cos\delta \sin h \\
\cos a \sin A &= \sin\delta \cos\phi - \cos\delta \cos h \sin\phi \\
\sin a &= \sin\delta \sin\phi + \cos\delta \cos h \cos\phi \\
\cos\delta \cos h &= \sin a \cos\phi - \cos a \cos A \sin\phi \\
\sin\delta &= \sin a \sin\phi + \cos a \cos A \cos\phi \ ,
\end{aligned} \tag{5.45}$$

where ϕ is the observer's latitude, and the hour angle, h, is related to the right ascension, α, by

$$h = \text{local sidereal time} - \alpha \ . \tag{5.46}$$

5.1.4 Precession, Nutation, Aberration and Refraction

Positions on the celestial sphere are not firmly fixed. The stars that we now see in mid-winter are not the ones seen by the ancients in mid-winter, although the shift is too slow to be noticed by the naked eye in an individual's lifetime. This shift is called the "precession of the equinoxes"; meaning that the locations of the spring and fall equinoxes are slowly moving, or precessing, around the stellar background.

The precessional motion is caused by the tidal action of the Moon and Sun on the spinning Earth. It is similar to the wobbling of a top. As a result, the Earth's axis of rotation sweeps out a cone in space, centered around the axis of the Earth's orbit and completing one circuit in about 26,000 years (Fig. 5.6). This cone has an angular radius, or opening angle, equal to the obliquity of the ecliptic, with a value of about $23°26.36'$.

The slow conical motion carries the Earth's equator with it, and as the equator moves, the two intersections between the celestial equator and the Sun's path, or ecliptic, move westward. One of these intersections is the Vernal, or Spring, Equinox, from which astronomical coordinates are measured. Precession therefore shifts positions on the celestial sphere, which was first observed by Hipparchus, a Greek astronomer living in the second century B.C. who charted the positions of bright stars (Hipparchus, 125 B.C.). The Vernal Equinox moves forward (westward) along the ecliptic at the rate of about $50''$ per year, where $1'' = 1$ second of arc, so right ascensions are continuously increasing with time at this rate.

Because of the precessional change in celestial positions, the equinox, or reference date, must be given in specifying right ascension or declination, the most common being 1950.0 and 2000.0. The standard epoch that is now recommended for use in new star catalogs and theories of motion is

$$\text{J2000.0} = \text{2000 January 1.5} = JD2451545.0 \ , \tag{5.47}$$

where JD denotes Julian Date and the prefix J denotes the current system of measuring time in Julian centuries of exactly 36525 days in 100 years. The old Besselian system, designated by B, used tropical centuries, defined by the motion of the Sun; the tropical year in the fundamental epoch of 1900 January 1.5 corresponds to a tropical century of 36524.21987817305 Julian days at 1900. Proper motions, precessional motions and other angular motions

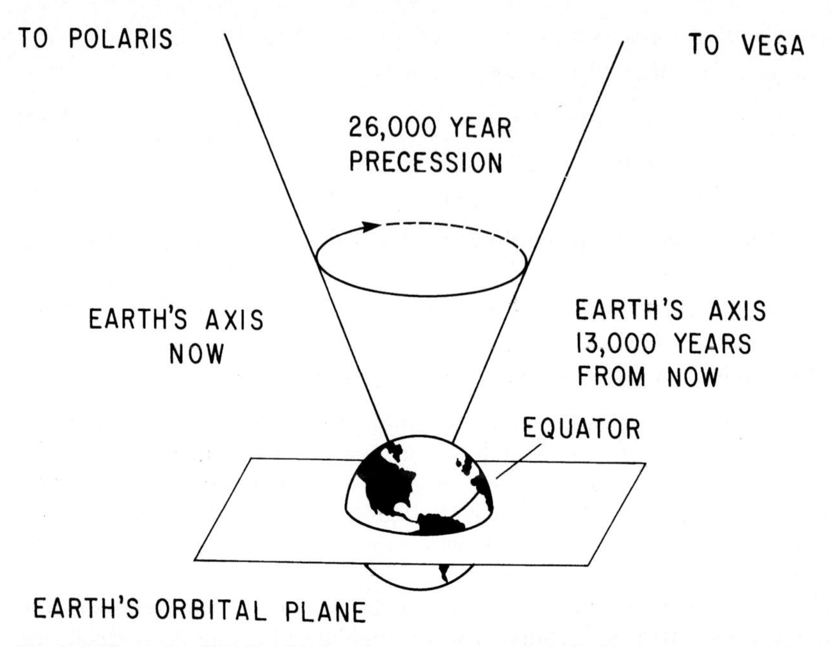

Fig. 5.6. The pole of the Earth's rotation traces out a circle in the sky once very 26,000 years, sweeping out a cone with an angular radius of about 23°26.4′. As the axis precesses, the equator and continents move with it. The north celestial pole now lies near the bright star Polaris, but in roughly 13,000 years the rotational axis will point towards another north star, Vega. The precessional motion of the Earth's rotational axis also causes a precession of the equinoxes; the Vernal Equinox will move through all the signs of the zodiac in about 26,000 years

expressed in units of tropical centuries can be transformed to units of Julian centuries by multiplication by the factor

$$F = 36525/36524.2198782 = 1.00002135903 \ , \tag{5.48}$$

where the tropical year in the fundamental epoch of 1900 January 1.5 corresponds to a tropical century of 36524.21987817305 Julian days at 1900.

We use the term lunisolar precession to describe the smooth, long-period (about 26,000 years) motion of the mean pole of the equator about the pole of the ecliptic at the rate of roughly 50″ per year. In addition, the plane of the ecliptic is subject to the gravitational action of the planets on the earth's orbit, making a contribution to precession known as planetary precession. If the equator were fixed, this motion would produce a precession of the equinox of about 12″ per century, and a decrease in the obliquity of the ecliptic of about 47″ per century. The combination of lunisolar and planetary precession is called general precession.

The detailed theory for computing the precessional correction to astronomical coordinates was given by Newcomb (1895) who obtained a precession constant of $p = 5025.64$ seconds of arc per tropical century at epoch B1900.0. The new definitions, adopted by the International Astronomical Union, or IAU, in 1976, are provided by Lieske et al. (1977) and modified by a small amount by Lieske (1979). The standard coordinate frame is now that of J2000.0, and the time unit for precessional motions and for proper motions of stars is now Julian centuries.

The general precession in longitude, p, at epoch J2000.0 is

$$p = 5029.0966 \text{ seconds of arc per Julian century.} \tag{5.49}$$

To the first approximation, we have the annual rates:

$$\begin{aligned}
&\text{General Precession in Longitude} = (50.290966 + 0.0222226\,T) \text{ seconds of arc per year} \\
&\text{Lunisolar Precession in Longitude} = (50.387784 + 0.0049263\,T) \text{ seconds of arc year} \\
&\text{Planetary Precession} = (-0.0188623 - 0.0476128\,T) \text{ seconds of arc per year}\ ,
\end{aligned} \tag{5.50}$$

where T is the time in Julian centuries from J2000.0. The general precession, p, may be resolved into a general precession in right ascension, m, and a general precession in declination, n, by

$$p = m \cos \varepsilon_0 + n \sin \varepsilon_0\ , \tag{5.51}$$

where the mean obliquity of the ecliptic, ε_0, has a value of

$$\varepsilon_0 = 23°26'\,21.448'' \text{ at J2000.0}\ . \tag{5.52}$$

If T denotes the number of Julian centuries since J2000.0, then the rates per Julian century of general precession in longitude, p, right ascension, m, and declination, n, are

$$\begin{aligned}
p &= (5029.0966 + 2.22226\,T - 0.000042\,T^2) \text{ seconds of arc per century} \\
m &= (4612.4362 + 2.79312\,T - 0.000278\,T^2) \text{ seconds of arc per century} \\
m &= (307.49574 + 0.186208\,T - 0.0000185\,T^2) \text{ seconds of time per century} \\
n &= (2004.3109 - 0.85330\,T - 0.000217\,T^2) \text{ seconds of arc per century}\ ,
\end{aligned} \tag{5.53}$$

where

$$T = [JD(\varepsilon_f) - JD(\varepsilon_0)]/36525 \tag{5.54}$$

for the Julian Date $JD(\varepsilon_f)$ at the fixed epoch, ε_f, and $JD(\varepsilon_0) = 2451545.0$ at J2000.0.

Formulas, valid over short periods, for reduction from the mean equinox of epoch t_1 to date $t = t_1 + \tau$, where τ is a fraction of a year and $|\tau| < 1$, are

$$\begin{aligned}
\alpha &= \alpha_1 + \tau(m' + n' \sin \alpha_1 \tan \delta_1) \\
\delta &= \delta_1 + \tau n' \cos \alpha_1
\end{aligned} \tag{5.55}$$

where the annual precession rates $m' = m/100.0$ and $n' = n/100.0$ are evaluated at t_1; these rates are given in *The Astronomical Almanac* for the relevant year.

More exact, but still approximate, formulae for the reduction of right ascension, α, and declination, δ, at the date, t, are:

For reduction to J2000.0

$$\begin{aligned} \alpha_0 &= \alpha - M - N \sin \alpha_m \tan \delta_m \\ \delta_0 &= \delta - N \cos \alpha_m \ . \end{aligned} \tag{5.56}$$

For reduction from J2000.0

$$\begin{aligned} \alpha &= \alpha_0 + M + N \sin \alpha_m \tan \delta_m \\ \delta &= \delta_0 + N \cos \alpha_m \ , \end{aligned} \tag{5.57}$$

where the subscript zero refers to epoch J2000.0, and α_m and δ_m refer to the mean epoch, with

$$\begin{aligned} \alpha_m &= \alpha - \tfrac{1}{2}(M + N \sin \alpha \tan \delta) = \alpha_0 + \tfrac{1}{2}(M + N \sin \alpha_0 \tan \delta_0) \\ \delta_m &= \delta - \tfrac{1}{2} N \cos \alpha_m = \delta_0 + \tfrac{1}{2} N \cos \alpha_m \end{aligned} \tag{5.58}$$

and the precessional constants M and N are given by

$$\begin{aligned} M &= 1.2812323\,t + 0.0003879\,t^2 + 0.0000101\,t^3 \text{ degrees} \\ N &= 0.5567530\,t - 0.0001185\,t^2 - 0.0000116\,t^3 \text{ degrees} \ , \end{aligned}$$

with

$$t = [JD(\varepsilon_D) - JD(\varepsilon_F)]/36525 = (T - 2000.0)/100 \ ,$$

where ε_F is a fixed epoch taken to be 2000.0 at $JD(2000.0) = 2451545.0$, ε_D is the epoch of date, and T is the date.

The precession matrix that enables one to precess to and from J2000.0 is given by Lieske (1979). The transformation of coordinates between the systems of B1950.0 and J2000.0, and the principal galactic axes referred to J2000.0, are found in Murray (1989). The matrix formulation for the transformation of the mean places and proper motions in astrometric catalogs between the old FK4 fundamental stellar reference system of B1950.0 and the new FK5 reference system of J2000.0 are provided by Standish (1982), Aoki et al. (1983), Smith et al. (1989), Soma and Aoki (1990), and Seidelmann (1992).

Rigorous formulae for the reduction of mean equatorial coordinates from an initial epoch to an epoch of date, and *vice versa*, are given in *The Astronomical Almanac*. It also includes a concise method of conversion from B1950.0 in the FK4 system to J2000.0 in the FK5 system and *vice versa*.

Superimposed on the precession is a superficially similar motion called nutation, which was first observed by Bradley (1748). It is a short-period motion of the Earth's rotation axis with respect to a spaced-fixed coordinate system, caused by the periodic gravitational torque of the Moon, Sun and planets on the Earth's equatorial bulge. Nutation has an amplitude of about 9″ and a variety of periods up to 18.6 years; the coefficient of the 18.6-year term is known as the constant of nutation, N, with a value at standard epoch J2000.0 of

$$N = 9.2025 \text{ seconds of arc at J2000.0} \ . \tag{5.59}$$

The nutation may be resolved into a correction, $\Delta\psi$, to the Sun's longitude, and a correction, $\Delta\varepsilon$, in obliquity (Kinoshita, 1977). Daily values for the nutations in longitude and obliquity are given in *The Astronomical Almanac* for the relevant year with a precision of about 1″. Approximate expressions for these nutations, also accurate to about 1″, are given by (Seidelmann, 1992)

$$\begin{aligned} \Delta\psi &= -0^\circ.0048 \sin(125^\circ.0 - 0^\circ.05295d) \\ &\quad - 0^\circ.0004 \sin(200^\circ.9 + 1^\circ.97129d) \\ \Delta\varepsilon &= +0^\circ.0026 \cos(125^\circ.0 - 0^\circ.05295d) \\ &\quad + 0^\circ.0002 \cos(200^\circ.9 + 1^\circ.97129d) \ , \end{aligned} \tag{5.60}$$

where d is the number of days from J2000.0, or JD 2451545.0. We also have the mean obliquity, ε_0, and true obliquity, ε, of date given by:

$$\begin{aligned} \varepsilon_0 &= 23^\circ\, 26'\, 21''.448 - 46''.8150\, T - 0''.00059\, T^2 + 0''.001813\, T^3 \ , \\ \varepsilon &= \varepsilon_0 + \Delta\varepsilon \ , \end{aligned} \tag{5.61}$$

where $T = [JD - 2451545.0]/36525$. Daily values of the obliquity of the ecliptic, ε, are given in *The Astronomical Almanac* for the relevant year. The 1980 IAU nutation series and fundamental arguments needed to evaluate $\Delta\Psi$ and $\Delta\varepsilon$ are given by Van Flandern (1981) and Seidelmann (1992); recent highly accurate observations by Very Long Baseline Interferometry (VLBI) suggest some systematic corrections to this series (Seidelmann, 1992).

First-order nutational corrections to right ascension, $\Delta\alpha$, and declination, $\Delta\delta$, are given by

$$\begin{aligned} \Delta\alpha &= (\cos\varepsilon + \sin\varepsilon \sin\alpha \tan\delta)\Delta\psi - \cos\alpha \tan\delta\Delta\varepsilon \ , \\ \Delta\delta &= \sin\varepsilon \cos\alpha\Delta\psi + \sin\alpha\Delta\varepsilon \ . \end{aligned} \tag{5.62}$$

Bradley (1728) first observed stellar aberration, which is a tilting of the apparent direction of a celestial object toward the direction of motion of the observer. That is, because the velocity of light is finite, the apparent direction of a moving celestial object from a moving observer is not the same as the geometric direction of the object from the observer at the same instant. For stars, the normal practice is to ignore the correction for the motion of the celestial object, and to compute the stellar aberration due to the motion of the observer. Aberration of planets and other members of the solar system include both their motions and that of the observer, with the designation planetary aberration.

The magnitude, $\Delta\theta$, of stellar aberration depends on the ratio of the velocity of the observer, v, to the velocity of light, c, and the angle, θ, between the direction of observation and the direction of motion. The displacement, $\Delta\theta$, in the sense of apparent minus mean place, is given by

$$\tan\Delta\theta = \frac{v\sin\theta}{c + v\cos\theta} \tag{5.63}$$

or

$$\Delta\theta \approx \frac{v}{c}\sin\theta - \frac{1}{2}\left(\frac{v}{c}\right)^2 \sin 2\theta + \left(\frac{v}{c}\right)^3 (\sin\theta\cos^2\theta - 0.33\sin^3\theta) + \cdots \ .$$

The term of order v/c is about 20 seconds of arc, and the term $(v/c)^2$ has a maximum value of 0.001 seconds of arc.

The motion of an observer on the Earth is the result of the diurnal rotation of the Earth, the orbital motion of the Earth about the center of mass of the solar system, and the motion of this center of mass in space. The stellar aberration is therefore made up of three components, which are referred to as diurnal aberration, annual aberration, and secular aberration.

The annual aberration corrections to the right ascension, $\Delta\alpha$, and declination, $\Delta\delta$, are given, to the first order in v/c, by

$$\Delta\alpha = \frac{1}{\cos\delta}[-\kappa\sin\lambda\sin\alpha - \kappa\cos\lambda\cos\varepsilon\cos\alpha] \tag{5.64}$$

and

$$\Delta\delta = -\kappa\sin\lambda\cos\alpha\sin\delta + \kappa\cos\lambda[\cos\varepsilon\sin\alpha\sin\delta - \sin\varepsilon\cos\delta] \ ,$$

where λ is the true geometric longitude of the Sun, ε is the mean obliquity of the ecliptic, and the constant of aberration is given by:

$$\kappa = \frac{2\pi a}{Pc}(1-e^2)^{-\frac{1}{2}} \tag{5.65}$$

or in the standard epoch

$$\kappa = 20.49552 \text{ seconds of arc for J2000.0} \ . \tag{5.66}$$

Here $a = 1.49598 \times 10^{13}$ cm is the astronomical unit (the mean distance from the Earth to the Sun), $e = 0.01671$ is the mean eccentricity of the Earth's orbit, $P = 3.1558 \times 10^7$ seconds is the length of the sidereal year, $c = 2.997925 \times 10^{10}$ cm sec^{-1} is the velocity of light, and it has been assumed that one radian = 2.062648×10^5 seconds of arc. The corrections to right ascension and declination are in the sense that $\Delta\alpha$ and $\Delta\delta$ are the apparent coordinate minus the mean coordinate, and the correction can be as large as 20.5 seconds of arc.

The approximate corrections to equatorial coordinates for annual aberration are given in *The Astronomical Almanac* in terms of the rectangular velocity components of the Earth, which are tabulated at daily intervals for the relevant year.

The diurnal aberration caused by the rotation of the Earth on its axis is described by the constant of diurnal aberration

$$\begin{aligned} \kappa_{\rm d} &= \frac{a\omega}{c}\frac{\rho}{a}\cos\phi' \\ \kappa_{\rm d} &= 0.02133\frac{\rho}{a}\cos\phi' \text{ seconds of time} \\ \kappa_{\rm d} &= 0.3200\frac{\rho}{a}\cos\phi' \text{ seconds of arc} \end{aligned} \tag{5.67}$$

where the rotation of the Earth on its axis carries the observer toward the east with a velocity $\omega\rho\cos\phi'$, the equatorial velocity of the Earth, is $\omega = 7.2921 \times 10^{-5}$ radians per second, $a_{\rm e}\omega = 0.465 \times 10^5$ cm sec^{-1} is the equatorial rotational velocity of the surface of the Earth, the equatorial radius of the Earth is $a = 6.37814 \times 10^8$ cm, and ρ and ϕ' are the geocentric distance, or radius, and latitude of the observer, respectively.

The diurnal aberration corrections to right ascension, $\Delta\alpha$, and declination, $\Delta\delta$, in the sense of apparent minus mean position, are given by

$$\Delta\alpha = \kappa_d \cos h \sec\delta = 0.02133\frac{\rho}{a}\cos\phi' \cos h \sec\delta \quad \text{seconds of time} \ , \tag{5.68}$$

and

$$\Delta\delta = \kappa_d \sin h \sin\delta = 0.3200\frac{\rho}{a}\cos\phi' \sin h \sin\delta \quad \text{seconds of arc} \ , \tag{5.69}$$

where the mean hour angle, h, is the local sidereal time minus the mean right ascension, α, and δ is the mean declination of the celestial object. For a star in transit, $h = 0°$ or $180°$, so $\Delta\delta$ is zero.

Formulae, examples and ephemerides for approximate reductions of equatorial coordinates using the day-number technique have been given in *The Astronomical Almanac*. The apparent right ascension and declination (α_1, δ_1), of a star at an epoch $t + \tau$, which includes the effects of precession, nutation, annual aberration, proper motion, and annual parallax, and includes the second-order day numbers (J and J'), may be calculated either using the Besselian day numbers A, B, C, D, E, and star constants, from

$$\begin{aligned} \alpha_1 &= \alpha + Aa + Bb + Cc + Dd + E + J\tan^2\delta + \tau\mu_\alpha + \pi(dX - cY) \ , \\ \delta_1 &= \delta + Aa' + Bb' + Cc' + Dd' + J'\tan\delta + \tau\mu_\delta + \pi(d'X - c'Y) \ , \end{aligned} \tag{5.70}$$

or using the independent day numbers f, g, G, h, H, from

$$\begin{aligned} \alpha_1 &= \alpha + f + g\sin(G+\alpha)\tan\delta + h\sin(H+\alpha)\sec\delta + J\tan^2\delta \\ &\quad + \tau\mu_\alpha + \pi(dX - cY) \ , \\ \delta_1 &= \delta + g\cos(G+\alpha) + h\cos(H+\alpha)\sin\delta + i\cos\delta + J'\tan\delta \\ &\quad + \tau\mu_\delta + \pi(d'X - c'Y) \ , \end{aligned} \tag{5.71}$$

where α, δ are the right ascension and declination of the star for the mean equinox and epoch of t, the μ_α and μ_δ are the annual proper motions in right ascension and declination, respectively, and π is the parallax. X and Y are the barycentric coordinates of the Earth. The day numbers are usually tabulated in seconds of arc; when used for calculating the star's right ascension, measured in time, either they or the star constants by which they are multiplied should be divided by 15.

The Besselian day numbers A, B and E correct for precession; C and D correct for aberration.

The quantities A, B, C, D, and E, in seconds or arc, and τ are tabulated daily in *The Astronomical Almanac*. The rectangular coordinates X, Y, Z of the Earth with respect to the barycenter of the solar system, referred to the mean equinox and equator J2000.0, and expressed in astronomical units, are given in *The Astronomical Almanac* for 0 hours TDB (Barycentric Dynamical Time) on each day of the relevant year.

The star constants, which are constant only for the mean equinox of the fixed epoch t, are defined by

$$\begin{aligned} a &= m/n + \sin\alpha\tan\delta, & a' &= \cos\alpha \ , \\ b &= \cos\alpha\tan\delta, & b' &= -\sin\alpha \ , \\ c &= \cos\alpha\sec\delta, & c' &= \tan\varepsilon_0\cos\delta - \sin\alpha\sin\delta \ , \\ d &= \sin\alpha\sec\delta, & d' &= \cos\alpha\sin\delta \ , \end{aligned} \tag{5.72}$$

where α and δ are the mean place of the star for epoch t. The precession rates m and n, and ε_0, also should be evaluated for epoch t.

The terms J and J' are called the second-order day numbers, providing second-order corrections to right ascension, $J \tan^2 \delta$, and declination $J' \tan \delta$. They are given by:

$$\begin{aligned} J &= +[g \sin(G+\alpha) \pm h \sin(H+\alpha)][g \cos(G+\alpha) \pm h \cos(H+\alpha)] \\ &= +[(A \pm D) \sin\alpha + (B \pm C) \cos\alpha][(A \pm D) \cos\alpha - (B \pm C) \sin\alpha] \end{aligned} \tag{5.73}$$

and

$$\begin{aligned} J' &= -\frac{1}{2}[g \sin(G+\alpha) \pm h \sin(H+\alpha)]^2 \\ &= -\frac{1}{2}[(A \pm D) \sin\alpha + (B \pm C) \cos\alpha]^2 \ , \end{aligned}$$

the upper sign being taken for positive declinations, and the lower sign for negative declinations. The expressions for second-order terms and J and J' assume that f, g, h are measured in radians, and J must be divided by 15 to be converted to seconds of time.

The independent day numbers f, g, and G correct for precession and nutation, and h, H, and i correct for annual aberration. They are given by

$$\begin{aligned} & f = (m/n)A + E = m\tau + \cos\varepsilon\Delta\psi, \qquad h \sin H = C \ , \\ & g \sin G = B, \qquad h \cos H = D \ , \\ & g \cos G = A, \qquad i = C \tan\varepsilon \end{aligned} \tag{5.74}$$

These day numbers are defined in terms of the Besselian day numbers.

To transform the right ascension and declination (α_0, δ_0) from the standard epoch and equinox of J2000.0 to the true equinox and equator of date, an approximate reduction for precession and nutation is:

$$\begin{aligned} \alpha &= \alpha_0 + f + g \sin(G + \alpha_0) \tan\delta_0 \ , \\ \delta &= \delta_0 + g \cos(G + \alpha_0) \ , \end{aligned} \tag{5.75}$$

where the units of right ascension are seconds of time, the units of declination are minutes of arc, and f, g and G are given by

$$\begin{aligned} f &= M + A(m/n) + E \ , \\ g &= \left(B^2 + (A+N)^2\right)^{\frac{1}{2}} \ , \\ G &= \tan^{-1}[B/(A+N)] + \frac{1}{2}M \ . \end{aligned} \tag{5.76}$$

The expressions for f, g, and G are an extension of the use of the day numbers, which apply precession and nutation from the epoch (e.g., the middle of the nearest year) only to the epoch of date $t + \tau$ where τ is the fraction of a year. Thus A, B, and E are the Besselian day numbers, and m and n are the annual rates of precession. The additional terms M and N, which are the accumulated precession angles, are required to apply precession from J2000.0 to epoch t.

Observed positions also differ from the true position due to the refraction in the atmosphere. The effect of refraction is removed from an observation at optical wavelengths by subtracting the constant of refraction, R, from the observed altitude.

$$R = \left(\frac{0.28P}{T+273}\right)\frac{0.0167}{\tan(H+7.31/(H+4.4))} \quad \text{degrees} \ , \tag{5.77}$$

where T is the temperature in degrees Celsius, or °C, P is the barometric pressure in millibars, or mb, and H is the apparent altitude. If T and P are unknown, assume the term in the first bracket is unity. The standard conditions used in *The Nautical Almanac* are $T = 10$ °C and $P = 1010$ mb; in *The Star Almanac* $T = 7$ °C and $P = 1005$ mb. The refraction decreases slightly for increased temperature and for decreasing atmospheric pressure, for example, in a low-pressure zone or in the mountains.

The apparent shift in altitude, R, in the sense of apparent minus the true altitude, is summarized in Table 5.1 for standard conditions.

At an altitude of 90° there is no correction for refraction. For rising and setting phenomena, a refraction constant of 34′ is used.

At optical wavelengths, the changes in right ascension, $\Delta\alpha$, and declination, $\Delta\delta$, caused by refraction are, for the observed true positions,

$$\begin{aligned} \Delta\alpha &= R \sec\delta \sin C \\ \Delta\delta &= R \cos C \ , \end{aligned} \tag{5.78}$$

where R is the constant of refraction, the parallactic angle, C, is given by $\sin C = \cos\phi \ \sin h/\sin z$, the observer's latitude is ϕ, the zenith distance is z, the hour angle is h, and δ is the declination. When an object is on the meridian, $C = 0°$ and the right ascension is unaffected by refraction.

At radio wavelengths, positions are displaced by refraction in the ionosphere. Detailed formulae for the apparent displacement of source positions due to ionospheric refraction are given by Komesaroff (1960).

The apparent direction of a celestial object may be significantly affected by the deflection of light in the gravitational field of the Sun, which is discussed in greater detail in Sect. 5.5.6. The angular deflection, ϕ, has a maximum value of 1.75 seconds of arc when the line of sight grazes the solar limb, and decreases for greater elongation angles, E, between the directions of the Sun center and the celestial object. That is:

$$\phi = \frac{0.00407}{\tan(E/2)} = 0.00407 \tan\left(\frac{180° - E}{2}\right) \quad \text{seconds of arc} \ . \tag{5.79}$$

Approximate values of the deflection angles, ϕ, for different elongations, E, are given in Table 5.2.

Table 5.1. The refraction R of a star at different altitudes H for standard conditions of atmospheric temperature and pressure

H	0°	5°	10°	20°	40°	60°
R	34′50″	9′45″	5′16″	2′37″	1′09″	33″

Table 5.2. Deflection, ϕ, by solar gravitation for varying angular distances, E, from Sun center

E	0°.25	0°.5	1°	2°	5°	10°	20°	50°	90°
ϕ	1″.866	0″.933	0″.466	0″.233	0″.093	0″.047	0″.023	0″.009	0″.004

The increments to be added to the calculated right ascension, α, and declination, δ, may be evaluated approximately from:

$$\cos E = \sin\delta \sin\delta_s + \cos\delta\cos\delta_s\cos(\alpha - \alpha_s) \ ,$$

$$\Delta\alpha = 0^s.000271 \frac{\cos\delta_s \sin(\alpha - \alpha_s)}{(1 - \cos E)\cos\delta} \ , \tag{5.80}$$

$$\Delta\delta = 0''.00407 \frac{\sin\delta\cos\delta_s\cos(\alpha - \alpha_s) - \cos\delta\sin\delta_s}{1 - \cos E} \ ,$$

where $\alpha, \delta, \alpha_s, \delta_s$ are the geocentric right ascensions and declinations of the star and the Sun, respectively. For stars, E, the geocentric elongation, approximates $180° - \psi$, where ψ is the heliocentric elongation.

5.2 Distance and Luminosity

5.2.1 Distance of the Sun, Moon and Planets

The distances to the Sun, Moon and planets can be measured by triangulation from different points on the Earth. These distances are inferred from their parallax, or the angular difference in the apparent direction of an object as seen from two different locations. The solar parallax, $\pi_\odot$, is, for example, defined as half the angular displacement of the Sun as viewed from opposite sides of the Earth, or:

$$\sin\pi_\odot = \mathrm{a_e}/\mathrm{A} \ , \tag{5.81}$$

where the equatorial radius of the Earth, $a_e = 6.378140 \times 10^8$ cm, and the unit distance, A, also called the astronomical unit, AU, is the semi-major axis of the Earth's orbit about the Sun with a value of (Lang, 1992):

$$A = 1 \text{ AU} = \text{ one astronomical unit} = 1.49597870 \times 10^{13} \text{cm} \ . \tag{5.82}$$

For these values of $\mathrm{a_e}$ and A we have a solar parallax of:

$$\pi_\odot = \arcsin(a_e/A) = 8.794148 \text{ seconds of arc} \ , \tag{5.83}$$

where one radian $= 2.062648 \times 10^5$ seconds of arc.

The equatorial horizontal parallax π_{object}, often called the horizontal parallax, of any object can be calculated from

$$\pi_{\mathrm{object}} = \pi_\odot/\Delta \ , \tag{5.84}$$

where Δ is the geocentric distance of the object in AU. For example, the equatorial horizontal parallax of the Moon is:

$$\pi_{\mathrm{moon}} = 57'\ 02.608''$$

$$\pi_{\mathrm{moon}} = 3422.608 \text{ seconds of arc} \ , \tag{5.85}$$

at the mean distance of the Moon from the Earth of

$$\Delta_{\rm moon} = 3.844 \times 10^{10}\ {\rm cm}$$
$$\Delta_{\rm moon} = 60.27\ \text{Earth radii} \tag{5.86}$$
$$\Delta_{\rm moon} = 0.002570\ {\rm AU}\ .$$

As the ocean tides flood and ebb, they create eddies in the water, producing friction and dissipating energy at the expense of the Earth's rotation. The tides therefore act as a brake on the spinning Earth, causing the day to become longer and the Moon to move steadily away from the Earth (also see Sect. 5.3.6). Laser signals to reflectors left on the Moon by astronauts indicate that the semimajor axis of the Moon's orbit is increasing at the rate of 3.82 ± 0.07 centimeters per year (Dickey et al., 1994).

As first observed by Kepler (1619) and explained by Newton (1687), the orbit of a planet describes an ellipse with the Sun at one focus, and a semi-major axis, a, that is related to the orbital period, P, by:

$$P^2 = 4\pi^2 a^3/[G(m_{\rm p} + M_\odot)]\ , \tag{5.87}$$

where the Newtonian gravitational constant $G = 6.672 \times 10^{-8}$ in c.g.s. units, $m_{\rm p}$ is the planet's mass, and the mass of the Sun is $M_\odot = 1.989 \times 10^{33}$ g. Since the planetary masses are much less than the solar mass, the unit distance, A, is given by:

$$A \approx (P_{\rm e}/P_{\rm p})^{2/3} a_{\rm p}\ , \tag{5.88}$$

where the subscripts e and p respectively denote the Earth and another planet.

Thus, once one establishes the mean distance of any planet from the Sun, the distances of all the others can be determined from their known periods. The current values for the semi-major axes, a, in astronomical units, AU, the orbital periods, P, in tropical years, and the average orbital speeds of the planets are given in Table 5.3.

The quest for accuracy in the mean distance of the Sun from the Earth has involved hundreds of trips to remote countries, tens of thousands of

Table 5.3. Mean orbital elements of the planets (adapted from Lang, 1992). For conversion purposes, 1 tropical year = $3.15569259747 \times 10^7$ seconds = 365.24219 days

Planet	Semi-Major Axis, a (AU)	Orbital Period (years)	Mean Orbital Velocity (km s^{-1})
Mercury	0.387099	0.2409	47.89
Venus	0.723332	0.6152	35.03
Earth	1.000000	1.0000	29.79
Mars	1.523688	1.8809	24.13
Jupiter	5.202834	11.8622	13.06
Saturn	9.538762	29.4577	9.64
Uranus	19.191391	84.0139	6.81
Neptune	30.061069	164.793	5.43
Pluto	39.529402	247.7	4.74

observations and photographs, and the lifetime work of several astronomers (Lang, 1985). It has been inferred with increasingly greater accuracy during the past two centuries by measuring the parallax of Venus, Mars and the nearby minor planet, or asteroid, Eros, during their closest approach to the Earth, but the values obtained often differed by several times their estimated errors (Fig. 5.7).

The controversy over the exact value of the solar parallax was not completely resolved until the 1960s when radio, or radar, pulses were used to accurately determine the distance to Venus, and to thereby measure the Sun's distance with an accuracy of about one kilometer (Ash, Shapiro, Smith, 1967; Muhleman, 1969).

Nowadays the accuracy of the Sun's distance is fixed by the exact value for the speed of light, c, defined to be a exactly:

$$c = 2.99792458 \times 10^{10} \text{ cm s}^{-1} \quad . \tag{5.89}$$

The International Astronomical Union, or IAU, specifies the light-time, τ_A, to travel the unit distance, A, as a primary constant:

$$\tau_A = 499.004782 \text{ seconds} \quad . \tag{5.90}$$

The IAU then specifies the unit distance, A, as a derived constant given by $A = c\tau_A = 1.49597870 \times 10^{13}$ cm.

The measured values of the planetary distances from the Sun are roughly approximated by the Titius (1766)–Bode (1772) law. If a_n denotes the mean distance of planet, n, then the Titius–Bode law states that:

$$a_n \approx 0.1[4 + 3 \times 2^n] \quad , \tag{5.91}$$

where $n = -\infty, 0, 1, \ldots, 7$, and the distances are given in astronomical units. Blagg (1913) gives a different relation. It is a progression in 1.73 (not 2) multiplied by a periodic fraction of n (also see Richardson, 1945, and Roy, 1953). A historical and theoretical discussion of the Titius–Bode law is given by Nieto (1972).

5.2.2 Distance to the Nearby Stars – Trigonometric Parallax

Even the closest stars are too far away for us to detect a shift in position from any two points on Earth. To triangulate the distances of nearby stars we need a wider baseline, the Earth's annual orbit. That is, as the Earth goes around the Sun, the apparent direction, or position, of all stars will change, and a nearby star will appear displaced relative to the backdrop of the more distant, seemingly stationary stars in the same part of the sky. When measured at intervals of six months from opposite sides of the Earth's orbit, then half the apparent angular displacement defines the trigonometric, or annual, parallax, π_t, given by:

$$\pi_t \approx \tan \pi_t = \frac{A}{D} \text{ radians} \quad , \tag{5.92}$$

where A is the astronomical unit and D is the distance of the star. The trigonometric parallax π_t is usually given in seconds of arc, which can be

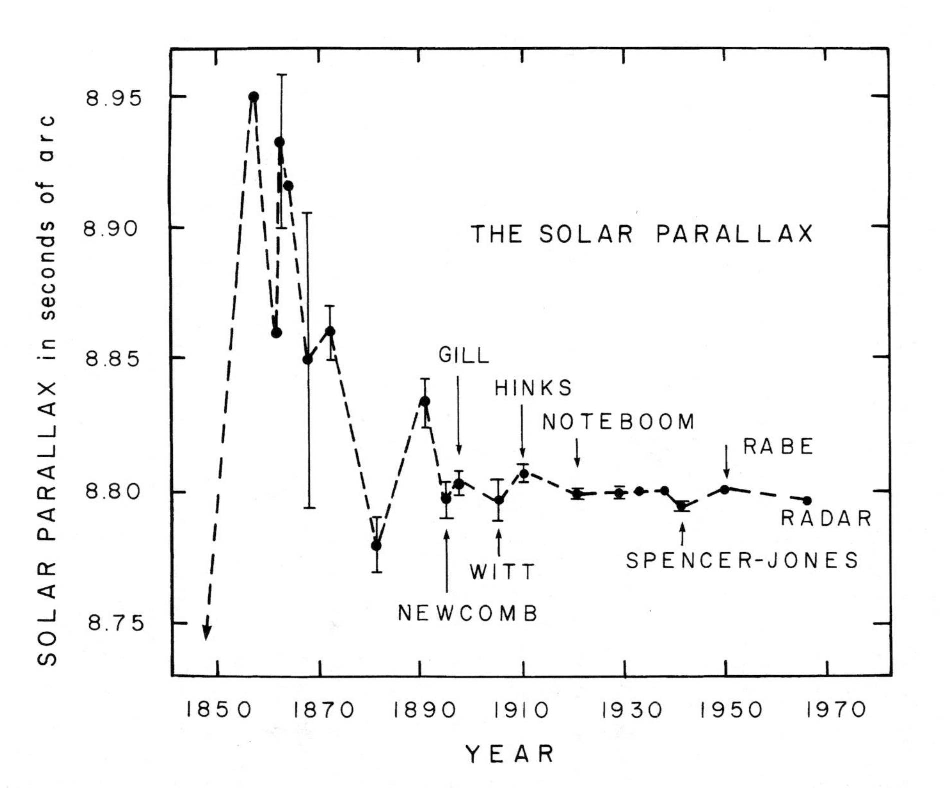

Fig. 5.7. Values of the solar parallax, $\pi_{\odot}$, obtained by different methods between 1850 and 1970. Here the error bars denote the probable errors in the determination. The data for 1941, 1950 and 1965 all have errors smaller than the plotted points. (Adapted from Lang, 1985)

obtained by multiplying by 2.062648×10^5 seconds of arc per radian. Because π_t is inversely proportional to the distance, the angular displacement is greater for nearby stars.

If the direction of a star lies in the plane of the Earth's orbit, or in the ecliptic, the star will move backwards and forwards in this plane during the year by an amount of π_t; when the star's direction is perpendicular to this plane it moves in a circle of angular radius π_t; and at other ecliptic latitudes the star appears to move in an ellipse with semi-major axis π_t.

A convenient unit of stellar distances is the parsec, abbreviated pc. The distance of a nearby star in parsecs is given by:

$$D = \frac{1}{\pi_t} \text{ parsccs} \quad , \tag{5.93}$$

when π_t is given in seconds of arc. A star whose parallax is one second of arc lies at a distance of one parsec, where:

$$\begin{aligned} \text{One parsec} &= 1\,\text{pc} = 3.085678 \times 10^{18} \text{ cm} \\ \text{One parsec} &= 1\,\text{pc} = 3.261633 \text{ light years} \\ \text{One parsec} &= 1\,\text{pc} = 206{,}265 \text{ AU} \quad . \end{aligned} \tag{5.94}$$

Distances comparable to the extent of our Galaxy are measured in units of kiloparsecs, or kpc, where 1 kpc $=$ 10^3 pc $=$ one thousand parsecs, and intergalactic distances are specified in units of Megaparsecs, or Mpc, with 1 Mpc $=$ 10^6 pc $=$ one million parsecs.

It was not until 1838/39 that Friedrich Wilhelm Bessel first measured the trigonometric parallax of a star, finding $\pi_t = 0.31 \pm 0.02$ seconds of arc for the star 61 Cygni (Bessel, 1839), close to the modern value of $\pi_t = 0.292 \pm 0.004$. He was closely followed by Henderson (1839) for α Centauri with $\pi_t = 1.16 \pm 0.11$ seconds of arc (modern value 0.760) – see Jackson (1956) for a historical review. Alpha Centauri is the nearest star with a distance of 1.316 parsecs or 4.29 light years.

Reliable distances for individual stars can only be determined for nearby stars with trigonometric parallaxes of $\pi_t \geq 0.045$ seconds of arc, or for stars with distances $D \leq 22$ parsecs. A catalogue giving the position, trigonometric parallax, proper motion and absolute visual magnitude for 2,241 stars nearer than 22 parsecs has been compiled by Lang (1992). Instruments aboard the Hipparcos satellite, named after the Greek astronomer Hipparchus, have been obtaining parallaxes of hundreds of thousands of stars with an unprecedented accuracy of one milliarcsecond, leading to dramatic improvements in our estimates for the distances of nearby stars.

5.2.3 Distance to the Nearby Stars – Stellar Motions

Everything in the Universe moves, and the stars are no exception. The Sun and other nearby stars, for example, revolve about a distant, massive galactic center at a speed of about 220 km s^{-1}, and a random stellar motion of about 10 km s^{-1} is superimposed on this circular motion. Nearby stars therefore appear to move across the sky relative to more distant stars. Such a change in position, caused by motions perpendicular to the line of sight, is now called the proper motion and designated by the symbol μ. It was first noticed by Edmond Halley in comparing the positions of bright stars, such as Sirius and Arcturus, with those measured by the ancient Greeks (Halley, 1718).

If we know the distance, D, to a star, as well as its proper motion, μ, then we can infer its velocity, V_t, perpendicular to the line of sight.

$$V_t = 4.74\,\mu D \text{ km s}^{-1} \quad , \tag{5.95}$$

where V_t is in units of km s^{-1} when μ is in seconds of arc per year and D is in parsecs. The components of the annual proper motion, μ, in right ascension and declination are given by Lang (1992) for 2,241 stars nearer than 22 parsecs.

The proper motions of stars within a nearby cluster of stars can be used to infer the distance of the cluster. If these stars are all moving on parallel tracks, with a common velocity greater than that of the random motions in the cluster, then they will appear to converge at a point, in much the same way that two parallel railroad tracks seem to meet at a distant point. If V_r denotes the radial component of velocity along the line of sight, and V_t indicates the transverse velocity component perpendicular to the line of sight, then the angle, θ, between the apparent direction of the cluster and that of the convergent point is given by:

$$\tan\theta = \frac{V_t}{V_r} = \frac{4.74\,\mu D}{V_r} \quad , \tag{5.96}$$

so the cluster distance is specified by:

$$D = V_r \frac{\tan\theta}{4.74\mu} \text{ parsecs} \quad , \tag{5.97}$$

when the radial velocity, determined from Doppler shifts of spectral features, is given in km s^{-1} and μ is in seconds of arc per year. This technique of inferring the distance of a nearby star cluster is called the convergent point method. A comparison of the convergent point and other methods for specifying the distance to the Hyades cluster has been given by Hanson (1980) who obtains a distance of $D = 45.7 \pm 0.8$ parsecs. Perryman et al. (1998) similarly use Hipparcos data to obtain a distance of 46.34 ± 0.27 parsecs for the center of mass of the Hyades cluster, corresponding to a distance modulus of $m - M = 3.33 \pm 0.01$ magnitude. The distances to about 400 open star clusters in our Galaxy are given by Lang (1992), many of these also have distances given in *The Astronomical Almanac*.

An alternative technique for inferring the distance of a nearby star cluster has been designated the moving-cluster method. It is based on the rate of change, $\dot{\theta}$, of the angular diameter, θ, of the cluster due to its radial motion at a velocity, V_r, relative to the Sun. The distance is given by:

$$D = \frac{V_r\theta}{\dot{\theta}} = \frac{V_r\theta\Delta t}{\Delta\theta} \quad , \tag{5.98}$$

where a change, $\Delta\theta$, in angular size has been measured over the time interval Δt.

5.2.4 Apparent and Absolute Luminosity

Celestial objects emit an intrinsic brightness, or absolute luminosity, and astronomers observe the apparent brightness which depends upon how much the radiation has dimmed on its way to us. According to the inverse-square law, the intrinsic brightness is reduced by the square of the distance as light travels outward and is dispersed into space. The apparent luminosity, l, is therefore given by:

$$l = L/D^2 \ \mathrm{erg\ s^{-1}\ cm^{-2}} \ , \tag{5.99}$$

where L is the absolute luminosity, in units of erg s^{-1}, and D is the distance in cm.

Thus, knowing an object's apparent luminosity and distance allows us to infer its absolute luminosity. Or, the distance of a celestial object can be gauged by comparing the absolute luminosity to the amount of radiation reaching the Earth. That is:

$$D = (L/l)^{1/2} \ . \tag{5.100}$$

For a given absolute luminosity, dimmer objects are farther away, and brighter ones closer. It is similar to judging the distance of a car's headlight or a street light from their apparent brightness.

The absolute luminosity of celestial objects is often specified in terms of the Sun's absolute luminosity, $L_{\odot}$, which can be determined from measurements of the solar radiation reaching the Earth and the distance between the Earth and the Sun.

Spacecraft have measured the Sun's total irradiance just outside the Earth's atmosphere. The ACRIM radiometer aboard the SMM spacecraft obtains, for example, a value of (Willson, 1993)

$$f_{\odot} = (1367.51 \pm 0.01) \times 10^3 \ \mathrm{erg\ s^{-1}\ cm^{-2}} \ . \tag{5.101}$$

This quantity is known as the solar constant. It is defined as the total amount of radiant solar energy per unit time per unit area reaching the top of the Earth's atmosphere at the Earth's mean distance from the Sun. The units of the solar constant sometimes include watts, where one watt $= 10^7$ erg s^{-1}.

The solar constant is not precisely known. Sackmann, Boothroyd and Kraemer (1993) average several spacecraft measurements to obtain:

$$f_{\odot} = (1370 \pm 2) \times 10^3 \ \mathrm{erg\ s^{-1}\ cm^{-2}} \ , \tag{5.102}$$

reflecting a disagreement in the absolute calibration of different spacecraft radiometers. We also know that the Sun's total radiative output changes by amounts of up to a few tenths of a percent on all time scales from 1 second to 10 years. Dark sunspots reduce the solar constant by up to 0.3 percent, or 0.003, for a few days, bright faculae enhance the Sun's radiative output by a comparable amount, and the solar constant varies by a few tenths of a percent in step with the decade-long solar activity cycle (Lang, 1995). So, the solar constant is not constant after all.

The Sun's absolute luminosity, $L_{\odot}$, is given by:

$$L_{\odot} = 4 \pi f_{\odot} A^2 \ , \tag{5.103}$$

which is the amount of energy per unit time crossing an imaginary sphere entered on the Sun with a radius equal to the Earth's semi-major axis $A =$ one astronomical unit $= 1.49597870 \times 10^{13}$ cm. Thus, we can conservatively estimate

$$L_\odot = (3.85 \pm 0.01) \times 10^{33} \text{ erg s}^{-1} . \tag{5.104}$$

By way of comparison, Van Den Bergh (1988) estimates that the absolute luminosity, L_G of our Galaxy is

$$L_G = (2.3 \pm 0.6) \times 10^{10} L_\odot . \tag{5.105}$$

The intrinsic brightness of a celestial object can also be described in terms of the magnitude scale invented by the ancient Greeks. The fourteen brightest stars visible from Greece were called stars of the "first magnitude", while the faintest stars visible to the naked eye were said to be of the "sixth magnitude". There are about 5,000 stars that we can see with the unaided eye across the celestial sphere. The 6th magnitude stars were about 1/100th the brightness of the 1st magnitude stars, and each of the six magnitude groups was 2.5 times brighter than the next faintest one. Even today, we retain their peculiar convention in which fainter celestial objects have larger magnitudes. Some examples of apparent magnitudes are the Sun at -27 mag, the full moon at -12 mag, Venus at about -4 mag when brightest, the naked eye limit for stars at about $+6$ mag, a visual limit of nearly $+15$ mag for a 12-inch telescope, and about $+30$ mag with an 18-hour exposure with the Hubble Space Telescope. This encompasses a range of fifty-seven magnitudes of apparent brightness, which is equivalent to a brightness ratio of nearly 10^{23}, or 100 billion trillion.

In the magnitude system, the measure of brightness is logarithmic, and the scale factor is such that a difference of five magnitudes corresponds to an intensity ratio of 100. For an object of apparent luminosity, l, the apparent magnitude, m, is given by:

$$m = -2.50 \log \, l + \text{constant} , \tag{5.106}$$

where log is the common logarithm to the base 10. The apparent magnitudes, m_1 and m_2, of two objects are therefore related by

$$m_1 - m_2 = -2.50 \log(l_1/l_2) , \tag{5.107}$$

or

$$l_1/l_2 = 2.512^{-(m_1 - m_2)} = 10^{-0.4(m_1 - m_2)} . \tag{5.108}$$

The constant $2.512 = 10^{2/5}$ has a common logarithm of 0.4; it was introduced by Pogson (1856) to calibrate the magnitude scale.

The absolute magnitude, M, of a celestial object is the magnitude the object would have at a distance of 10 parsecs. Using the previous equation with the inverse square law, we can define a distance modulus, μ_o, by:

$$\mu_o = m - M = 5 \log D - 5 \quad \text{for } D \text{ in parsecs} , \tag{5.109}$$

or

$$\mu_o = m - M = -5(1 + \log \pi) , \tag{5.110}$$

where π is the trigonometric (annual) parallax in seconds of arc. Stellar astronomers usually measure distances in parsecs, but the distances to galaxies are often specified in Megaparsecs. The distance modulus is therefore alternatively defined by:

$$\mu_o = m - M = 5\log D + 25 \quad \text{for } D \text{ in Megaparsecs} \ . \tag{5.111}$$

As light travels through the space between the stars it is absorbed and scattered by interstellar dust. The absorption, or extinction, of light by interstellar dust diminishes the light intensity and increases the apparent magnitude by an amount A, and the distance modulus has to be corrected for this to yield the true distance modulus:

$$\begin{aligned} \mu_o &= m - M - A = 5\log D - 5 \quad \text{for } D \text{ in parsecs} \ . \\ \mu_o &= m - M - A = 5\log D + 25 \quad \text{for } D \text{ in Megaparsecs} \ . \end{aligned} \tag{5.112}$$

Interstellar dust is confined primarily to the plane of our Galaxy, typically increasing stellar magnitudes by about a magnitude per kiloparsec. The amount of interstellar absorption, A_λ, is a function of wavelength, λ, falling sharply as the wavelength increases in the standard photometric regions.

The apparent and absolute magnitudes depend on what wavelength or color is being observed, for they are actually measured over a small portion of the electromagnetic spectrum. Filter characteristics of astronomical photometric systems are given in Table 5.4. Quantities measured in a given wavelength region are usually denoted by a subscript that specifies the region. An example is the UBV, or ultraviolet-blue-visual, system using filters centered at 3650, 4400 and 5500 Angstroms. For the Sun we have (Lang, 1992)

$$\begin{aligned} m_{V\odot} &= -26.78 \text{ mag} \\ M_{V\odot} &= +4.82 \text{ mag} \\ \mu_{o,\odot} &= (m - M)_\odot = -31.6 \text{ mag} \\ m_B - m_V &= B - V = 0.68 \pm 0.005 \\ m_U - m_B &= U - B = 0.17 \pm 0.01 . \end{aligned} \tag{5.113}$$

Table 5.4. Central wavelengths and passbands of photometric filters used in stellar astronomy. The passband is the full-width to half intensity. One Angstrom is equivalent to 10^{-8} cm

System-Filter	Wavelength (Angstroms)	Passband (Angstroms)	System-Filter	Wavelength (Angstroms)	Passband (Angstroms)
UBV-			Infrared-		
U	3650	700	R	7000	2200
B	4400	1000	I	8800	2400
V	5500	900	J	12000	3800
			K	22000	4800
Six Colors-			L	34000	7000
			M	50000	12000
U	3550	500	N	104000	57000
V	4200	800			
B	4900	800	uvby β		
G	5700	800			
R	7200	1800	u	3500	340
I	10300	1800	v	4100	200
			b	4700	160
			y	5500	240
			β	4860	30150

Interstellar absorption also depends on the spectral region used, for interstellar dust scatters blue light more than red light. This means that the light of a celestial object is reddened during its voyage through space; the same effect in the Earth's atmosphere explains the red color of sunsets. The interstellar reddening of starlight can be described by the absorption, A_{V}, in the visual band, the color excess E(B – V) due to the difference in interstellar absorption in the blue, B, and visual, V, passbands, and the difference $m_{\mathrm{B}} - m_{\mathrm{V}} = \mathrm{B} - \mathrm{V}$ of the apparent magnitudes observed through the two bands. Thus, we have:

$$A_{\mathrm{V}} = R \times \mathrm{E(B - V)} \ , \tag{5.114}$$

where the ratio of total to selective absorption, or the reddening factor R, is about 3.0. Once R is known, a measurement of the color excess can be converted into the visual extinction, $\mathrm{A_V}$, in magnitudes. The reddening factor is given by (Olson, 1975)

$$R = 3.25 + 0.25(\mathrm{B - V})_{\mathrm{o}} + 0.05\mathrm{E(B - V)} \ , \tag{5.115}$$

where the subscript o for B – V refers to the intrinsic, unreddened color. The formula for R is valid for $(\mathrm{B - V})_{\mathrm{o}} < 1.4$ mag and $\mathrm{E(B - V)} < 1.5$ mag. Burstein and Heiles (1982) provide contour maps of E(B – V) that can be combined with R to give A_{V}.

Alternatively, the extinction can be related to the column density of neutral hydrogen, N_{H}, which is detected by radio astronomers through its emission near 21 cm wavelength. This is because the amount of interstellar dust is proportional to the amount of hydrogen in the space between the stars. Burstein and Heiles (1978) give:

$$A_{\mathrm{B}} = (R + 1)\mathrm{E(B - V)} = A_{\mathrm{V}} + \mathrm{E(B - V)} = -0.149 + 6.41 \times 10^{-4} N_{\mathrm{H}} \ , \tag{5.116}$$

where N_{H} is in units of 10^{18} atoms per square centimeter, and can be determined as a function of galactic longitude, l, and latitude, b, by:

$$N_{\mathrm{H}} = \begin{cases} 323\csc|b| & b < 0 \\ 323\csc|b| + [105 - 3.8b - 89\cos(l - 140°)], & 2b > 0 \ . \end{cases} \tag{5.117}$$

Average values of $\mathrm{A}_{\lambda}/\mathrm{E(B - V)}$ at other wavelengths, λ, from 0.001 to 0.0001 cm are given by Savage and Mathis (1979) and reproduced by Rowan-Robinson (1985).

The color indices, CI, that specify the apparent magnitude differences $m_{\mathrm{B}} - m_{\mathrm{V}} = \mathrm{B} - \mathrm{V}$ or $m_{\mathrm{U}} - m_{\mathrm{B}} = \mathrm{U} - \mathrm{B}$ are tabulated by Lang (1992) for main-sequence, giant and supergiant stars of different spectral class. The positions, apparent magnitudes and color indices U – B and B – V for about 1,500 bright stars are given in *The Astronomical Almanac*; it also contains the color indices U – B, B – V, V – R and V – I for 107 photometric (UBVRI) standard stars. The two-color diagram of U – B and B – V for over 29,000 stars is given in Fig. 5.8.

The total radiant flux integrated over all wavelengths defines the apparent bolometric magnitude, m_{bol}, and the absolute bolometric magnitude, M_{bol}. The bolometric correction required to obtain this total output from the visual one is:

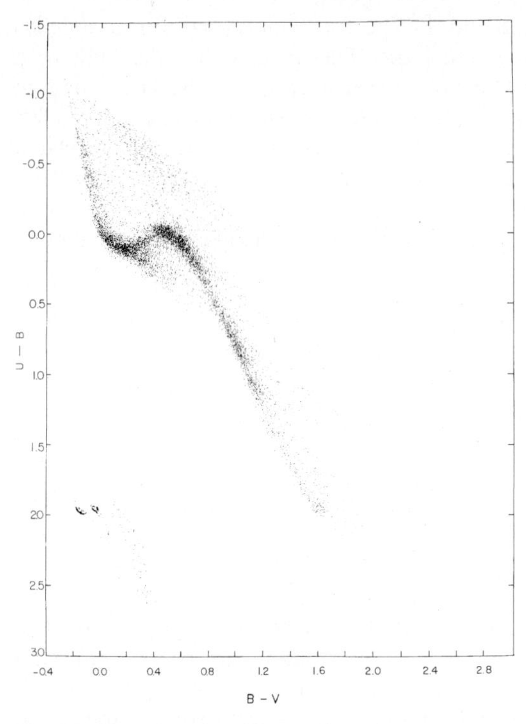

Fig. 5.8. The two-color diagram for over 29,000 stars. The difference between the apparent magnitudes, m, are plotted for the ultraviolet, U, blue, B, and visual, V, spectral regions with $U - B = m_U - m_B$ and $B - V = m_B - m_V$. (Courtesy of the Geneva Observatory)

$$\text{Bolometric Correction} = \text{BC} = M_V - M_{\text{bol}} = m_v - m_{\text{bol}}, \tag{5.118}$$

but a definition with the reversed sign, or BC $= M_{\text{bol}} - M_V$ is also used. For the Sun we have (Lang, 1992):

$$\begin{aligned} m_{\text{bol},\odot} &= -26.85 \text{ mag} \\ M_{\text{bol},\odot} &= 4.75 \text{ mag} \\ \text{BC}_\odot &= +0.17 \text{ mag} . \end{aligned} \tag{5.119}$$

Once the bolometric magnitude of a celestial object is known, we can infer the absolute luminosity from:

$$\log(L/L_\odot) = 0.4(4.75 - M_{\text{bol}}) = 0.4(M_{\text{bol},\odot} - M_{\text{bol}}) \tag{5.120}$$

or equivalently

$$L = 10^{0.4(4.75 - M_{\text{bol}})} L_\odot , \tag{5.121}$$

where

$$L_\odot = 3.85 \times 10^{33} \text{ erg s}^{-1} .$$

The absolute luminosity, L, of a star can be used to specify its effective temperature, T_{eff}, using the Stefan–Boltzmann law

$$T_{\mathrm{eff}} = [L/(4\pi\sigma R^2)]^{1/4} \ , \tag{5.122}$$

where the Stefan–Boltzmann constant $\sigma = 5.670 \times 10^{-5}$ erg s^{-1} cm^{-2} K^{-4}, and R is the radius of the celestial object. The effective temperature is the surface temperature that the object would have if it were a perfect black body radiating its given absolute luminosity. For the Sun we have:

$$\begin{aligned} R_{\odot} &= 6.96 \times 10^{10} \text{ cm} \\ T_{\mathrm{eff},\odot} &= 5780\mathrm{K} \ , \end{aligned} \tag{5.123}$$

and for any other star

$$\log(L/L_{\odot}) = 4\log(T_{\mathrm{eff}}/T_{\mathrm{eff},\odot}) + 2\log(R/R_{\odot}) \ .$$

Radio astronomers measure the flux density, $S(\nu)$, at frequency, ν, where

$$S(\nu) = \frac{f(\nu)}{\Delta\nu} \text{ erg s}^{-1} \text{ cm}^{-2} \text{ Hz}^{-1} \ . \tag{5.124}$$

Here $f(\nu)$ is the energy flux at frequency, ν, and $\Delta\nu$ is the bandwidth of the filter used. The conventional unit for radio flux density is the flux unit, f.u., or Jansky, Jy, given by:

$$1 \text{ f.u.} = 1 \text{ Jy} = 10^{-26} \text{ watt m}^{-2} \text{ Hz}^{-1} = 10^{-23} \text{ erg s}^{-1} \text{ cm}^{-2} \text{ Hz}^{-1} \ . \tag{5.125}$$

Solar radio astronomers often use the solar flux unit, s.f.u., given by:

$$1 \text{ s.f.u.} = 10^4 \text{ f.u.} = 10^4 \text{ Jy} \ . \tag{5.126}$$

The observed flux density, $S_{\mathrm{o}}(\nu)$, depends on the source flux density, $S_{\mathrm{s}}(\nu)$, the antenna efficiency, η_{A}, the antenna solid angle, Ω_{A}, and the solid angle extent of the source, Ω_{s}.

$$S_{\mathrm{o}}(\nu) = \eta_{\mathrm{A}} S_{\mathrm{s}}(\nu) \qquad \text{for } \Omega_{\mathrm{s}} \ll \Omega_{\mathrm{A}} \tag{5.127}$$

$$S_{\mathrm{o}}(\nu) = \eta_{\mathrm{A}} \frac{\Omega_{\mathrm{A}}}{\Omega_{\mathrm{s}}} S_{\mathrm{s}}(\nu) \quad \text{for } \Omega_{\mathrm{s}} \geq \Omega_{\mathrm{A}}. \tag{5.128}$$

For a nonthermal radiator of spectral index, α, the total luminosity, L, is given by

$$L = 4\pi D^2 S_{\mathrm{s}}(\nu_{\mathrm{o}}) \nu_{\mathrm{o}}^{\alpha} \int_{\nu_1}^{\nu_2} \nu^{-\alpha} \, d\nu \ , \tag{5.129}$$

where the source distance is D, the source flux density at frequency, ν_{o}, is $S_{\mathrm{s}}(\nu_{\mathrm{o}})$, and the source flux density is assumed to be negligible for frequencies, ν, such that $\nu \leq \nu_1$ and $\nu \geq \nu_2$.

When a radiator is thermal, its brightness, $B_\nu(T)$, at frequency, ν, is given by (Planck, 1901, 1913; Wien, 1893, 1894; Rayleigh, 1900, 1905; Jeans, 1905, 1909; Milne, 1930).

$$\begin{aligned} B_\nu(T) &= \frac{S_{\mathrm{s}}(\nu)}{\Omega_{\mathrm{s}}} = \frac{2h\nu^3}{c^2} \frac{1}{\left[\exp\left(\frac{h\nu}{kT}\right) - 1\right]} \text{ erg s}^{-1} \text{ cm}^{-2} \text{ Hz}^{-1} \text{ ster}^{-1} \ , \\ &= \frac{2h\nu^3}{c^2} \exp(-h\nu/kT) \text{ if } h\nu > kT \text{ Wien's Law} \\ &= \frac{2\nu^2 kT}{c^2} \text{ if } h\nu < kT \text{ Rayleigh–Jeans Law} \end{aligned} \tag{5.130}$$

where Ω_s is the solid angle extent of the source, T is the brightness temperature of the source, c is the velocity of light, and h and k are, respectively, Planck's and Boltzmann's constants. The Rayleigh–Jeans approximation is used at radio frequencies, $\nu \approx 10^9$ Hz, for the criterion $h\nu < kT$, or $\nu < 2 \times 10^{10}\, T$, is valid. Wien's approximation is sometimes used at optical frequencies, where $\nu \approx 10^{15}$ Hz.

The observed brightness, $B_o(\nu)$, is related to the source brightness, $B_s(\nu)$, by the relations

$$\begin{aligned} B_o(\nu) &= \eta_A \frac{\Omega_s}{\Omega_A} B_s(\nu) = \frac{S_o(\nu)}{\Omega_A} \quad \text{for } \Omega_s \ll \Omega_A \\ B_o(\nu) &= \eta_A B_s(\nu) = \frac{S_o(\nu)}{\Omega_A} \quad \text{for } \Omega_s \geq \Omega_A \ , \end{aligned} \tag{5.131}$$

where Ω_s and Ω_A denote, respectively, the solid angle extent of the source and the antenna, and η_A is the antenna efficiency. The temperature calculated from these equations is called the equivalent brightness temperature. It is the temperature of a black body radiator which has the same relative intensity distribution as the source under observation.

Accurate distances can only be inferred from the apparent magnitudes, or apparent luminosity, if we know the absolute magnitude, or absolute luminosity. However, not all celestial objects shine with the same brightness. Unless we understand the energizing process, we cannot tell if dim objects are really farther away than bright ones, or if their internal fires are simply cooler, throwing off less light. The problem of measuring astronomical distances is therefore reduced to a search for a "standard candle" that always shines with the same intrinsic brightness.

The absolute luminosities and distances of stars can, for example, be determined from their spectral type. Secchi (1868) and Rutherfurd (1863) first noted that stars of different colors have very different spectra. Secchi divided the stars into four groups with the colors white, yellow, orange, and red, and with the respective spectral features of strong absorption lines of hydrogen, strong calcium lines, strong metallic lines and broad absorption lines which are more luminous on the violet side. When larger numbers of stars were observed, letters were introduced to denote a wider variety of color and spectral types (Pickering and Fleming, 1897; Pickering, 1889). The development of the Henry Draper Catalogue of 225,330 stars (Maury and Pickering, 1897; Cannon and Pickering, 1918–1924), together with additions and improvements by Morgan, Keenan and Kellman (1943), has led to the spectral classification, Sp, of:

$$\text{Q, P, W, O, B, A, F, G, K, M, S, R, and N} \ . \tag{5.132}$$

Within the standard sequence beginning with O, the bluest stars are labeled O, white stars A, yellow G, orange K and red M. The spectral characteristics of stars of different spectral class, including the luminosity classes I, II, and III are given by Lang (1992).

The absolute visual magnitudes, M_V, bolometric corrections, BC, absolute bolometric magnitudes, M_{bol}, absolute luminosities, L, and effective temperatures, T_{eff}, for stars of different spectral type, Sp, are tabulated by Lang (1992) for main-sequence, giant and supergiant stars.

So, the apparent and absolute luminosities can be combined with the inverse-square law to determine distance, and for ordinary stars we can infer the absolute luminosity from their spectral characteristics. As next discussed, another technique is used to determine the absolute luminosity of certain variable stars from their period of brightness variation.

5.2.5 Absolute Luminosity of Variable Stars

Cepheid variable stars, named after their prototype δ Cephei, are stars whose intrinsic brightness changes periodically with time, producing regular fluctuations in their apparent luminosity with a period of 2 to 150 days, depending on the star. They are also very bright stars with absolute visual magnitudes of $M_V = -2$ to -7. Since they are so luminous, they can be seen at exceptionally large distances, including nearby galaxies.

At the turn of the twentieth century, Henrietta Leavitt examined Cepheid variables in the Magellanic Clouds, which are the nearest galaxies to our own, orbiting it in close companionship. By 1908 she reported that the brighter variable stars in the Magellanic Clouds tended to have the longer cycles of variation, and four years later Leavitt showed that the variation period of Cepheids in the Small Magellanic Cloud increase linearly with decreasing apparent magnitude (Leavitt, 1908, 1912). Because the extent of the Magellanic Clouds is small compared to their total distance, the relation of period to apparent brightness implied also a real connection with absolute brightness or luminosity, establishing the important period-luminosity relation. Once this relation is suitably calibrated, observation of the period, P, of a Cepheid variable star leads to a determination of its absolute magnitude, M; then using the observed apparent magnitudes, the distance of the star can be calculated using the inverse-square law or our expression for the distance modulus.

For more than 80 years, researchers have sought to exactly calibrate the period-luminosity (PL) relation for Cepheid variables. At visible wavelengths, V, astronomers establish the zero point, a, and slope, b, in the PL relation

$$M_V = -a - b \log P \ , \qquad (5.133)$$

where M_V is the absolute visual magnitude and P is the period in days. The fascinating early history of the calibration has been reviewed by Fernie (1969), but also see Sandage (1972). A few highlights include Ejnar Hertzsprung's initial calibration of Leavitt's relation (Hertzsprung, 1913), Harlow Shapley's use of a similar zero point to establish the vast size of the Milky Way (Shapley, 1918), and Edwin Hubble's use of Cepheids to demonstrate the extragalactic nature of nearby spiral nebulae (Hubble, 1925). The calibration was dramatically corrected in 1952 by Walter Baade, who effectively doubled the size of the known Universe and showed that there are two classes of variable stars, with different period-luminosity relations (Baade, 1952).

More recently, Gieren, Barnes and Moffett (1993) determined the distance to 100 galactic classical Cepheids, and obtained a period-luminosity, PL(V), relation in the V bandpass of

$$M_V = -1.371 - 2.986 \log P \ , \tag{5.134}$$

with respective uncertainties of ± 0.095 and ± 0.094 in the zero point and slope. This relation is shown in Fig. 5.9 together with the observed data. By way of comparison, Caldwell and Coulson (1986) and Feast and Walker(1987) find that

$$\langle M_V \rangle = -1.35 - 2.78 \log P \tag{5.135}$$

for the Magellanic Clouds, where the brackets $\langle \ \rangle$ denote an average value, and that one can use

$$\langle M_B \rangle = -1.16 - 2.24 \log P \tag{5.136}$$

for extragalactic objects in the blue, B, band.

More recently, Feast and Catchpole (1997) used trigonometric parallaxes of Cepheid variable stars, obtained from the Hipparcos satellite, to derive a zero point for the period-luminosity relation. Applying a slope from the Large Magellanic Cloud, they find the relation:

$$\langle M_V \rangle = -1.43 - 2.81 \log P \ , \tag{5.137}$$

with a standard error in the zero point of 0.10 mag (also see Sandage and Tammann, 1998). That means that a Cepheid variable of a given period is slightly brighter and therefore about 10 percent further away than was previously thought.

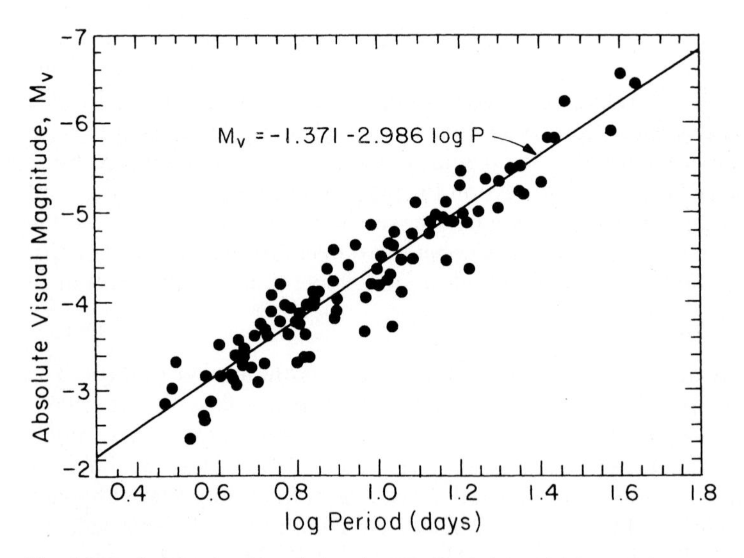

Fig. 5.9. Period-luminosity relation for 100 Cepheid variable stars in our Galaxy. Here M_V is the absolute visual magnitude, P is the period in days, and the line denotes the least-squares fit to the data $M_V = -1.371 - 2.986 \log P$, with an uncertainty of ± 0.095 in the numerical coefficients. (Adapted from Gieren, Barnes and Moffett, 1993)

The infrared (JHK) period-luminosity relation for the Magellanic Clouds is given by (Laney and Stobie, 1986; Feast and Walker, 1987)

$$\langle M_{\rm H} \rangle = -2.14 - 3.42 \log P \ . \tag{5.138}$$

At a wavelength of 0.000105 cm (IV band) Cepheid observations in the Magellanic Clouds yield (Visvanathan, 1985; Feast and Walker, 1987)

$$\langle M_{\rm IV} \rangle = -1.88 - 3.14 \log P \ . \tag{5.139}$$

Cepheids are important distance indicators both because they are luminous, and also because their variations are well understood theoretically. Massive stars become pulsationally unstable in the later stages of their evolution, after exhausting the hydrogen fuel in their cores, with a pulsation period, P, that is proportional to $\rho^{-1/2}$ where ρ is the mean density of the star.

$$P = Q_{\rm o}(\rho_{\odot}/\rho)^{1/2} \ , \tag{5.140}$$

where the pulsation constant $Q \approx 0.116$ days and the mean density of the Sun is $\rho_{\odot} = 1.41$ g cm^{-3}.

As shown by Sandage (1958), we can obtain a period-luminosity-color (PVC) relation that depends on the color index B – V:

$$M_{\rm V} = n - m \log P + q \ ({\rm B - V}) \ . \tag{5.141}$$

Different determinations of the constants n, m and q have been tabulated by Rowan-Robinson (1985). As an example, Iben and Tuggle (1972) used detailed pulsation and evolution theory for Cepheids to obtain

$$\langle M_{\rm V} \rangle = -2.61 - 3.76 \log P + 2.60 \ ({\rm B - V}) \ . \tag{5.142}$$

Feast and Walker (1987) have reviewed the use of Cepheids as distance indicators, obtaining:

$$\langle M_{\rm V} \rangle = -2.27 - 3.80 \log P + 2.70 \ (\langle {\rm B_o} \rangle - \langle {\rm V_o} \rangle) \tag{5.143}$$

for the Magellanic Clouds where the brackets $\langle \ \rangle$ denote an average value and the subscript o denotes the intrinsic value (also see Martin, Warren and Feast, 1979). For metal-poor Cepheids in the Magellanic Clouds, we have (Feast and Walker, 1987; Caldwell and Coulson, 1986)

$$\langle M_{\rm V} \rangle = -2.13 - 3.53 \log P + 2.13 \ (\langle {\rm B_o} \rangle - \langle {\rm V_o} \rangle) \ . \tag{5.144}$$

Feast and Walker also give

$$\langle M_{\rm V} \rangle = -3.13 - 3.66 \log P + 3.71 \ ({\rm V - I})_{\rm o} \tag{5.145}$$

for a PLC relation in V and I, and

$$\langle M_{\rm H} \rangle = -3.59 - 3.64 \log P + 3.28 \langle {\rm J - K} \rangle_{\rm o} \tag{5.146}$$

for an infrared PLC relation.

One of the methods of calibrating the Cepheid period-luminosity relation is to obtain the radii, and hence the absolute magnitudes of the stars. For a star of radius, R, and effective temperature, $T_{\rm eff}$, the bolometric magnitude, $M_{\rm bol}$, is given by

$$M_{\rm bol} = -5 \log R - 10 \log T_{\rm eff} + M_{{\rm bol},\odot} + 10 \log T_{{\rm eff},\odot} + 5 \log R_{\odot} \ , \tag{5.147}$$

where the solar values are $M_{bol,\odot} = 4.75$ mag, $T_{eff,\odot} = 5780$ K, and $R_\odot = 6.955 \times 10^{10}$ cm and the effective temperature can be inferred from the spectral features or color of the star.

As suggested by Baade (1926), and modified by Wesselink (1946) – also see Becker (1940), the linear radius of a pulsating star can be inferred from measurements of its changing apparent brightness and radial velocity, as determined from the Doppler shift of spectral lines. This technique, now known as the Baade–Wesselink method, leads to a period-radius relation for variable stars.

It follows from the Stefan–Boltzmann law that the change, Δl, in apparent luminosity, l, is given by

$$\frac{\Delta l}{l} = \frac{2\Delta R}{R} = k\frac{\Delta V_r \Delta t}{R} \ , \tag{5.148}$$

where ΔR is the change in radius, R, the $\Delta V_r \Delta t$ denotes the time integral of the change ΔV_r in the observed radial velocity V_r, and the constant k lies between 3/2 and 4/3.

The formulation given by Balona (1977) is

$$\Delta M = a\Delta C + b\Delta R + c \ , \tag{5.149}$$

where a, b and c are constants, ΔM is the change in magnitude, ΔC the change in color $(B - V)$, and the radius $R = \langle R \rangle + \Delta R$, where $\langle R \rangle$ is the mean radius and the radius variation, ΔR, is obtained by integrating the observed radial velocity of the star over the variability cycle. That is,

$$\Delta R = kP \int V_r \, d\theta \ , \tag{5.150}$$

where P is the pulsation period, θ is the phase, and the constant $k \approx 1.31$ accounts for geometrical projection and limb darkening. The constant a is the slope of the color-surface brightness curve, the constant b is related to the mean radius $\langle R \rangle$ of the star, and given by

$$b = 5 \log_{10} e / \langle R \rangle \ . \tag{5.151}$$

Bolona (1977) shows how to obtain the constants a, b and c, as well as $\langle R \rangle$, using a maximum-likelihood fit to the observables ΔM, ΔC and ΔR.

The visual surface brightness method for inferring stellar radii may be considered as a modern version of the classical Baade–Wesselink technique. It was first developed by Barnes and Evans (1976), and later improved by Barnes, Evans and Moffett (1978), who found an empirical correlation between the visual surface brightness and the visual-to-red (V – R) color index for stars which have had their angular radii measured by interferometric methods.

The visual surface brightness parameter, F_V, is defined as

$$F_V = 4.2207 - 0.1 V_o - 0.5 \log \phi \ . \tag{5.152}$$

V_o is the apparent visual magnitude, corrected for interstellar reddening, and ϕ is the stellar angular diameter in milliseconds of arc. F_V can also be expressed as

$$F_V = \log T_{eff} + 0.1 \ \mathrm{BC} \ , \tag{5.153}$$

where T_{eff} is the effective temperature and BC the bolometric correction with respect to the V bandpass. Using stars with known angular diameters, Barnes and Evans (1976) obtained the empirical relation

$$F_V = b + m(V - R) \ . \tag{5.154}$$

If the values of the constants b and m are known, then F_V can be obtained from the (V – R) color of a star; this yields its angular diameter.

In the case of a Cepheid, the instantaneous linear displacements $2\Delta R$ of the pulsating stellar surface can be obtained by integrating the observed radial velocity curve. The linear diameter of the star, $2R$, in astronomical units, is related to its angular diameter at the same phase in milliseconds of arc by

$$2R = 10^{-3} D\phi \ , \tag{5.155}$$

where D is the distance of the star in parsecs. For a radial pulsator like a Cepheid, the instantaneous linear diameter $2R$ can be written as the sum of the mean diameter $2\langle R \rangle$, and the instantaneous displacement from the mean, $2\Delta R$, or

$$2(\Delta R + \langle R \rangle) = 10^{-3} D\phi \ . \tag{5.156}$$

A regression analysis of ϕ against $2\Delta R$ thus yields both the distance and the mean diameter of the star. While the mean diameter depends only on m, the slope of the F_V equation, the distance of the star obtained this way is dependent on both the slope m and the zero point b of this relation.

Period-radius, or PR, relations for Cepheid variable stars have been reviewed by Fernie (1984), who obtained a weighted mean theoretical relation

$$\log R = 1.179 + 0.692 \log P \ , \tag{5.157}$$

where R is the radius in solar radii, P is the period in days, and the uncertainty in the numerical coefficients is ± 0.006. Measured values of these coefficients have been tabulated by Moffett and Barnes (1987) for different observational results, including those obtained by the Baade–Wesselink and the surface-brightness methods. The theoretical PR relation lies in the middle of these solutions. For example, the theory relation predicts a radius of 266 solar radii for a 63 day Cepheid, and a weighted mean of eight recent measurements yields a radius of 256 solar radii (only a 3.15% difference). Gieren (1988) discussed the observed Cepheid period-luminosity relation obtained from the visual surface brightness method, and Gieren, Barnes and Moffett (1989) discuss the galactic period-radius relation for classical Cepheids. The theoretical period-radius relation of classical Cepheids is discussed by Bono, Caputo and Marconi (1998).

Finally, there are the RR Lyrae pulsating stars that have short variation periods of a few hours to about one day and are less luminous than the longer period (1 to 150 days) Cepheid variables. The mean absolute visual magnitude, $\langle M_{\mathrm{v}} \rangle$, of RR Lyrae stars is $\langle M_{\mathrm{v}} \rangle = 0.6$ to 0.7 mag (Tsujimoto, Miyamoto and Yoshii, 1998), so these variable stars can only be used to determine relatively nearby distances. In contrast the long-period Cepheid variables with $\langle M_{\mathrm{v}} \rangle = -2$ to -7 can be seen to much greater distances. However, RR Lyrae stars can be used to estimate the distances of the globular star clusters in which they are found. The distances of about 150 globular star clusters in our Galaxy are given by Lang (1992) and also in *The Astronomical Almanac*. RR Lyrae variables are relatively old, Population II stars, while the Cepheids belong to the younger Population I.

5.2.6 Kinematic Distance and Galactic Structure

The stars in our Milky Way Galaxy, and the gas and dust between them, are rotating differentially about a distant galactic center, following Kepler's third law in which rotation velocity decreases as the square root of distance from the center. Slipher (1917) observed rotation in spiral nebulae, now known to be galaxies that resemble the Milky Way. Shapley (1918) showed that the system of globular clusters is centered at a remote galactic center rather than the Sun. Lindblad (1925) then suggested that the high apparent velocities of clusters could be accounted for if both the Sun and globular star clusters were rotating about the galactic center. Oort (1927, 1928) and Joy (1939) provided some observational evidence for differential galactic rotation; Oort put forth the following formulae for circular motion about the galactic center (also see Burton, 1988).

The Doppler-shift velocity along the line-of-sight, or the radial velocity V_r, of a galactic object at the distance, R, from the galactic center is given by

$$V_r = R_o[\omega(R) - \omega(R_o)] \sin l \qquad (5.158)$$

or

$$V_r \approx R_o \left(\frac{d\omega}{dR}\right)_{R=R_o} (R - R_o) \sin l \quad \text{for } R - R_o \ll R_o \qquad (5.159)$$

where R_o is the distance of the local standard of rest from the galactic center, $\omega(R)$ is the circular angular velocity of the Galaxy at R, and l denotes the galactic longitude. The proper motion μ, of the celestial object in galactic longitude is

$$\mu = \frac{1}{4.74}\left[-\frac{1}{2}R_o\left(\frac{d\omega}{dR}\right)_{R=R_o} \cos 2l - \frac{1}{2}R_o\left(\frac{d\omega}{dR}\right)_{R=R_o} - \omega(R_o)\right] , \qquad (5.160)$$

where μ is in second of arc per year. Using Oort's constants of differential galactic rotation, A and B, given by

$$A = -\frac{1}{2}R_o\left(\frac{d\omega}{dR}\right)_{R=R_o} , \qquad (5.161)$$

which is a measure of the local rate of shear of galactic rotation, and

$$B = -\frac{1}{2}R_o\left(\frac{d\omega}{dR}\right)_{R=R_o} - \omega(R_o) \qquad (5.162)$$

we obtain

$$V_r = -2A(R - R_o) \sin l \text{ for } R - R_o \ll R_o , \qquad (5.163)$$

and

$$\mu = \frac{1}{4.74}[B + A \cos 2l] . \qquad (5.164)$$

A review of galactic constants gives (Kerr and Lynden-Bell, 1986)

$$\begin{aligned} R_o &= 8.5 \pm 1 \text{ kpc} \\ \theta_o &= 222 \pm 20 \text{ km s}^{-1} = R_o\, \omega(R_o) \\ A &= 14.4 \pm 1.2 \text{ km s}^{-1} \text{ kpc}^{-1} \\ B &= -12.0 \pm 2.8 \text{ km s}^{-1} \text{ kpc}^{-1} \end{aligned} \quad (5.165)$$

and

$$A - B = 26.4 \pm 1.9 \text{ km s}^{-1} \text{ kpc}^{-1} \ . \quad (5.166)$$

Reid (1989) obtains

$$R_o = 7.7 \pm 0.9 \text{ kpc} \ . \quad (5.167)$$

Estimates for the distance to the center of the Galaxy are reviewed by Reid (1993). Galactic kinematics of Cepheids from Hipparcos proper motions indicate that $A = 14.82 \pm 0.84$ km s^{-1} kpc^{-1} and $B = -12.7 \pm 0.64$ km s^{-1} kpc^{-1}, with $R_o = 8.5 \pm 0.5$ kpc (Feast and Whitelock, 1997).

The International Astronomical Union has adopted the standard values of (Kerr and Lynden-Bell, 1986)

$$\begin{aligned} R_o &= 8.5 \text{ kpc} \\ \theta_o &= 220 \text{ km s}^{-1} = R_o\, \omega(R_o) \end{aligned} \quad (5.168)$$

which implies

$$A - B = 25.9 \text{ km s}^{-1} \text{ kpc}^{-1} \ . \quad (5.169)$$

A measurement of the radial velocity, V_r, may lead to a kinematic measurement of the distance, D, of a galactic object from the local standard of rest. For an object in the galactic plane and within one kiloparsec of the Sun, for example, the radial component of velocity due to differential galactic rotation is:

$$V_r = A\, D \sin 2l \quad (5.170)$$

where

$$R^2 = R_0^2 + D^2 - 2R_o D \cos l \ . \quad (5.171)$$

The tangential component, V_t, of velocity due to differential rotation for a nearby object within about a kiloparsec from the Sun is given by

$$V_t = D[A \cos(2l) + B] \ . \quad (5.172)$$

For an object within the galactic plane, outside the central galactic bulge, and inside the Sun's orbit around the center, for 3 kpc $< R < R_o = 8.5$ kpc, the galactic rotation curve, that specifies angular rotation velocity as a function of the distance from the galactic center, R, can be determined from Doppler-shift observations of the 21-cm spectral line of interstellar atomic hydrogen. Within this region, the distance from the galactic center reaches a minimum, and the rotation velocity has a maximum, at the "tangent point". This point is called the "tangent point", or the "subcentral point", because it lies where the line-of-sight is tangent to a galactocentric circle. At this point, $R = R_{\min} = R_o|\sin l|$ and the radial velocity, $V_{r,\max}$, is given by

$$V_{r,\max} = R_o[\omega(R_{\min}) - \omega(R_o)] \sin l \quad \text{for } 3 \text{ kpc} < R < R_o$$

or

$$\begin{aligned} V_{r,\max} &= 2\,AR_o \sin l(1 - |\sin l|) \\ &= R_o\left[\frac{\theta_c(R)}{R} - \frac{\theta_c(R_o)}{R_o}\right] \sin l \quad \text{for } 3 \text{ kpc } < R < R_o \end{aligned} \tag{5.173}$$

from which we can obtain a rotation curve $R\omega(R) = \theta_c(R)$ for different R.

For $R < R_o$, a single value of radial velocity, V_r, corresponds to two values of R, and there is an ambiguity in determining the distance to a galactic object. This ambiguity is resolved by obtaining $\omega(R)$ from V_r, and then comparing $\omega(R)$ to models which give $\omega(R)$ as a function of R. A simple analytic expression of this rotation curve is (Burton, 1988):

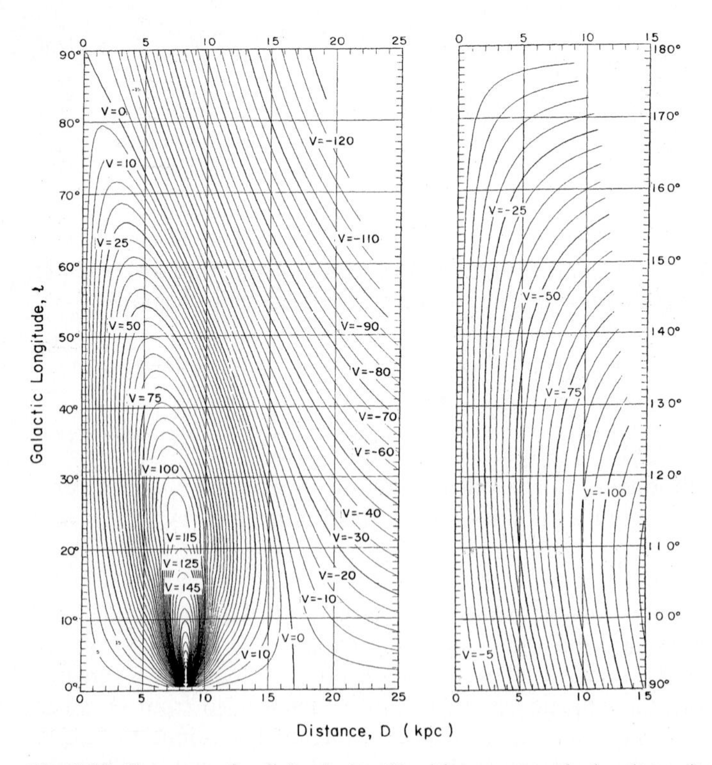

Fig. 5.10. Contours of radial velocity, V, with respect to the local standard of rest, plotted as a function of distance, D, from the Sun in kiloparsecs, where 1 kpc $= 3.0857 \times 10^{21}$ cm, and the galactic longitude, ℓ. (Adapted from Burton, 1988)

$$\frac{\theta_c(R)}{\theta_o} = \frac{R\omega(R)}{\theta_o} = 1.0074\left(\frac{R}{R_o}\right)^{0.0382} + 0.00698 \quad \text{for } 3 \text{ kpc} < R < R_o , \tag{5.174}$$

where $R_o = 8.5$ kpc and $\theta_o = 220$ km s^{-1}. Velocities with respect to the local standard of rest, using this rotation curve, are plotted as a function of distance, D, from the Sun in Fig. 5.10. This figure can be used to assign a kinematic distance for objects of known radial velocity and at low galactic latitude, or within the galactic plane.

Kinematic distances cannot be obtained in the outer-Galaxy where $R > R_o$. There is no unique velocity corresponding to a known unique distance, as there is for the direction through the inner Galaxy where $R < R_o$.

As illustrated in Fig. 5.11, the luminous O and B stars, and their associated regions of ionized hydrogen or H II regions, delineate the spiral shape of our Galaxy. Unlike the other stars, these comparatively ephemeral, hot O and B stars are young enough to still be located near the dense interstellar concentrations where they were formed and hence outline the nearby sections of the spiral arms (Morgan, Sharpless and Osterbrock, 1951). These features are shown in greater detail, and at greater distances, using 21-cm observations of unionized, or neutral hydrogen, called H I regions (Oort, Kerr and Westerhout, 1958). The Sun, for example, lies within one of these localized concentrations of neutral hydrogen.

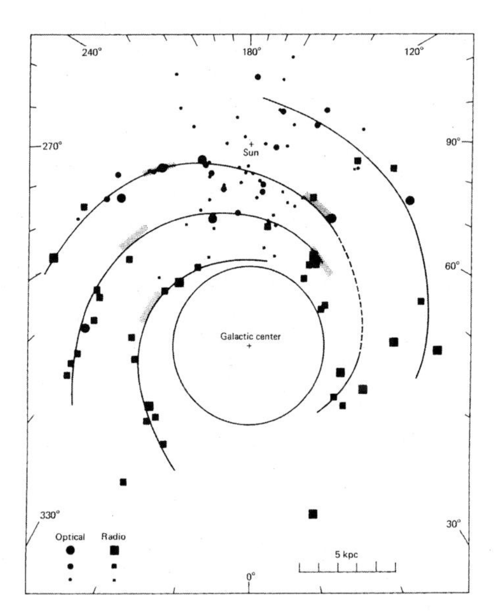

Fig. 5.11. Spiral model of our Galaxy obtained from young, ionized hydrogen, or H II, regions. The filled circles are for data taken at optical wavelengths; the filled squares are for radio-wavelength data. Hatched areas denote intensity maxima in the radio continuum and in neutral hydrogen. The galactic longitude, ℓ, as viewed from the Sun is indicated along the edge of the figure, and the scale is shown at the lower right. (Adapted from Georgelin and Georgelin, 1976)

A persistent dilemma has been understanding why the spiral arms do not wind up to form a featureless ball of gas and dust. For example, the period, P_o, for one rotation of the Sun about the galactic center is $P_o = 2\pi R_o/\theta_o = 2.4 \times 10^8$ years, where 1 kpc $= 3.086 \times 10^{21}$ cm and 1 year $= 3.1557 \times 10^7$ seconds. The Sun is about 4.6×10^9 years old, and our Galaxy is about three times as old as the Sun. So, the spiral arms should have been wrapped around the galactic center many times during the lifetime of either the Sun or our older Galaxy.

Lindblad (1927) first suggested that the observed persistence of galactic spiral arms against differential rotation would be explained if the material arms coexist with a density wave pattern propagating around the Galaxy with a pattern speed, $\Omega_p(R)$, given by

$$\Omega_c(R) - \frac{\kappa(R)}{m} < \Omega_p(R) < \Omega_c(R) + \frac{\kappa(R)}{m} \tag{5.175}$$

where $\Omega_c(R)$ and $\Omega_p(R)$ are the respective angular velocities of the material and the density wave around the galactic center at a distance, R, therefrom, $\kappa(R)$ is the epicyclical frequency, and m is an integer denoting the number of spiral arms. The epicyclical frequency is defined by the relation

$$[\kappa(R)]^2 = [2\Omega_c(R)]^2 \left\{ 1 + \frac{R}{2\Omega_c(R)} \frac{d\Omega_c(R)}{dR} \right\}. \tag{5.176}$$

and is related to the spacing λ_* between the spiral arms by the equation

$$\lambda_* \approx \frac{4\pi^2 G\sigma(R)}{[\kappa(R)]^2}, \tag{5.177}$$

where G is the Newtonian gravitational constant, and $\sigma(R)$ is the projected mass surface density. In the presence of a density wave, the outer and inner sides of a concentration of hydrogen will have, respectively, higher and lower speeds than would have been observed in its absence. The rotation curve will then exhibit oscillations which are in fact observed, and kinematic distances must be corrected for the density wave effect. Such corrections may be made using the density wave theory of Lin and Shu (1967) and Lin, Yuan, and Shu (1969). They derive a dispersion relation between the arm spacing, λ, and the angular frequency, ν, at which stars see this pattern

$$\nu = \frac{m[\Omega_c(R) - \Omega_p(R)]}{\kappa(R)}. \tag{5.178}$$

They show that λ varies smoothly from 0.55 λ_* to 0.05 λ_* as $|\nu|$ varies from zero to one, and indicate that $\Omega_p(R_o) = 13.5$ km s^{-1} kpc^{-1}. The spiral density wave theory is reviewed by Shu (1992) and by Bertin and Lin (1996).

Infrared photographs of spiral galaxies reveal their density wave structure (Elmegreen and Elmegreen, 1984), and optical tracers have been used to trace sprial wave resonances in them (Elmegreen and Elmegreen, 1990; Elmegreen, Elmegreen and Montenegro, 1992). Images of the stars, gas and dust of galaxies are compared with expectations of density wave theory by Knapen and Beckman (1996). The properties of barred spiral galaxies are discussed by Elmegreen and Elmegreen (1985) and in Buta, Crocker and Elmegreen (1996).

5.2.7 Luminosity, Distance, Luminosity Function and Mass Density of Galaxies

Estimates for the absolute luminosity of our Galaxy have been briefly reviewed by Van Den Bergh (1988). When corrected for absorption by interstellar dust, the optical data indicates an absolute blue magnitude, M_{BG}, and absolute blue luminosity, L_{BG}, of

$$M_{\mathrm{BG}} = -20.5 \pm 0.2 \tag{5.179}$$

and

$$L_{\mathrm{BG}} = (2.5 \pm 0.5) \times 10^{10} L_{\mathrm{B}\odot} \tag{5.180}$$

where the absolute blue luminosity of the Sun is $L_{\mathrm{B}\odot} \approx 3.0 \times 10^{33}$ erg s^{-1}. Spacecraft photometry and determinations of the velocity widths of interstellar hydrogen imply a Galactic luminosity of

$$L_{\mathrm{BG}} = (2.3 \pm 0.6) \times 10^{10} L_{\mathrm{B}\odot} \ . \tag{5.181}$$

The absolute magnitude, M_{G}, and absolute luminosity, L_{G} of a galaxy are related by:

$$\log\left(\frac{L}{L_{\odot}}\right) = \frac{1}{2.5}\left[M_{\mathrm{G}} - M_{\mathrm{bol},\odot}\right] \tag{5.182}$$

where the absolute luminosity of the Sun is $L_{\odot} = 3.85 \times 10^{33}$erg s^{-1}, and the absolute bolometric magnitude of the Sun is $M_{\mathrm{bol},\odot} = 4.75$.

Once the intrinsic brightness of an extragalactic object is known, its apparent brightness can be used to gauge its distance, since the luminosity falls off as the inverse square of the distance. The distance, D, of a galaxy of apparent magnitude, m_{G}, and absolute magnitude, M_{G}, is given by the distance modulus, μ_{G}:

$$\mu_{\mathrm{G}} = m_{\mathrm{G}} - M_{\mathrm{G}} = 5 \log D + 25 \quad \text{for } D \text{ in Megaparsecs} \ , \tag{5.183}$$

where the magnitudes are bolometric values corrected for interstellar absorption, and the extragalactic distances are given in units of Megaparsecs, or Mpc, where 1 Mpc $= 3.086 \times 10^{24}$ cm.

Extragalactic distances are inferred in a step-like process, beginning with the nearest galaxies in which bright stars can be viewed and extending to more remote ones whose distances are gauged by brighter "standard candles" such as supernovae or the galaxies themselves. Rowan-Robinson (1985) wrote an excellent summary of this cosmological distance ladder, updating it in Rowan-Robinson (1988). The status of the cosmic distance scale was reviewed by Van Den Bergh (1989, 1992), and a critical review of extragalactic distance-measuring techniques was provided by Jacoby et al. (1992).

A compilation of distance estimates to key nearby galaxies is given in Table 5.5. It includes the closest galaxies, the Large and Small Magellanic Clouds (LMC and SMC). They are small companions to our own Galaxy, in

Table 5.5. Distance estimates to nearby galaxies and the Virgo and Coma Clusters of galaxies.*

Galaxy or Cluster	Distance	Reference
Large Magellanic Cloud (LMC)	49 ± 2 kpc	Van Den Bergh (1992)
Large Magellanic Cloud (LMC)	55.0 ± 2.5 kpc	Feast & Catchpole (1997)
Small Magellanic Cloud (SMC)	58 ± 4 kpc	Van Den Bergh (1992)
Andromeda nebula (M31)	725 ± 35 kpc	Van Den Bergh (1992)
Andromeda nebula (M31)	899 ± 47 kpc	Feast & Catchpole (1997)
Andromeda nebula (M31)	784 ± 17 kpc	Stanek and Garnavich (1998)
Triangulum nebula (M33)	795 ± 75 kpc	Van Den Bergh (1992)
NGC 300	1.6 ± 0.15 Mpc	Van Den Bergh (1992)
NGC 5128	3.45 ± 0.25 Mpc	Van Den Bergh (1992)
NGC 3031 (M81)	3.63 ± 0.34 Mpc	Freedman et al. (1994)
M96 (Leo Group)	11.6 ± 0.9 Mpc	Tanvir et al. (1995)
Virgo cluster (NGC 4571)	14.9 ± 1.2 Mpc	Pierce et al. (1994)
Virgo cluster (average core)	15.8 ± 1.1 Mpc	Van Den Bergh (1992)
Virgo cluster (M 100)	17.1 ± 1.8 Mpc	Freedman et al. (1994)
Virgo cluster (four methods core)	18.3 ± 1.1 Mpc	Tanvir et al. (1995)
Virgo cluster (globular clusters)	21.3 ± 2.7 Mpc	Sandage and Tammann (1995)
Coma cluster	87 ± 6 Mpc	Van Den Bergh (1992)
Coma cluster	105 ± 12 Mpc	Tanvir et al. (1995)
Coma cluster	93 ± 13 Mpc	Whitmore et al. (1995)

*One kiloparsec = 1 kpc = 3.086×10^{21} cm; one Megaparsec = 1 Mpc = 3.086×10^{24} cm.

relatively close orbit around it, once thought to be at respective distances of 49 ± 2 kpc and 58 ± 4 kpc for LMC and SMC, where 1 kpc = 3.086×10^{21} cm. The nearest large spiral galaxy, the Andromeda nebula or, M31, is located at 725 ± 35 kpc. It is similar in size and structure to our own Galaxy and has several dwarf satellite galaxies as well as a more substantial companion, the spiral galaxy M33 in the Triangulum.

Recent parallax measurements of Cepheid variable stars using instruments aboard the Hipparcos satellite indicate that they are a bit brighter and therefore slightly farther away than astronomers had believed (Feast and Catchpole, 1997). This correction leads to a distance modulus of 18.70 ± 0.10 for the Large Magellanic Cloud and 24.77 ± 0.11 for M31. The distances with this calibration are 55.0 ± 2.5 kpc for the LMC and 899 ± 47 kpc for M31, so the Large Magellanic Cloud is about 10 percent farther away than previously thought. However, a distance of 784 ± 17 kpc is obtained for M31 using Hubble Space Telescope and Hipparcos red clump stars (Stanek and Garnavich, 1998; also see Stanek, Zaritsky and Harris, 1998).

Our Galaxy, M31, M33 and at least 39 small galaxies make up the Local Group of galaxies that interact gravitationally and are located within about

1.5 Mpc or 1,500 kpc. A similar, nearby group of galaxies, the M81 group, is about 3.6 Mpc away, and the Leo group, around M96, is located at about 12 Mpc. Further out, between 14 and 23 Mpc, is a vast cluster of thousands of galaxies, called the Virgo cluster since it is centered on the constellation Virgo.

Cepheid variable stars provide a foundation for the extragalactic distance scale, establishing the first rung of the cosmological distance ladder. They are among the brightest stars, and can be easily identified in nearby galaxies. Cepheids pulsate with extraordinary regularity, providing a measure of their innate brightness through the period-luminosity relation. A historical review of that relation was provided by Fernie (1969), and its zero-point calibration was discussed by Sandage (1972). Contemporary investigations of the period-luminosity relation for variable stars were summarized in Sect. 5.2.5.

The quest for accuracy in the distances to nearby galaxies has led to increasingly precise estimates using Cepheid variable stars. The extragalactic nature of spiral nebulae was, for example, conclusively demonstrated by Hubble (1925) using Cepheids in M31 and M33, but nowadays we accept distances to them of two and a half times Hubble's value. This was mainly the result of Baade's (1944) resolution of stars in the nuclear region of the spiral galaxy M31 and of the two elliptical companions, M32 and NGC 205, followed by discovery of two stellar populations and a recalibration of the zero point in the period-luminosity relation for classical Cepheids in M31 (Baade, 1952). Öpik (1922) obtained a more nearly correct distance for M31 from its mass, itself inferred from rotation, and the assumption that the mass-to-luminosity ratio was stellar.

The relative distances to nearby galaxies within the Local Group are now inferred using Cepheid variable stars to better than 5%. Moreover, as illustrated in Table 5.5, the Hubble Space Telescope has viewed fainter Cepheid variable stars in more remote galaxies out to the Virgo cluster (Freedman et al., 1994, Tanvir et al., 1995), while ground-based observations of Cepheid variables in another Virgo cluster galaxy provide a similar distance of about 15 Mpc (Pierce et al., 1994).

The distance to the Coma cluster of galaxies is obtained in leapfrog fashion from the distance of the Virgo cluster, assuming a difference in distance modulus, μ, or a distance, D, ratio of (Sandage and Tammann, 1990; Van Den Berg, 1992; Tanvir et al., 1995).

$$\mu(\mathrm{Coma}) - \mu(\mathrm{Virgo}) = 3.75 \pm 0.1 \text{ mag} . \tag{5.184}$$

and

$$\frac{D(\mathrm{Coma})}{D(\mathrm{Virgo})} = 5.6 \pm 0.3 . \tag{5.185}$$

The extragalactic Cepheid distance scale is itself calibrated using independent measurements of the distance of the Large Magellanic Cloud, or LMC, and there is now excellent internal agreement between diverse methods for gauging its distance (Van Den Bergh, 1992). An example is the geometric measurement using the light echo ring of the supernova SN 1987A in the LMC. The light travel time to the ring gives its absolute size, by multiplication with the velocity of light, and this can be combined with measurements of the ring's angular size to infer a LMC distance modulus of 18.55 ± 0.13 and a distance of 50.1 ± 2.4 kpc (Panagia et al., 1991; Madore and Freeman, 1991; Mc Call, 1993). As already mentioned, Hipparcos satellite results give a distance modulus of 18.70 ± 0.10 and a distance of 55.0 ± 2.5 kpc for the Large Magellanic Cloud (Feast and Catchpole, 1997).

Distances out to the Virgo cluster can be independently determined from bright regions of ionized hydrogen, or H II regions, that envelop massive, new, nonvariable O or B stars in galaxies. Gum and De Vaucoulecurs (1952) showed that large ring-like H II regions could be used as extragalactic distance indicators, and Sérsic (1959, 1960) inferred the distances of nearby galaxies from the diameters of their bright H II regions. One can measure the angular diameters and combine these with the expected linear size to infer the distance to H II regions. Allan Sandage and Gustav Tammann subsequently used the technique in a classic series of paper on the extragalactic distance ladder (Sandage and Tammann, 1974). This work was refined by improved diameter measurements (Kennicutt, 1979), and by distance estimates of H II regions based on their hydrogen alpha flux (Kennicutt, 1979, 1981) or their velocity dispersions (Melnick, 1977).

A galaxy's globular star clusters provide another secondary distance indicator. The absolute luminosity of hundreds of globular clusters in our Galaxy, as well as in the nearby M31, describe a luminosity function with mean peak (turnover) absolute magnitudes of (Sandage and Tammann, 1995)

$$\begin{aligned} \langle M_{\mathrm{B}} \rangle_{\mathrm{o}} &= -6.93 \pm 0.08 \text{ mag} \\ \langle M_{\mathrm{v}} \rangle_{\mathrm{o}} &= -7.62 \pm 0.08 \text{ mag} \quad . \end{aligned} \tag{5.186}$$

When this local calibration is applied to the E galaxy Virgo cluster core, a distance modulus of $(m - M)_{\mathrm{o}} = 31.64 \pm 0.25$ and a distance of $D = 21.3 \pm 2.7$ Mpc is obtained, which is consistent, within the uncertainties, with that obtained by Hanes (1979) for the Virgo cluster $(m - M)_{\mathrm{o}} = 30.7 \pm 0.3$. Whitmore et al. (1995) have used Hubble Space Telescope observations of over 1,000 globular clusters in M87, located at the core of the Virgo cluster, to similarly obtain $(m - M) = 31.12$ and $D = 16.75$ Mpc assuming $\langle M_{\mathrm{v}} \rangle_{\mathrm{o}} = -7.4 \pm 0.25$ (Secker, 1992).

Planetary nebulae have recently been used as distance indicators for nearby galaxies. These stars eject a shell of matter during a late stage in their evolution to white dwarfs. The central stars, which are essentially burnt-out cores with temperatures of up to 125,000 K, ionize the expanding shells. There is a remarkably abrupt upper limit to the absolute luminosity of the line emission from the ionized shell. This cutoff luminosity provides the distance estimator (Ciardullo et al., 1989, 1991; Jacoby et al., 1990, 1992).

Novae also provide "standard candles" for nearby galaxies. They are easily identified as luminous stars that abruptly brighten by as much as 100,000 times, and then slowly decline in luminosity. At maximum amplitude, novae reach an absolute visual magnitude of -5 to -10, which is more luminous than all but the longest period Cepheids. Novae are neither new nor temporary stars, but instead interacting binary systems (Kraft, 1964) in which mass from a cool, main-sequence star flows onto the surface of a white dwarf companion, resulting in a nuclear explosion. Because they belong to an old stellar population, novae are found predominantly in elliptical galaxies and the bulges of spiral galaxies.

Mc Laughlin (1945) realized that classical novae obey an empirical maximum luminosity-rate of decline relationship that provides a powerful tool for the determination of nearby galaxy distances (Schmidt, 1957). The distance scale using novae has been calibrated by Cohen (1985, 1988) who determined the distances of old novae within our Galaxy using spatially resolved observations of their expanding shells. If the angular diameter and expansion velocity of a nova shell are known, one can infer distance from an expansion parallax. The observational data obey the linear relationship given by:

$$M_{\mathrm{v}}(\max) = -10.66(\pm 0.33) + 2.31(\pm 0.26)\log(t_2) \ , \tag{5.187}$$

where $M_{\mathrm{v}}(\max)$ is the maximum absolute visual magnitude and t_2 is the time in days to decline two magnitudes below maximum light, or $t_2 = 1/\dot{m}$ where $\dot{m}$ is the rate of decline, in magnitudes per day, over the first 2 magnitudes. If one assumes that all novae reach the same absolute magnitude 15 days after maximum light, equal to the mean observed value of $M_{\mathrm{v}} = -5.60$ magnitudes after 15 days, then Cohen (1985, 1988) also obtains

$$M_{\mathrm{v}} = (\max, \mathrm{corr}) = -10.70(\pm 0.30) + 2.41(\pm 0.23)\log(t_2) \ . \tag{5.188}$$

Additional discussions of novae as distance indicators are given by Van Den Bergh and Pritchet (1986) and in the general review of extragalactic distance techniques provided by Jacoby et al. (1992).

Supernovae permit us to estimate distances to galaxies that are far too distant to see individual stars, including the Cepheids (see Fig. 5.12). A supernova is a massive star that explodes at the end stages of stellar evolution when all its thermonuclear fuel has been consumed. At maximum intensity, supernovae are millions of times brighter than Cepheids and can briefly outshine all of the stars in the galaxy combined. If we can assume that a certain kind of supernova flares to the same brightness no matter where or when it explodes, then we can use it as a "standard candle" to infer distance from the apparent brightness out to enormous distances of 1,000 Mpc.

Statistical studies of supernovae in galaxies indicate that there are two types, denoted type I and type II, distinguished by the absence or presence of hydrogen in their spectra (Lang, 1992). Supernovae of type Ia lack hydrogen lines and helium lines in their spectra; during the first month after maximum light they have a strong red absorption line due to singly ionized silicon. Type II supernovae have a wide range in peak absolute magnitude and cannot be treated as standard candles. In contrast, the intrinsic luminosity of type Ia supernovae at peak brightness, a few days after the explosion, is thought to vary relatively little from case to case. This is because they are all believed to result from the nuclear detonation of a white dwarf that is at or near the

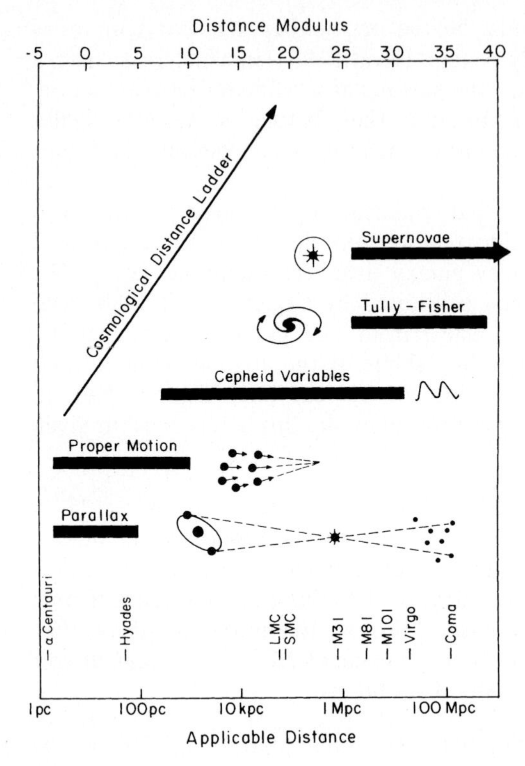

Fig. 5.12. Different techniques have been used to infer the distances of increasingly remote objects, describing a cosmological distance ladder. Here the units are in parsecs, pc, where 1 pc = 3.0857×10^{18} cm, kiloparsecs, kpc = 10^3 pc = a thousand parsecs, and megaparsecs, Mpc = 10^6 pc = a million parsecs. Alpha Centauri is the nearest star, Hyades is a nearby open star cluster, LMC and SMC respectively denote the Large and Small Magellanic clouds, M31 is the nearest spiral galaxy, also called Andromeda, and the Virgo cluster is the nearest large cluster of galaxies. (Adapted from Rowan-Robinson, 1985, and Roth and Primack, 1996)

Chandrasekar limit of 1.4 solar masses; the white dwarf accretes additional mass from a nearby companion star until it blows up (Arnett, Branch and Wheeler, 1985; Wheeler and Harkness, 1990).

The observational status and calibration of type Ia supernovae, or SNe Ia, was reviewed by Branch and Tammann (1992). Their best estimate for the mean absolute blue magnitude, $\langle M_{\mathrm{B}} \rangle$, of type Ia supernovae at maximum light is

$$\langle M_{\mathrm{B}}(\mathrm{max}) \rangle = -19.6 \pm 0.2 \text{ mag} \ , \tag{5.189}$$

including historical supernovae in our Galaxy and SNe Ia in nearby galaxies and the Virgo cluster (also see Tammann and Leibundgut, 1990). Subsequent calibration of the absolute magnitudes of 34 SNe Ia results in (Sandage and Tammann, 1993)

$$\langle M_{\mathrm{B}}(\mathrm{max}) \rangle = -19.55 \pm 0.14 \text{ mag} \ , \tag{5.190}$$

and

$$\langle M_{\rm v}({\rm max})\rangle = -19.71 \pm 0.13 \text{ mag} . \tag{5.191}$$

Sandage et al. (1994) provided additional calibration, and Sandage et al. (1996) obtain a peak brightness of type Ia supernovae of

$$\langle M_{\rm B}({\rm max})\rangle = -19.47 \pm 0.07 \text{ mag} , \tag{5.192}$$

and

$$\langle M_{\rm v}({\rm max})\rangle = -19.48 \pm 0.07 \text{ mag} . \tag{5.193}$$

The dispersion in the maximum absolute magnitudes of type Ia supernovae is discussed in the next Sect. 5.2.8.

The distances of remote galaxies can also be inferred from rotation measurements, or line widths, of galaxies. The line width is a measure of the rotation velocity, and the mass of a galaxy can be calculated from the square of the velocity using the virial theorem (also see Sect. 5.4.5.). The distance is then determined from the known mass-absolute luminosity ratio and the measured apparent luminosity through the inverse square law. Indeed, Öpik (1922) used optical measurements of the rotation velocity of M31 with the virial theorem mass to show that its distance had to be about 480 kpc if its mass to luminosity ratio is comparable to stars in our Galaxy, which was much more nearly correct than Hubble's (1925) distance using Cepheid variables with an incorrect period-luminosity relation.

In practice, correlation between the line width and the absolute magnitudes is established, and calibrated empirically using data from nearby galaxies. Such a correlation is called a Tully–Fisher relation after the first demonstration of it (Tully and Fisher, 1977). Their calibration was reanalyzed by Sandage and Tammann (1976) and Bottinelli et al. (1980), and the observations extended to a large number of galaxies by Fisher and Tully (1981) and Bottinelli et al. (1985). An example of such relations is given by Rowan-Robinson (1985), where the absolute photographic magnitude, $M^{\rm o}_{\rm pg}$, corrected for the effects of inclination and extinction, is related to the radio 21-cm line width, $W_{\rm R}$, by:

$$M^{\rm o}_{\rm pg} = -5.90(\pm 0.34) \log\left(\frac{W_{\rm R}}{\sin i}\right) - 4.39(\pm 0.08) , \tag{5.194}$$

where $W_{\rm R}$, is in kilometers per second and i is the inclination angle between the normal to the plane of the galaxy and the line of sight.

Aaronson et al. (1980, 1983, 1986) have proposed that the 21-cm line width should be correlated with the infrared absolute magnitude rather than the optical one, to reduce uncertainties caused by obscuration. Optical rotation curve measurements provide an alternative to the radio ones, and may apply to distances beyond the range of radio telescopes (Dressler and Faber, 1990; Mathewson, Ford and Buchhorn, 1992).

Calibrations of the Tully–Fisher relations using nearby galaxies yields (Pierce and Tully, 1988):

$$\langle M_{\rm B}\rangle = -6.86 \log\left(W^{i}_{\rm R}\right) - 2.27(\pm 0.25) \tag{5.195}$$

for the mean absolute blue magnitude, $\langle M_{\mathrm{B}} \rangle$, and the neutral hydrogen, radio line width, W_{R}^{i}, has been deprojected to edge-on orientation, or inclination corrected, and

$$\langle M_{\mathrm{H}} \rangle = -9.25 \log\left(W_{\mathrm{R}}^{i}\right) + 1.40(\pm 0.14) \tag{5.196}$$

for the infrared Tully-Fisher relation at $1.6\,\mu$m (H band). The absolute calibration in B adopted by Richter and Huchtmeier (1984) is

$$\langle M_{\mathrm{B}} \rangle = (-7.1 \pm 0.2) \log\left(W_{\mathrm{R}}^{i}\right) - 2.12(\pm 0.10) \ , \tag{5.197}$$

where the inclination-corrected 21-cm line width, W_{R}^{i}, is read at the 20% intensity level. Additional calibrations, with an uncertainty of about 7% in the zero point, are provided by Freedman (1990) and Pierce and Tully (1992).

The relative distance estimate of a single galaxy using the Tully–Fisher relation is accurate to 15%, and if enough galaxies in a cluster are observed, the method should give an ensemble distance with a precision better than 5% (Tully, 1993).

Allan Sandage, Gustav Tammann and their coworkers argue that Tully-Fisher relations can result in incorrect distances because of a selection bias at large distances in favor of atypically bright galaxies that are easier to detect. That is, in a flux-limited or magnitude-limited sample, the less luminous galaxies are missing at larger distances, an effect that is often called the Malmquist bias after Malmquist (1920). Since the absolute magnitudes of galaxies exhibit a dispersion about the mean $\langle M \rangle$, the true absolute magnitude of any particular galaxy is either brighter or fainter than the mean, and the intrinsically brighter galaxies that are favored at larger distances will have too small a calculated distance using $\langle M \rangle$. If the data are from a flux-limited sample, the Tully–Fisher relations may shift to higher magnitudes with increasing distance and the error will become larger with larger true distance, distorting the cosmological distance scale (Sandage, 1988; Kraan-Korteweg, Cameron and Tammann, 1988; Sandage, 1994; Sandage, Tammann and Federspiel, 1995).

These authors argue that the Malmquist bias can be removed by using a distance-limited sample for which the Tully–Fisher relations become (Kraan-Korteweg, Cameron and Tammann, 1988):

$$\langle M_{\mathrm{B}} \rangle = -6.69 \log\left(W_{\mathrm{R}}^{i}\right) - (2.77 \pm 0.10) \tag{5.198}$$

and

$$\langle M_{\mathrm{H}} \rangle = -10.12 \log\left(W_{\mathrm{R}}^{i}\right) + (4.12 \pm 0.14) \tag{5.199}$$

or Sandage (1988, 1994)

$$\langle M_{\mathrm{B}} \rangle = -7.416 \log\left(W_{\mathrm{R}}^{i}\right) - 1.239(\pm 0.15) \ . \tag{5.200}$$

All of these Tully–Fisher relations apply to spiral galaxies which are rich in the interstellar gas that emits the 21-cm radio radiation, and which have extended global structures whose rotation produces the observed line broadening.

In the gas-poor elliptical galaxies, the velocity broadening of stellar absorption lines can be used to infer the core velocity dispersion, σ_v, which describes the random motions of the stars. Faber and Jackson (1976) found that the total absolute luminosity, L, of an elliptical galaxy increases with the fourth power of the velocity spread in the optical spectrum, or

$$L = \text{constant} \times \sigma_v^4 , \tag{5.201}$$

which has become known as the Faber–Jackson relation. They also demonstrated that the ratio of mass, M, to absolute luminosity, L, is correlated with absolute magnitude for normal elliptical galaxies, obtaining $(M/L) = \text{constant} \times L^{0.5}$.

The correlation between the velocity dispersion and absolute luminosity of elliptical galaxies has been confirmed by subsequent investigations (Binney, 1982), and related to their mass-to-light ratio by (Van Der Marel, 1991)

$$\frac{M}{L} = \text{constant} \times L^{0.35 \pm 0.05} , \tag{5.202}$$

where M denotes the mass of an elliptical galaxy with absolute luminosity L.

A three-parameter correlation, often abbreviated by $D_{\mathrm{n}} - \sigma$, provides a more precise relationship between the luminosity, dimension, and internal motions of elliptical galaxies (Dressler et al., 1987; Djorgovski and Davis, 1987). The D_{n} is the diameter of an aperture at a specified surface brightness, hence combining luminosity and dimension information, and σ is the root-mean-square, or rms, velocity dispersion. The $D_{\mathrm{n}} - \sigma$ relation for elliptical galaxies has been reviewed by Jacoby et al. (1992). It provides relative distance estimates with an rms uncertainty of 20%; but since there are no large, nearby elliptical galaxies to act as calibrators, the fiducial zero point is set by some nearby entity like the Virgo cluster (Tully 1993).

Yet another distance estimator involves the surface brightness fluctuations of galaxies (Tonry et al., 1988, 1989; Tonry, 1991). This technique has also been reviewed by Jacoby et al. (1992). It involves the power spectra of luminosity fluctuations in elliptical galaxies and in the spheroidal components of lenticular and early spirals, which are apparently immune to Malmquist bias.

Enormous amounts of effort continue to be invested in the precise determination of the distances to nearby galaxies. This is because the stakes are high, establishing the size of the Universe. As far as cosmological distances are concerned, the ongoing controversy reduces to a choice between the long and short extragalactic distance scale (also see Sect. 5.2.8). Advocates for larger distances to nearby galaxies, and therefore a bigger Universe, include Sandage and Tammann in work previously mentioned. De Vaucouleurs (1982, 1993) has long argued for a smaller Universe and shorter distance scale, reasoning that there is a large zero-point error in the long scale with an unexplained discontinuity between galaxies inside and outside the Local Group. The controversy is intimately related to the determination of the Hubble constant, H_{o}, which is discussed in the next Sect. 5.2.8. It specifies the current rate of expansion of the Universe, and is a crucial parameter in determining its eventual fate.

But in the meantime, we turn to discussions of the luminosity function and present mass density of galaxies, which are of comparable cosmological importance.

It is often useful to describe an optical, or visible, luminosity function, $\phi(L)$, for galaxies of a particular type T. This is a distribution function for which the number of galaxies per unit volume, $dN(L)$, with absolute luminosity in the interval L to $L + dL$, is given by

$$dN(L) = \phi(L)dL \ . \tag{5.203}$$

When summed over all galaxy types, $\phi(L)$ describes a general luminosity function for galaxies. If the absolute magnitude, M, is used instead of L, then $\phi(L)dL = \phi(M)dM$ and $dM/M = -2.5\ln(dL/L)$.

For each type of galaxy, the luminosity function $\phi(L)$ may depend upon the evolution and space distribution, and integration over space and time are required to predict various observable quantities. We might, for example, assume that the galaxies in question are uniformly distributed in space, so that the probability of finding a galaxy in a given volume of space is just proportional to that volume. Then the number of galaxies, $N(r, L)$, that lie in the absolute luminosity range L to $L + dL$ within distance r to $r + dr$ is given by:

$$N(r, L) = 4\pi nr^2\phi(L)dL\,dr \tag{5.204}$$

where n is the space density of these objects.

If space is Euclidean and galaxies are uniformly distributed in that space, then the apparent luminosity, l, of the radiation from any spherical shell at a distance, r, and of thickness, dr, is

$$l \approx \frac{LN(r, L)}{4\pi r^2} \approx nL\phi(L)dL\,dr \ . \tag{5.205}$$

When the contributions of all spherical shells are added, l becomes infinite, contradicting the observed fact that the sky is dark at night. This paradox, often called Olbers' paradox, was first noted by Halley (1720), De Cheseaux (1744), and Olbers (1826). Dickson (1968), Jaki (1969), and Tipler (1988) give historical reviews of this paradox.

Olbers' paradox was first applied to stars before galaxies were discovered; since we now know that the local stellar Universe has finite edges at the boundaries of our Milky Way Galaxy, the modern version applies to the larger Universe of galaxies that has no visible edge and no preferred center. Olbers' paradox may be resolved by the expansion of the Universe, which redshifts the light of distant galaxies out the visible band of wavelengths.

The luminosity function of galaxies has been reviewed by Binggeli, Sandage and Tammann (1988). Efstathiou, Ellis and Peterson (1988) have reviewed methods for estimating the field-galaxy luminosity function, $\phi(L)$, for magnitude-limited redshift surveys. A useful representation is the Schechter luminosity function, $\phi_s(L)$, given by Schechter (1976):

$$\phi_s(L) = \phi^*\left(\frac{L}{L_*}\right)^\alpha \exp\left(-\frac{L}{L_*}\right)\frac{dL}{L_*} \ , \tag{5.206}$$

where ϕ^* is a normalization constant, α is the slope of the luminosity function at low luminosity, and L^* is the absolute luminosity corresponding to the "knee" of the luminosity function. The absolute magnitude, M^*, corresponds to the absolute luminosity, L^*. For $L >> L^*$, the $\phi_s(L)$ decreases exponentially, while for faint luminosity $L << L^*$ it obeys a power law with the exponent α.

The number of galaxies in the absolute magnitude interval M to $M + dM$ is:

$$\Phi(M)dM \propto \Xi^{\alpha+1}e^{-\Xi}dM \ , \tag{5.207}$$

with

$$\Xi = 10^{-0.4(M-M^*)} \ .$$

The field luminosity functions derived from five redshift surveys are compatible with each other, and described by a Schecter function with parameters (Efstathiou, Ellis and Peterson, 1988):

$$\begin{aligned} \Phi^* &= (1.56 \pm 0.34) \times 10^{-2}h^3 \ \text{Mpc}^{-3} \\ \alpha &= -1.07 \pm 0.05 \\ \mathrm{M}^*_{\mathrm{B_T}} &= -19.68 \pm 0.10 \end{aligned} \tag{5.208}$$

where the Hubble constant $H_o = 100h \text{ km sec}^{-1}\text{ Mpc}^{-1}$. and the subscript B_T denotes the blue magnitude, and for the Sun $M_{\mathrm{B}_\odot} = 5.48$. Loveday, Peterson, Efstathiou and Maddox (1992) similarly estimate the Schecter luminosity function from a redshift survey of southern galaxies with

$$\begin{aligned} \Phi^* &= (1.40 \pm 0.17) \times 10^{-2}h^3 \ \text{Mpc}^{-3} \\ \alpha &= -0.97 \pm 0.15 \\ \mathrm{M}^*_{\mathrm{BJ}} &= -19.50 \end{aligned} \tag{5.209}$$

and a mean space density, $\bar{n}$, of galaxies of:

$$\bar{n} = 5.52 \times 10^{-2}h^3 \ \text{Mpc}^{-3} \ , \tag{5.210}$$

where the Hubble constant $H_o = 100h \text{ km s}^{-1}\text{ Mpc}^{-1}$.

The mean luminosity density, $\langle L_{\mathrm{B}} \rangle$, of field galaxies in the blue band is given by (Efstathiou, Ellis and Peterson, 1988)

$$\langle L_{\mathrm{B}} \rangle = \phi^* \Gamma(\alpha + 2)L^*$$

and

$$\langle L_{\mathrm{B}} \rangle \approx (1.93 \pm 0.8) \times 10^8 hL_\odot \ \text{Mpc}^{-3} \ , \tag{5.211}$$

where Γ is the incomplete gamma function.

Oort (1958) introduced the idea that the present mass density, ρ_{mo}, of visible galaxies, which is of particular cosmological importance, could be determined from

$$\rho_{\mathrm{mo}} = \sum_{\mathrm{T}} \int \phi(L) \mathscr{L} \left(\frac{M}{L} \right) dL$$

or

$$\rho_{\mathrm{mo}} = \sum_{\mathrm{T}} \mathscr{L} \left(\frac{M}{L} \right) \tag{5.212}$$

where the summation is over different galaxy types, T, of luminosity density, $\mathscr{L}$, and mass-to-light ratio, M/L. Average values of these parameters, determined from 1970 to 1990, have been summarized by Persic and Salucci (1992), with typical values that do not differ substantially from that given by Kiang (1961)

$$\langle \mathcal{L} \rangle \approx 3 \times 10^8 h L_\odot \quad \mathrm{Mpc}^{-3} \tag{5.213}$$

and

$$\left\langle \frac{M}{L} \right\rangle \leq 20\, h\, M_\odot / L_\odot \tag{5.214}$$

where $M_\odot = 1.989 \times 10^{33}$ is the Sun's mass and $L_\odot = 3.85 \times 10^{33}\mathrm{erg\ s}^{-1}$ for the Sun.

That is, the present mass density of the visible components of galaxies is

$$\rho_{\mathrm{mo}} \leq 6 \times 10^9 h^2 L_\odot (M_\odot / L_\odot) \tag{5.215}$$

or

$$\rho_{\mathrm{mo}} \leq 0.4 \times 10^{-30} h^2 \quad \mathrm{g\ cm}^{-3} \tag{5.216}$$

where we have used 1 Mpc $= 3.086 \times 10^{24}$ cm. This is often compared with the critical mass density, ρ_c, needed to close the Universe.

$$\rho_\mathrm{c} = \frac{3H_\mathrm{o}^2}{8\pi G} = 1.9 \times 10^{-29} h^2 \quad \mathrm{g\ cm}^{-3} \ , \tag{5.217}$$

or with the density parameter

$$\Omega_\mathrm{o} = \frac{\rho_{\mathrm{mo}}}{\rho_\mathrm{c}} \leq 0.02 \ . \tag{5.218}$$

For specific galaxy types, Persic and Salucci (1992) obtain:

$$\begin{aligned} &\Omega_\mathrm{o} \approx 0.0015 \qquad (\mathrm{E+S0\ galaxies}) \\ &\Omega_\mathrm{o} \approx 0.007 \qquad (\mathrm{S\ galaxies}) \ . \end{aligned} \tag{5.219}$$

Discussions of the non-visible mass component of galaxies is given in Section 5.4.5.

5.2.8 The Hubble Constant and the Expanding Universe

Observations of the spectral lines of extragalactic objects indicate that the observed wavelength, λ_o, is longer than the emitted wavelength, λ_e. This difference is quantified by the redshift, z, given by:

$$z = \frac{\lambda_\mathrm{o} - \lambda_\mathrm{e}}{\lambda_\mathrm{e}} \ . \tag{5.220}$$

It is attributed to the Doppler effect of the radial motion at a recession velocity, V_r, away from the observer and along the line of sight to the emitting object.

$$V_\mathrm{r} = cz \quad \text{for } V_\mathrm{r} \ll c \ , \tag{5.221}$$

where $c = 2.9979 \times 10^5$ km s^{-1} is the velocity of light, and

$$V_\mathrm{r} = c \left[\frac{(z+1)^2 - 1}{(z+1)^2 + 1} \right] \ , \tag{5.222}$$

for large velocities approaching the speed of light.

As early as 1917, Vesto Slipher had shown that many bright spiral nebulae are receding from our Galaxy with tremendous speeds of $V_r \geq 1,000$ km s^{-1} (Slipher, 1917). More than a decade later, Edwin Hubble demonstrated that the velocity, V_r, increases with the distance, D, of the spiral nebulae, which are now called galaxies (Hubble, 1929, also see Wirtz 1921, 1924). Although Hubble did not specifically state it, the velocity and distance of extragalactic objects are now related by Hubble's law:

$$V_r = H_o D = cz \ , \tag{5.223}$$

where H_o is the Hubble constant. The standard practice is to write this constant as

$$H_o = 100h \text{ km s}^{-1} \text{ Mpc}^{-1} \ , \tag{5.224}$$

where h lies between 0.5 and 1.0. Measured values range between $H_o = 50$ km s^{-1} Mpc^{-1} and $H_o = 100$ km s^{-1} Mpc^{-1}. Thus, when the recession velocity is given in units of km s^{-1}, Hubble's law provides the distance in Mpc, where 1 Mpc = onc Megaparsec = 3.085678×10^{24} cm = 3.26 million light years.

According to Hubble's law, the farther a galaxy is from Earth, the faster it is receding from us and the more its light waves are stretched toward the red end of the spectrum. Thus, if a galaxy is twice as far away from us as a closer one, it will be seen to be moving at twice the speed with twice the measured redshift. A historical review of the origins of the velocity-distance relation, now known as Hubble's law, is given by Smith (1979).

Hubble's law is now explained by the expansion of the Universe that began with the big bang. The resulting Hubble flow is characteristic of explosions; the farther apart one particle of debris is from another, the faster they move apart from each other. Such an expansion was realized as a consequence of Einstein's General Theory of Relativity even before the discovery of Hubble's law (Friedmann, 1922, 1924; Lemaitre, 1927). In a perfectly uniform and isotropic expanding Universe, the relative expansion or recession velocity, V_R, of two particles is proportional to their separation, D, such that $V_R = H_o \times D$ where H_o denotes the value of the Hubble constant. As the Universe expands, the ratio V_R/D of recession velocity to separation changes, so the Hubble constant drifts slowly with time and the subscript o in H_o denotes the present value.

The Hubble constant specifies the current rate of expansion of the Universe. That is, it is a measure of the rate at which the expansion velocity increases with increasing separation between our Galaxy and others, thereby establishing the shape and age of the Universe. If the Universe is expanding at a relatively fast rate, with a comparatively large value of H_o, then the Universe must contain more mass to hold itself within a spherical shape and to eventually halt its expansion in the future. Smaller values of H_o require less mass to achieve such a balance between gravity and expansion. If the mass density of the Universe is known, then a precise value of H_o determines whether or not the Universe will expand forever.

A low value of the Hubble constant implies that the Universe is expanding slowly and that the Universe must be relatively old to have reached its current size. By contrast, a high value of the Hubble constant implies a rapid expansion and a relatively young Universe. Indeed, the current age of the Universe, t_o, is given roughly by the reciprocal of the Hubble constant, or $t_o \approx 1/H_o$. A discussion of this expansion age of the Universe is provided in Sects. 5.3.11 and 5.3.12, where it is compared with the ages of the oldest known astronomical objects (Sects. 5.3.9 and 5.3.10).

For a uniform and isotropic Universe, the Hubble constant H_o is just the slope of a plot of recession velocity versus distance. Such a plot is called the Hubble diagram. The example given in Fig. 5.13 for nearby galaxies indicates that H_o lies between 50 and 100 km s^{-1} Mpc^{-1} and that $H_o = 73 \pm 11$ km s^{-1} Mpc^{-1}. A similar velocity-distance plot, extending to larger distances, is given in Fig. 5.14 for the brightest galaxies in clusters of galaxies; this data provides similar bounds to H_o with a value of $H_o = 67 \pm 15$ km s^{-1} Mpc^{-1}. That value would mean that there is an increase in the recessional velocity of 67 kilometers per second for every Megaparsec of distance from the Earth or from any other observer position.

Astronomers have been participating in a contentious and sometimes acrimonious debate over the exact value of the Hubble constant for nearly half a century, and its measured variation between different observers is often much larger than their claimed internal uncertainties. Hubble and Humason (1931) found a H_o of 558 km s^{-1} Mpc^{-1}, but corrections for the calibration of the Cepheid period-luminosity relation (Baade, 1952) led to a reduction to about 250. Ernst Öpik had previously obtained a more nearly correct distance for

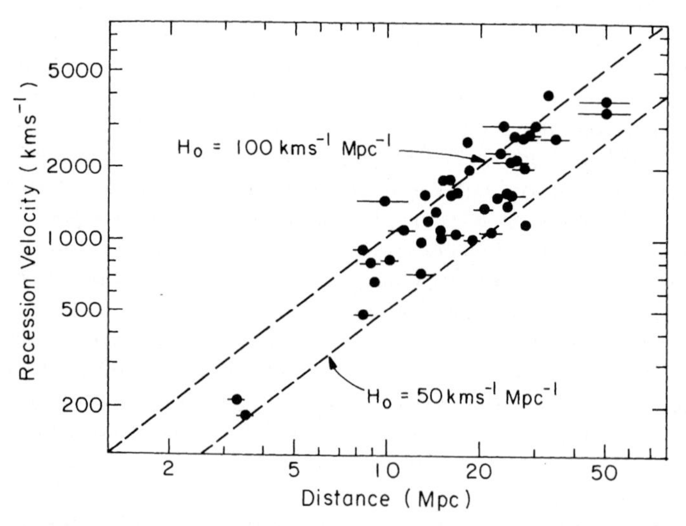

Fig. 5.13. Velocity-distance plot, or Hubble diagram, for nearby galaxies. The two dashed lines denote a Hubble constant of $H_o = 50$ and 100 km s^{-1} Mpc^{-1}. These lines are displaced slightly toward the right when corrected for the effect of the 300 km s^{-1} infall towards the Virgo cluster of galaxies. Adapted from Jacoby et al. (1992) who obtain a value of $H_o = 73 \pm 11$ km s^{-1} Mpc^{-1} from this data using an unweighted average of the distance to the Virgo cluster, at 17.6 ± 2.2 Mpc, to bootstrap to the Coma cluster

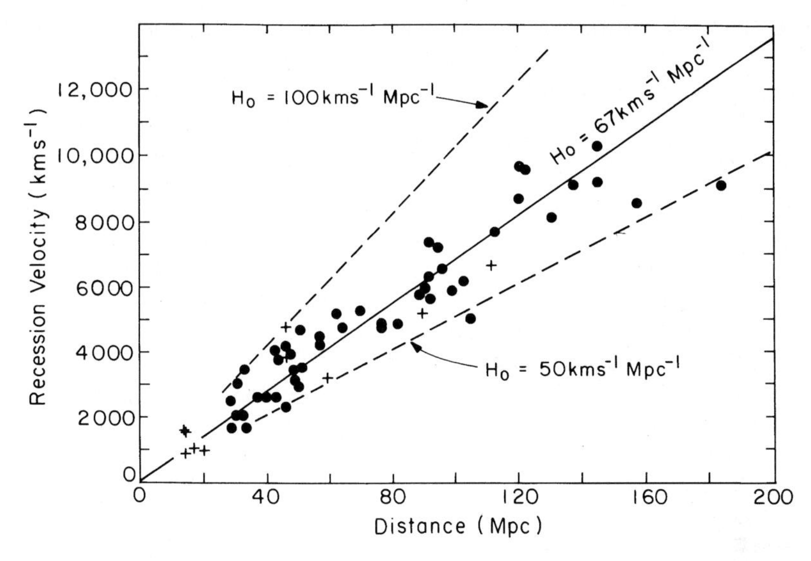

Fig. 5.14. Velocity-distance plot, or Hubble diagram, for the brightest cluster galaxies. The two dashed lines denote a Hubble constant of $H_0 = 50$ and 100 km s^{-1} Mpc^{-1}, whereas the solid line corresponds to $H_0 = 67$ km s^{-1} Mpc^{-1}. The recession velocities have been corrected for galactic rotation and the Galaxy's absolute motion in space. Adapted from Rowan-Robinson (1985) who obtains a value of $H_0 = 67 \pm 15$ km s^{-1} Mpc^{-1} from his examination of all the available evidence

Andromeda under the assumption that its mass-to-luminosity ratio was stellar (Öpik, 1922). Sandage (1958) then showed that the brightest "stars" in Virgo spiral galaxies were, in fact, H II regions; correcting for this error he found that $H_o \approx 75$ km s^{-1} Mpc^{-1} with a possible uncertainty of two. Sandage and Tammann (1975) obtained $H_o = 57 \pm 3$ for nearby spiral galaxies, closely the same as $H_o = 55 \pm 6$ for remote spirals; whereas De Vaucouleurs and Bollinger (1979) used the velocity-distance relation, or Hubble diagram, for remote galaxies to obtain a Hubble expansion rate of $H_o = 100 \pm 10$ km s^{-1} Mpc^{-1}. As illustrated in Table 5.6, the factor of two dispute over the value of the long-contested Hubble constant has continued to recent times.

Why has there been recurrent conflict and confusion over the exact value of the Hubble constant? There are two sources of uncertainty in determining its precise value. The expansion or recession velocities are uncertain for relatively nearby objects, and the distances are imprecise for remote ones. In other words, the discord results from local velocity anomalies and the lack of a single yardstick by which all distances can be measured.

The observed distribution of galaxies is not smooth and uniform. The nearby galaxies are distributed in a decidedly inhomogeneous way, appearing in small groups, such as the Local Group, in larger clusters of galaxies, such as the Virgo and Coma clusters, and in bigger superclusters like the Local Supercluster. The gravitational forces of these concentrations of matter pull surrounding galaxies toward them, producing "peculiar" motions that have

Table 5.6. Forty years of measurements of the Hubble constant H_o, given in units of km s^{-1} Mpc^{-1}

Long Distance Scale (H_o in km s^{-1} Mpc^{-1})	Short Distance Scale (H_o in km s^{-1} Mpc^{-1})
$H_o \approx 50$ Sandage (1971)	$H_o \approx 100$ Baade and Swope (1955)
$H_o = 57 \pm 3$ Sandage and Tammann (1975)	$H_o = 100 \pm 10$ De Vaucouleurs and Bollinger (1979)
$H_o = 50 \pm 7$ Sandage and Tammann (1984)	$H_o = 95 \pm 4$ Aaronson et al. (1980)
$H_o = 57 \pm 1$ Kraan-Korteweg Cameron and Tammann (1988)	$H_o = 95 \pm 1$ De Vaucouleurs (1982)
$H_o = 67 \pm 8$ Van Den Bergh (1989)	$H_o = 87 \pm 10$ Tully (1990)
$H_o = 47 \pm 5$ Sandage and Tammann (1993)	$H_o = 76 \pm 9$ Van Den Bergh (1992)
$H_o = 69 \pm 8$ Tanvir et al. (1995)	$H_o = 85 \pm 10$ Pierce and Tully (1988)
$H_o = 55 \pm 7$ Sandage and Tammann (1995)	$H_o = 86 \pm 1$ De Vaucouleurs (1993)
$H_o = 67 \pm 7$ Reiss, Press and Kirshner (1995)	$H_o = 90 \pm 10$ Tully (1993)
$H_o = 58 \pm 4$ Sandage et al. (1996)	$H_o = 87 \pm 7$ Pierce et al. (1994)

nothing to do with the expansion of the Universe. The observed velocities only become dominated by a uniform Hubble flow, or free expansion of the Universe, at very remote distances where the homogeneous and isotropic mass distribution assumed in the cosmology of the expanding Universe becomes valid. And, although the distances to nearby galaxies are in good absolute agreement, there is still disagreement between different techniques used to connect the local distances to remote galaxies where even the brightest objects of standard luminosity are dim and difficult to observe.

The observed Doppler shift of a galaxy is due to the sum of a cosmological part, called Hubble's law, that is proportional to the galaxy distance, D, and a Doppler shift due to the line-of-sight component, V_P, of the peculiar velocity of the galaxy relative to the mean Hubble flow. That is, the redshift, z, is given by:

$$cz = H_o D + V_P \ , \tag{5.225}$$

so the radial peculiar velocity, V_P, is the difference between the observed quantity, cz, and the Hubble velocity, $H_o D$, where c is the velocity of light. Thus, although redshifts and their associated velocity, cz, can be determined accurately, not all of the velocity can be attributed to the expansion of the Universe. As was illustrated in Figs. 5.13 and 5.14, this results in considerable scatter in the velocity-distance plots, or Hubble diagrams, for galaxies.

The expansion or recession speed, $H_o D$, only becomes much greater than the localized peculiar velocities at very large distances where the velocities are expected to converge to the uniform expansion characteristic of an isotropic, homogeneous Universe. This expansion is known as the smooth or uniform

Hubble flow, and the value of the Hubble constant inferred from it is called the global value of H_0. Widely different values of H_0 have resulted, in part, from controversies about the velocities of various components of the Universe. These peculiar motions are discussed in Sect. 5.7.6.2 where their implications for large-scale mass concentrations are also mentioned.

The uncertainty caused by bulk flows and other peculiar motions can be overcome by probing the Universe to great depths and larger redshifts where the expansion or recession speed greatly exceeds the speed of all other motions. The Hubble constant, H_0, can then be inferred from the measured redshifts and distances using Hubble's law; but imprecision in absolute distances then confuse the precise determination of H_0.

There is a consensus regarding the distances of nearby galaxies using Cepheid variable stars and other methods out to distances of about 5 Megaparsecs (Van Den Bergh, 1992, 1994, Sect. 5.2.7). Astronomers can also reliably measure the relative distances of remote galaxies (Fukugita, Hogan and Peebles, 1993). Nevertheless attempts to calibrate the absolute cosmic distance scale at large distances remain controversial.

The Hubble Space Telescope can be used to pick out and monitor Cepheid variable stars in galaxies as far away as 25 Megaparsecs, thereby improving the extragalactic distance scale and estimates of H_0 (Tanvir et al., 1995; Freedman et al., 1994; Sandage et al., 1996; also see Table 5.6). Still, even at these distances, the expansion is significantly distorted by flows toward local mass concentrations. In other words, Cepheids are much too faint to be observed out to distances where the expansion or recession velocities are substantially larger than those due to gravitational interactions with nearby galaxies or to large-scale flows. Only when we reach the Coma cluster, at a distance of about 100 Megaparsecs, does the recessional velocity, about 7,000 km s^{-1} there (Fukugita et al., 1991), become large enough for a clear view of the Hubble flow. So, a different type of "standard candle" must be used to measure distances to remote galaxies where departures from the idealized expansion of the Universe introduce only small uncertainties. Supernovae of type Ia may provide such a long-range yardstick; at peak brightness they shine with a million times more light then a Cepheid variable star and therefore can be identified at much greater distances.

A type Ia supernova, or SNe Ia, is an exploding star whose distinguishing signature is the absence of hydrogen in its spectrum (Lang, 1992). Such a supernova is thought to be a unique standard candle – an object whose intrinsic brightness is accurately known. Branch (1988) discusses supernova estimates of the Hubble constant that were then available, and Branch and Tammann (1992) review the use of type Ia supernovae as standard candles. Jacoby et al. (1992) discuss the dispersion in their absolute magnitudes and the physical basis for their explosion. Estimates for the absolute magnitudes of type Ia supernovae at maximum light are provided in the preceding Sect. 5.2.7.

Some type Ia supernovae may appear brighter at maximum than others, but their intrinsic luminosity at peak brightness varies little from case to case no matter when or where they occur. For example, Kowal (1968) showed that the dispersion in their maximum absolute magnitude, a few days after the explosion, is 0.6 magnitudes or less. Phillips (1993) inferred a similar intrinsic dispersion in the maximum light of type Ia supernovae, and Vaughan et al. (1995) reduced the uncertainty to about 0.3 magnitudes by removing faint, spectroscopically peculiar examples. The way a supernova brightens and then fades, called its light curve, can also be used to diminish uncertainty in their absolute brightness (Phillips, 1993; Hamuy et al., 1995).

The peak absolute luminosity and the light curves of type Ia supernovae can also be calculated using detailed physical models of the explosion. They are attributed to the thermonuclear disruption of a carbon-oxygen white dwarf star that has accreted enough mass from a companion star to exceed the Chandrasekhar limit and suddenly blow up (Wheeler and Harkness, 1990). The nuclear energy released in the explosion provides the kinetic energy of the ejected matter, and the observable light output is powered by the delayed energy input from the radioactive decay of ^{56}Ni and ^{56}Co. Nugent et al. (1995) provide physical models of the expanding atmospheres and light curves.

Relatively low values of the Hubble constant, H_o, are inferred from the Hubble diagrams, light curves and physics of type Ia supernovae. Since these stellar explosions do not occur very often, perhaps every 50 years or so for a typical galaxy, nearby supernovae are quite rare and reliable calibration of their distances has been difficult. However, Sandage and his colleagues have been able to gauge the absolute distances and calibrate the intrinsic peak luminosity of a few of these exploding beacons using Cepheid variable stars in the parent galaxies. The resulting Hubble diagram, or velocity-distance plot, of type Ia supernovae at distances far beyond any peculiar velocity anomalies gives global values of $H_o = 47 \pm 5$ (Sandage and Tammann, 1993), $H_o = 55 \pm 8$ (Sandage et al., 1994), $H_o \leq 59 \pm 4$ (Tammann and Sandage, 1995), and $H_o = 58 \pm 4$ km s^{-1} Mpc^{-1} (Sandage et al., 1996).

A relation between the peak luminosity and the decline rate of type Ia supernovae (Phillips, 1993) has been used to obtain $H_o = 62$ to 67 km s^{-1} Mpc^{-1} (Hamuy et al., 1995). The light curve shapes have been used by Reiss, Press and Kirshner (1995, 1996) to derive $H_o = 67 \pm 7$ and 64 ± 3 (statistical) km s^{-1} Mpc^{-1}. Explosion models have been used to calculate their light curves that are then compared to observations, leading to values of H_o that depend on the physics rather than astronomical calibration. Such a technique provides $H_o = 55$ to 60 (Van Den Bergh, 1995), a value of $H_o = 50 \pm 12$ (Nugent et al., 1995) and $H_o = 67 \pm 9$ km s^{-1} Mpc^{-1} out to a distance of about 1,300 Megaparsecs (Höflich and Khokhlov, 1996). The determination of cosmological parameters from high-redshift supernovae has been discussed by Garnavich et al. (1998) and Perlmutter et al. (1997, 1998).

Nevertheless, long-standing disputes still delay consensus, and controversy remains over the precise value of the Hubble constant. For instance, one can also infer the distances of remote galaxies using the Tully–Fisher relation that links a spiral galaxy's intrinsic luminosity to its rotation velocity, or line width, as measured by Doppler broadening (also see Sect. 5.2.7). This method yields global values of $H_o = 87 \pm 10$ km s^{-1} Mpc^{-1} (Tully, 1990), or well above the supernova value. When applied to the Coma cluster, the Tully–Fisher method gives $H_o = 92 \pm 21$ km s^{-1} Mpc^{-1} (Fukugita et al., 1991); and a high value of

$H_o = 80 \pm 17$ km s^{-1} Mpc^{-1} is similarly inferred from Hubble Space Telescope (HST) measurements of Cepheid variable stars in M100, a member of the Virgo cluster (Freedman et al., 1994). However, the HST value uses the Large Magellanic Cloud as a benchmark, and a subsequent 10 percent increase in its estimated distance suggests a decrease in the Hubble constant by a similar amount (Feast and Catchpole, 1997). Galaxy distances using type Ia supernovae have been compared with those using the Tully–Fisher relation (Pierce, 1994), finding no evidence for any discrepancy between relative distances using the two methods. When the Tully–Fisher distances were calibrated using variable stars, and then applied to the parent galaxies of the supernovae, giving an absolute distance calibration for the more distant supernovae, a global estimate of the Hubble constant of $H_o = 86 \pm 7$ km s^{-1} Mpc^{-1} was obtained (Pierce, 1994). In their critical review of determining extragalactic distances, Jacoby et al. (1992) favor a value of $H_o \approx 85$ km s^{-1} Mpc^{-1}, with lower values being perhaps attributed to an infall to the Virgo cluster. Observations of Cepheid distances, the Tully–Fisher relation and corrections for peculiar motions leads to $H_o = 69 \pm 5$ km s^{-1} Mpc^{-1} (Giovanelli et al., 1997).

In contrast, Sandage (1994) and Sandage, Tammann and Federspiel (1995) argue that the Malmquist bias introduced in flux-limited samples (Malmquist, 1920, Sect. 5.2.7) overestimates H_o by favoring atypically bright galaxies at large distances, reducing the value from $H_o = 80$ to 90 km s^{-1} Mpc^{-1} – as inferred by the Tully–Fisher method – to unbiased values of $H_o = 40$ to 55 km s^{-1} Mpc^{-1}. Observational selection bias affecting the determination of the extragalactic distance scale is reviewed by Teerikorpi (1997).

The critical differences in measurements of the Hubble constant seem to arise in estimating distances on scales of tens of Megaparsecs where a controversy has persisted for at least two decades (also see Table 5.6). There is the long distance scale, with $H_o = 50$ to 60 km s^{-1} Mpc^{-1}, and the short distance scale leading to $H_o = 80$ to 90 km s^{-1} Mpc^{-1}. De Vaucouleurs (1993), a persistent champion of the short scale, summed up the discrepancy, arguing that it is not so much due to a progressive Malmquist bias in the short scale, but the result of a large zero-point error in the long scale with an unexplained dichotomy between galaxies inside and outside the Local Group. The disagreements appear to be narrowing, perhaps leading to a value that lies somewhere between the two extremes. But the controversy will no doubt continue to rage, for a precise value of H_o is intimately related to cosmological models, as well as to the age and eventual fate of the Universe (see Sects. 5.3.11 and 5.3.12).

Since distances can only be directly measured for relatively nearby extragalactic objects, Hubble's law for remote objects is sometimes expressed by the redshift-magnitude relation in which the apparent bolometric magnitude, m_{bol}, is given by:

$$m_{\text{bol}} = 5\log(cz/H_o) + M_{\text{bol}} + 25 \ , \tag{5.226}$$

where M_{bol} is the absolute bolometric magnitude and the Hubble constant H_o is given in units of km s^{-1} Mpc^{-1}. This version of Hubble's law, known as the redshift-magnitude relation, was first tested by Hubble and Humason (1934) for relatively nearby galaxies. It is valid, within the observational errors, for normal galaxies, radio galaxies, and quasars (Lang et al., 1975).

5.2.9 Extragalactic Distances and Cosmological Models

The recession of the galaxies is attributed to the expansion of the Universe, in which its scale factor, $R(t)$, increases with time, t. In this theoretical context, Hubble's constant, $H(t)$, varies with time, t, and is defined by

$$H(t) = \frac{\dot{R}(t)}{R(t)} \ , \tag{5.227}$$

where the dot denotes differentiation with respect to time.

At the present epoch, denoted by subscript o, we have

$$H(t_o) = H_o = \frac{\dot{R}(t_o)}{R(t_o)} = \frac{\dot{R}_o}{R_o} \ . \tag{5.228}$$

So, the Hubble constant H_o specifies the current rate of expansion of the Universe.

The behavior of $R(t)$ depends upon the Hubble constant, H_o, the present mass density of the Universe, ρ_o, which tends to decelerate the expansion, and a possible cosmological constant, Λ, that provides an apparent acceleration in opposition to gravity even in a vacuum. These three constants, H_o, ρ_o and Λ, therefore specify a deceleration parameter, q_o, given by:

$$q_o = \frac{-\ddot{R}_o}{R_o H_o^2} = \frac{-\ddot{R}_o R_o}{\dot{R}_o^2} = \frac{\Omega_o}{2} - \frac{\Lambda c^2}{3H_o^2} \tag{5.229}$$

where the density parameter, Ω_o, is given by:

$$\Omega_o = \frac{\rho_o}{\rho_c} , \tag{5.230}$$

and the critical mass density, ρ_c, is

$$\rho_c = \frac{3H_o^2}{8\pi G} = 1.879 \times 10^{-29} h^2 \ \text{g cm}^{-3} \ , \tag{5.231}$$

where $H_o = 100h$ km s^{-1} Mpc^{-1}, and the Newtonian gravitational constant $G = 6.672 \times 10^{-8}$ in c.g.s. units.

For a model Universe in which the cosmological constant $\Lambda = 0$, the curvature of space is directly related to the density of the Universe by:

$$\frac{kc^2}{R_c^2} = H_o^2(2q_o - 1) \quad \text{for } \Lambda = 0 \ , \tag{5.232}$$

where $R_c(t)$ is the radius of curvature of curved space at time t, and the space curvature constant, k, has the critical values:

$$\begin{aligned} k &= +1 \quad \text{for } q_o > 0.5 \quad \text{and } \Lambda = 0 \\ k &= 0.5 \quad \text{for } q_o = 0.5 \quad \text{and } \Lambda = 0 \\ k &= -1 \quad \text{for } 0.0 < q_o < 0.5 \quad \text{and } \Lambda = 0 \ . \end{aligned} \tag{5.233}$$

The scale factor $R(t)$ is plotted as a function of time, t, in Fig. 5.15 for the three values of the space curvature constant k. For $k = +1$, space is elliptical and closed

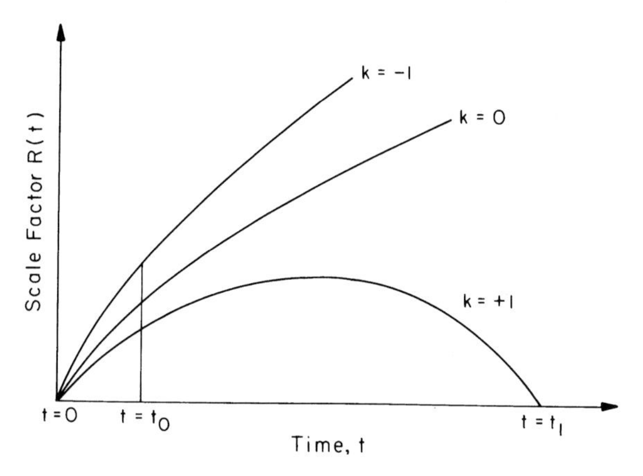

Fig. 5.15. Scale factor, $R(t)$, of the Universe as a function of time, t, for three model universes in which the cosmological constant is zero. They are for a space curvature constant $k = -1$ (open hyperbolic space, ever-expanding Universe), $k = 0$ (flat, Euclidean space, Einstein – De Sitter Universe), and $k = +1$ (closed, spherical space, oscillating Universe). The time $t = 0$ corresponds to the big bang that started the expansion of the Universe. At the present time, t_o, the scale factor $R(t)$ is increasing in all three models. At a future time, t_1, the $R(t)$ will reach zero again if the space curvature constant $k = +1$, and there is enough matter present in the Universe to eventually stop its expansion

with $\rho_o > \rho_c$ in an oscillating Universe that will contract in the future (Friedmann, 1922). When $k = 0$, space is "flat" and Euclidean in an ever expanding Einstein – De Sitter Universe for which the current mass density $\rho_o = \rho_c$ (Einstein and De Sitter, 1932). If $k = -1$, space is open and hyperbolic with $\rho_o < \rho_c$ in an ever-expanding Universe (Friedmann, 1924, Lemaitre, 1927, Milne, 1935).

For the special case of a "flat" Euclidean Universe without space curvature and with zero cosmological constant,

$$q_o = \frac{\Omega_o}{2} = \frac{\rho_o}{2\rho_c} = 0.5 \quad \text{for } k = 0 \quad \text{and } \Lambda = 0 \ . \tag{5.234}$$

Such a cosmological model is called the Einstein – De Sitter model (Einstein and de Sitter, 1932).

The fate of the Universe, and the apparent distances in it, are determined by the total amount of matter in the Universe and its curvature of space. It will also depend upon any repulsive terms quantified by the cosmological constant Λ. Thus, we describe different cosmological models by the density parameter, $\Omega_o = \rho_o/\rho_c$, and the deceleration parameter $q_o = (\Omega_o/2) - \Lambda c^2/(3H_o^2)$, where ρ_o is the current mass density of the Universe, the critical mass density, ρ_c, is given by $\rho_c = 3H_o^2/(8\pi G) = 1.879 \times 10^{-29} h^2$ g cm^{-3} for $H_o = 100h$ km s^{-1} Mpc^{-1}, the Newtonian gravitational constant $G = 6.672 \times 10^{-8}$ in c.g.s. units, and the velocity of light $c = 2.9979 \times 10^5$ km s^{-1}. For zero cosmological constant, we have $\Omega_o = 2q_o$ for $\Lambda = 0$.

When the cosmological constant $\Lambda = 0$, we can also describe a luminosity distance, D_L, for which the apparent luminosity, l, of light falls off as the inverse square of the distance, or

$$\text{TDT} = \text{TAI} + 32.184 \text{ seconds} . \tag{5.245}$$

An approximate formula that is usually sufficient for converting TDT to Barycentric Dynamical Time (TDB) is:

$$\text{TDB} = \text{TDT} + 0^{\text{s}}.001658 \sin g + 0^{\text{s}}.000014 \sin 2g \tag{5.246}$$

where

$$g = 357^{\circ}.53 + 0.98560028(\text{JD} - 2451545.0) ,$$

and JD is the Julian date to two decimals of a day. The value of g is given in terms of the day number in *The Astronomical Almanac* for the relevant year. More accurate expressions are given by Moyer (1981) and Seidelmann (1992).

In 1991 the International Astronomical Union, or IAU, adopted new resolutions in which Terrestrial Dynamical Time (TDT) was renamed Terrestrial Time (TT), and two new time scales were defined – Geocentric Coordinate Time (TCG) and Barycentric Coordinate Time (TCB). They are coordinate times in coordinate systems having their spatial origins at the center of mass of the Earth and at the solar system barycenter, respectively. These time scales will be introduced into *The Astronomical Almanac* when new fundamental theories and ephemerides based on them are adopted by the IAU.

5.3.3 Julian Date, Tropical Year, Synodic Month and Calendars

To facilitate chronological data, astronomical days, beginning at noon in Greenwich, England, are numbered consecutively from an epoch that precedes the historical period. This Julian day number, expressed as an integer, denotes the number of complete days that have elapsed since the beginning of the Julian period defined as Julian day 0 commencing at Greenwich noon on 1 January 4713 B.C., Julian proleptic calendar. (The Julian calendar, introduced by Julius Caesar in –45, was a solar calendar with months of fixed length, serving as a standard for European civilization until the Gregorian reform of +1582). The Julian proleptic calendar is formed by applying the rules of the Julian calendar to times before Caesar's reform, and the Julian date (JD) specifies the particular instant of a day by ending the Julian day number with the fraction of the day elapsed since the preceding Greenwich noon. The new year at 2000 January 1, 12^{h} UT1 occurs, for example, at JD 2451545.0. The Julian dates, JD, for different Julian Year epochs are:

$$\begin{aligned} &1900 \text{ January } 0.5 = \text{JD } 2415020.0 \\ &1925 \text{ January } 0.5 = \text{JD } 2424151.0 \\ &1950 \text{ January } 0.5 = \text{JD } 2433282.0 \\ &2000 \text{ January } 0.5 = \text{JD } 2451544.0 \\ &2050 \text{ January } 0.5 = \text{JD } 2469807.0 \\ &2100 \text{ January } 0.5 = \text{JD } 2488069.0 \end{aligned} \tag{5.247}$$

By definition, one Julian century has 36,525 days and one day is defined as 86,400 seconds of International Atomic Time (TAI). So,

$$
\begin{aligned}
1 \text{ Julian year} &= 365.25 \text{ days} = 8766 \text{ hours} \\
&= 525960 \text{ minutes} = 31557600 \text{ seconds} ,
\end{aligned}
\tag{5.248}
$$

where

$$1 \text{ day} = 24 \text{ hours} = 1440 \text{ minutes} = 86400 \text{ seconds} . \tag{5.249}$$

The principal astronomical cycles are the day, the year and the month, respectively based on the Earth's rotation about its axis, the Earth's revolution about the Sun, and the revolution of the Moon about the Earth. The tropical year, based on the orbital period of the Earth around the Sun, is defined as the mean interval between vernal equinoxes; it corresponds to the cycle of the seasons and is given by (Laskar, 1986):

$$
\begin{aligned}
\text{tropical year} = 365^{\text{d}}.2421896698 - 0.00000615359T - 7.29 \times 10^{-10}T^2 \\
+ 2.64 \times 10^{-10}T^3
\end{aligned}
\tag{5.250}
$$

where $T = (\text{JD} - 2451545.0)/36525$ and JD is the Julian day number. However, the interval from a particular vernal equinox to the next may vary from this mean by several minutes.

The old Besselian system, B, used tropical centuries defined by the motion of the Sun. The Julian Dates for the beginning of key Besselian years, BY, are:

B.Y.	Julian Date	B.Y.	Julian Date
B1850.0	2396758.203	B2000.0	2451544.533
B1900.0	2415020.313	B2025.0	2460675.588
B1950.0	2433282.423	B2050.0	2469806.643
B1975.0	2442413.478	B2100.0	2488068.753

(5.251)

The lengths of various types of years, at epoch 1900.0, are:

Tropical (equinox to equinox)	365.2421897 days
Sidereal (fixed star to fixed star)	365.25636 days
Anomalistic (perihelion to perihelion)	365.25964 days
Eclipse (Moon's node to Moon's node)	346.62005 days
Gaussian (Kepler's law for $a = 1$)	365.25690 days
Julian	365.25 days

(5.252)

The synodic month, the mean interval between conjunctions of the Moon and Sun, corresponds to the cycle of lunar phases. The following expression for the synodic month is based on the lunar theory of Chapront-Touze and Chapront (1988):

$$\text{synodic month} = 29^{\text{d}}.5305888531 + 0.00000021621T - 3.64 \times 10^{-10}T^2 . \tag{5.253}$$

Again $T = (\text{JD} - 2451545.0)/36525$ and JD is the Julian day number. Any particular phase cycle may vary from the mean by up to seven hours. For convenience, it is common to speak of a lunar year of twelve synodic months, or 354.35707 days.

The lengths of various types of months are:

Synodic month (new Moon to new Moon)	29.53059	
Tropical (equinox to equinox)	27.32158	
Sidereal (fixed star to fixed star)	27.32166	(5.254)
Anomalistic (perigee to perigee)	27.55455	
Draconic (node to node)	27.21222	

A solar calendar is designed to maintain synchrony with the tropical year; an example is the Gregorian calendar that serves as an international standard for civil use, with common years of 365 days in length and leap years of 366 days in length; it is based on a cycle of 400 years of average length of 365.2425 days per calendar year, which is a close approximation to the length of the tropical year. Each cycle of 400 years corresponds to 146097 days, which is exactly divisible by 7. Every year that is exactly divisible by four is a leap year, except for years that are exactly divisible by 100; these centurial years are leap years only if they are exactly divisible by 400, so the year 2000 is a leap year and 1900 is not.

A lunar calendar, such as the Islamic calendar, follows the cycle of lunar phases; its months therefore systematically shift with respect to the months of the Gregorian calendar. The Hebrew and Chinese calendars are lunisolar calendars, with months based on the lunar phase cycle; but every few years a whole month is intercalculated to bring the calendar back in phase with the tropical year.

A careful choice of origin gives simple formulae for Julian day numbers from Gregorian or Julian calendar dates, and calendar dates from day numbers (Hatcher, 1984). The two calendars require different formulae for computing both the number of complete years and complete months that have elapsed. These complexities vanish if the calculations are performed in a year beginning March 1. If A, M, and D denote the year, month and day in the usual way, with new year on January 1, and A', M', D', the same variable for the year beginning March 1, then

$$A' = A - [(12 - M)/10]\mathrm{INT}$$
$$M' = (M - 3)\mathrm{MOD}\ 12 \quad (5.255)$$

and

$$A = A' + (M'/10)\mathrm{INT}$$
$$M = (M' + 2)\mathrm{MOD}\ 12 + 1\ . \quad (5.256)$$

Here, for the variables x and y, (x) INT is the integral part of x, and x is assumed to be positive; (x) MOD y is the positive remainder on dividing x by y. For example, from March to December A' and A are numerically the same; for January and February A' will be less than A. The day of the month remains the same in both systems, so $D = D'$.

The Julian day number, JD, can be computed for either the Gregorian or the Julian calendar dates by evaluating (Hatcher, 1984):

$$
\begin{aligned}
y &= [365.25(A' + 4712)] \text{ INT} \\
d &= (30.6\, M' + 0.5) \text{ INT} \\
N &= y + d + D + 59 \ . \\
g &= \{[(A'/100) + 49] \text{ INT} \times 0.75\} \text{ INT} - 38 \ .
\end{aligned}
\tag{5.257}
$$

Then

$$
\begin{aligned}
&\text{JD} = N - g \quad \text{for Gregorian calendar dates} \\
&\text{JD} = N \quad \text{for Julian calendar dates}
\end{aligned}
\tag{5.258}
$$

Hatcher (1984) gives similar formulae for computing the calendar dates from the Julian day number, and Hatcher (1985) provides generalized equations relating Julian day numbers and calendar dates for a variety of calendars, including the Gregorian solar calendar and the schematic Islamic lunar calendars.

The Julian Day numbers for the beginning of each month from the calendar year 1950 to the calendar year 2050 are given in *The Astronomical Almanac*. It also contains tables for compiling the Julian Date for Gregorian calendar dates for any day from 1600 to 2200, as well as tabulations of the Julian date at 0 hours UT against Gregorian calendar dates for Universal and Sidereal Times of the year of the *Almanac*.

5.3.4 Sidereal Time (ST), or Star Time

Local Sidereal Time (LST) specifies the orientation of the celestial (equatorial) coordinate system with respect to the local terrestrial coordinate system in which the sidereal time is measured. The unit of duration of Sidereal Time (ST) is the period of the Earth's rotation with respect to a point nearly fixed with respect to the stars. The value of the Local Sidereal Time is equal to the right ascension of the local meridian, and is most accurately determined by interferometer observations of radio sources.

For a celestial object X,

Local Hour Angle of X = Local Sidereal Time − Right Ascension of X

or

$$
\text{LHA } X = \text{LST} - \text{RA } X \ . \tag{5.259}
$$

To each local meridian on the Earth there corresponds a Local Sidereal Time, LST, corrected to the Greenwich Sidereal Time, GST, of the prime meridian (Greenwich) when the geographic longitude, λ, is known; thus:

Local Sidereal Time = Greenwich Sidereal Time + East Longitude

or

$$
\text{LST} = \text{GST} + \lambda \ . \tag{5.260}
$$

Local sidereal time is subject to the motion of the equinox, itself due to precession and nutation; otherwise it is a direct measure of the diurnal rotation of the Earth. The measure of the rotation of the Earth with respect to the true equinox is called the apparent sidereal time. That is, the apparent sidereal time is defined as the hour angle of the true vernal equinox (first point of Aries).

When the position of the true equinox is corrected for nutation of the Earth's axis, the resultant sidereal time is called the mean sidereal time. The difference between the apparent and mean sidereal times is called the equation of the equinoxes.

5.3.5 Universal Time (UT), or Sun Time, Time Zones and the Equation of Time

The Earth's rotation with respect to the Sun is the basis of Universal Time, also called Solar Time. The unit of duration of Universal Time represents the solar day, defined to be as uniform as possible despite variations in the rotation of the Earth. Universal Time (UT) is the measure of time used as the basis for all civil time keeping. As the Earth rotates about its axis, it moves in orbit around the Sun, so Universal Time involves both the nonuniform diurnal rotation of the Earth and the Earth's orbital motion around the Sun. The Universal Times (UT) of the principal astronomical phenomena involving configurations of the Sun, Moon and planets are given in *The Astronomical Almanac*.

Universal Time is equivalent to the standard Civil Time for 0 degrees longitude, which is defined to be the meridian through Greenwich, England. By specifying the longitude of any other location on the globe, we can exactly infer the local Sun time. For example, the Sun will be overhead at noon in Boston about 4 hours 46 minutes later than noon in Greenwich; because Boston is about 73 degrees 33 minutes of arc west of Greenwich. And in the same way, noon in New York City will occur about 10 minutes later than in Boston because New York City is slightly west of Boston.

The world has been divided into standard time zones based on about one hour, or roughly 15 degree, increments in longitude, so our watches differ by others in hourly increments and are slightly out of synchronism with the Sun. Fig. 5.16 is a world map of the different time zones.

The standard time zones are identified with letter designations, in addition to appropriate names. Thus the standard time zone for 0 degrees longitude is labeled Z. The standard time zone for 15 degrees east longitude is labeled A. Subsequent letters designate zones increasing eastward in 15 degree increments, with M designating the zone for 180 degrees east. The time zone for west 15 degrees is labeled N, with letters progressing to Y for 180 degrees west longitude.

In practice, however, each country selects the appropriate time zone, or zones, for itself. The exact zone boundaries within each country are determined through that country's internal political process. In addition, some countries disregard the aforementioned convention by adopting a zone on a fraction of an hour. Many countries also adopt daylight saving time, sometimes called summer time or advanced time, which is generally one hour in advance of the standard time normally kept in that zone. The dates for changing to and from daylight saving time vary among the countries; a few countries retain advanced time all year.

Due to the eccentricity and inclination of the Earth's orbit around the Sun, the apparent motion of the Sun around the ecliptic is not perfectly uniform through the year. The Civil Time, or clock time, that we use in daily life must therefore be adjusted to provide the average apparent motion of the Sun

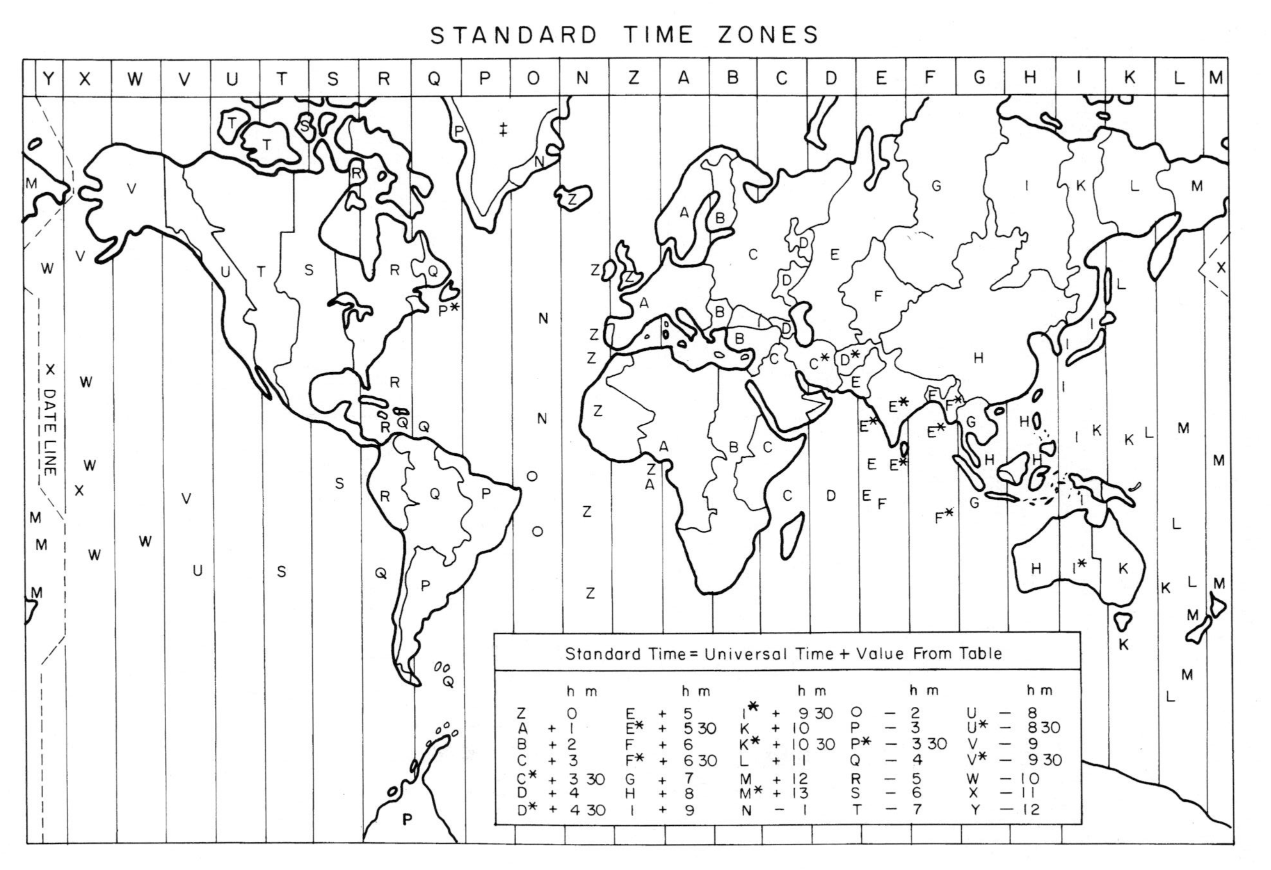

Fig. 5.16. Standard time zones

around the Earth's equator. The true rate of the Sun's apparent motion changes in the course of the year, and the difference between the True Sun Time and Civil Time is called the Equation of Time.

$$\text{True Sun Time} = \text{Civil Time} + \text{Equation of Time} \quad (5.261)$$

or

$$\text{LAT} = \text{LMT} + \text{Equation of Time} \ , \quad (5.262)$$

where LAT is the local apparent solar time and LMT is the local mean time. Adding the equation of time to Civil Time will give the True Sun Time. The equation of time varies through the year in a continuous manner by up to 16 minutes, as shown in Fig. 5.17. The equation of time is also useful for erecting sundials and determining the transit time of the Sun.

The following algorithm gives the equation of time, E, the Greenwich hour angle, GHA, declination, δ, and semidiameter, SD of the Sun, in degrees, to a precision of better than 1.0 minutes of arc (Seidelmann, 1992).

(1) Using the Julian date, JD, and the universal time, UT, in hours, calculate T, the number of centuries from J2000:

$$T = (\text{JD} + \text{UT}/24 - 2451545.0)/36525 \ . \quad (5.263)$$

(2) Calculate the mean longitude corrected for aberration, L; the mean anomaly, G; the ecliptic longitude, λ; and the obliquity of the ecliptic, ϵ:

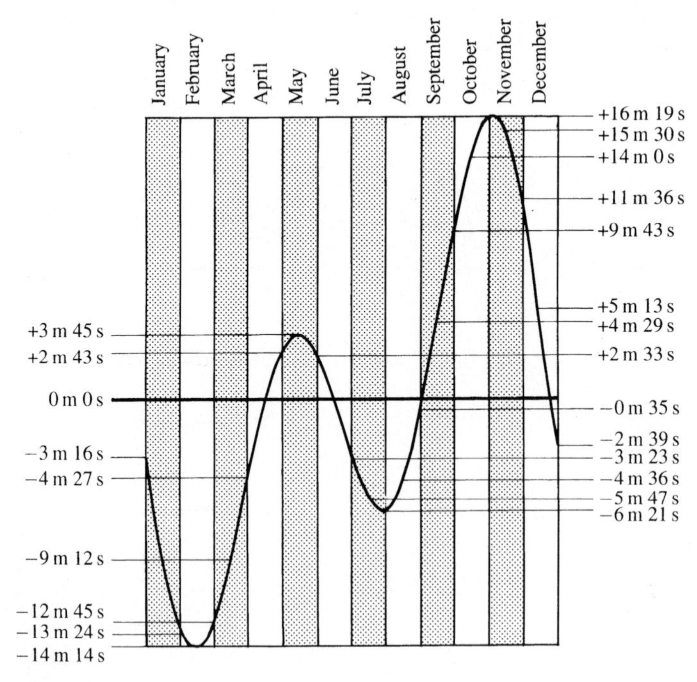

Fig. 5.17. Variation of the Equation of Time through the year. Add the time given here to Civil Time, or clock time, to obtain the True Sun Time

$$
\begin{aligned}
L &= 280^\circ.460 + 36000^\circ.770T, \text{ remove multiples of } 360^\circ \ , \\
G &= 357^\circ.528 + 35999^\circ.050T \ , \\
\lambda &= L + 1^\circ.915 \sin G + 0^\circ.020 \sin 2G \ , \\
\epsilon &= 23^\circ.4393 - 0^\circ.01300T \ .
\end{aligned}
\tag{5.264}
$$

(3) The equation of time E, the GHA, δ, and SD are given by

$$
\begin{aligned}
E &= -1^\circ.915 \sin G - 0^\circ.020 \sin 2G + 2^\circ.466 \sin 2\lambda - 0^\circ.053 \sin 4\lambda \ , \\
\text{GHA} &= 15 \text{ UT} - 180^\circ + E \ , \\
\delta &= \sin^{-1}(\sin \epsilon \sin \lambda) \ , \\
\text{SD} &= 0^\circ.267/(1 - 0.017 \cos G) \ .
\end{aligned}
\tag{5.265}
$$

Plotting the position of the apparent Sun relative to the mean Sun (which moves uniformly along the equator) produces the analemmic curve, see Fig. 5.18. The displacement of the apparent Sun in longitude is given by the equation of time, E, and the displacement in latitude is given by the declination, δ. The progress of the apparent Sun throughout the year is indicated at the beginning of each month on the closed curve. The analemmic curve plays an important role in certain types of sundial.

The Local Apparent Solar Time, LAT, is given by:

$$
\text{LAT} = \text{LST} - \text{RA Sun} - 12^{\text{h}} \ , \tag{5.266}
$$

where LST is the local sidereal time, defined by the rotation of the Earth with respect to the equinox, and RA Sun is the right ascension of the Sun. Universal time is similarly defined in terms of Greenwich Sidereal Time (GST) by:

$$
\text{UT} = \text{GST} - \text{RA U} - 12^{\text{h}} \ , \tag{5.267}
$$

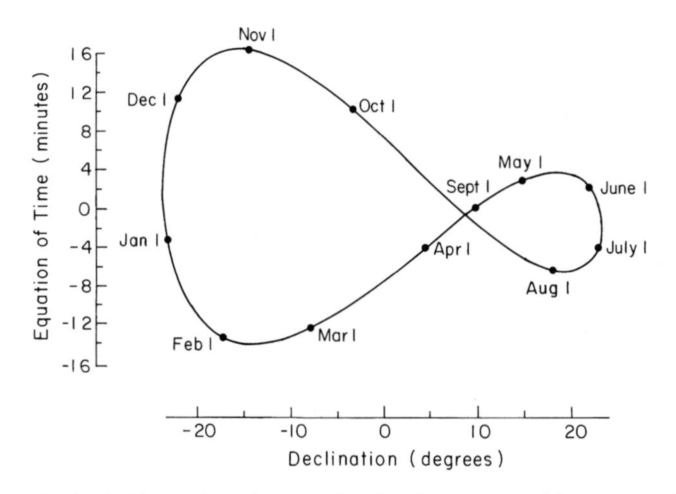

Fig. 5.18. The analemmic curve showing the progress of the apparent Sun throughout the year

where the hypothetical point U moves around the celestial equator at a uniform rate, and the coefficients for RA U are chosen so that UT may for most purposes be regarded as mean solar time on the Greenwich meridian.

5.3.6 Variation in the Earth's Rotation

There is an uncertainty in measurements of time that depends upon the Earth's rotation, whose variation has two main components. The first component, called polar motion, is the change in the orientation of the Earth's rotation axis relative to its crust. The second component is a variation in the rate of rotation, or changes in the length of the day.

Universal Time that has not been corrected for any variations in the Earth's rotation axis is designated UT0, and depends on the place of observation. The UT1 time scale has been corrected for the shift in longitude of the observing station caused by polar motion, and is thus independent of observing location, but is influenced by the slightly variable rotation of the Earth. Indeed, accurate measurements of UT1 are now used to determine variations in the Earth's rotation. A tabulation of the difference TDT – UT1 between Terrestrial Dynamic Time and UT1 is given in *The Astronomical Almanac* for the beginning of each year since 1984.

The modern definition of UT1 was established in order to fulfill the following conditions:

(a) UT1 is proportional to the angle of rotation of the Earth in space, reckoned around the true position of the rotation axis. In other words, UT1 is proportional to the integral of the modulus of the rotation vector of the Earth.
(b) The rate of UT1 is chosen so that the day of UT1 is close to the mean duration of the solar day.
(c) The phase of UT1 is chosen so that 12^{h} UT1 corresponds approximately, in the average, to the instant when the Sun crosses the meridian of Greenwich.

Fluctuations in UT1 are sometimes expressed in terms of the length of the day (LOD), which is the negative of the time derivative of UT1. All broadcast services currently distribute timescales based on the redefined Coordinated Universal Time (UTC), which differs from International Atomic Time (TAI) by an integer number of seconds, and is maintained within 0.90 second of UT1 by the introduction of one-second steps (leap seconds) when necessary, usually at the end of June or December.

Current techniques for determining variations in the Earth's rotation, and the time differences UT0 – UTC or UT1 – UTC, include laser ranging to geodetic satellites such as the Laser Geodynamics Satellite, or LAGEOS (Cohen et al., 1985), with a ranging precision of 1 to 15 centimeters; laser ranging to cube corners left on the Moon, determining UT0 – UTC with an accuracy of 0.3 to 0.7 milliseconds (Mc Carthy and Luzum, 1991); and Very Long Baseline Interferometry, or VLBI, of extragalactic radio sources with an accuracy of 0.05 milliseconds for UT1 – UTC (Robertson, 1991).

The motion of the Earth's rotation axis within our planet causes the North Pole to move roughly in a circle of about 9 meters in radius every year or so.

This polar motion, also known as the Chandler wobble, was inferred by Seth Carlo Chandler in 1891 using observations of the astronomical latitude at Cambridge, Massachusetts (Chandler 1891, 1892, 1893, 1898; Lang, 1985). Chandler found that the astronomical latitude exhibits both an annual and a 14-month variation of about 0.3 seconds of arc, corresponding to a 9-meter displacement measured at the Earth's surface. (The astronomical latitude is the complement of the angle between the Earth's rotation axis and the local vertical or the angle between the horizon and the average position of the Pole Star.)

The advent of the International Latitude Service in the 1890s brought about monitoring of polar motion that has continued to the present day (Fedorov et al., 1972; Yumi and Yokoyama, 1980). More recently, Very Long Baseline Interferometry (VLBI) of distant radio sources has reduced the formal errors of the determinations of the pole position to 0.0005 seconds of arc (0.5 milliseconds of arc), or about a centimeter of displacement on the Earth's surface (Robertson, 1991).

If the Earth was completely rigid, the wobble period would be 10 months, but the nonrigidity of the elastic Earth can lengthen the period to 14 months (Newcomb, 1892). The excitation of the Earth's wobble results from mass shifts, perhaps due to meteorological effects such as the seasonal distribution of atmospheric masses or to solid-Earth effects, but the exact nature of the excitation mechanism for the Chandler wobble remains one of the major unsolved problems of geophysics in spite of the accuracy of measuring the motion.

Variations in the Earth's rate of rotation were first inferred from observations of the Moon and planets. John Couch Adams showed that the Moon is apparently speeding up in its motion (Adams, 1853, 1860); Simon Newcomb then showed that the Moon is also apparently moving in fits and starts, at first speeding up in its orbit and then slowing down on timescales of decades (Newcomb, 1909). Variations in the Earth's rate of rotation were conclusively demonstrated a few years later (Glauert, 1915). In 1939 Harold Spencer Jones showed that similar accelerations and fluctuations occur in the apparent motions of the Moon, Venus and the Sun (Spencer Jones, 1939), showing that the observed motions are actually due to variations in the Earth's rotation. Laminated tidal sediments on Earth have confirmed that its rotation is slowing down, and shown that the length of the terrestrial day 900 million years ago was about 18 hours (Sonett et al., 1996).

The apparent accelerations of the Moon, Venus and the Sun are produced by a deceleration in the rotation of the Earth, corresponding to a small, slow increase in the length of the day. The apparent fluctuations in their motions are due to irregular changes in the Earth's rate of rotation. Thus, the Earth's broken clock is gradually running down, while also alternately running slow and fast in a seemingly random fashion.

The most complete current references on historical variations in the Earth's rate of rotation, based on ancient sightings of eclipses and lunar occultations, are given by Stephenson and Morrison (1984) and Mc Carthy and Babcock (1986). The overall increase in the length of the day over the centuries is explained by a tidal deceleration of the Earth's rotation. As the ocean tides flood and ebb, they create eddies in the water, producing friction and dissipating energy at the expense of the Earth's rotation (Jeffreys, 1920, 1975; Munk and Mac Donald, 1960; Lambeck, 1978, 1980). The tides therefore act as a brake on the spinning Earth, and the day becomes longer at a rate of about 2 milliseconds per year. This decrease in the Earth's rotational angular momentum is compensated by an increase in the Moon's orbital angular momentum, and as a result the Moon steadily moves away from the Earth at the rate of a few centimeters per year.

In recent years, the lunar tidal deceleration has been estimated using lunar laser ranging, analysis of the orbits of artificial satellites, numerical tidal models and astronomical models. The lunar laser ranging data indicate that the secular acceleration of the Moon is -25.88 ± 0.5 seconds of arc per (century)^2, which is equivalent to an increase in the length-of-day at the rate of about two milliseconds per century and causes the Moon to move away from the Earth at a rate of 3.82 ± 0.07 centimeters per year (Dickey et al., 1994). The long-term increase in the length-of-day has been documented over the past 2,700 years from records of occultations of stars by the Moon and solar and lunar eclipses (Stephenson and Morrison, 1984), with a rate of 2.43 ± 0.07 milliseconds per century between 390 B.C. and A.D. 948 and 1.40 ± 0.04 milliseconds per century between A.D. 948 and 1800.

More accurate telescopic observations for the recent historical period from 1656 (Stephenson and Morrison, 1984; Mc Carthy and Babcock, 1986; Stephenson and Lieske, 1988) indicated that pronounced fluctuations in the length of the day are superimposed on the long-term increase; they have amplitudes up to a millisecond of time and occur with periods from a few days to a year or more. Stormy weather is jerking the Earth around (Fig. 5.19), or in more technical terms, extensive meteorological effects produce variations in angular momentum accompanied by fluctuations in the solid Earth's rotation (Chao, 1989). Sinusoidal variations with amplitudes of up to one millisecond and periods ranging from a few days to months might also be explained by periodic changes of the tidal bulge in the solid Earth caused by the Moon and the Sun (Robertson, 1991).

5.3.7 Relations Between Sidereal Time and Universal Time

As the Earth rotates about its axis, it also moves in its orbit around the Sun. The rotation of the Earth with respect to a fixed star, called Sidereal Time (ST) or star time, is therefore different from the Earth's rotation with respect to the Sun, called Universal Time (UT) or Sun time, due to the Earth's motion in its orbit and the resultant additional rotation with respect to the Sun.

Greenwich Mean Sidereal Time (GMST), or star time, is related to the corrected Universal Time (UT1), or Sun time, by (Aoki et al., 1982; Seidelmann, 1992):

$$\text{GMST of } 0^{\text{h}}\text{UT1} = 24110.54841 + 8640184.812866 T_u + 0.093104 T_u^2 - 6.2 \times 10^{-6} T_u^3 \text{ seconds} \tag{5.268}$$

where 24110.54841 seconds = 6 hours 41 minutes 50.54841 seconds, and $T_u = d_u/36525$, d_u is the number of days of Universal Time elapsed since JD

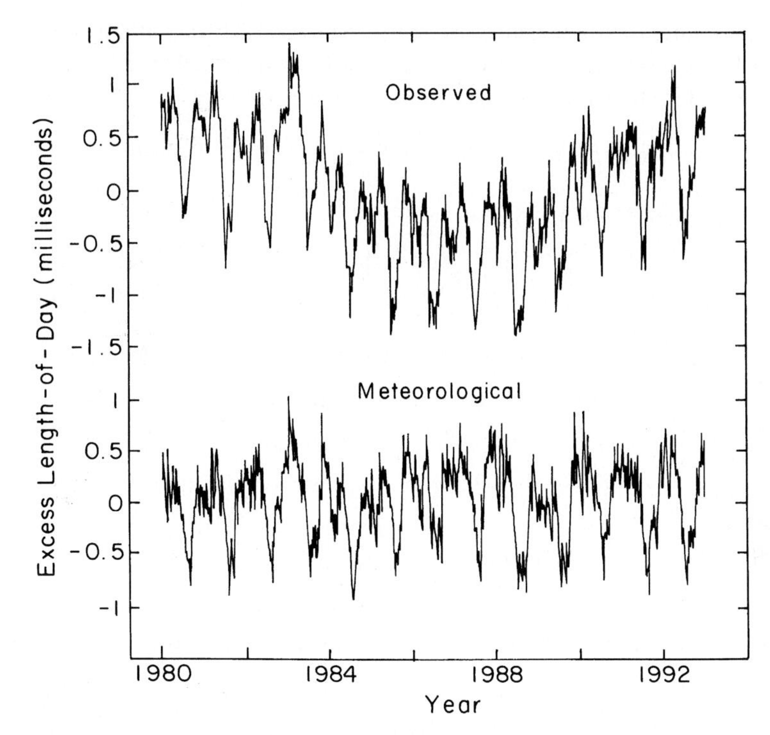

Fig. 5.19. Geodetically observed excess length-of-day (top) compared with that predicted by meteorologically derived data (bottom). The mean and zonal tidal signals have been removed to give the excess length-of-day beyond that caused by tidal effects. (Courtesy of Benjamin Fong Chao)

2451545.0 UT1 (2000 January 1, 12^h UT1), taking on values of $\pm 0.5, \pm 1.5, \pm 2.5, \pm 3.5, \ldots$. This relation conforms to the position and motion of the equinox specified in the IAU 1976 System of Astronomical Constants, the 1980 IAU Theory of Nutation, and the positions and proper motions of stars in the FK5 catalog.

The relationship between Greenwich Sidereal Time (GST) and Universal Time (UT) for the year is given in the relevant volume of *The Astronomical Almanac*. It also tabulates GST for 0 hours UT of each day of the year and provides examples for conversion between Local Sidereal Time (LST) and Universal Time (UT) and *vice versa*.

The number of seconds of sidereal time in a Julian century may be obtained by differentiation of the preceding equation to obtain

$$s' = 8640184.812866 + 0.186208T - 1.86 \times 10^{-5}T^2 \text{ seconds} . \tag{5.269}$$

Dividing by 36525 (the number of days in a Julian century) gives the number of sidereal seconds, s, in a day of Universal Time, UT1.

$$s = 86636.55536790872 + 5.098097 \times 10^{-6}T - 5.09 \times 10^{-10}T^2 \text{ seconds} , \tag{5.270}$$

where T is the number of Julian centuries elapsed since JD 2451545.0. Dividing this equation by 86400 gives the ratio of mean sidereal time to UT1:

$$r' = 1.002737909350795 + 5.9006 \times 10^{-11}T - 5.9 \times 10^{-15}T^2 \ . \quad (5.271)$$

Taking the inverse of this expression gives the ratio of UT1 to mean sidereal time:

$$1/r' = 0.997269566329084 - 5.8684 \times 10^{-11}T + 5.9 \times 10^{-15}T^2 \ . \quad (5.272)$$

Although the lengths of a day of UT1 and a day of mean sidereal time vary slightly with variations in the Earth's rotation, the ratio of UT1 to mean sidereal time is unaffected by rotational variations.

The equivalent measures of the lengths of the days in the two times are:

$$\begin{aligned} &\text{mean sidereal day} = 0.99726956633 \text{ days of mean solar time} \\ &\text{mean sidereal day} = 23^{\text{h}}\,56^{\text{m}}\,04^{\text{s}}.090524 \text{ of mean solar time} \ , \\ &\text{mean sidereal day} = 86164.090524 \text{ seconds of mean solar time} \ , \end{aligned} \quad (5.273)$$

and

$$\begin{aligned} &\text{mean solar day} = 1.00273790935 \text{ days of mean sidereal time} \ , \\ &\text{mean solar day} = 24^{\text{h}}\,03^{\text{m}}\,56^{\text{s}}.5553678 \text{ of mean sidereal time} \ . \\ &\text{mean solar day} = 86636.5553678 \text{ seconds of mean sidereal time} \ , \end{aligned} \quad (5.274)$$

where mean solar time is equivalent to Universal Time. Thus, mean sidereal time gains $03^{\text{m}}\,56.5553678^{\text{s}}$ per day on mean solar time, which means that relative to the Sun the stars rise and set four minutes earlier each day.

The rotational period of the Earth is 86164.09890369732 seconds of UT1 or $23^{\text{h}}\,56^{\text{m}}\,04.^{\text{s}}09890369732$. This gives a ratio of a UT1-day to the period of rotation of 1.002737811906. The rate of rotation is 15.04106717866910 seconds of arc per second of time.

5.3.8 Age of the Earth, Moon and Meteorites

Under the assumption that the Earth was initially molten and cooled from the outside in, Lord Kelvin (then William Thomson) calculated that it would take 20 to 100 million years for the Earth's surface temperature gradient to reach that observed today (Kelvin, 1862, 1899). This age agreed with calculations based on the sodium content of the oceans. Assuming that the oceans were initially precipitated salt free, and that sodium, Na, has been removed by rivers from the land and deposited into the oceans at the current rate, then it takes about 100 million years to obtain the ocean's present salt, NaCl, content (Joly 1899, 1908). Around 100 million years are also required to account for all geological history if the present rates of erosion applied in the past (Geikie, 1871).

The discovery of radioactivity led to older estimates for the Earth's age. Rapid, energetic particles emitted by radioactive elements heat the interior of our planet (Rutherford, 1905). The Earth's internal heat is therefore provided by radioactivity, and it is not due to cooling from an initially molten state.

Radioactive elements also clocked the Earth's age, leading to the conclusion that it is more than a billion years old, or roughly ten times older than had been thought (Rayleigh, 1905; Boltwood, 1907; Aston, 1929; Rutherford, 1929). This age is determined by measuring the relative amounts of radioactive materials and their non-radioactive products. When this ratio is combined with the known rates of radioactive decay, the time since the rock solidified and locked in the radioactive atoms is obtained.

The law of radioactive decay, which Rutherford formulated, states that the number of radioactive atoms, N_t, at time t will be less than the number, N_0, at time $t = 0$ according to the relation:

$$N_t = N_0 \exp(-\lambda t) = N_0 \exp(-0.693t/\tau_{1/2}) \; , \tag{5.275}$$

where the radioactive decay constant $\lambda = 0.693/\tau_{1/2} = \ln(2)/\tau_{1/2}$ and $\tau_{1/2}$ is the half-life of the radioactive species. Half-lives of radioactive decay chains are given in Table 5.7.

The number of radioactive atoms in the rock will be halved in a time equal to the half-life. Radioactive uranium, ^{238}U, decays, for example, into lead, ^{206}Pb, (which is stable), with a half-life of about 4.47 billion years; so every 4.47 billion years the amount of uranium-238 in a rock will be halved. In contrast, the other radioactive isotope of uranium, ^{235}U, has a short half life (704 million years) and is now much less abundant than it was in the early solar system.

Since the time a rock solidified, the sum of the number of uranium-238 and lead-206 atoms must be a constant, or $N(^{238}U) + N(^{206}Pb) =$ constant. We may find out the value of the constant by measuring the amounts of ^{238}U and ^{206}Pb currently present in the rock sample and then use the previous equation to deduce the amount of uranium-238 contained in past times, inferring the same for lead-206 since their sum is constant. If we assume that all the ^{206}Pb comes from the radioactive decay of ^{238}U, we can find the time in the past when the amount of lead, $N(^{206}Pb)$, was zero. This time is the age of the rock sample when it solidified.

If a terrestrial rock, lunar sample or a non-terrestrial meteorite became a closed system at time $t = 0$, then the present abundances of lead and uranium are related by the equation:

Table 5.7. Radioactive isotopes (parents), stable daughter isotopes, and half-lives of elements commonly used for dating (Adapted from Arnould and Takahashi, 1990, and Birck, 1990)

Parent	Daughter	Half-life (years)
Rubidium (Rb) 87	Strontium (Sr) 87	48.8 billion
Rhenium (Re) 187	Osmium (Os) 187	44 billion
Lutetium (Lu) 176	Halfnium (Hf) 176	35.7 billion
Thorium (Th) 232	Lead (Pb) 208	14.05 billion
Uranium (U) 238	Lead (Pb) 206	4.47 billion
Potassium (K) 40	Argon (Ar) 40	1.27 billion
Samarium (Sm) 146	Neodymium (Nd) 142	0.10 billion
Uranium (U) 235	Lead (Pb) 207	704 million
Plutonium (Pu) 244	Thorium (Th) 232	83 million
Iodine (I) 129	Xenon (Xe) 129	16 million
Palladium (Pd) 107	Silver (Ag) 107	6.5 million
Manganese (Mn) 53	Chromium (Cr) 53	3.7 million
Aluminum (Al) 26	Magnesium (Mg) 26	0.72 million

$$\left(\frac{\mathrm{Pb}^{206}}{\mathrm{Pb}^{204}}\right)_t = \left(\frac{\mathrm{U}^{238}}{\mathrm{Pb}^{204}}\right)_t [\exp(\lambda_{238}t) - 1] + \left(\frac{\mathrm{Pb}^{206}}{\mathrm{Pb}^{204}}\right)_0 , \tag{5.276}$$

where the subscripts t and 0 denote the present and initial abundances, respectively. If all of the rock samples have the same initial $\mathrm{Pb}^{206}/\mathrm{Pb}^{204}$ abundances, and if all of them have the same age, t, then a plot of $(\mathrm{Pb}^{206}/\mathrm{Pb}^{204})_t$ against $(\mathrm{U}^{238}/\mathrm{Pb}^{204})_t$ should lie in a straight line of slope $[\exp(\lambda_{238}t) - 1]$; such a plot is called an isochron. If a system formed t years ago and initially contained no lead, then a curve of the ratios $^{207}\mathrm{Pb}/^{206}\mathrm{Pb}$ and $^{238}\mathrm{U}/^{206}\mathrm{Pb}$ also provides the age t.

In general, the system will have both uranium and lead present, and this lead will consist of both primordial Pb that was initially present when the system formed and radiogenic Pb produced by U and Th decay prior to time t. Data arrays of natural systems, such as the lunar highlands, show both a lower and upper intersection with the theoretical curve, denoting a formation time and a time of a major disturbance (see Fig. 5.20).

Holmes and Lawson (1926) pointed out that the observed lead to uranium abundance ratio established a lower limit to the age of the Earth of about 1 Gyr; and Rutherford (1929) and Aston (1929) noticed that if uranium was formed with equal isotopic abundances of $^{236}\mathrm{U}$ and $^{238}\mathrm{U}$, then the present abundance ratio indicated a time of about 3 Gyr since the rocks solidified. One Gyr is equivalent to 10^9 years or a billion years. No rock older than 3.9 Gyr has been found on the Earth's surface (Birck, 1990), but indirect evidence for the first 600 million years of the Earth is found from lunar samples and meteorites.

Radioactive dating of samples returned from the lunar surface describe the Moon's early history (Wasserburg et al., 1977; Taylor, 1982; Lang and Whitney, 1991). They indicate that the Moon formed about 4.5 Gyr ago, and that an intense bombardment created the large impact basins and most of the highland craters about 4.0 Gyr ago (see Fig. 5.20). There followed an era of volcanism, from 4.2 to 3.1 Gyr ago, as molten basaltic lava welled up from the lunar interior, flooding the great impact basins and producing the dark circular maria that can be seen today. Except for the occasional large impacts that produced the rayed craters, and churning by small meteorites, the Moon has remained essentially unchanged for the past 3 Gyr.

Meteorites provide the most accurate estimate for the age of the solar system. Patterson, Tilton and Inghram (1955) and Patterson (1956) showed that the ratios for the lead isotopes $^{206}\mathrm{Pb}/^{204}\mathrm{Pb}$ and $^{207}\mathrm{Pb}/^{204}\mathrm{Pb}$ in three stone and two iron meteorites lie on a straight line whose slope corresponds to an age of $t_{\mathrm{m}} = 4.55 \pm 0.07$ Gyr, which was taken to be the time since the Earth attained its present mass. This age estimate assumed that all meteorites formed at the same time, originally containing lead of the same isotopic composition, and that they now contain uranium with the same isotopic composition as that in the Earth. Similar data for the carbonaceous chondrites have more recently resulted in an age of $t_{\mathrm{m}} = 4.566 \pm 0.002$ Gyr (Birck, 1990); these meteorites are about 10 million years older than those of other types (chondrites and achondrites).

Examination of the isotopic abundance of other elements, such as $^{40}\mathrm{K}$ and $^{40}\mathrm{Ar}$ or $^{87}\mathrm{Rb}$ and $^{87}\mathrm{Sr}$, lead to similar meteorite age estimates (Anders 1962,

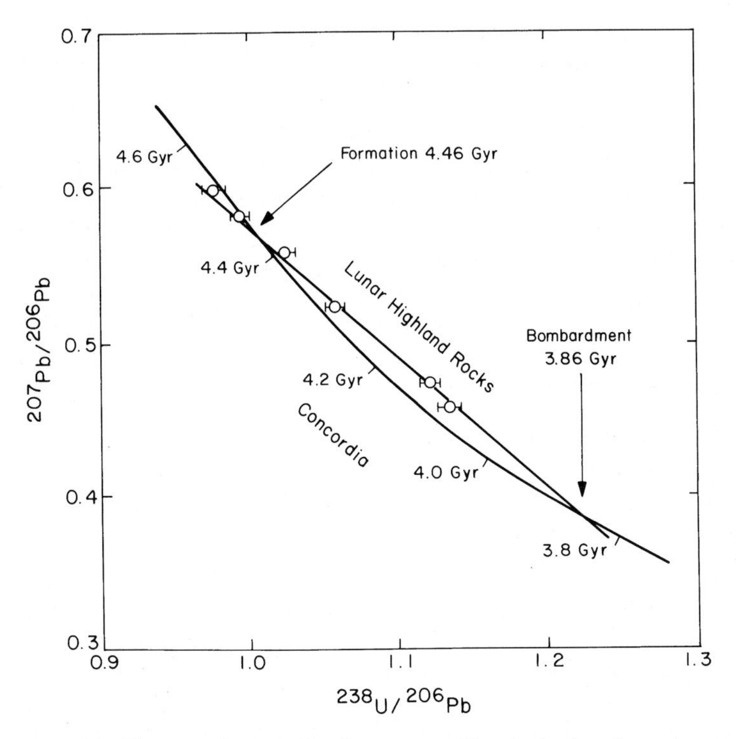

Fig. 5.20. The curved concordia diagram specifies the lead and uranium concentrations for times denoted along the curve in Gyr = 10^9 years = one billion years. Rocks from the lunar highlands describe a straight-line isochron (data points with error bars) indicating a formation age of 4.46 Gyr followed by an intense bombardment 3.86 Gyr ago. (Adapted from Birck, 1990 and Wasserburg et al., 1977)

1964; Birck, 1990). The terrestrial abundance ratios of lead and uranium indicate that primordial lead on the Earth had the same isotopic composition as that of the meteorites 4.54 ± 0.02 Gyr ago (Ostic, Russell, and Reynolds 1963). Thus we can assume that the major solidification of the Earth, Moon and meteorites occurred at the same time, t_{m}, given by

$$t_{\mathrm{m}} = 4.566 \pm 0.002 \text{ Gyr} . \tag{5.277}$$

Rounding off the number and allowing for possible systematic errors, we obtain an age, t_{ss}, for the solid bodies in solar system of

$$t_{\mathrm{ss}} = 4.6 \pm 0.1 \text{ Gyr} . \tag{5.278}$$

5.3.9 Age of the Sun, Stars, and Star Clusters

The stars have formed, and continue to be formed, through the collapse of interstellar clouds. The gravitational free-fall time, $t_{\text{f-f}}$, required for an isolated interstellar cloud of radius r to collapse under gravity, with no restraining forces, is given by twice integrating the equation $d^2r/dt^2 = -GM/r^2$ with respect to time, t, obtaining (Harwit, 1988):

$$t_{\text{f-f}} = \frac{\pi\sqrt{2}}{4}\left(\frac{R^3}{GM}\right)^{1/2} = \left(\frac{3\pi}{32G\rho_{\text{i}}}\right)^{1/2} \tag{5.279}$$

where $\rho_{\text{i}} = 3M/4\pi R^3$ is the initial mass density for a mass M and initial radius R. Note that for an interstellar cloud of a mass density ρ_{i}, this expression is independent of the size, R. With mass and radius measured in solar units, the free-fall collapse time is:

$$t_{\text{f-f}} \approx 1.77 \times 10^3 \left(\frac{M_\odot}{M}\right)^{1/2} \left(\frac{R}{R_\odot}\right)^{3/2} \text{ seconds} \tag{5.280}$$

or

$$t_{\text{f-f}} \approx 2.10 \times 10^3 (\rho/\rho_\odot)^{-1/2} \text{ seconds} \tag{5.281}$$

where the solar values are $M_\odot = 1.989 \times 10^{33}$ g, $L_\odot = 3.85 \times 10^{33}$ erg s^{-1}, $R_\odot = 6.96 \times 10^{10}$ cm, and the mean density of the Sun $\rho_\odot = 3M_\odot/4\pi R_\odot^3 = 1.41$ g cm^{-3}.

Once formed by gravitational collapse, a star will begin to shine. The Sun, for example, most likely formed together with the planets, moons and meteorites and has since continued to shine for 4.5 to 4.6 billion years. However, the Sun has not always been thought to be so old; its estimated age grew roughly in parallel with the discovery of radioactivity and the increase in the measured age of the Earth. The difficulty was explaining how the Sun could supply light and heat with the current intensity for long periods of time. A normal fire of the Sun's intensity would burn out in a few thousand years.

William Thomson, later Lord Kelvin, showed that the Sun might have illuminated the Earth at its present rate for a much longer time, about 10 million years but up to 100 and perhaps 500 million years, by slowly contracting (Kelvin, 1862). If the Sun, or any other star, was gradually shrinking, then its gravitational potential energy, Ω, would be slowly converted into kinetic energy, T_{k}, heating the stellar gas to incandescence with the luminosity, L. If the star is in hydrostatic equilibrium, in which outward gas pressure balances the inward pull of gravity, then from the virial theorem, $T_{\text{k}} = -\Omega/2$, and for a sphere of mass, M, and radius, R, the $\Omega = -3GM^2/(5R)$.

When the star shines by this store of energy, then its lifetime is given by the Kelvin-Helmholtz time (Kelvin, 1863, Helmholtz, 1854):

$$t_{\text{K-H}} = \frac{T_{\text{k}}}{L} \approx \frac{3GM^2}{10RL} \approx 10^7 \left(\frac{M}{M_\odot}\right)^2 \left(\frac{L_\odot}{L}\right) \left(\frac{R_\odot}{R}\right) \text{ years }, \tag{5.282}$$

where 1 year $= 3.1557 \times 10^7$ seconds. This age is, of course, incompatible with radioactive dating of rocks and fossils that indicate that the Sun was supplying the Earth with its heat and light for billions of years.

The problem of the source of stellar energy was brought into sharp relief by Arthur Stanley Eddington, whose calculations showed that gravitational contraction could suffice for only about 100,000 years for giant stars. Eddington pointed out that nuclear processes were a likely candidate for the stellar energy source, noting that nuclear fusion in the Sun's hot, dense core can fuel the Sun at the observed rate for billions of years (Eddington, 1920). The basic idea relies on Einstein's famous mass-energy equivalence (Einstein, 1905, 1906). For an energy, E, and mass, M, the $E = Mc^2$, where $c = 2.9979 \times 10^{10}$ cm s^{-1} is the velocity of light. For the Sun, $E_{\odot} = M_{\odot}c^2 = 1.79 \times 10^{54}$ erg. If all of this energy could be converted into radiation, it could make the Sun shine for a thermonuclear time $t_N = E_{\odot}/L_{\odot} = 1.4 \times 10^{13}$ years.

Eddington argued that hydrogen might be transformed into helium inside stars, with the resultant mass difference released as energy to power the Sun; a similar idea was advanced at about the same time by the French physicist Jean Perrin, but with less effect on the astronomical community (Eddington, 1920; Perrin, 1920). The mass of the helium nucleus, m_{He}, is 0.007, or 0.7 per cent, less than the mass of the four protons, 4 m_{H}, that formed it, so the efficiency of conversion of mass to energy by this process is $m_{\mathrm{He}}/4m_{\mathrm{H}} = 0.007$. Moreover, thermonuclear reactions are limited to the hot, dense core of a star, and the hydrogen fuel within that core will be totally consumed when the star has exhausted about 0.12, or 12 percent, of its original energy (Schönberg and Chandrasekhar, 1942). So the main-sequence thermonuclear lifetime, t_{ms}, when a star shines by converting hydrogen into helium is approximated by:

$$t_{\mathrm{ms}} = 0.12\left(\frac{m_{\mathrm{He}}}{4m_{\mathrm{H}}}\right)\frac{Mc^2}{L} \approx 10\left(\frac{M}{M_{\odot}}\right)\left(\frac{L_{\odot}}{L}\right) \text{ Gyr} \tag{5.283}$$

for a star of mass, M, and absolute luminosity, L. Here the solar values are denoted by $M_{\odot}$ and $L_{\odot}$. The mass-luminosity relation for hydrogen-burning stars near a solar mass is roughly where $L = L_{\odot}(M/M_{\odot})^n$. Taking $n = 3.5$, the thermonuclear lifetime for converting hydrogen into helium is:

$$t_{\mathrm{ms}} \approx 10\left(\frac{L_{\odot}}{L}\right)^{5/7} \text{ Gyr} \approx 10\left(\frac{M}{M_{\odot}}\right)^{-2.5} \text{ Gyr} . \tag{5.284}$$

where 1 Gyr $= 10^9$ years = a billion years. However, the index n decreases significantly as the mass, M, increases beyond about 10 $M_{\odot}$.

The paths of stellar evolution can be traced in the Hertzsprung–Russell, or H-R, diagram in which the ordinate is the absolute magnitude, M, and the abscissa is the color B-V. Since stars of different colors also exhibit different spectral lines and have varying temperatures, the abscissa of the H-R diagram also describes the spectral class, O, B, A, F, G, K or M, and the stellar surface temperature, T_s, or effective temperature, T_{eff}. The ordinate of the H-R diagram can also specify the stellar absolute luminosity, L, often specified in solar units, $L_{\odot} = 3.85 \times 10^{33}$ erg s^{-1}.

Hertzsprung (1905) and Russell (1914) were able to specify the first H-R diagrams once the absolute luminosities of individual stars were inferred from their distances, obtained by parallax measurements and apparent luminosity, and their spectral class was obtained from the absorption lines in their spectra. Russell's plot showed two well-defined classes of stars. These he called dwarf and giant stars on the basis of size. The giant stars are bigger and therefore more luminous than the dwarf stars for a given temperature, just as we would infer from the Stefan-Boltzmann relation $L = 7.1258 \times 10^{-4} R^2 T^4$ erg s^{-1} for a star of radius, R, and effective or surface temperature, T. Dwarf stars on the main sequence include the Sun (spectral type G2V) whose radius, $R_\odot = 6.96 \times 10^{10}$ cm, and range in size from 0.2 $R_\odot$ for spectral class M7 to 15 $R_\odot$ for spectral class O3, while giant stars can be hundreds of times larger than the Sun (Lang, 1992).

Russell's dwarf stars are also called the main-sequence stars, since most stars occupy this part of the H-R diagram (Fig. 5.21). It was later realized (Adams, 1914, 1915) that an anomalous star, located in the lower left-hand corner of Russell's original plot is a representative of yet another class of stars of high surface temperature and low luminosity; these stars are now called white dwarfs because of their initially white color and exceedingly small sizes. The white dwarfs are comparable to the Earth in size, with a radius, R_{wd}, about equal to one hundredth that of the Sun, or $R_{\mathrm{wd}} = 0.01 R_\odot$.

The beginning of the main sequence is defined as that stage when the nuclear energy production just begins to balance the luminosity. That is, when a star first contracts from the interstellar medium, the generation of energy by hydrogen burning develops internal pressures which oppose gravitational

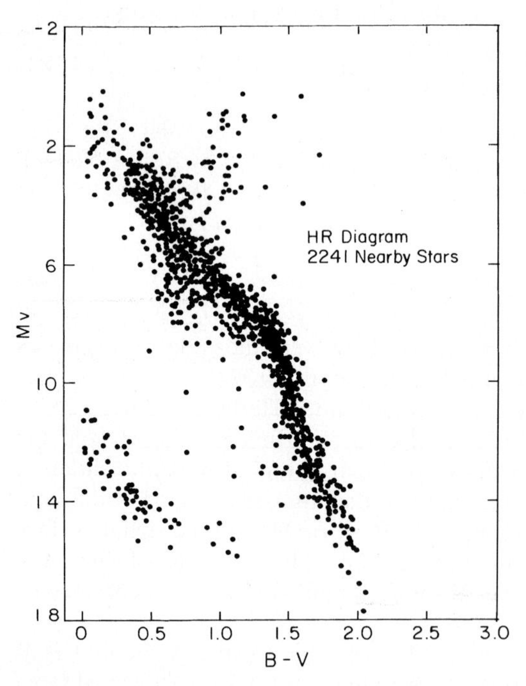

Fig. 5.21. Hertzsprung–Russell, or H-R, diagram for 2,241 nearby stars. Here the absolute visual magnitudes, M_{V}, are plotted against the color, B-V. The vast majority (roughly 90%) of the stars occupy the main sequence in the H-R diagram, running diagonally from the hot, luminous upper left to the cooler, less-luminous lower right. The Sun is a main-sequence star with $M_{\mathrm{V}} = 4.82$ and B-V $= 0.68$. The white dwarf stars, in the lower left, are low-luminosity stars about the size of the Earth. (Adapted from Lang, 1992)

contraction, and the star is stabilized on the main sequence at the point appropriate to its mass.

According to the Vogt–Russell theorem (Vogt, 1926; Russell, Dugan, and Stewart, 1926), a star of a given mass, age, and initial composition occupies a unique position on the H-R diagram. The equations which determine this position are:

The mass equation:

$$\frac{dM(r)}{dr} = 4\pi r^2 \rho \text{ or } \frac{dr}{dM(r)} = \frac{1}{4\pi r^2 \rho} , \tag{5.285}$$

where $M(r)$ is the mass within radius, r, and ρ is the mass density given by:

$$\rho = \left(2X + \tfrac{3}{4}Y + \tfrac{1}{2}Z\right)^{-1} N m_{\mathrm{H}}$$

for a fully ionized gas of total number density, N, respective mass fractions X, Y, and Z for hydrogen, helium and the heavier elements, and where $m_{\mathrm{H}} = 1.673 \times 10^{-24}$ grams is the mass of the hydrogen atom. Representative values of X, Y, and Z are $X = 0.61$, $Y = 0.37$, and $Z = 0.02$ for population I stars, and $X = 0.90$, $Y = 0.10$, and $Z = 0.001$ for population II stars. The presently accepted values for the solar system are $X = 0.70683 \pm 0.025$, $Y = 0.27431 \pm 0.06$ and $Z = 0.01886 \pm 0.00085$ (Anders and Grevesse, 1989).

The equation of hydrostatic equilibrium:

$$\frac{dP}{dr} = -\rho \frac{GM(r)}{r^2} \quad \text{or} \quad \frac{dP}{dM(r)} = -\frac{GM(r)}{4\pi r^4} , \tag{5.286}$$

where P is the gas pressure and G is the gravitation constant.

The equation for energy balance:

$$\frac{dL(r)}{dr} = 4\pi r^2 \rho \varepsilon \quad \text{or} \quad \frac{dL(r)}{dM(r)} = \varepsilon \tag{5.287}$$

where $L(r)$ is the energy flux from within r, and ε is the rate of energy generation per unit mass per unit time. Values of ε are discussed in Volume I. Hydrogen burning by the proton–proton reaction or the CNO cycle is responsible for all the energy production of stars on the main sequence.

The equation of radiation energy transfer:

$$\frac{dT}{dr} = \frac{-3\kappa\rho L(r)}{4acT^3 4\pi r^2} \quad \text{or} \quad \frac{dT}{dM(r)} = \frac{-3\kappa L(r)}{64\pi^2 acT^3 r^4} , \tag{5.288}$$

where T is the temperature, the radiation density constant $a = 7.566 \times 10^{-15}$ erg cm^{-3} K^{-4}, and κ is the opacity to radiation in area per unit mass.

The equation of convective energy transport:

$$\frac{dT}{dr} = \left(\frac{\Gamma - 1}{\Gamma}\right) \frac{T}{P} \frac{dP}{dr} \quad \text{or} \quad \frac{dT}{dM(r)} = \left(\frac{\Gamma - 1}{\Gamma}\right) \frac{T}{P} \frac{dP}{dM(r)} , \tag{5.289}$$

where Γ is the adiabatic exponent.

Given an initial composition and a mass, the previous five equations determine an age for an observed luminosity and temperature. The current main sequence age of the Sun is, for example, 4.50 to 4.52 Gyr, with a total age including pre-main-sequence collapse of 4.55 Gyr (Sackmann, Boothroyd and Kraemer, 1993). The total main-sequence lifetime, t_{ms}, is specified in Table 5.8 for stars of different spectral class.

The Sun is slowly growing in luminous intensity as it ages. This steady luminosity increase is a consequence of increasing amounts of helium in the Sun's core; the greater mean density produces higher core temperatures, faster nuclear reactions and a steady increase in luminosity. Calculations indicate that the Sun began its life about 4.6 Gyr ago when its luminosity was 0.70, or 70 percent, of its present luminosity, $L_\odot$, and that the Sun's luminosity will grow to $2.2L_\odot$ in about 6.5 Gyr from now, when the Sun will deplete its core hydrogen and begin expanding to become a giant star. The Sun eventually reaches a luminosity of $2300L_\odot$ and a radius of $170R_\odot$ as a giant star, shedding $0.275M_\odot$ and engulfing the planet Mercury (Sackmann, Boothroyd and Kraemer, 1993).

The faint early luminosity of the Sun suggests an early cold climate on the Earth, which is in disagreement with the ancient warm climate record suggested by terrestrial fossils and sedimentary rocks. These rocks, which must have been deposited in liquid water, date from 3.8 Gyr ago, and there is fossil evidence for life around 3.5 Gyr ago. This discrepancy, known as the faint-young-Sun paradox, might be resolved if the Earth's primitive atmosphere contained about a thousand times more carbon dioxide than it does now; the enhanced greenhouse effect would have kept the oceans from freezing and protected Earth's young surface from the chilling effects of a faint Sun (Lang, 1995).

Öpik (1938) showed that both the giant and main-sequence stars shine by thermonuclear processes following a well-defined series of nuclear reactions as the gas at the core gets hotter and denser under the continued action of gravitation. Each reaction ignites near the center of the star and feeds on the ashes of its predecessors, which continue to occur further out. Once the central

Table 5.8. The main-sequence stars and their total main-sequence lifetime, t_{ms} (Schaller et al., 1992 and Lang, 1992).*

Spectral type	Effective temperature (°K)	Mass ($M/M_\odot$)	Luminosity ($L/L_\odot$)	Main-sequence Lifetime, t_{ms} (years)
O5 V	44 500	60	7.9×10^5	3.7×10^6
B0 V	30 000	17	5.2×10^4	1.1×10^7
B5 V	15 400	6	8.3×10^2	6.5×10^7
A0 V	9 520	3	5.4×10	2.9×10^8
F0 V	7 200	1.6	6.5	1.5×10^9
G0 V	6 030	1.1	1.5	5.1×10^9
K0 V	5 250	0.8	0.42	1.4×10^{10}
M0 V	3 850	0.5	0.077	4.8×10^{10}
M5 V	3 240	0.2	0.011	1.4×10^{11}

* The lifetimes for the hydrogen-burning phase with $Z = 0.01$ for stars of different mass are from Schaller et al. (1992), who also provide the lifetimes for the helium-burning and carbon-burning phases. The spectral types, temperatures and luminosities are from Lang (1992).

hydrogen of a main-sequence star is converted into helium, for example, the core must undergo gravitational contraction until it is hot enough for helium burning; in the meantime the hydrogen envelope expands and the former main-sequence star evolves into a giant star. Such an evolution is portrayed in the H-R diagram of a globular cluster of stars with varying mass but a common birthdate (Fig. 5.22). Theoretical tracks in the H-R diagram for individual stars evolving from the main sequence to the giant regions were provided by Sandage and Schwarzschild (1952) and Sandage (1957), and more recently by Iben and Renzini (1984), Jimenez et al. (1996) and Jiminez and Mac Donald (1996).

As a star ages, it precedes through successive core contraction and static core burning stages, in which heavier elements are synthesized from lighter ones. The basic element producing reactions were outlined by Burbidge, Burbidge, Fowler and Hoyle (1957), often called B^2FH. As shown in Table 5.9, the successive element-producing reactions proceed at a progressively hotter, denser and faster pace. The relative abundances of the different elements and isotopes is related to their nuclear stability convoluted with astrophysical processes.

More than 90% of all main-sequence stars will end up as white dwarf stars, after shedding most of their mass in the giant phases. Their initially high temperatures and low luminosity imply that white dwarf stars are small, with sizes comparable to that of the Earth, or about 0.012 that of the Sun. These

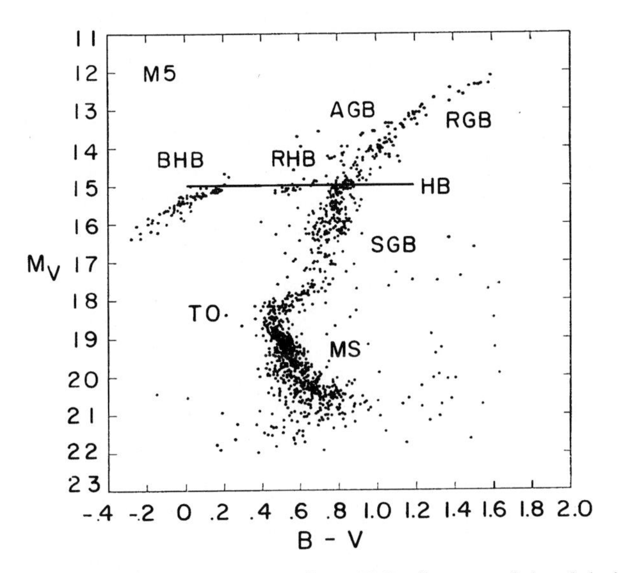

Fig. 5.22. Hertzsprung–Russell, or H-R, diagram of the globular cluster, M5 (adapted from Arp, 1962). The absolute visual magnitudes, M_V, are plotted as a function of the color, B-V. Evolutionary phases are the main sequence, MS, the turnoff, TO, point from the main sequence, the subgiant branch, SGB, the horizontal branch, HB, often subdivided into the red, RHB, and blue, BHB, parts that are separated by an instability strip, and the giant branch, GB, often subdivided into the red, RGB, or first, FGB, and asymptotic, AGB parts

Table 5.9. Central temperature, density and duration of thermonuclear reactions for a Population I star of initial mass, M, equal to 25 solar masses, or $M = 25M_{\odot}$ (Weaver, Zimmerman and Woosley, 1978)

Evolutionary Phase	Central Temperature (10^6 K)	Central Density (g cm^{-3})	Duration (seconds)*
Hydrogen burning (H → He)	37	3.8	2.3×10^{14}
Helium burning (He → C and O)	180	620	2.1×10^{13}
Carbon burning (C → Ne)	720	6.4×10^5	5.2×10^9
Neon burning (Ne → Mg and Si)	1400	3.7×10^6	3.9×10^7
Oxygen burning (O → Si)	1800	1.3×10^7	1.6×10^7
Silicon burning (Si → Fe)	3400	1.1×10^8	1.2×10^5

* For conversion one year = 3.156×10^7 seconds.

stellar remnants have masses, M, of about 0.6 of the solar mass, or $M = 0.6M_{\odot}$, and high densities of about 100,000 g cm^{-3}. Degenerate electron pressure stabilizes these stars against further collapse. Other, more massive stars will end up as neutron stars or black holes, which are discussed in other parts of this book (Sect. 5.6).

The light of white-dwarfs, the embers of low-mass stars, comes from the leftover heat of their former lives, and since they cannot sustain thermonuclear reactions there is nothing left to fuel their fires. White dwarf stars therefore slowly cool off and fade into darkness as they age. However, the faintest white dwarfs, observable in nearby space, have not yet faded to invisibility, and they can be used to infer the age of the galactic disk within the local solar neighborhood. That nearby white dwarfs have not been seen with surface temperatures of less than about 4,000 degrees Kelvin suggests that they formed between 7 and 15 Gyr ago.

Since hotter, more luminous white dwarf stars cool more rapidly, with a cooling rate that is roughly proportional to $L^{-5/7}$ for absolute luminosity L (Mestel, 1952), the observed number of white dwarf stars in a given volume of space increases monotonically with decreasing luminosity. That is, there are many more faint white dwarfs than bright ones, so the low-luminosity objects dominate the observed luminosity function of nearby white dwarf stars. The luminosity function is defined as the number of stars per unit volume per unit absolute bolometric magnitude interval specified as a function of luminosity.

Observations indicate that the numbers of white dwarfs per unit volume increases continuously at fainter luminosity up to log $(L/L_{\odot}) = -4.5$, and then shows an extremely sharp decline at very low luminosity (Liebert, 1980; Liebert, Dahn and Monet, 1988). The oldest white dwarf stars, which are the remnants of the earliest stars, have therefore not yet had time to cool to the lowest luminosity, at least in the local solar neighborhood where they can be observed. These stars are so faint that they must be within about 200 pc to be detected, and those used in the luminosity function are generally within 50 pc, where 1 pc $= 3.0856 \times 10^{18}$ cm.

The luminosity of the turndown in the white-dwarf luminosity function can be used to infer the cooling age, t_{wd}, of the oldest white dwarfs in the local part of the galactic disk (see Fig. 5.23). Winget et al. (1987) have combined the observed turning point between $\log(L/L_{\odot}) = -4.3$ and -4.7 with theoretical cooling rates, which depend upon model composition and mass, to obtain a cooling age of:

$$t_{wd} = 9.3 \pm 1.8 \text{ Gyr} . \qquad (5.290)$$

Recent estimates for the local age of the galactic disk, t_{gd}, inferred from the white dwarf data and theoretical models, are (Hernanz et al., 1990; Wood, 1992):

$$t_{gd} = t_{wd} + t_{ms} = 8.3 \text{ to } 15 \text{ Gyr} ,$$

and (5.291)

$$t_{gd} = t_{wd} + t_{ms} = 7.5 \text{ to } 11 \text{ Gyr} ,$$

where the main-sequence evolution time, $t_{ms} = 10(M/M_{\odot})^{-2.5}$ Gyr $= 0.5$ to 1.0 Gyr where 1 Gyr $= 10^9$ years = one billion years. A spread of about 3 Gyr for t_{wd} is due to uncertainty in the turndown luminosity for the local population of white dwarfs.

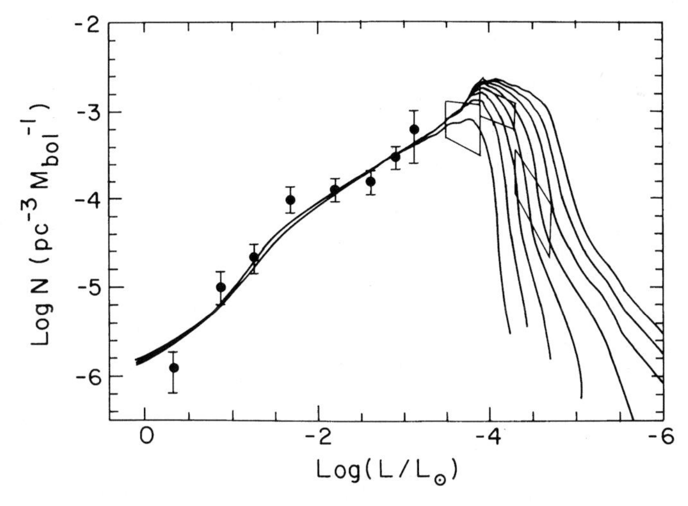

Fig. 5.23. Observed luminosity function of nearby white dwarf stars (data points with error bars) compared to theoretical models with cooling ages, t_{wd}, from 7 to 13 Gyr (solid curves), providing an estimated age for the local galactic disk of $t_{gd} = t_{wd} + t_{ms} = 7.5$ to 11.0 Gyr, where the main-sequence evolution time $t_{ms} = 0.5$ to 1.0 Gyr and 1 Gyr $= 10^9$ years = one billion years. Here N is the white dwarf density per cubic parsec, pc^3, per unit absolute bolometric magnitude, M_{bol}, the absolute luminosity L is given in solar units $L_{\odot} = 3.85 \times 10^{33}$ erg s^{-1}, and 1 pc $= 3.0857 \times 10^{18}$ cm. (Adapted from Wood, 1992)

Most stars, and perhaps all stars, were born in clusters. Yet, many of them are not now in star clusters, including the Sun. These stars must move fast enough to eventually escape the gravitational pull of the cluster (Ambartzumian, 1938; Spitzer, 1940).

The stars within a cluster are analogous to atoms within a gas, and characterized by a mean random velocity, V, with a Maxwellian distribution of stellar velocities. That is, the sizes of stars are small compared to their separations, even in a dense cluster, and a star cluster can be described in terms of an individual star's motion and the gravitational pull of the other stars on it. Most stars have random speeds comparable to V, but a few of them move with exceptionally high velocities; after a series of weak, distant gravitational encounters with other stars, these high-speed stars can move even faster, enabling them to escape the combined gravitational pull of all of the stars in the cluster. This escape requires the remaining stars to readjust their random distribution in a relaxation time, t_r, given by (Spitzer and Härm, 1958; Shu, 1982):

$$t_r = \frac{(3/2)^{3/2} V^3}{2\pi G^2 m^2 n \ln N} \approx \frac{R}{V} \frac{N}{12 \ln(N/2)} \tag{5.292}$$

where G is the gravitational constant and there are a total of N stars of mean stellar mass m with $N = 4\pi n R^3/3$ for a number density of stars, n, and a cluster radius R. The total mass of the cluster is Nm and a typical speed $V = (GNm/R)^{1/2}$. If the crossing time, $t_c = R/V$, then the relaxation time $t_r = n_r \times t_c$, where $n_r = 0.1N/\ln(N/2)$ is the number of crossings that are required for a star's velocity to change by the order of itself.

For a system whose median radius is r_h with stars of mean-square speed $\langle V \rangle^2 = 0.4\ GM/r_h$ and a total mass $M = Nm$, the median relaxation time, t_{rh}, is (Spitzer and Hart, 1971; Binney and Tremaine, 1987):

$$\begin{aligned} t_{rh} &= \frac{0.14N}{\ln(0.4N)} \left(\frac{r_h^3}{GM} \right)^{1/2} \\ &= \frac{0.65}{\ln(0.4N)} \left(\frac{M}{10^5\ M_\odot} \right)^{1/2} \left(\frac{M_\odot}{m} \right) \left(\frac{r_h}{1\ \mathrm{pc}} \right)^{3/2} \mathrm{Gyr}\ . \end{aligned} \tag{5.293}$$

where $1\ \mathrm{pc} = 3.086 \times 10^{18}$ cm, the Sun's mass, $M_\odot = 1.989 \times 10^{33}$ g, and $1\ \mathrm{Gyr} = 10^9$ years.

The evaporation time, t_{evap}, in which a significant fraction of the stars leave a cluster is:

$$t_{evap} \approx 100 t_r \approx 100 t_{rh}\ , \tag{5.294}$$

which sets an upper limit to the lifetime of any bound stellar system. (In an evaporation time, the cluster reduces its number of stars by a factor of $e = 2.718\ldots$.) For a typical open star cluster t_{evap} is about 3 Gyr, and for a typical globular star cluster t_{evap} is about 80 Gyr. Thus, the open star clusters that we see today are dispersing with time, and were created after our Galaxy formed, while the visible globular star clusters date back to the formation of our Galaxy.

When the statistical properties of N-body gravitating systems are taken into account, the collective relaxation time, t_{cr}, is (Gurzadyan and Savvidy, 1986):

$$t_{cr} = \left(\frac{15}{4}\right)^{\frac{2}{3}} \frac{1}{2\pi\sqrt{2}} \frac{\langle V \rangle}{G\langle M \rangle n^{\frac{2}{3}}} \ , \tag{5.295}$$

for an average stellar velocity $\langle V \rangle$, mean stellar mass $\langle M \rangle$, and stellar density, n. When normalized to values of stellar systems like globular clusters and galaxies, this expression becomes

$$t_{cr} \approx 0.1 \frac{\langle V \rangle}{10 \text{ km s}^{-1}} \left(\frac{n}{1 \text{ pc}^{-3}}\right)^{-\frac{2}{3}} \frac{M_{\odot}}{\langle M \rangle} \text{ Gyr} \ . \tag{5.296}$$

For clusters of galaxies, $t_{cr} \approx 10 - 1000$ Gyr.

Gurzadyan and Savvidy (1986) compare this collective relaxation time to the binary relaxation time, t_{br}, due to stellar binary encounters:

$$\frac{t_{br}}{t_{cr}} \approx \frac{\langle V^2 \rangle}{GMn^{\frac{1}{3}}} \frac{1}{\ln N} \approx \frac{d}{r_*} \frac{1}{\ln N} \ , \tag{5.297}$$

where d is the mean distance between stars, and $r_* = GM/\langle V^2 \rangle$ is the radius of gravitational influence. Here

$$t_{br} = \frac{\sqrt{2}\langle V^3 \rangle}{\pi G^2 \langle M^2 \rangle n \ln(N/2)} \approx t_r, \tag{5.298}$$

which has been given previously. With increasing stellar density, the distance, d, between stars decreases and approaches r_*, and the binary encounters become the dominant relaxation mechanism.

The dynamical time scale, t_{dyn}, for a stellar system of characteristic size, D, is given by:

$$t_{dyn} = \frac{D^{\frac{3}{2}}}{(GMN)^{\frac{1}{2}}} \approx \frac{d}{D} t_{cr} \ . \tag{5.299}$$

Under the hypothesis that all stars which are in a cluster of stars were formed at the same time, the age of a cluster can be found by matching a family of evolutionary tracks of various stellar masses, initial compositions and ages to the observed Hertzsprung–Russell diagram of the cluster. That is, the theory of stellar evolution permits us to compute the ages of star clusters from the turn-off point on the main sequence; the details depend on the initial composition of the stellar material and the fraction of the stellar mass that is exhausted at the turn-off point. Using plausible stellar models, we can infer the ages of stars of different masses, and therefore luminosity, and compare this age scale to the Hertzsprung–Russell diagrams of star clusters. As illustrated in Fig. 5.24, such techniques have been used to determine the ages of open star clusters, including well-observed clusters such as the Pleiades (77.6 million years) and the Hyades (about 660 million years). More recently, Perryman et al. (1998) obtain an age of 625 ± 50 million years for the Hyades cluster using

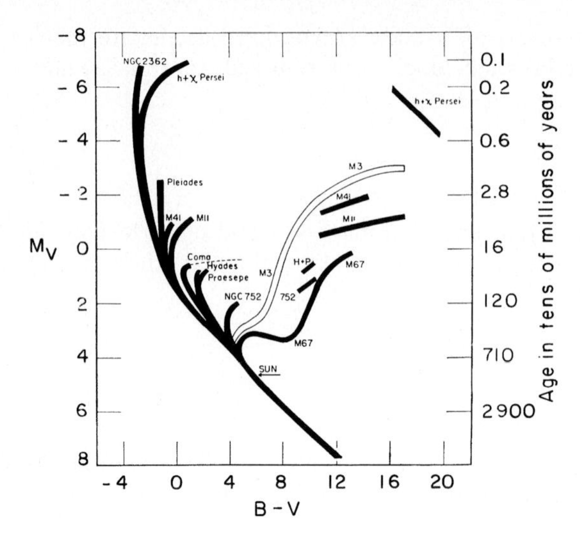

Fig. 5.24. The Hertzsprung–Russell, or H-R, diagrams for the stars in ten open, or galactic, clusters are compared with that of one globular cluster, M3. Here M_V is the absolute visual magnitude, the color is B-V, and the ordinate at the right gives estimates of the ages of the open clusters derived from the magnitude of the turn-off point at the upper end of the main sequence. (Adapted from Sandage, 1957). For comparison, Mermilliod (1981) obtains ages of 77.6 million years for the Pleiades and about 660 million years for the Hyades using their turn-off points from the main sequence. More recently, Perryman et al. (1998) obtain an age of 625 ± 50 million years for the Hyades cluster using Hipparcos data.

Hipparcos data. The open star clusters are located within the disk of our Galaxy and so appear to lie in the Milky Way.

Open star clusters contain several hundred to several thousand stars distributed in a region about a parsec across, often including hot, highly luminous stars. They are therefore relatively young, and because of their weak internal gravitation tend to disperse in time as the result of encounters with giant molecular clouds and galactic tidal effects. In the solar vicinity, these disruptive forces result in the destruction of a typical open cluster in a few hundred million years or less. The vast majority are therefore less than a few hundred million years old, and ages inferred from their Hertzsprung–Russell diagrams tend to confirm this (Sandage, 1957; Mermilliod, 1981; Lang, 1992).

Nevertheless, a few open clusters are relatively rich, massive, and born in low gas density areas, remaining gravitationally bound and surviving disruptive forces for long periods of time. These older systems have ages that span the lifetime of the galactic disk and can probe the earlier stages of its formation. Their ages are best determined by choosing parameters that characterize their color-magnitude (Hertzsprung–Russell) diagrams, thereby bypassing both observational and theoretical calibration problems. (Since the cluster Hertzsprung–Russell diagrams commonly use color instead of spectral class, they are often called color-magnitude diagrams.) An example of such a parameter is the ratio of the difference between the luminosity of the main-sequence turnoff and that of the horizontal branch to the color difference between the turnoff and the giant branch (Anthony-Twarog and Twarog, 1985; Twarog and Anthony-Twarog, 1989).

Old open clusters lead to estimates of the minimum age of the galactic disk, t_{gd}. Janes and Phelps (1994) have identified 72 open star clusters older than the Hyades (about 700 million years), and 19 older than M 87 (about 5 Gyr). Four of the oldest open clusters seen within 1 kiloparsec (3.0857×10^{21} cm) from the Sun are M 67, Mel 66, NGC 188 and NGC 6791 with respective ages of 4.7, 6.8, 7.8 and 7.8 Gyr, where 1 Gyr $= 10^9$ years = one billion years (Janes and Phelps, 1994). Thus, the minimum age of the galactic disk, t_{gd}, inferred from the oldest known open star clusters, t_{oc}, is:

$$t_{gd} \geq t_{oc} = 8 \text{ Gyr} . \tag{5.300}$$

The oldest presently observable stellar systems are the globular star clusters. They are roughly spherical, densely packed clusters of up to millions of stars. The high stellar density suggests that globular clusters have the gravitational cohesion to survive for very long periods of time. Moreover, the stars in globular clusters have very low levels of heavier elements, made inside previous generations of stars, suggesting that the globular clusters formed early before the interstellar medium was enriched by the supernova explosions of dying massive stars.

The globular clusters are also distributed within a large, roughly-spherical halo around the galactic center, in contrast with the open clusters that are found only in the galactic disk (see Figs. 5.3, 5.4). This is what would be expected if the globular clusters were older than the open clusters, and if galaxy formation proceeded by collapse. The Hertzsprung–Russell diagrams of globular clusters do, in fact, illustrate the later stages of stellar evolution with the giant branches and one horizontal branch (see Figs. 5.22, 5.25); in contrast the H-R diagrams of the open clusters show only the main-sequence and the beginning of an evolutionary turnoff from it (Fig. 5.24).

The observed H-R diagrams of globular clusters include stars which connect the main sequence to the red giant branch, providing an observational basis for understanding a star's evolution (Sandage and Schwarzschild, 1952; Sandage, 1957). After spending most of its life on the main sequence at essentially constant surface temperature and luminosity, a star consumes the available hydrogen near its center. The core then contracts, liberating gravitational energy, and the envelope expands, creating a giant star with a lower surface temperature but a central temperature high enough to eventually ignite helium in the core. The point of initial helium ignition is at the top end of the giant branch and is known as the "helium flash". On the initial evolution up the giant branch to the helium flash, the nuclear burning occurs in the hydrogen shell surrounding the collapsing helium core. This transition from the main-sequence to giant phase of evolution is marked by a distinctive bend in the color-magnitude, or H-R, diagram of globular clusters.

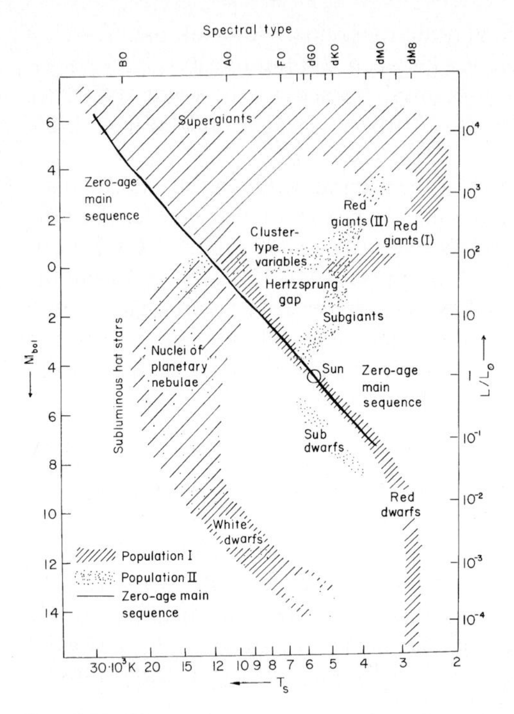

Fig. 5.25. Composite Hertzsprung–Russell, or H-R, diagram. Stars of different absolute luminosity, L–right axis in solar units $L_{\odot} = 3.85 \times 10^{33}$ erg s^{-1}, or absolute bolometric magnitude, M_{bol}–left axis, are plotted as a function of surface temperature, T_s–bottom axis, or spectral type–top axis. Stars on the left of the diagram are hot, those on the right are cool. Bright stars are found at the top of the diagram and faint stars on the bottom. Running diagonally from top left to bottom right is the main sequence along which roughly 90% of the stars fall. Stars on the subgiant and red giant branch are sometimes called Population II stars; they are delineated in the H-R diagrams of the relatively old stars found in globular clusters (see Fig. 5.22)

The globular clusters contain the oldest known stars in our Galaxy, with ages that provide limits to the minimum age of the Universe. The ages, t_{gc}, of globular clusters are inferred from the observed color-magnitude, or H-R, diagrams by computing a set of theoretical curves called isochrones. Each isochrone shows the evolutionary tracks of stars of different mass and the same age on the H-R diagram. In this comparison, it is assumed that all stars currently observed in the globular cluster originated at the same time from matter of the same composition, so one uncertainty involves the assumed initial composition of the material from which the stars formed. Since theoretical models determine absolute luminosity and temperature, they must be transformed into absolute magnitude and color for comparison with the observed data. Such a transformation involves additional uncertainties in determining the distances, used to obtain the intrinsic brightness from the apparent one, and reddening by the interstellar medium.

The classical method of determining the age of a globular cluster is based upon the luminosity of the main-sequence turnoff point, or by determining how long it takes for stars to burn their core hydrogen and thereby move off the main sequence. Theoretical calculations of the main-sequence lifetimes of stars of the appropriate composition, as a function of luminosity, then permits a direct measure of the cluster age by comparison with the observed color-magnitude, or H-R, diagram.

Using the largest telescope then available, the 200-inch Hale telescope, with fast photographic emulsions and photoelectric techniques, Sandage (1953) obtained the colors and magnitudes of thousands of stars, including the very faint ones, for the globular cluster M3, inferring a time of 5 Gyr since the cluster stars were on the main sequence. Sandage subsequently obtained an average absolute age, t_{gc}, for four globular clusters of:

$$t_{gc} = 11.5 \pm 0.3 \text{ Gyr for M3, M13, M15 and M}\mathit{92} \text{ (Sandage, 1970)} , \tag{5.301}$$

where the uncertainties were due to observational errors, the uncertainties of the assumed helium and metal abundance, and to differences in theoretical models – Sandage used the theoretical evolutionary tracks of Iben and Rood (1970). In the meantime, an older age had been obtained for another globular cluster:

$$t_{gc} = 20 \pm 4 \text{ Gyr for M5 (Arp, 1962)} , \tag{5.302}$$

using the stellar evolutionary models of Haselgrove and Hoyle (1959) to obtain the absolute luminosity of the bluest stars.

More recently, globular cluster dating involves either the direct main-sequence fitting of the observed color-magnitude diagrams to theoretical isochrones, or the determination of the magnitude difference between horizontal-branch stars and the main-sequence turnoff at the color of the turnoff. Progress in determining theoretical isochrones and their fits to observational data include work by Ciardullo and Demarque (1977), Vandenberg (1983), Iben and Renzini (1984), Green, Demarque and King (1987), Proffitt and Vandenberg (1991), Jiminez et al. (1996) and Jiminez and Mac Donald (1996). Modern electronic detectors, such as the Charge Coupled Device, or CCD, have resulted in an order-of-magnitude improvement in the precision of the observational data, enabling observers to obtain higher precision color-magnitude diagrams extending to much fainter levels than were previously possible (Fig. 5.26). Such techniques have, for example, led to an accurately-determined age for the globular cluster 47 Tucanae, or NGC 104, of (Hesser et al., 1987):

$$t_{gc} = 13.5 \pm 0.5 \text{ Gyr for 47 Tucanae} . \tag{5.303}$$

Yet, the uncertainties of specific ages for most globular clusters remain large (Chaboyer, 1995). Our present inability to accurately determine their distances can lead to changes of 25% in the determination of absolute ages, and uncertainties in chemical composition result in comparable errors. Even if the distances and compositions of globular clusters were known exactly, the total uncertainties in the age estimates of globular clusters would be ±15% due to errors arising from uncertainties in the physics of stellar evolution models.

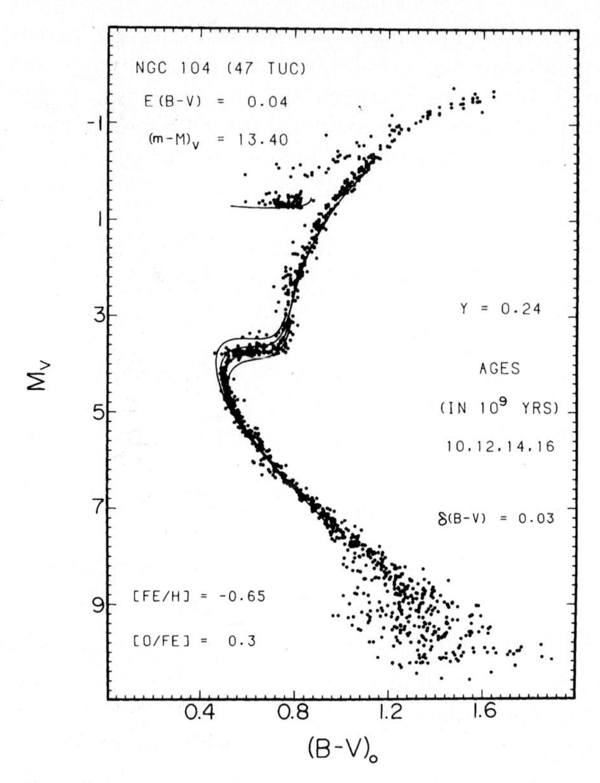

Fig. 5.26. The Hertzsprung–Russell, or H-R, diagram for the globular star cluster 47 Tucanae, or NGC 104. Here theoretical evolutionary tracks, called isochrones, of ages 10, 12, 14 and 16 Gyr have been fit to the observed plot of absolute visual magnitude, M_V, versus the intrinsic, or unreddened, color $(B\text{-}V)_0$, covering more than eleven magnitudes. An age of $t_{gc} = 13.5$ Gyr, with an uncertainty of 0.5 Gyr, is inferred from this diagram, where 1 Gyr $= 10^9$ years = one billion years. (Adapted from Hesser et al., 1987, courtesy of James E. Hesser and the Publications of the Astronomical Society of the Pacific)

Nevertheless, relative ages of globular clusters can be obtained with greater accuracy (Sandage, 1970; Peterson, 1987).

So, it is perhaps not surprising that individual authors obtain age estimates for the same globular clusters that can differ as much as 3 Gyr. The globular cluster M92 is a good example. It is at a high galactic latitude, where interstellar reddening is less, and it has a low metallicity, where the stars may be the oldest. Moreover, CCD photometry has been used to obtain detailed color-magnitude diagrams for M92. Comparisons of this data with theoretical isochrones lead to:

$$\begin{aligned} t_{gc} &= 18 \pm 2 \text{ Gyr for M92 (Sandage, 1983)} \ , \\ t_{gc} &= 16 \text{ to } 17 \text{ Gyr for M92 (Stetson and Harris, 1988)} \ , \\ t_{gc} &= 14 \pm 2 \text{ Gyr for M92 (Proffitt and Vandenberg, 1991)} \ , \\ t_{gc} &= 17 \pm 2 \text{ Gyr for M92 (Chaboyer et al., 1992)} \ , \end{aligned} \tag{5.304}$$

$$t_{gc} = 15.8 \pm 2.1 \text{ Gyr for M92 (Bolte and Hogan, 1995)} .$$

Other estimates indicate:

$$\begin{aligned} t_{gc} &= 13.5 \pm 2 \text{ Gyr for the oldest globular clusters (Jiminez et al., 1996)} \\ t_{gc} &\approx 11.8 \pm 2.5 \text{ Gyr for the oldest globular clusters} \\ &\quad \text{(Feast and Catchpole, 1997; Gratton et al., 1997)} . \end{aligned} \tag{5.305}$$

The above error estimates do not include complete estimates of systematics between techniques and assumptions, as is dramatically illustrated by the fact that some age estimates are outside of the stated error bars of others.

Chaboyer et al. (1996) have attempted to examine the absolute ages and uncertainties in ages of the oldest globular clusters in our Galaxy using a Monte Carlo analysis of input parameters with their code, obtaining a lower limit, at the 95 percent confidence level, of

$$t_{gc} \geq 12.07 \text{ Gyr} ,$$

a median age for the distribution of

$$t_{gc}(\text{median}) = 14.56 \text{ Gyr} , \tag{5.306}$$

and a 95 percent range of ages of

$$11.6 \text{ Gyr} \leq t_{gc} \leq 18.1 \text{ Gyr} .$$

Shi (1995) used extreme systematic assumptions to show that a very robust lower boundary exits for the age of the oldest globular clusters of:

$$t_{gc} \geq 10 \text{ Gyr} , \tag{5.307}$$

and showed how the age inferred from observations depends on assumptions such as the initial helium abundance, using high abundance values to estimate an age for metal poor globular clusters of:

$$t_{gc} = 11 \pm 1 \text{ Gyr} . \tag{5.308}$$

To obtain the age of the galactic halo, one would have to add a formation time of 0.1 to 2.0 Gyr to these values of the globular cluster ages. The age of the galactic globular cluster system has been reviewed by Vandenberg, Stetson and Bolte (1996).

Some of the globular star clusters are several billion years younger than the majority, which is important for models of galaxy formation. That is, the difference in ages between the oldest and youngest globular clusters provides an estimate for the time in which the galactic halo formed, constraining models for the early stages of our Galaxy and apparently ruling out the rapid collapse of the galactic halo. The total range of the true, or absolute, ages of the globular clusters lie in the range:

$$\begin{aligned} t_{gc} &= 10 \text{ to } 19 \text{ Gyr (Carney, 1980)} \\ t_{gc} &= 11 \text{ to } 21 \text{ Gyr (Chaboyer, 1995)} . \end{aligned} \tag{5.309}$$

All of these age estimates are compatible with those inferred for M92, and they all indicate that globular clusters are several Gyr older than the oldest known open cluster ($t_{oc} = 8$ Gyr), which is as it should be since the globular clusters have progressed further along the evolutionary paths in the H-R diagram. The oldest globular clusters may also predate the local galactic disk ($t_{gd} = 7$ to 15 Gyr) which would be expected if the Galaxy formed by collapse from a spherical halo into a flattened disk.

The available data also indicate a significant age spread of several Gyr among the globular clusters (Sarajedini and King, 1989; Vandenberg, Bolte and Stetson, 1990; Chaboyer, Sarajedini and Demarque, 1992).

Hence, globular cluster formation appears to have taken place over an extended period of time. They therefore argue in favor of a more chaotic, long-lasting collapse advocated by Searle and Zinn (1978) perhaps involving merging of smaller systems rather than the uniform, rapid collapse (≤ 1 Gyr) suggested by Eggen, Lynden-Bell and Sandage (1962). Nevertheless, the central bulge of our Galaxy has roughly the same age as the oldest globular clusters in the galactic halo, suggesting the near-coeval formation of the galactic bulge and halo (Ortolani et al., 1995) or even that the center of our Galaxy served as a nucleus around which the rest of the Galaxy was built (Lee, 1992).

5.3.10 Nucleochronology and the Age of our Galaxy

The stars which synthesized the heavy elements in the solar system were formed, evolved and eventually ejected the ashes of their nuclear fires into the interstellar medium before the solar system formed. So, our Galaxy must be much older than the solar system, and the difference in age can be inferred from the time it takes to create the heavy elements that were present when the nascent solar system contracted from the local, enriched interstellar gas.

The approximate age of our Galaxy, t_G, can be inferred from the presence of radioactive parent elements on the surface of the Earth and Moon and in meteorites. If Δ denotes the duration of nucleosynthesis, in which these elements were made, then

$$t_G = \Delta + t_{ss} \quad , \tag{5.310}$$

where $t_{ss} = 4.6 \pm 0.1$ Gyr is the age of the solar system. In this section, all ages are given in Gyr, where 1 Gyr $= 10^9$ years $=$ one billion years.

Nucleochronology utilizes our knowledge of the current meteoritic abundances and the production rates of radioactive nuclear species in stellar explosions to determine Δ in the context of models of the chemical evolution of the Galaxy (see Symbalisty and Schramm, 1981 for a review). Half-lives of long-lived radioactive elements and their nucleosynthesis sites are given by Cowan, Thielemann and Truran (1991). The half-lives of the more frequently used ones, including the long-lived parent elements thorium, uranium and rhenium, were also given in Table 5.7. Meteoritic abundance ratios, production ratios, galactic chemical evolution models, observational constraints, and age determinations from nucleochronology are also reviewed by Cowan, Thielemann and Truran (1991).

Rutherford (1929) realized that the uranium we now detect on the Earth must have been created before our planet formed, and that the production of uranium ceased thereafter. He estimated that it would take $\Delta \approx 3.4$ Gyr for the production of the presently observed abundance of uranium isotopes, although he incorrectly supposed that they were produced in the Sun.

We now realize that elements heavier than iron cannot be produced in successive static burning stages within stars. This is because any nuclear reaction involving the iron group of nuclei, with atomic weight $A \approx 56$, cannot provide fuel for the thermonuclear fires that support a star and make it shine. Instead, the iron-group elements act as the seeds for the synthesis of heavier elements by neutron capture. Such processes were first suggested by George Gamow for nonequilibrium nucleosynthesis during the early stages of the expansion of the Universe (Gamow, 1948; Alpher, Bethe and Gamow, 1948), and applied to the later stages of stellar evolution by Burbidge, Burbidge, Fowler and Hoyle (1957), often called B^2FH, and independently by Cameron (1957).

Double-peaked features in the abundance curve (Fig. 5.27) indicate that two neutron capture processes, called the *r*-process and the *s*-process, must synthesize elements with atomic weights A greater than 60. The rapid (*r*-process) neutron capture occurs on time scales of about 100 seconds, which is rapid (r) compared to electron beta decay in the synthesis networks, while the *s*-process is much slower (s), occurring over time scales of 10^2 to 10^5 years. All naturally occurring radioactive elements with $A \geq 209$, including the long-lived uranium, U, and thorium, Th, parents, ^{238}U, ^{235}U and ^{232}Th, require the *r*-process, which builds beyond mass 238 to nuclei that decay back to these radioactive parents. The *r*-process probably occurs during stellar explosions, called supernovae, that rapidly provide a large neutron flux with a short duration.

B^2FH estimated the duration, Δ, of *r*-process nucleosynthesis from its beginning in the first stars in our Galaxy to the last events before the formation of the solar system. They were able to calculate the age of uranium isotopes using estimates of the production ratio of $^{235}U/^{238}U$ in the *r*-process, the lifetimes of their radioactive decay, and their present-day terrestrial abundance ratio. B^2FH used different models to obtain:

$$t_G = 6.6 \text{ to } 18 \text{ Gyr} \ ,$$

or (5.311)

$$\Delta = 2 \text{ to } 11 \text{ Gyr} \ ,$$

for $t_{ss} = 4.55$ Gyr. If all the uranium was produced by a single *r*-process, then $t_G = 6.6$ Gyr; uniform nucleosynthesis from a time t_G ago to a time t_1 before the formation of the solar system resulted in $t_G = 18$ Gyr for $t_1 = 0$ and $t_G = 11.5$ Gyr for $t_1 = 0.5$ Gyr. Taking the relative abundances of radioactive elements in meteorites into account, the best value for B^2FH was

$$t_G \approx 10.0 \text{ Gyr} \quad \text{or} \quad \Delta \approx 5.4 \text{ Gyr} \ . \tag{5.312}$$

Fowler and Hoyle (1960) extended the work of B^2FH, providing a framework for nucleochronology, or nuclear cosmochronology, in which the observed and derived relative abundances of radioactive nuclei are used to determine the time scale, Δ, of nucleosynthesis before the origin of the solar system, t_{ss} years ago. The basic inputs of these calculations are the present abundances of the radioactive elements, their half lives, their production rates by the *r*-process,

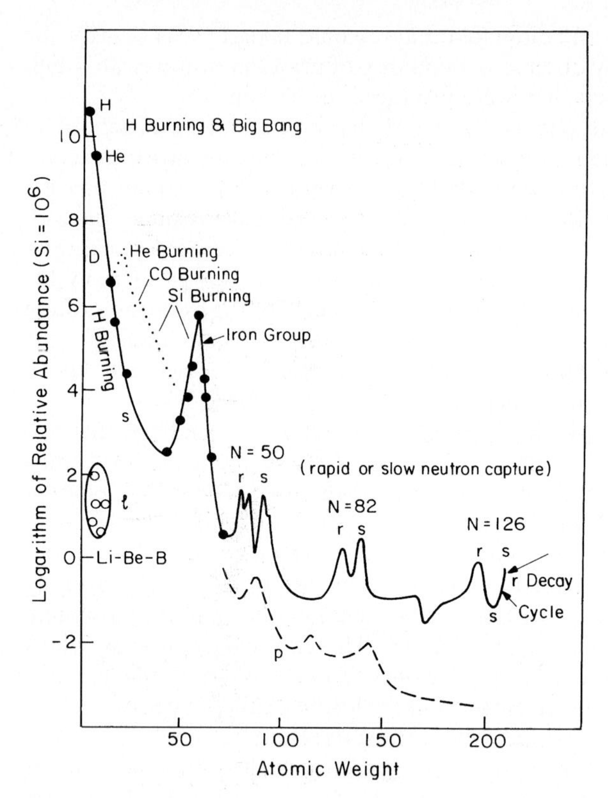

Fig. 5.27. Schematic abundances for elements in the solar system and in main sequence stars similar in mass and age to the Sun. Here atomic abundances relative to silicon Si $= 10^6$, are plotted as a function of atomic weight, $A = N + Z$, for N neutrons and Z protons, using the data of Suess and Urey (1956), with specification of the relevant nucleosynthesis processes by Burbidge, Burbidge, Fowler and Hoyle (1957). Note the overabundances relative to their neighbors of the alpha-particle nuclei with $A = 16, 20, \ldots, 40$, the peak at the iron group nuclei near $A = 56$, and the twin peaks at magic neutron numbers $N = 50, 82$ and 126. The double-peaked features are due to two neutron capture processes, a rapid one (r-process) for atomic weights $A = 80, 130$ and 194, and a slow (s) process for $A = 90, 138$ and 208

and a model for the chemical evolution of our Galaxy. Fowler and Hoyle assumed that star formation and evolution has decreased exponentially since the origin of the Galaxy to the present time, obtaining

$$t_G = \Delta + t_{ss} + t_e \approx 15 \text{ Gyr}$$

with

$$\Delta \approx 7 \text{ Gyr} , \tag{5.313}$$

where thorium, Th, and uranium, U, synthesis started at an evolution time $t_e = 2.8$ to 3.8 Gyr that preceded the first supernova and $t_{ss} = 4.7 \pm 0.1$ Gyr assuming that it took the Sun and planets 0.1 to 0.2 Gyr to contract from the local interstellar medium. They obtained r-process production ratios of

1.65 ± 0.15 for $^{235}U/^{238}U$ and $^{232}Th/^{238}U$, and respective abundance ratios of $0.34 + 0.03$ and $2.33 + 0.2$ at t_{ss} years ago, assuming subsequent natural radioactive decay to the current abundance ratios of $^{235}U/^{238}U = 0.00723$ and $^{232}Th/^{238}U = 3.8 \pm 0.3$.

Fowler (1987) subsequently used *r*-process production ratios of 1.34 ± 0.19 for $^{235}U/^{238}U$ and 1.71 ± 0.01 for $^{232}Th/^{238}U$ to assert that:

$$t_G = 10.0 \pm 1.6 \text{ Gyr}$$

and

$$\Delta = 5.4 \pm 1.5 \text{ Gyr} \ . \tag{5.314}$$

This conclusion was severely criticized by Clayton (1988) who reasoned that "one can get almost any answer [for the age of the Galaxy] one wants from nuclear cosmochronology by choosing parameters well within their range of uncertainties; especially with U and Th, and by following an unproductive practice of arbitrarily specifying the nucleosynthesis history." His examination of five cosmoradiogenic chronologies, including the ^{187}Re to ^{187}Os method he introduced (Clayton, 1964) resulted in (Clayton, 1988):

$$12 \text{ Gyr} < t_G < 20 \text{ Gyr} \tag{5.315}$$

as the most unbiased interpretation of the nuclear data alone.

The difficulty is that nucleochronology must be evaluated within the context of the chemical evolution of our Galaxy if it is to tightly constrain its age. This was realized by Tinsley (1975, 1977), who introduced sophisticated chemical evolution models, using the notation of $\langle \tau \rangle$,

$$\langle \tau \rangle = \frac{\int_0^T t\psi(t)dt}{\int_0^T \psi(t)dt} \tag{5.316}$$

for the mean time of nucleosynthesis with a total duration, T, where $\psi(t)$ is the mass of stars formed per unit time. Schramm and Wasserburg (1970) showed that the only model independent quantity is a lower limit to the total duration of nucleosynthesis:

$$T = \Delta^{\max} - \Delta_d + \langle \tau \rangle \geq \Delta^{\max} - \Delta_d \ , \tag{5.317}$$

where Δ_d is the time interval between the termination of nucleosynthesis and the solidification of solid bodies in the solar system, the period of free decay of the elements, and the model-independent parameter, $\Delta^{\max}$, introduced by Schramm and Wasserburg (1970) is given by the age parameter

$$\Delta_{ij}^{\max} = \frac{\ln R(i,j)}{\lambda_i - \lambda_j} \ , \tag{5.318}$$

where λ_i and λ_j are the decay constants of two long-lived isotopes,

$$R(i,j) = \frac{P_i/P_j}{N_i(T + \Delta)N_j(T + \Delta)} \ ;$$

P_i/P_j is the production ratio in stars, and N_i, N_j are the number of atoms of each element present in the proto-solar system material.

For long-lived radioactive elements, where $\lambda_i T$ and $\lambda_j T \ll 1$, the Δ_{ij}^{max} tend to the same value for all pairs providing a mean age Δ^{max} that is independent of the rate of nucleosynthesis and the model of chemical evolution. Schramm (1990) obtains:

$$\Delta^{max} = 1.8 \text{ to } 13.4 \text{ Gyr for } {}^{187}\text{Re}/{}^{187}\text{Os}$$

$$\Delta^{max} = 4.1 \pm 1.4 \text{ Gyr for } {}^{232}\text{Th}/{}^{238}\text{U} \tag{5.319}$$

and

$$\Delta^{max} = 1.9 \pm 0.3 \text{ Gyr for } {}^{235}\text{U}/{}^{238}\text{U} \ .$$

The age of the galaxy, t_G, is given by (Schramm and Wasserburg, 1970):

$$t_G = T + \Delta_d + t_{ss} = \Delta^{max} + \langle\tau\rangle + t_{ss} \ ,$$

so it depends on chemical evolution through the parameter $\langle\tau\rangle$, and on the age of the solar system, t_{ss}. Nevertheless, Meyer and Schramm (1986) were able to use nuclear physics arguments to provide constraints on the ratio $\langle\tau\rangle/T$, or in their notation t_{υ}/T to obtain

$$0.43 \le \langle\tau\rangle/T \le 0.59 \ ,$$

and

$$9 \text{ Gyr} \le t_G \le 28 \text{ Gyr} \tag{5.320}$$

A summary of age estimates for the duration, Δ, of nucleosynthesis prior to the formation of our Galaxy, and the age of the Galaxy, t_G, is given in Table 5.10.

Schramm (1990) reaffirms this definite lower limit of $t_G \ge 9$ Gyr, and states that assuming standard galactic evolution assumptions yields the best galactic evolution age of:

$$t_G = 12 \text{ to } 18 \text{ Gyr} \tag{5.321}$$

with a firm, model independent lower bound at 9.6 Gyr.

Table 5.10. Estimates for the duration, Δ, in Gyr of nucleosynthesis in our Galaxy prior to the formation of the solar system $t_{ss} = 4.6 \pm 0.1$ Gyr ago. The associated nucleochronologic values for the age of our Galaxy, t_G, is also given in Gyr assuming $t_G = t_{ss} + \Delta$. Here 1 Gyr = 10^9 years = one billion years

Authors	Duration, Δ, of Nucleosynthesis (Gyr)	Age of Our Galaxy, t_G. (Gyr)
B^2FH (1957)	5.4	10.0
Fowler and Hoyle (1960)	7.0	11.6
Schramm and Wasserburg (1970)	>3.2	>7.8
Meyer and Schramm (1986)	4.4 to 23.4	9 to 28
Cowan *et al.* (1987)	7.8 to 10.1	12.4 to 14.7
Fowler (1987)	5.4 ± 1.5	10.0 ± 1.6
Clayton (1988)	7.4 to 15.4	12 to 20
Schramm (1990)	>5	>9.6
Cowan *et al.* (1991)	5.4 to 15.4	10 to 20

5.3.11 Approximate Expansion Age of the Universe

As early as 1917, Vesto Slipher had shown that spiral nebulae are receding from our Galaxy with tremendous velocities, sometimes exceeding 1,000 km s^{-1} (Slipher, 1917). More than a decade later, Edwin Hubble convincingly demonstrated that the velocities, V, increase with the distance, D, of the spiral nebulae, which are now called galaxies (Hubble, 1929). In the meantime, Alexander Friedmann and Georges Lemaître had shown that such an expanding Universe is a mathematical consequence of Einstein's General Theory of Relativity (Friedmann, 1922, 1924; Lemaître, 1927).

The systematic recession of the galaxies is explained by the Big Bang theory in which the Universe we now observe expanded from a dense, hot beginning. A wide range of observational evidence, including the cosmic abundances of the light elements and the precise black–body character of the microwave background radiation, provide strong evidence for the Big-Bang model (Peebles, Schramm, Turner and Kron, 1991). In this theory, the expanding Universe began with a big bang in a single explosive moment a finite time, t_o, ago, and the recessional velocities of the galaxies are the cosmical consequence of this expansion.

The current rate of expansion of the Universe is quantified by the Hubble constant, H_o, the ratio of the speed with which distant galaxies appear to be receding from us to their distance. That is, the velocity, V, for a galaxy at distance, D, is given by Hubble's law, $V = H_o \times D$, although Hubble did not specifically state this relation. So, a measurement of the Hubble constant, H_o, sets the physical scale for the Universe, and provides a rough estimate for the expansion age, t_o. This may be quantified by the

$$\text{Hubble time} = t_H = H_o^{-1} = 9.778h^{-1}\ \text{Gyr}\ , \qquad (5.322)$$

where $H_o = 100h$ km s^{-1} Mpc^{-1} and 1 Gyr $= 10^9$ years $=$ one billion years.

That is, the expansion of the Universe began roughly t_H years ago when the Universe was in an intensely hot and compressed state. A faster rate of expansion, or a larger value of H_o, means that less time has elapsed since the expansion began with the big bang, and a slower expansion with a smaller value of Hubble's constant suggests a greater age.

The subscript o in t_o and H_o, as well as other cosmological parameters, indicates their value today. However, the age of the expanding Universe obviously increases as time goes on, and the Hubble constant, H(t), probably decreases with increasing time, t. The expansion rate is, for example, slowed with time by the mutual gravitation of all the matter in the Universe. Since gravity acts to decelerate the Universe, the expansion was faster in the past than it is now, and the current age of the Universe, t_o, must be less than the Hubble time, t_H (at least for zero cosmological constant – which will be discussed later).

Moreover, the use of the Hubble constant to infer the age of the expanding Universe has been fraught with uncertainty since its first use in that way, and it remains one of the more uncertain methods of establishing the age of the Universe. For instance, when we use Hubble's value of $H_o = 558$ km s^{-1}

Mpc^{-1} (Hubble and Humason, 1931), then the approximate age of the Universe is $t_{\mathrm{H}} = 1/H_{\mathrm{o}} = 1.8$ Gyr. However, radioactive dating had already shown that the oldest terrestrial rocks are at least 3.4 Gyr old (Rutherford, 1929). Thus, the Earth seemed to be older than the Universe (Lang, 1985).

The paradoxical discrepancy between the age of the Earth and the spiral galaxies provided support for the steady state cosmology in which the Universe expands but does not evolve, with new matter being created out of empty space to keep the mass density constant (Bondi and Gold, 1948; Hoyle, 1948). Then, in 1952, Walter Baade showed that Edwin Hubble's distance measurements for nearby spiral nebulae were underestimated by at least a factor of two (Baade, 1952). Baade's new value of $H_{\mathrm{o}} \approx 100$ km s^{-1} Mpc^{-1} (Baade and Swope, 1955) results in $t_{\mathrm{H}} = 1/H_{\mathrm{o}} \approx 10$ Gyr, bringing the expansion age of the Universe into rough accord with the estimated age of the Earth. Strong, observational evidence for a dense, primeval origin of an evolving expanding Universe, including the present deuterium and helium abundance and the remnant, three-degree microwave background radiation, eventually led Hoyle to recant the steady state theory (Hoyle, 1968).

The use of the Hubble constant, H_{o}, to infer an expansion age of the Universe remains an inaccurate one, in part because of the ongoing controversy over the exact value of H_{o}. There are now values lying around $H_{\mathrm{o}} = 50$ km s^{-1} Mpc^{-1} and other estimates almost twice as large, in the region of $H_{\mathrm{o}} = 80$ km s^{-1} Mpc^{-1}(see Section 5.2.8). We probably do not know either H_{o} or the Hubble time, t_{H}, better than $\pm 15\%$. Indeed, a recent upward revision in the distance to the Large Magellanic Cloud by ten percent indicates that the Universe is slightly larger and about a billion years older than was previously thought on the basis of Cepheid calibration of this distance (Feast and Catchpole, 1997). The range in opinion is illustrated by:

$$t_{\mathrm{H}} = 11.2 \pm 1.0 \text{ Gyr for } H_{\mathrm{o}} = 87 \pm 7 \text{ km s}^{-1} \text{ Mpc}^{-1} \quad \text{(Pierce et al., 1994)} \ , \tag{5.323}$$

and

$$t_{\mathrm{H}} = 17.8 \pm 2.6 \text{ Gyr for } H_{\mathrm{o}} = 55 \pm 7 \text{ km s}^{-1} \text{ Mpc}^{-1} \quad \text{(Sandage and Tammann, 1995)} \ . \tag{5.324}$$

So, we can now only say with certainty that the Hubble time lies somewhere between 10 and 20 Gyr, with one camp arguing for a young, rapidly evolving Universe and the other reasoning that the Universe is about twice as old. Nevertheless, some cosmologists now think that the number is likely to settle somewhere between 12 billion and 14 billion years. Either range in ages is in approximate agreement with the oldest known stellar ages, confirming the basic premise of a big bang at the beginning of an evolving Universe.

Any estimate for the actual age of the Universe from the Hubble time involves additional assumptions about the mean density of the Universe and the possibility of a cosmological constant, Λ, that opposes gravity. In other words, a precise value of the expansion age, t_{o}, depends on the theoretical model of choice, as well as on the uncertainties in the Hubble constant. For instance, in a matter-dominated Universe with zero cosmological constant, $\Lambda = 0$, the current expansion age, t_o, is given by:

$$0 < t_o < 2t_H/3 \text{ for a bound Universe },$$
$$2t_H/3 < t_o < t_H \text{ for an unbound Universe }. \quad (5.325)$$

In a bound Universe, there is enough matter to stop the expansion in the future, whereas an unbound Universe continues to expand and thin out forever.

A particularly simple cosmological assumption is the Einstein – De Sitter model in which the cosmological constant is set to zero, and the mass density is just enough to keep the Universe expanding at precisely the velocity needed to sustain that expansion indefinitely (Einstein and DeSitter, 1932). In such a model, space curvature has a negligible influence on the expansion, and the "flat", matter-dominated Universe has an age:

$$t_o = 2t_H/3 = 6.6h^{-1} \text{ Gyr for the Einstein–De Sitter model with } \Lambda = 0$$
$$= 8 \text{ Gyr for } H_o = 80 \text{ km s}^{-1} \text{ Mpc}^{-1}$$
$$= 13 \text{ Gyr for } H_o = 50 \text{ km s}^{-1} \text{ Mpc}^{-1}. \quad (5.326)$$

The Einstein – De Sitter model is also in accord with the inflationary cosmology (see Guth, 1981 and Sect. 5.7.2) which predicts such an equilibrium between the restraining pull of gravity and the momentum of expansion, provided that the cosmological constant is zero.

Thus, if one believes in the Einstein – De Sitter model, and its inflationary successor, then the Universe is 8 to 13 Gyr old. In contrast, the oldest globular star clusters in our Galaxy, have a median age of 14.56 Gyr and a lower limit of 12.07 Gyr (Chaboyer et al., 1996; also see Sect. 5.3.9), not including an additional formation time of 0.1 to 2.0 Gyr. Thus, the value of $H_o = 58 \pm 4$ km s^{-1} Mpc^{-1} determined by Sandage et al. (1996) is just barely consistent with the inflationary big-bang cosmology and the lower limits to the ages of the oldest globular clusters. Any larger value of H_o and/or greater globular cluster ages would be inconsistent with inflation, or with any other scenario with critical closure density and zero cosmological constant. Some astronomers therefore argue that there is an age crisis in which the Universe appears younger than some of its components.

However, the globular cluster ages could be as low as 10 Gyr (Shi, 1995), so there may not be any conflict. Of course, it is also possible that the Universe is not at the critical density or even that the cosmological constant Λ is not equal to zero; in this latter case any plausible value for the Hubble constant H_o can be accommodated.

5.3.12 Expansion Age from Cosmological Models

The value of the expansion rate, or the Hubble constant H_o, must be combined with certain assumptions about the mass and other properties of the Universe to estimate the time, t_o, since the big bang. The low expansion ages inferred from the Einstein – De Sitter model may be incorrect because these assumptions, made by cosmologists, are wrong. An older Universe, obtained under different assumptions, would bring the expansion age into accord with the age of the oldest known astronomical objects, and also help to explain how all the large structures, such as clusters of galaxies, could have developed from the presumably smooth and homogeneous beginnings after the big bang.

There may be less matter than assumed in the simplest "flat" Universe scenario, and the mass density may not be high enough to keep the Universe balanced between collapse and expansion into nothing. That is, the expansion rate is slowed by the mutual gravitational attraction of all the matter of the Universe, so a lower mass density would result in older age estimates.

Another possibility involves Einstein's cosmological constant Λ, included in some solutions of Einstein's equations used to describe the evolution of the Universe. The Universe is possibly accelerated by the energy density of a physical vacuum, quantified by a non-zero cosmological constant $\Lambda \neq 0$. It provides a force that opposes gravity, even in empty space, and makes a Universe with a given amount of matter older than it would otherwise be.

In today's terminology, we specify the current age of the Universe in terms of the density parameter, Ω_o, the ratio of the present-day actual mass density to the critical amount required to forever slow the Universe's expansion. That is, the

$$\text{Density parameter} = \Omega_o = 2q_o = \rho_o/\rho_c \ , \tag{5.327}$$

where the deceleration parameter is q_o, the present mass density of the Universe is ρ_o and $\rho_c = 3H_o^2/(8\pi G) = 1.879 \times 10^{-29}h^2$ g cm^{-3} is the critical mass density of the Universe, where $H_o = 100h$ km s^{-1}Mpc^{-1} and G is the Newtonian gravitational constant.

The present age of the Universe, t_o expressed as a function of the Hubble time, $t_H = 1/H_o$, is, for zero cosmological constant $\Lambda = 0$:

$$t_o = \tfrac{2}{3}t_H \quad \text{for } \Omega_o = 1 \text{ or } k = 0, \quad \text{and } \Lambda = 0 \ ,$$
$$0 < t_o < \tfrac{2}{3}t_H \quad \text{for } \Omega_o > 1 \text{ or } k = +1, \quad \text{and } \Lambda = 0 \ , \tag{5.328}$$

and

$$\tfrac{2}{3}t_H < t_o < t_H \quad \text{for } 0 < \Omega_o < 1 \text{ or } k = -1, \quad \text{and } \Lambda = 0 \ ,$$

where we have introduced the space curvature constant k as an equivalent way of specifying the density parameter Ω_o.

A Universe on the edge between perpetual expansion and eventual contraction is described by the Einstein – De Sitter and inflation models in which $\Omega_o = 1$, the space curvature constant $k = 0$, and space is flat and Euclidean in an ever-expanding Universe for which $\rho_o = \rho_c$ (Einstein and De Sitter, 1932). When $k = +1$, then $\rho_o > \rho_c$ for a closed, spherical Universe with positive curvature that will stop expanding in the future and eventually recollapse. For $k = -1$, the $\rho_o < \rho_c$ in an open, hyperbolic ever-expanding Universe with negative curvature.

Observational constraints indicate that:

$$0.50 < h < 0.85 \ ,$$
$$0.1 < \Omega_o < 2 \ , \tag{5.329}$$

and

$$t_o = 15 \pm 5\,\mathrm{Gyr} \; .$$

An inventory of the amount of visible and invisible matter indicates that the current mass density of the Universe may be only a few per cent of the critical density, so observations suggest that $\Omega_o \leq 0.3$, but there remains the possibility of dark, non-baryonic matter to bring Ω_o up to 1.0 or more (see Sect. 5.4.6). The present observational data would seem to favor general relativistic cosmological models with either low density ($\Omega_o < 1$ for $\Lambda = 0$) or a dominant cosmological constant term with $\Lambda \neq 0$.

Detailed calculations give (Sandage, 1961; Krisciunas, 1993):

$$\frac{t_o}{t_H} = \frac{1}{1-\Omega_o} + \frac{\Omega_o}{(\Omega_o - 1)^{\frac{3}{2}}} \sin^{-1}\left[\left(\frac{\Omega_o - 1}{\Omega_o}\right)^{\frac{1}{2}}\right] \quad \text{for } \Omega_o > 1 \quad \text{and } \Lambda = 0 \; ,$$

and (5.330)

$$\frac{t_o}{t_H} = \frac{1}{1-\Omega_o} - \frac{\Omega_o}{2(1-\Omega_o)^{\frac{3}{2}}} \ln\left[\frac{2-\Omega_o}{\Omega_o} + \frac{2(1-\Omega_o)^{\frac{1}{2}}}{\Omega_o}\right] \quad \text{for } 0 < \Omega_o < 1 \; ,$$

and $\Lambda = 0$. For a completely empty Universe, $\Omega_o = 0$, and zero cosmological constant $\Lambda = 0$, we have $t_o = t_H$ so the age of the Universe is always less than the Hubble time when the cosmological constant $\Lambda = 0$, unless there is no matter in the Universe. Graphs of the age of the Universe, t_o, for different values of $H_o = 100h$ km s^{-1} Mpc^{-1} and Ω_o are given in Figs. 5.28 and 5.29.

The cosmological constant Λ was introduced by Einstein into the equations of motion of general relativity in order to keep a static, or non-expanding, Universe from contracting to a point under its mutual gravitation (Einstein, 1917, 1919). When the expansion of the Universe was subsequently discovered, Einstein withdrew his cosmological constant and proposed the

Table 5.11. Age of the Universe, t_o, in units of the Hubble time $t_H = 9.778h^{-1}$ Gyr, with the Hubble constant $H_o = 100h$ km s^{-1} Mpc^{-1}, for zero cosmological constant $\Lambda = 0$, and different values of the density parameter $\Omega_o = \rho_o/\rho_c$, the ratio of the present mass density, ρ_o, to the critical mass density, ρ_c, needed to close the Universe (adapted from Krisciunas, 1993)

$\Omega_o = \rho_o/\rho_c$	t_o/t_H
0.00	1.000
0.03	0.954
0.06	0.927
0.10	0.898
0.15	0.870
0.30	0.809
0.50	0.754
1.00	0.667
1.5	0.611
2.00	0.571
3.00	0.513
10.00	0.352

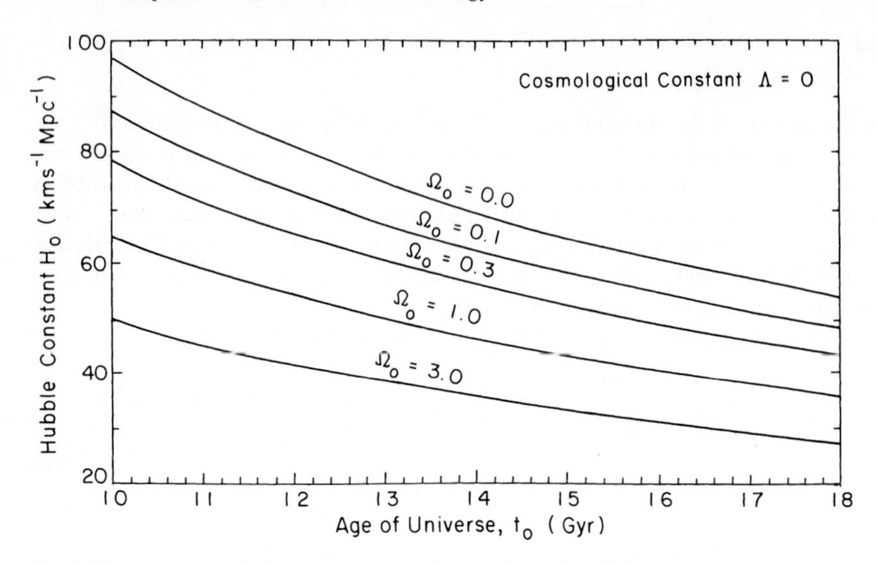

Fig. 5.28. Contours of constant density parameter Ω_o for different values of the Hubble constant, H_o, in km s^{-1} Mpc^{-1}, and ages of the Universe, t_o, in Gyr, where 1 Gyr = one billion years = 10^9 years. These curves are for model universes in which the cosmological constant $\Lambda = 0$. (Adapted from Krisciunas, 1993)

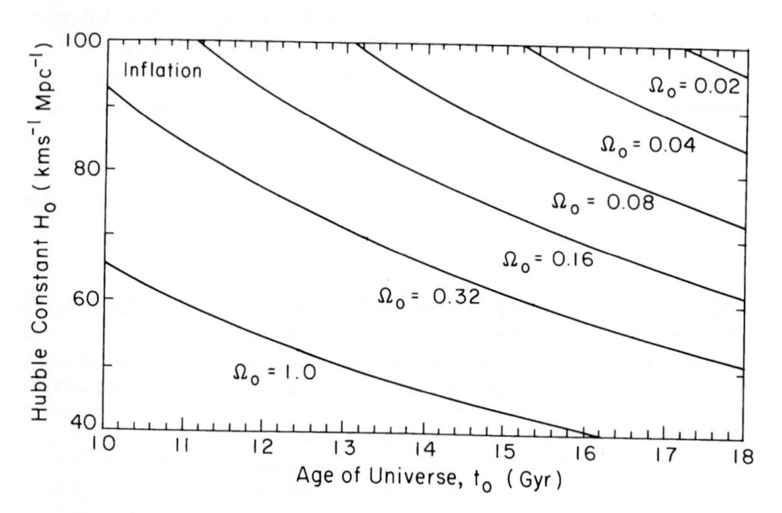

Fig. 5.29. Contours of constant density parameter Ω_o for different values of the Hubble constant, H_o, in km s^{-1} Mpc^{-1}, and ages of the Universe, t_o, in Gyr, where 1 Gyr = one billion years = 10^9 years. These curves are for model universes that assume the correctness of the inflationary scenario for which $\Omega_o + \lambda = 1$

Einstein – De Sitter model (Einstein and De Sitter, 1932). However, the term has recently been revived, in part, because of concerns over the apparent age of the Universe. The cosmological constant Λ allows regions devoid of matter to exert anti-gravity or repulsive forces, effectively accelerating the expansion and making the Universe older than it seems based on a given value of the Hubble constant H_o and the current mass density ρ_o.

For nonzero cosmological constant, $\Lambda \neq 0$, it is convenient to use the dimensionless, or reduced, cosmological constant λ, where

$$\lambda = \frac{\Lambda c^2}{3H_o^2} , \tag{5.331}$$

and assume the special case

$$\Omega_o + \lambda = 1 , \tag{5.332}$$

for an inflationary Universe with $0 < \Omega_o < 1$. Then

$$\frac{t_o}{t_H} = \frac{2}{3}\frac{1}{(1-\Omega_o)^{\frac{1}{2}}}\ln\left[\frac{1+(1-\Omega_o)^{\frac{1}{2}}}{\Omega_o^{\frac{1}{2}}}\right] \quad \text{for } \Omega_o + \lambda = 1 \quad \text{and } 0 < \Omega_o < 1 . \tag{5.333}$$

For an object of redshift, z, we can specify the look-back time, t_L, at which the radiation was emitted. Looking at objects at larger and larger redshift is looking further and further into the past. For small redshifts and low velocities $v = cz \ll c$, where c is the velocity of light,

$$t_L = t_H z = 9.778 h^{-1} z \text{ Gyr} \quad \text{for } z < 0.1 . \tag{5.334}$$

For larger redshifts, and a completely bound Universe we have:

$$\frac{t_L}{t_H} = \frac{2}{3}\left[1 - \frac{1}{(1+z)^{\frac{3}{2}}}\right] \quad \text{for } \Omega_o = 1 \quad \text{and } \Lambda = 0 , \tag{5.335}$$

and for an empty Universe:

$$\frac{t_L}{t_H} = \frac{z}{(1+z)} \quad \text{for } \Omega_o = 0 \quad \text{and } \Lambda = 0 . \tag{5.336}$$

For $\Omega_o > 1$ and $\Lambda = 0$ detailed calculations give Krisciunas (1993):

Table 5.12. Age of the Universe, t_o, in units of the Hubble time $t_H = 9.778 h^{-1}$ Gyr, with the Hubble constant $H_o = 100\,h$ km s^{-1} Mpc^{-1}, for nonzero cosmological constant $\Lambda \neq 0$, assuming $\lambda = \Lambda c^2/3H_o^2 = 1$ and for different values of the density parameter $\Omega_o = \rho_o/\rho_c$, the ratio of the present mass density, ρ_o, to the critical mass density, ρ_c, needed to close the Universe (adapted from Krisciunas, 1993)

$\Omega_o = \rho_o/\rho_c$	t_o/t_H
0.00	∞
0.03	1.727
0.10	1.375
0.30	1.077
1.00	0.785
3.00	0.550

$$\frac{t_L}{t_H} = \frac{(1+\Omega_o z)^{\frac{1}{2}}}{(\Omega_o - 1)(1+z)} + \frac{\Omega_o}{(\Omega_o - 1)^{\frac{3}{2}}} \tan^{-1}\left[\left(\frac{1+\Omega_o z}{\Omega_o - 1}\right)^{\frac{1}{2}}\right]$$
$$- \frac{1}{(\Omega_o - 1)} - \frac{\Omega_o}{(\Omega_o - 1)^{\frac{3}{2}}} \tan^{-1}\left[\left(\frac{1}{\Omega_o - 1}\right)^{1/2}\right] . \qquad (5.337)$$

and for $0 < \Omega_o < 1$ and $\Lambda = 0$:

$$\frac{t_L}{t_H} = - \frac{(1+\Omega_o z)^{\frac{1}{2}}}{(1-\Omega_o)(1+z)} - \frac{\Omega_o}{2(1-\Omega_o)^{\frac{3}{2}}} \ln\left[\frac{(1+\Omega_o z)^{\frac{1}{2}} - (1-\Omega_o)^{\frac{1}{2}}}{(1+\Omega_o z)^{\frac{1}{2}} + (1-\Omega_o)^{\frac{1}{2}}}\right]$$
$$+ \frac{1}{(1-\Omega_o)} + \frac{\Omega_o}{2(1-\Omega_o)^{\frac{3}{2}}} \ln\left[\frac{1-(1-\Omega_o)^{\frac{1}{2}}}{1+(1-\Omega_o)^{\frac{1}{2}}}\right] . \qquad (5.338)$$

For a large redshift $z = 5.0$, corresponding to the largest ones observed to date, the look-back time, t_L, ranges from 9.7 to 15.5 Gyr depending on the choice of the Hubble constant, H_o, and the density parameter, Ω_o; but the difference $t_o - t_L$ between the age of the Universe, t_o, and the look-back time is between 1 and 2 Gyr.

For nonzero cosmological constant $\Lambda \neq 0$, the look-back time can be solved by numerical integration of the equation (Krisciunas, 1993)

$$\frac{t_L}{t_H} = \int_o^z \frac{dz'}{(1+z')^2 \left\{\Omega_o z' + 1 - \lambda \left[\frac{2z'+(z')^2}{(1+z')^2}\right]\right\}^{\frac{1}{2}}} , \qquad (5.339)$$

where

$$\lambda = \frac{\Lambda c^2}{3H_o^2} .$$

5.4 Mass

5.4.1 Inertial Mass, Gravitational Mass and the Newtonian Gravitational Constant

The equation of motion of a particle of inertial mass, M_i, is given by (Newton, 1687)

$$F = M_i a , \qquad (5.340)$$

whereas the law of gravitation is given by

$$F = -M_g g , \qquad (5.341)$$

where M_g is the gravitational mass. Here F is the force producing an acceleration a, or the force on a mass caused by the gravitational acceleration, g. For a distance, R, from a spherical mass, we have (Newton, 1687)

$$g = \frac{M_g G}{R^2} , \qquad (5.342)$$

where M_g is the gravitational mass, and G is the Newtonian constant of gravitation.

Laboratory measurements provide (Cavendish, 1798; Heyl, 1930; Rose et al., 1969)

$$G = (6.674 \pm 0.012) \times 10^{-8}\ \mathrm{cm^3\ s^{-2} g^{-1}} \tag{5.343}$$

where 0.012 represents three standard deviations. A more recent value by Luther and Towler (1982) is:

$$G = (6.6726 \pm 0.0005) \times 10^{-8}\ \mathrm{cm^3\ s^{-2}\ g^{-1}} \tag{5.344}$$

where the uncertainty is one standard deviation. This accuracy has been questioned by Kuroda (1995). Submarine measurements of $G = (6.677 \pm 0.013) \times 10^{-8}$ are, within their accuracy, consistent with the laboratory measurements (Zumberge et al., 1991).

In the opening paragraphs of his Principia, Newton (1687) stated that the inertial "mass" of a body, which regulates its response to an applied force, is equal to its "weight", that property that regulates its response to gravitation. These two quantities have since become known as the inertial mass and gravitational mass. The equivalence of these two types of mass is now generally referred to as the Weak Equivalence Principle, which essentially means that all bodies fall in a gravitational field with the same acceleration regardless of their mass or external structure. The Strong Equivalence Principle extends the weak one to all self-gravitating bodies, and demands that the laws of physics apply regardless of place or time in the Universe.

A direct test of the Weak Equivalence Principle is the Eötvös experiment, the comparison of the acceleration from rest of two laboratory-sized bodies of different composition in an equal gravitational field. Thus, Eötvös (1889) showed that the difference between M_i and M_g for wood and platinum was less than 10^{-9}. Roll, Krotov, and Dicke (1964) and Braginsky and Panov (1972) have improved this limit to respective values of 10^{-11} and 10^{-12}. It has also been shown that molecules (Estermann et al., 1938) and neutrons (Dabbs et al., 1965) fall with the same acceleration as atoms, and that the gravitational force on electrons in copper is the same as that on free electrons (Witteborn and Fairbank, 1967). The Weak Equivalence Principle has also been tested for neutrinos and photons from the supernova SN 1987A (Longo, 1988; Krauss and Tremaine, 1988), updated in the classical way by Adelberger et al. (1990), and reviewed by Will (1993).

Nordtvedt (1968) suggested that lunar laser ranging could test, at a scientific level, whether Earth and Moon accelerate equally toward the Sun. Such observations have confirmed the universality of free–fall for two different bodies (Dickey et al., 1994; Nordtvedt, 1995, 1996), any difference being fractionally less than 5×10^{-13}. Laboratory experiments are slightly less precise, but do have the versatility of comparing acceleration rates among a variety of different objects (Su et al., 1994).

Most theories of gravity that violate the Strong Equivalence Principle predict that the locally measured Newtonian gravitational constant, G, may vary with time as the Universe evolves. Observations of the binary pulsar limit this variation to $\dot{G}/G \leq 10^{-11}$ yr^{-1} (Damour, Gibbons and Taylor, 1988; Taylor and Weisberg, 1989; Damour and Taylor, 1991). Similar stringent experimental bounds on the time variation of G, and therefore on the Strong Equivalence Principle, have been obtained from radar range data to the Viking lander on Mars (Hellings et al., 1983) and from lunar laser ranging (Müller et al., 1991; Dickey et al., 1994). Possible variations of G with time are also discussed in Sect. 5.7.3 within the context of cosmological models.

All of the available experimental evidence, then, allows us to follow Newton's example and write the equation of motion of a system of particles interacting gravitationally as

$$M_{\mathrm{n}} \frac{d^2 X_{\mathrm{n}}}{dt^2} = G \sum\nolimits_{\mathrm{m}} \frac{M_{\mathrm{n}} M_{\mathrm{m}} (X_{\mathrm{m}} - X_{\mathrm{n}})}{|X_{\mathrm{m}} - X_{\mathrm{n}}|^3} , \tag{5.345}$$

where M_{n} is the mass of the n particle, and X_{n} is its Cartesian position vector in an inertial reference frame at time t.

5.4.2 Mass of the Sun, Earth, Moon and Planets

A planet has an elliptical orbit with the Sun at one focus, and an orbital period, P, and semimajor axis, a, related by Kepler's third law (Kepler, 1619)

$$\frac{P^2}{a^3} = \frac{4\pi^2}{G(M_{\mathrm{p}} + M_{\odot})} \approx \frac{4\pi^2}{GM_{\odot}} \tag{5.346}$$

where the mass of the planet is M_{P}, the Newtonian gravitational constant $G = 6.672 \times 10^{-8}$ in c.g.s. units, and the mass of the Sun is given by:

$$M_{\odot} = \frac{4\pi^2 a^3}{GP^2} = 1.989 \times 10^{33} \text{ grams} , \tag{5.347}$$

using the values for the Earth's orbit $a = A = 1.4959787 \times 10^{13}$ cm and P = one year = 3.15569×10^7 seconds.

The motions of artificial satellites about the Earth also obey Kepler's third law, with the Sun's mass replaced by the mass of the Earth, M_{E}, and observations of these satellites indicate that the geocentric gravitational constant is

$$GM_{\mathrm{E}} = 3.986005 \times 10^{20} \text{ cm}^3 \text{ s}^{-2} , \tag{5.348}$$

and that the mass of the Earth is

$$M_{\mathrm{E}} = 5.9742 \times 10^{27} \text{ grams} . \tag{5.349}$$

The ratio of the mass of the Sun to that of the Earth, or the Earth's reciprocal mass, is:

$$M_{\odot}/M_{\mathrm{E}} = 332946.0 . \tag{5.350}$$

The ratio of the mass of the Moon, M_{M}, to that of the Earth is:

$$\mu = M_{\mathrm{M}}/M_{\mathrm{E}} = 0.0123002 = 1/81.300587 , \tag{5.351}$$

or

$$M_{\mathrm{M}} = 7.3483 \times 10^{25} \text{ grams} . \tag{5.352}$$

The reciprocal mass of the Moon is:

$$M_\odot/M_\mathrm{M} = 27068708.7 \ , \tag{5.353}$$

and the reciprocal mass of the Earth–Moon system is:

$$M_\odot/(M_\mathrm{M}+M_\mathrm{E}) = 328900.5 \ . \tag{5.354}$$

Because the constant, G, is only known to three significant figures, the orbits of the planets are calculated using the Gaussian constant of gravitation, k, given by

$$k^2 = \frac{4\pi^2 a_\mathrm{E}^3}{P_\mathrm{E}^2(M_\mathrm{E}+M_\odot)} \ . \tag{5.355}$$

Gauss (1857) took $a_\mathrm{E} = 1.0$, $P_\mathrm{E} = 365.2563835$ days, and $M_\mathrm{E}/M_\odot = 1/354,710$ to obtain

$$k = 0.01720209895(\mathrm{a.u.})^{3/2}(\text{ephemeris day})^{-1}(\text{solar mass})^{-1/2} \ . \tag{5.356}$$

This value of k is presently used as a defining constant for the ephemeris, and it is customary to give the planet distance in astronomical units (a.u.) and the planet mass, M_p, in inverse mass units, $M_\odot/M_\mathrm{p}$.

Masses of the planets are given in Table 5.13.

Table 5.13. Radius, mass and mean density of the planets (Lang, 1992). Pluto has a radius of 1,123 km and a mass of 1.36×10^{25} grams

	Mercury	Venus	Earth	Mars
Equatorial Radius, R_e (km)	2,439	6,051	6,378.140	3,397
Equatorial Radius, ($R_\mathrm{E} = 1.0$)	0.382	0.949	1.000	0.533
Reciprocal Mass ($M_\odot/M_\mathrm{p}$)	6023600	408525.1	332946.043	3098710
Mass, M_p (grams)	3.3022×10^{26}	4.8690×10^{27}	5.9742×10^{27}	6.4191×10^{26}
Mass M_p ($M_\mathrm{E} = 1.0$)	0.05527	0.81499	1.00000	0.10745
Mean Density (grams cm^{-3})	5.43	5.25	5.52	3.93
	Jupiter	**Saturn**	**Uranus**	**Neptune**
Equatorial Radius, R_e (km)	71,492	60,268	25,559	24,764
Equatorial Radius, ($R_\mathrm{E} = 1.0$)	11.19	9.46	3.98	3.81
Reciprocal Mass ($M_\odot/M_\mathrm{p}$)	1047.3492	3497.91	22902.94	19434
Mass, M_p (grams)	1.8992×10^{30}	5.6865×10^{29}	8.6849×10^{28}	1.0235×10^{29}
Mass M_p ($M_\mathrm{E} = 1.0$)	317.894	95.1843	14.5373	17.1321
Mean Density (grams cm^{-3})	1.33	0.71	1.24	1.67

5.4.3 Mass and Mass Loss of Stars

The majority of stellar masses, M, lie between 0.3 and 3.0 solar masses, or $0.3\ M_\odot \leq M \leq 3.0\ M_\odot$, where the mass of the Sun is $M_\odot = 1.989 \times 10^{33}$ grams. The largest accurately determined stellar mass is 26.9 $M_\odot$ for V382 Cyg. Theoretical calculations indicate that a star must have a mass greater than 0.07 solar masses, or $M \geq 0.07\ M_\odot$, to shine and support itself by fusing hydrogen into helium.

Stellar masses can be inferred from observations of binary star systems that come in three varieties – visual, spectroscopic and eclipsing binaries. In a visual binary, the two stars are seen as separate images that orbit one another with the passage of time. For a spectroscopic binary, thc orbital motion is inferred from a periodic variation of the Doppler shift of spectral lines. If the lines of only one component are observed, it is called a single-lined spectroscopic binary; the lines from both components are observed for a double-lined spectroscopic binary. In an eclipsing binary, periodic changes in the total light from the system are interpreted in terms of eclipses of one star by another.

The sum of the two masses, M_1 and M_2, of the two stars in a visual binary system can be inferred from Kepler's third law:

$$M_1 + M_2 = \frac{4\pi^2 a^3}{GP^2} \tag{5.357}$$

where a is the semimajor axis of the absolute elliptical orbit, P is the orbital period, and the Newtonian gravitational constant $G = 6.672 \times 10^{-8}$ in c.g.s. units. With a suitable choice of units,

$$M_1 + M_2 = \frac{a^3}{P^2} = \frac{\alpha^3}{\pi^3 P^2} = \frac{\alpha^3 D^3}{P^2} , \tag{5.358}$$

when the masses are in solar masses, a is in astronomical units, P is in years, α is the angular size of the semimajor axis in seconds of arc, π is the trigonometric parallax of the binary star system in seconds of arc, and D is the distance of the visual binary in parsecs.

Since the masses of the two components of the system will often be comparable, each star will follow an elliptical orbit about the center of mass, with respective semimajor axes, a_1 and a_2, given by:

$$a = a_1 + a_2$$

or

$$\alpha = \alpha_1 + \alpha_2 \tag{5.359}$$

and

$$\frac{a_1}{a_2} = \frac{\alpha_1}{\alpha_2} = \frac{M_2}{M_1}$$

so that individual masses can be calculated.

When the fainter component is not directly observable, its presence can be inferred from the motion of the brighter component 1, for which we have

$a_1/a = M_2/(M_1 + M_2)$ and

$$(M_1 + M_2)\left(\frac{M_2}{M_1 + M_2}\right)^3 = \frac{\alpha_1^3}{\pi^3 P^2} \tag{5.360}$$

where α_1 and π are in seconds of arc and P is in years.

We usually see an apparent orbit which is the projection of the absolute orbit onto the plane of the sky. If the orbital plane is not seen face on, but instead inclined at an angle, i, we observe $\alpha' = \alpha \sin i$ and

$$(M_1 + M_2)\sin^3 i = \frac{(\alpha')^3}{\pi^3 P^2} \tag{5.361}$$

so we know the mass of the system except for some unknown inclination factor.

For a double-lined spectroscopic binary, we can determine the quantities $V_1 \sin i$ and $V_2 \sin i$ from the maximum red and blue Doppler shifts of the spectral lines of each star, where

$$\begin{aligned} V_1 &= 2\pi a_1/P \ , \\ V_2 &= 2\pi a_2/P \ , \end{aligned} \tag{5.362}$$

and from Kepler's third law:

$$(M_1 + M_2)\sin^3 i = \frac{P(V_1 \sin i + V_2 \sin i)^3}{2\pi G} \ . \tag{5.363}$$

If only one spectrum is observed, then we can use:

$$\begin{aligned} \frac{4\pi^2}{G}\frac{(a_1 \sin i)^3}{P^2} &= (M_1 + M_2)\left(\frac{M_2}{M_1 + M_2}\right)^3 \sin^3 i \\ &= \frac{M_2^3 \sin^3 i}{(M_1 + M_2)^2} \ . \end{aligned} \tag{5.364}$$

The quantity on the right is called the mass function.

If the binary is also an eclipsing system, then we know that i is close to 90 degrees, and from the duration and depth of the eclipses we can also deduce the radii of the stars relative to the orbital radii.

Popper (1980) reviewed eclipsing as well as visual binary star systems, and found masses ranging from 16.7 to 0.25 solar masses, with the mass decreasing systematically along the spectral sequence from O B A F G K M. These measured masses are reproduced by Lang (1992) who also gives tables of the mass and radius for stars of different spectral type and luminosity class. An example is shown in Table 5.14. Accurate masses and radii of detached, double-lined eclipsing binary systems have also been provided by Anderson (1991).

The spectral class, Sp, also establishes the effective temperature, T_{eff}, which can be combined with the stellar radius, R, to obtain the absolute luminosity, L, from the Stefan-Boltzmann law, $L = 4\pi\sigma R^2 (T_{\mathrm{eff}})^4$, where the Stefan-Boltzmann constant $\sigma = 5.67 \times 10^{-5}$ erg s^{-1} cm^{-2} K^{-4}. As indicated in the composite Hertzsprung-Russell diagram (see Sect. 5.3.9, Fig. 5.25), the spectral

Table 5.14. Stellar mass, M, and radius, R, in units of the Sun's values, $M_\odot$ and $R_\odot$, for main sequence stars of different spectral type, Sp. Adapted from Lang (1992)

Sp	M ($M_\odot$)	R ($R_\odot$)
O3	120	15
O5	60	12
O8	23	8.5
O9	19	7.8
B0	17.5	7.4
B1	13	6.4
B2	9.8	5.6
B3	7.6	4.8
B5	5.9	3.9
B8	3.8	3.0
A0	2.9	2.4
A5	2.0	1.7
F0	1.6	1.5
F5	1.3	1.3
G0	1.05	1.1
G5	0.92	0.92
K0	0.79	0.85
K5	0.67	0.72
M0	0.51	0.60
M3	0.33	0.45
M5	0.21	0.27
M7	0.12	0.18
M8	0.06	0.1

sequence also corresponds to a systematic decrease in absolute luminosity for main-sequence stars from O to M.

Thus, more massive main-sequence stars are hotter and shine with greater luminosity. As illustrated in Fig. 5.30, the observational data are consistent with an absolute luminosity, L, that increases with the fourth power of the mass, M, or

$$L \propto M^4 \tag{5.365}$$

and

$$\log\left(\frac{L}{L_\odot}\right) = 4\log\left(\frac{M}{M_\odot}\right) \tag{5.366}$$

where the Sun's absolute luminosity $L_\odot = 3.85\times10^{33}$erg s^{-1} and the solar mass $M_\odot = 1.989 \times 10^{33}$ grams. Mc Cluskey and Kondo (1972) obtain the relation

$$\log\left(\frac{L}{L_\odot}\right) = 3.8\log\left(\frac{M}{M_\odot}\right) + 0.08 \text{ for } M > 0.2\ M_\odot\ . \tag{5.367}$$

For a main-sequence star of absolute bolometric magnitude, M_{bol}, this empirical relation becomes:

$$\log(M/M_\odot) = 0.48 - 0.105\ M_{\text{bol}} \text{ for } -8 < M_{\text{bol}} < 10.5\ . \tag{5.368}$$

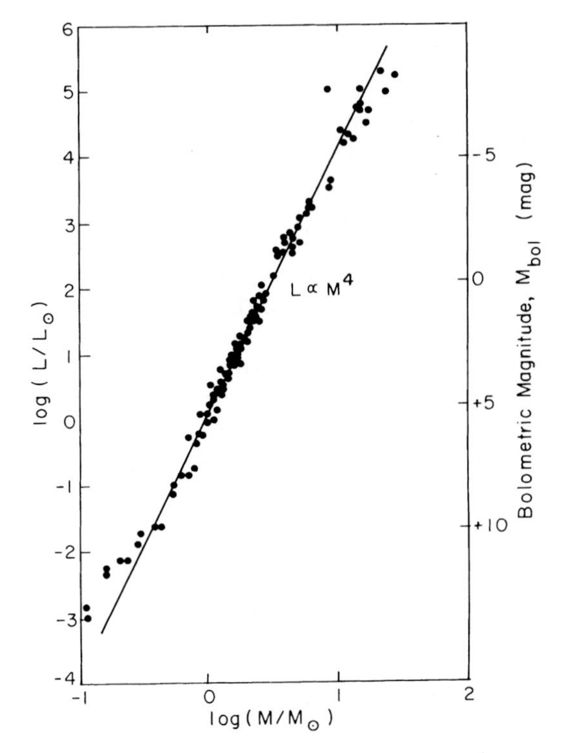

Fig. 5.30. Empirical mass-luminosity relation for main-sequence stars of absolute luminosity, L, in units of the solar luminosity, $L_{\odot} = 3.85 \times 10^{33}$ erg s^{-1}, and mass, M, in units of the solar mass $M_{\odot} = 1.989 \times 10^{33}$ grams. Adapted from data given by Popper (1980) who reviewed mass determinations for binary star systems and inferred luminosities from the radius and effective temperature. The straight line corresponds to a luminosity proportional to the fourth power of the mass

These mass-luminosity relations do not apply to stars that have evolved away from the main sequence, so they cannot be used to infer the masses of giant or white dwarf stars.

From the theoretical point of view, we would expect more massive main-sequence stars to have hotter central temperatures to support their greater weight, and that thermonuclear reactions in the stellar cores would proceed at a faster rate at the higher temperatures, producing a more luminous output.

A fully theoretical mass-luminosity relation can be derived only if the law of energy generation is known. However, as shown by Eddington (1924), a simple mass-luminosity relation can be derived from basic assumptions. The mass of main-sequence stars is supported by gas pressure, so their internal temperatures, T, scale inversely with their radius, R, and directly with their mass, M, or $T =$ constant $\times M/R$. From the assumptions that the internal heat is transported by radiation and that the stellar opacity is relatively insensitive to mass, it follows that the stellar luminosity must increase roughly with the fourth power of temperature and the third power of the mass, or that:

$$L \propto \frac{R^4 T^4}{\langle \kappa \rangle M} \propto M^3 \tag{5.369}$$

for a main-sequence star with average opacity $\langle \kappa \rangle$.

A closely related theorem asserts that main-sequence stars of the same compositions (with therefore the same laws of opacity and energy generation) have radii, luminosities, effective temperatures and mean densities that are determined solely by the stars' masses. Vogt (1926) first derived this theorem and Russell et al. (1926) included it in their textbook, apparently having derived it independently, and hence it is generally called the Vogt–Russell theorem. This theorem implies that a main-sequence star is uniquely described by the equations of hydrostatic equilibrium (for giant stars radiative equilibrium applies), the perfect gas law, the equation of energy balance and the equation of radiation energy transfer; these equations were given in Sect. 5.3.9. They imply that a star of a given mass, age and initial composition occupies a unique position, related to the star's evolutionary history, on the Hertzsprung–Russell diagram.

Stars gradually lose mass as stellar winds carry their outer atmospheres into the surrounding interstellar space. Such mass loss from cool main-sequence stars has been reviewed by Dupree (1986); the effect of mass loss on the evolution of massive stars is reviewed by Chiosi and Maeder (1986).

The Sun loses mass at the rate of

$$\dot{M}_{\odot} = 2 \times 10^{-14}\, M_{\odot} \text{ per year} \tag{5.370}$$

but the Sun will have evolved into a giant star long before it is blown away by the solar wind.

As illustrated in Figure 5.31, more luminous stars, and hence stars with greater mass, have much greater mass loss rates of up to 10^{-4} solar masses per year.

Empirical expressions that relate mass loss, $\dot{M}$, to stellar mass, M, absolute luminosity, L, radius, R, and/or effective temperature, $T_{\rm eff}$, have been tabulated by Chiosi and Maeder (1986). For main-sequence stars, De Jager et al. (1988) obtain:

$$\log(-\dot{M}) = 1.769 \log(L/L_{\odot}) - 1.676 \log(T_{\rm eff}) - 8.158 \ , \tag{5.371}$$

so the mass loss rate increases with increasing stellar luminosity and with decreasing stellar effective temperature along the main-sequence from spectral class O to M.

Mass loss rates are greatest for supergiant stars, either from the blue OB supergiants or the red M ones. The Doppler shifts of spectral line profiles, observed at visible and ultraviolet wavelengths, indicate that material is expanding at speeds well in excess of their escape velocity.

The mass loss rate can be accurately inferred from radio techniques, since the observed radio emission originates in a region where the outflow has already reached the terminal velocity. The radio flux density, S_ν, arising from the bremsstrahlung of the ionized, expanding stellar atmosphere is given by (Panagia and Felli, 1975)

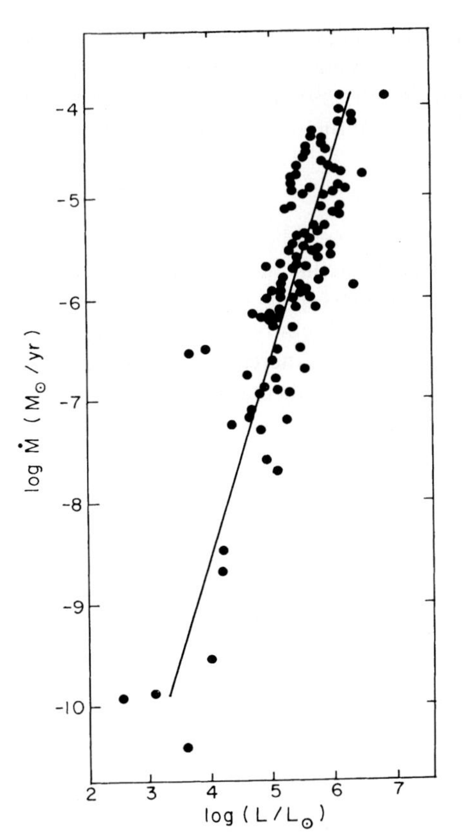

Fig. 5.31. The relation between mass loss rate, $\dot{M}$, in solar masses per year and absolute luminosity, L, in units of the Sun's luminosity $L_{\odot} = 3.85 \times 10^{33}$ erg s^{-1}. (Adapted from Chiosi and Maeder, 1986)

$$S_\nu = 5.12 \left[\frac{\nu}{10\ \mathrm{GHz}}\right]^{0.6} \left[\frac{T_\mathrm{e}}{10^4\mathrm{K}}\right]^{0.1} \left[\frac{\dot{M}}{10^{-5} M_{\odot}\ \mathrm{yr}^{-1}}\right]^{4/3}$$
$$\times \left(\frac{\mu}{1.2}\right)^{-4/3} \left[\frac{V_{\exp}}{10^3 \mathrm{km\ s}^{-1}}\right]^{-4/3} \bar{Z}^{2/3} \left[\frac{D}{\mathrm{kpc}}\right]^{-2} \mathrm{mJy} \ , \qquad (5.372)$$

where the units of milliJansky are 1 mJy = 10^{-26} erg s^{-1} cm^{-2} Hz^{-1}, the observing frequency is ν, the electron temperature is T_e, the mean atomic weight per electron is μ, the expansion velocity is $V_{\exp}$, the average ionic charge is $\bar{Z}$, and D is the source distance.

5.4.4 Mass of Galaxies – Light Emitting Regions

If the Sun's circular orbit about the galactic center at distance R_o is controlled by a dominant nuclear mass, M_N, concentrated within R_o, then it follows from Kepler's third law that

$$M_\mathrm{N} = R_\mathrm{o}\theta_\mathrm{o}^2/G = 1.9 \times 10^{44}\ \mathrm{grams} = 0.95 \times 10^{11}\ M_{\odot} \ , \qquad (5.373)$$

where (Kerr and Lynden-Bell, 1986) $R_\mathrm{o} = 8.5$ kpc $= 2.6 \times 10^{22}$ cm and $\theta_\mathrm{o} = 2\pi R_\mathrm{o}/P = 220$ km s^{-1} for a rotation period P, the Newtonian gravitational constant $G = 6.6726 \times 10^{-8}$, and the solar mass $M_{\odot} = 1.989 \times 10^{33}$ grams. Since the total luminosity, L_{BG}, of our Galaxy in the blue band is $L_{BG} = (2.3 \pm 0.6) \times 10^{10}\ L_{\odot}$ (Van Den Bergh, 1988), for an absolute solar luminosity of $L_{\odot} = 3.85 \times 10^{33}$ erg s^{-1}, we have the nuclear mass-to-luminosity ratio of:

$$M_N/L_{BG} = 4.3\,M_{\odot}/L_{\odot}\ . \tag{5.374}$$

The mass of the luminous, visible disks of spiral galaxies can be inferred from their rotation (Öpik, 1922). The circular stellar motions about a central, nuclear region are inferred from the Doppler shifts of spectral lines, obtaining a rotation velocity, V_{HO}, at the luminous edge, or Holmberg radius, R_{HO}. For a radius $R \leq R_{HO}$, the rotation velocity increases with increasing distance R from the center, and the central mass, M_{CS}, of the luminous matter of spiral galaxies is estimated from:

$$M_{CS} = R_{HO}(V_{HO})^2/G\ , \tag{5.375}$$

where the gravitational constant $G = 6.672 \times 10^{-8}$ cm^3 g^{-1} s^{-2} and typical values are (Roberts, 1969; Burbidge and Burbidge, 1975)

$$\begin{aligned} &R_{HO} \leq 20\,\text{kpc} \\ &M_{CS} = 10^{10} \text{ to } 5 \times 10^{11}\,M_{\odot} \end{aligned} \tag{5.376}$$

and

$$M_{CS}/L_{BS} = (1 \text{ to } 10)\,M_{\odot}/L_{B\odot}$$

where the solar mass $M_{\odot} = 1.989 \times 10^{33}$ grams, L_{BS} denotes the absolute blue luminosity of the spiral galaxy and the solar blue luminosity is $L_{B\odot} \approx 3 \times 10^{33}$ erg s^{-1}.

More detailed formulae for the rotation velocity, $V_{rot}(R)$, and the disk mass, M_{DS}, of the flattened disk of a spiral galaxy are (Brandt, 1960; Brandt and Belton, 1962):

$$V_{rot}(R) = \frac{V_T R}{R_T\left[\frac{1}{3} + \frac{2}{3}\left(\frac{R}{R_T}\right)^n\right]^{3/2n}} \tag{5.377}$$

and

$$M_{DS} = \left(\frac{3}{2}\right)^{3/n} \frac{R_T V_T^2}{G}\ , \tag{5.378}$$

where the integer n is determined by fitting to the rotation curve, and the rotation velocity is assumed to increase with distance, R, from the center to a maximum value of V_T at radius R_T, and to fall off as $R^{-1/2}$ beyond this turnoff point. The visible mass and mass-to-light ratios obtained with these formulae are similar to the previous simpler expression.

The mass-to-light ratio for the visible cores of elliptical galaxies can be obtained from the central line-of-sight velocity dispersion, σ_v, and spatially-resolved surface photometry of the galaxy. This can be done by fitting a King model to the central brightness distribution to obtain the core radius, R_c, and central surface luminosity, I_o, of the fit, and using the formula (King and

Minkowski, 1972; Binney, 1982)

$$\frac{M_{\rm CE}}{L_{\rm BE}} = \frac{9\sigma_{\rm v}^2}{2\pi G I_{\rm o} R_{\rm c}} \ . \tag{5.379}$$

Alternatively, one can fit a $R^{1/4}$ profile to the photometry to obtain an effective radius $R_{\rm e}$ and the associated surface luminosity $I_{\rm e}$ at $R_{\rm e}$, and infer the core mass to-light-ratio using (Binney, 1982; Lauer, 1985)

$$\frac{M_{\rm CE}}{L_{\rm BE}} = \frac{0.201\sigma_{\rm v}^2}{G I_{\rm e} R_{\rm e}} \ . \tag{5.380}$$

The blue bolometric magnitude is given by:

$$M_{\rm B} = -28.92 - 5.38 \log R_{\rm e} + 1.36\, I_{\rm e} \ . \tag{5.381}$$

Typical values for the cores of elliptical galaxies are (Morton and Chevalier, 1973; Faber and Jackson, 1976; Lauer, 1985; Van Der Marel, 1991):

$$R_{\rm c} = 0.5 \text{ to } 20\,\text{kpc}$$

$$M_{\rm CE} = (10^{10} \text{ to } 10^{12})\, M_{\odot} \tag{5.382}$$

and

$$M_{\rm CE}/L_{\rm BE} = (5 \text{ to } 20)\, M_{\odot}/L_{\rm B\odot} \ ,$$

where $L_{\rm BE}$ is the absolute blue luminosity of the elliptical galaxy.

5.4.5 Mass of Galaxies – Beyond the Luminous Boundaries

Astronomers have recently had to confront the fact that they may have been spending their lifetimes studying only a small part of the Universe that shines in the dark, and that the familiar, shimmering night sky may therefore tell us little about the nature of the Universe. They already knew about one sort of invisible Universe that is not seen visually with optical telescopes, that detected by radio or X-ray telescopes. But now there was evidence for truly dark and invisible matter that does not emit any kind of electromagnetic radiation, at least any kind detectable so far. This unfamiliar dark material cannot be seen, but it can be felt. Its presence is inferred from its gravitational effect on visible matter, suggesting vast amounts of enigmatic, unseen material hiding in the dark beyond the luminous boundaries of galaxies. Here we provide estimates for its mass and mass-to-light ratio; the possible nature of the dark matter is discussed within the broader cosmological context in the next Sect. 5.4.6.

Nonluminous matter has been inferred from the way its gravitational tug affects the spins of nearby spiral galaxies, including our own Milky Way, and the motions of galaxies within small groups and rich clusters. The elusive dark material keeps individual galaxies or groups of them from flying apart at high speeds. It is required to hold them together, and suggests that there is ten or twenty times more dark stuff than the luminous stellar kind. For a spiral galaxy one determines the orbital velocity, V, as a function of orbital distance, R, from the center, using Kepler's third law to infer the mass $M = V^2 R/G$ interior to the orbit. The statistical equivalent, called the virial theorem, is used to establish the mass, M, of a gravitationally bound group or cluster of galaxies;

Table 5.15. Distance, luminosity, mass and velocity of nearby galaxies (adapted from Peebles, 1995).*

Object	Distance (Mpc)	Luminosity (10^{10} $L_\odot$)	Mass** (10^{11} $M_\odot$)	Velocity*** (km s^{-1})
Milky Way	0.0	2	20.	0
Fornax	0.14	0.0015	0.01	−34
Leo I	0.22	0.0003	0.003	177
Leo II	0.22	0.0001	0.001	16
NGC 6822	0.52	0.03	0.8	33
M31	0.75	3	26.	−123
I1613	0.77	0.01	0.3	−155
WLM	0.95	0.005	0.1	−64
Sex AB	1.1	0.01	0.3	172
NGC 3109	1.2	0.02	0.5	203
NGC 300	2.10	0.3	5.	110
NGC 55	2.10	0.06	2.	115

* Velocities and distances of remote satellites of the Milky Way are also given by Zaritsky et al. (1989) and by Fich and Tremaine (1991).
** Often assuming $M/L = 100$ in solar units.
*** Velocity referred to the center of our Galaxy.

where K_R is the force per unit mass in the R direction, R is the distance from the galactic center, Oort's constants are $A = 14.4 \pm 1.2$ km s^{-1} kpc^{-1} and $B = -12.0 \pm 2.8$ km s^{-1} kpc^{-1} (Kerr and Lynden-Bell, 1986), and the force per unit mass, K_z, exerted by our Galaxy in the direction, z, perpendicular to the galactic plane is (Oort, 1932; Hill, 1960; Oort, 1960):

$$\frac{\partial K_z}{\partial z} = -0.3 \times 10^{-10} \text{ cm sec}^{-2}\text{pc}^{-1} , \quad (5.393)$$

where 1 pc $= 3.0857 \times 10^{18}$ cm.

Near the galactic plane and near the Sun, Oort (1932) obtained

$$\rho(0) = 10^{-23} \text{ g cm}^{-3} = 0.16\, M_\odot \text{ pc}^{-3} \quad (5.394)$$

from his analysis of the numbers and velocities of stars. Oort concluded that the mass density in visible stars, $\rho_s = 0.10\, M_\odot$ pc^{-3}, fell short by 30–50% of adding up to the amount of gravitating matter, $\rho(0)$, implied by the velocities. The quantity ρ (0) is sometimes called the "Oort limit" in honor of his early work.

However, there is some uncertainty in the determination of $\rho(0)$, with measured values ranging from (Gilmore, Wyse and Kuijken, 1989) $\rho(0) = 0.1$ to 0.2 $M_\odot$pc^{-3}, or between the approximate mass density of visible stars and twice that amount. The distribution of nearby stars mapped by Hipparcos indicate that Crézé, Chereul, Bienaymé and Pichon, 1998)

$$\rho(0) = 0.076 \pm 0.015\, M_\odot \text{ pc}^{-3} ,$$

a value that leaves no room for any disk shaped component of dark matter.

Additional detailed models used to estimate mass in various components of our Galaxy are given by Caldwell and Ostriker (1981), Bahcall, Schmidt and Soneira (1983) and Bahcall (1984).

More recently, Amaral et al. (1996) have compiled velocity data for interstellar gas, planetary nebulae, and asymptotic-giant-branch stars in the Milky Way's disk, concluding that their motion does not require the presence of any invisible matter within the Milky Way out to a radius of about 12 kpc. That does not preclude substantial amounts of unseen material in a more extensive halo. Still, the ongoing controversial possibility of invisible disk material is important for determining, and perhaps detecting, the elusive substance.

Mass estimates for nearby, spiral galaxies, seen more or less edge on, have been determined from measurements of their orbital, or rotational, velocity, V, as a function of distance, R, from their center. Such rotation curves are inferred from the Doppler shifts of the spectral lines of bright stars or emission nebulae, seen at optical wavelengths, or interstellar hydrogen gas detected at radio wavelengths near 21 cm. If the dominant mass, M, is centrally located, the more distant the stars or gas the lesser the gravitational pull and the slower the orbital speed, all in accordance with Kepler's third law $V^2 = GM/R$, where G is the Newtonian constant of gravitation. However, after an initial rise, a typical spiral rotation velocity remains constant with increasing radius (see Fig. 5.32). Stars or gas at the periphery of a spiral galaxy move at speeds as great or greater than those close to the center.

The non-Keplerian rotation of spiral galaxies was noted by Freeman (1970), who attributed it to a gravitating mass outside the visible regions. In the 1970s and 1980s, radio astronomers discovered that remote, invisible, neutral hydrogen gas clouds, detected at 21-cm wavelength in some spiral galaxies, are spinning about their centers unexpectedly fast. The 21-cm rotation velocity

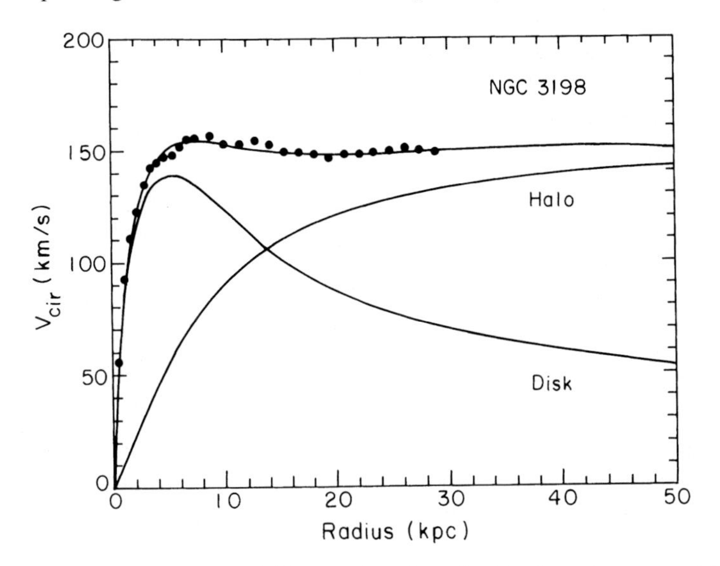

Fig. 5.32. Rotation speed, V_{cir}, as a function of radius from the center of the spiral, Sc, galaxy NGC 3198. The observed neutral hydrogen 21-cm data (top) are attributed to an optically luminous disk, and a "dark" radio halo which contributes most of the mass at distant regions from the center. The linear scale assumes a Hubble constant of $H_o = 75$ km s^{-1} Mpc^{-1}. (Adapted from Van Albada, Bahcall, Begeman and Sancisi, 1985).

remains constant with increasing distance from the center of the galaxy well beyond the visible stars (Rogstad and Shostak, 1972; Roberts and Rots, 1973; Roberts, 1975; Faber and Gallagher, 1979; Bosma, 1981; Kent, 1986, 1987). The mass of this gas usually amounts to less than 10 percent of the total mass of the galaxy (Roberts, 1975), but its rotation velocities imply much greater amounts of unseen mass beyond the luminous boundaries with a total mass of about 10 times the mass in visible stars.

During the same period, Vera Rubin showed that the visible stellar matter out at the edge of nearby spiral galaxies also moves just as fast as that located in the inner parts. She used optical spectroscopy to measure how fast spiral galaxies are spinning, from their luminous cores to their faint outer regions, arguing that they are spinning so fast that they must be embedded in substantial quantities of unseen matter to keep the stars from moving out of control (Rubin and Ford, 1970; Faber and Gallagher, 1979; Rubin, Ford and Thonnard, 1980; Rubin, 1979, 1983, 1987, 1993).

This also meant that there is more to a spiral galaxy than meets the eye; some unseen force is holding them together. Remote gas clouds and stars must be moving in response to the gravitational attraction of mass we cannot see. If this were not so, the galaxies could not retain their high-speed, outer parts; they could not stay together and would fly apart. Unlike luminosity, the unseen matter is not concentrated near the center of spiral galaxies, and instead it amounts to an increasing fraction of the total mass at larger distances from the nucleus. So, the distribution of light in a spiral galaxy might not be a guide to all the matter.

In important early papers, Ostriker, Peebles and Yahil (1974) and Einasto, Kaasik and Saar (1974) tabulated galaxy masses, M, as a function of the radius, R, to which they applied and found M increasing linearly with R out to at least 100 kpc and up to 10^{12} solar masses, with

$$M(R) = V^2R/G \approx 10^{12}\ M_\odot \text{ at } 1 \text{ Mpc} \ . \tag{5.395}$$

This mass of about 10^{12} solar masses is about ten times the amount of mass that had previously been estimated for the visible stellar component. In other words, about 90 percent of the mass of spiral galaxies had to be dark and invisible (Rubin, 1993), with a mass to light ratio in solar units of:

$$(M/L)_s \approx 200(M_\odot/L_\odot) \ , \tag{5.396}$$

where M is the mass of unseen matter and L is the total optical luminosity of the visible galaxy. Thus, each luminous disk of a spiral galaxy appears to be immersed in a halo of dark, invisible matter, somewhat like our own Galaxy (Trimble, 1987).

Estimates of the total mass of galaxies exceed that of their visible stellar disks by amounts that depend on the assumed extent. The halo mass, $M_{\rm H}(R_{\rm H})$, within radial distance, $R_{\rm H}$, from the center for both our Galaxy and other spiral galaxies is given by (Carr, 1994)

$$M_{\rm H}(R_{\rm H}) \approx 10M_{\rm VIS}(R_{\rm H}/100 \text{ kpc}) \ , \tag{5.397}$$

where $M_{\rm VIS}$ is the mass in visible stars, and $R_{\rm H}$ is in units of 100 kpc. So there appears to be several times as much matter outside the visible disks as inside, and invisible matter seems to constitute the dominant mass of these galaxies.

The motions of faint satellites in apparent orbit about spiral galaxies can be used to probe their distant regions, and to extend the estimated galaxy mass well beyond either the optically luminous or the gas-dominated radio regions. The results provide total mass estimates, M_S, for spiral galaxies of (Bahcall and Tremaine, 1981; Zaritsky et al., 1993)

$$M_S = 2 \times 10^{12}\ M_\odot \text{ within 200 kpc} . \tag{5.398}$$

Although elliptical galaxies do not seem to be supported by rotation, the presence of a nonluminous component is inferred from their high-velocity dispersions and extended X-ray halos of hot gas (Faber and Gallagher, 1979; Trimble, 1987). The mass determinations for the optically-invisible portions of elliptical galaxies can be inferred from their X-ray radiation, interpreted as the thermal bremsstrahlung of a hot multi-million degree gas. Assuming that the X-ray emitting material is in hydrostatic equilibrium, is spherically symmetric, and obeys the ideal gas law, we have a mass, $M(R)$ within radius, R, given by (Fabricant and Gorenstein, 1983; Forman, Jones and Tucker, 1985):

$$M(R) = \frac{-kT_{gas}}{G\mu m_H}\left(\frac{d \log \rho_{gas}}{d \log R} + \frac{d \log T_{gas}}{d \log R}\right) R . \tag{5.399}$$

where ρ_{gas} and T_{gas} are the gas density and temperature, G is the constant of gravitation, m_H is the mass of the hydrogen atom, μ is the mean molecular weight of the gas, and $M(R)$ is the total gravitating mass within a radius R, which will, in general, be much greater than the X-ray emitting gas.

To determine M(R) unambiguously, we need to determine the gas temperature and gas density gradients in the galaxy corona. The temperature profile can be parameterized by

$$T(R) = T_o[1 + (R/R_T)]^{-\alpha} \propto R^{-\alpha} , \tag{5.400}$$

where $T(R)$ is the temperature at radius R, and R_T is some characteristic radius which is small compared to the density core radius R_c. The density distribution, ρ_{gas}(R) for an isothermal gas is given by

$$\rho_{gas}(R) = \rho_{gas}(0)\left[1 + (R/R_c)^2\right]^{-3\beta/2} \tag{5.401}$$

so the electron density, N_e(R) at radius R is

$$N_e(R) \propto R^{-3\beta} . \tag{5.402}$$

Thus, assuming hydrostatic equilibrium

$$M(R) = 3.2 \times 10^{11}(3\beta + \alpha)\left(\frac{T}{10^7}\right)\left(\frac{R}{10\text{ kpc}}\right) M_\odot , \tag{5.403}$$

where the gas temperature T is in 10^7 degrees Kelvin, and is measured at a radius R in units of 10 kpc. An equivalent expression is:

$$M(R) = 1.8 \times 10^{12}(3\beta + \alpha)\left(\frac{T}{1\text{kev}}\right)\left(\frac{R}{10^3\text{ arc sec}}\right)\left(\frac{D}{10\text{ Mpc}}\right) M_\odot , \tag{5.404}$$

where the temperature is in kev, the angular radius is in 10^3 arcseconds, and the distance is in 10 Mpc.

Measurements with X-ray telescopes aboard the Einstein Observatory, Rosat and Asca indicate a total mass, M_E, for elliptical galaxies of up to (Forman, Jones and Tucker, 1985; Mushotzky et al., 1994; Kim and Fabbiano, 1995):

$$M_E = 5 \times 10^{12} M_\odot \tag{5.405}$$

and

$$M_E/L_B = 100\ M_\odot/L_\odot \tag{5.406}$$

out to 100 kpc, which is a factor of ten larger than the mass-to-light ratio of the visible parts of bright elliptical galaxies.

For spirals of linear scale, R, in Megaparsecs, or Mpc, the mass-to-light ratio increases with radius according to the relation (Bahcall, Lubin and Dorman, 1995)

$$\frac{M_S}{L_B} = (60 \pm 10)h\left(\frac{R}{0.1\text{Mpc}}\right)\frac{M_\odot}{L_\odot} \text{ for spirals} \tag{5.407}$$

and for ellipticals

$$\frac{M_E}{L_B} = (200 \pm 50)h\left(\frac{R}{0.1\text{Mpc}}\right)\frac{M_\odot}{L_\odot} \text{ for ellipticals .} \tag{5.408}$$

These relations are shown in Fig. 5.33, which also indicates that the mass-to-light ratios do not increase without bound, extending to about 200 h^{-1} where the Hubble constant $H_o = 100h$ km $s^{-1}Mpc^{-1}$.

Galaxies are clumped together on various scales, from small groups to rich clusters and superclusters, and velocity dispersion measurements indicate that the dynamical mass exceeds the visible mass on all of these scales. The virial theorem is used to determine the total group or cluster mass from the random motions of the component galaxies under the assumption that they are gravitationally bound (Zwicky, 1937). The method assumes that the kinetic energy, T, of the random motions of the galaxies is just balanced by the gravitational potential energy, Ω, of the cluster, or from the virial theorem

$$2T + \Omega = 0 \ . \tag{5.409}$$

For a spherically symmetric cluster that is dynamically stable (not expanding or contracting), the cluster mass, M_{CL}, is

$$M_{CL} = R_{CL}\sigma_V^2/G \tag{5.410}$$

where σ_V is the velocity dispersion and R_{CL} is the cluster radius. Masses inferred from the relative motions of galaxies in apparently-bound groups and clusters of galaxies exceed by a factor of about 10 times the sum of those inferred from the internal dynamics of the luminous parts of their component galaxies (Faber and Gallagher, 1979; Rood, 1981). However, the apparent discrepancy could be resolved by the total mass of the member galaxies if their extensive dark halos are included (Bahcall, Lubin and Dorman, 1995).

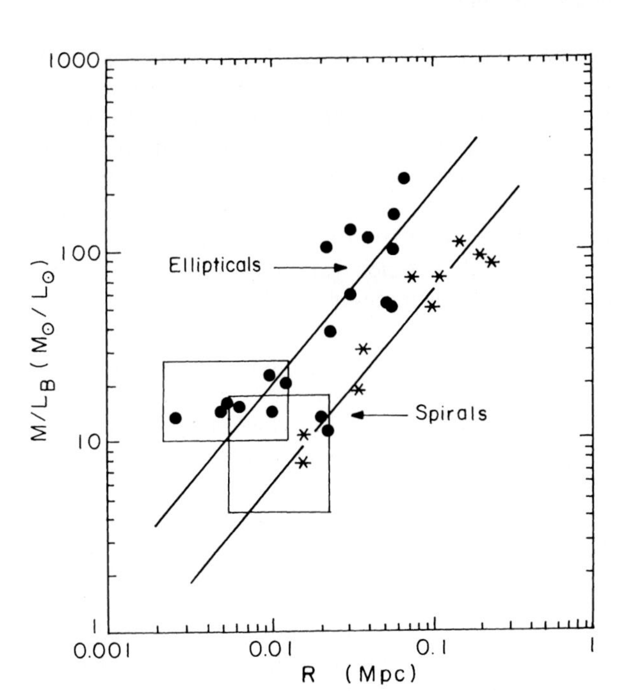

Fig. 5.33. The observed mass-to-light ratio, M/L_B, for bright spiral galaxies (asterisk symbols) and bright elliptical galaxies (solid dots) as a function of linear scale, R, assuming a Hubble constant of $H_o = 100$ km s^{-1} Mpc^{-1}. Here M is the total mass, L_B is the total corrected absolute blue luminosity, and the solid lines denote the best fit to $M/L_B =$ constant $\times$ R, which for spirals is $M/L_B = 60$ (R/0.1 Mpc) and for ellipticals is $M/L_B = 200$ (R/0.1 Mpc). (Adapted from Bahcall, Lubin and Dorman, 1995).

The mass determinations for binary galaxies were pioneered by Page (1960, 1961), extended by Turner (1976) and Peterson (1978) and reviewed by Faber and Gallagher (1979). Measurements of the velocity differences and projected separations of binary galaxies result in a mass $M(R)_{\mathrm{BI}}$ within radius, R, of (Ostriker, 1987)

$$M(R)_{\mathrm{BI}} = 1.3 \times 10^{12}(R/100\ \mathrm{kpc})M_\odot \ , \tag{5.411}$$

where R is in kiloparsecs or kpc, and the mass of the Sun is $M_\odot$. In terms of the mass-to-light ratio, the binary results give:

$$(M/L_{\mathrm{B}})_{\mathrm{BI}} \approx (70 \pm 20)h(M_\odot/L_{\mathrm{B}\odot}) \tag{5.412}$$

within a radius of about $100h^{-1}$ kpc, where the Hubble constant $H_o = 100h$ km s^{-1}Mpc^{-1}.

When the virial theorem is applied to groups of galaxies containing three to ten comparatively bright galaxies, one obtains (Huchra and Geller, 1982):

$$(M/L_{\mathrm{B}})_{\mathrm{G}} \approx 170h(M_\odot/L_{B\odot}) \ , \tag{5.413}$$

assuming that the groups are bound and not flying apart. As we have already seen, the mass of our own small group of galaxies, called the Local Group, is $M_{\rm LG} \approx 3 \times 10^{12}\ M_\odot$ with a mass-to-light ratio in solar units of $(M/L_{\rm B})_{\rm LG} \approx 100\ M_\odot/L_{\rm B\odot}$, inferred under the assumption that the self-gravitation of the Milky Way and Andromeda is enough to turn around their initial Hubble separation velocity, due to the expansion of the Universe, by the present age of the Universe (Kahn and Woltjer, 1959; Gunn, 1974).

In the 1930s Fritz Zwicky showed that rich galaxy clusters, such as the Coma cluster, require some unseen "dunkle materie", or dark material, to keep them gravitationally bound (Zwicky, 1933, 1937). A rich cluster is one containing hundreds or thousands of galaxies. Zwicky inferred the total cluster mass, $M_{\rm C}$, from the virial theorem and the velocity dispersion, $\sigma_{\rm V}$, of the galaxies using $M_{\rm C} = R_{\rm C}\sigma_{\rm V}^2/G$ for a cluster radius $R_{\rm C}$, showing that the total mass is at least 10 times greater than can be explained by the starlight seen in the thousands of member galaxies. The luminous galaxies are moving around at fast speed, and there isn't nearly enough mass associated with them to hold the Coma cluster together. In 1936, Sinclair Smith showed that a similar discrepancy exists between the gravitationally inferred mass of the Virgo cluster and the combined mass of its individual members (Smith, 1936).

Over the decades, the redshifts and velocity dispersions of hundreds of rich clusters of galaxies have been measured (Struble and Rood, 1991), and their core radii inferred (Sarazin, 1986). Since the radius $R = \theta D$ depends on the angular size, θ, and the distance $D = cz/H_{\rm o}$ at redshift z for velocity of light c, the radius depends on the Hubble constant $H_{\rm o} = 100\ h$ km s^{-1}Mpc^{-1}, as does the total cluster mass inferred from the virial theorem:

$$M_{\rm C} = R_{\rm C}\sigma_{\rm V}^2/G = \theta cz\sigma_{\rm V}^2/(GH_{\rm o}) \ . \tag{5.414}$$

So the total cluster mass is proportional to h^{-1}, or $M_{\rm C} \propto h^{-1}$. The results are usually expressed as a mass-to-light ratio, $(M/L)_{\rm C}$, and since the absolute luminosity $L = l/D^2$ for apparent luminosity, l, and distance, D, we have $L \propto h^{-2}$ and $M/L \propto h$.

The average mass-to-light ratio of $\langle M/L \rangle \geq 100h\ M_\odot/L_\odot$ has now been demonstrated for Coma, Virgo, Perseus and a number of other rich clusters (Rood, 1981; Trimble, 1987). Since this is much larger than the mass-to-light ratio of the starlight from individual galaxies, it suggests the presence of substantial amounts of unseen material. Such nonluminous material was often called the missing mass, but really it is the light that is missing.

The Coma cluster of galaxies serves as a representative example (Zwicky, 1937; Rood et al., 1972; Briel, Henry and Böhringer, 1992; White et al., 1993). Using a velocity dispersion of $\sigma_{\rm V} = 1{,}000$ km s^{-1} and a cluster radius of $R_{\rm C} = 5$ Mpc, with the virial theorem, we obtain a total cluster mass of

$$M_{\rm C}({\rm Coma}) \approx 10^{15} h^{-1} M_\odot \tag{5.415}$$

solar masses. The absolute blue luminosity inferred from photometry of the visible galaxies is:

$$L_{\rm BC}({\rm Coma}) = 2 \times 10^{12} h^{-2} L_\odot \ , \tag{5.416}$$

in units of the solar luminosity $L_\odot$, so one infers a mass-to-light ratio of the Coma cluster of galaxies of:

$$(M/L_{\rm B})_{\rm C} \approx 500hM_\odot/L_\odot \text{ for the Coma cluster of galaxies} . \tag{5.417}$$

Similar mass-to-light ratios are obtained for other rich clusters of galaxies (Faber and Gallagher, 1979; Rood, 1981), and since the luminous portion of the galaxies have $(M/L_{\rm B}) \approx 10h\,M_\odot/L_\odot$, the visible members of the component galaxies provide less than 10 percent of the mass of the cluster.

Clusters of galaxies are grouped together in larger superclusters with typical masses of $M_{\rm SC} \approx 10^{16}$ to 10^{17} solar masses in a radius of $R_{\rm SC} \approx 50h^{-1}$ Mpc (Abell, 1961). If bound structures, they imply a mass-to-light ratio of (Abell, 1961; Ford et al., 1981):

$$(M/L)_{\rm SC} = (200 \pm 100)h(M_\odot/L_\odot) , \tag{5.418}$$

where the uncertainty denotes a total range.

The thermal X-ray emission (bremsstrahlung) of the hot, 100 million-degree gas in clusters of galaxies has also been used to probe their gravitational mass. For rich clusters that appear to be in virial equilibrium, or bound, the X-ray data indicate (Sarazin, 1986, 1988; White and Fabian, 1995)

$$(M/L)_{\rm C} \approx 200h(M_\odot/L_\odot) . \tag{5.419}$$

The X-ray emitting gas permeates the space between the galaxies. Although the amount of cluster gas can be comparable to the mass content in all of the visible stars in the cluster of galaxies, it accounts for only 5 to 10 percent of the total cluster mass, $M_{\rm C}$. That is (Sarazin, 1988; Jones and Forman, 1992; Böhringer, 1994)

$$M_{\rm gas}(\text{cluster of galaxies}) = (0.05 \text{ to } 0.10)h^{-3/2}M_{\rm CL} \tag{5.420}$$

for a Hubble constant $H_{\rm o} = 100h$ km s^{-1}Mpc^{-1}.

Hot X-ray emitting gas has also been discovered in a compact group of four galaxies, which extends well beyond the galaxies themselves (Ponman and Bertram, 1993). As in the case of rich clusters, the multi-million degree gas suggests the presence of dark matter, for visible material in both the gas and galaxies does not provide enough gravity to contain the high-temperature gas.

So, 80 to 90 percent of the mass of rich clusters of galaxies is in some nonluminous form, neither in ordinary stars nor in the gas that emits X-rays, which together comprise only 10 to 20 percent of the total cluster mass. The unknown material, detected by its gravitational effect may be mainly caused by the high fraction (about 80%) of elliptical galaxies within them. Indeed, the mass-to-light ratios of groups and clusters of galaxies is comparable to those of the largest elliptical galaxies, and the ratio does not appear to increase with linear scale beyond 0.2 Mpc (see Fig. 5.34). The virial theorem mass for clusters may be accounted for by the intercluster gas and member galaxies, including their halo dark matter, which may be stripped off in dense clusters but remain there (Bahcall, Lubin and Dorman, 1995). So, we therefore turn to the nature of this unknown dark matter and its cosmological implications.

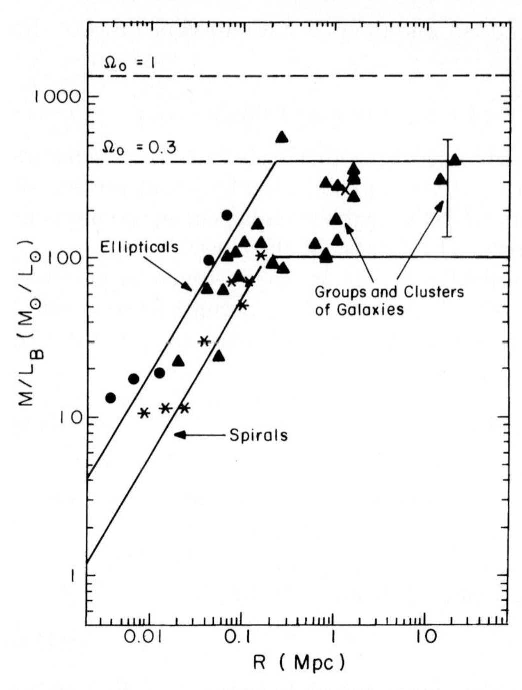

Fig. 5.34. Median values for the mass-to-light ratio M/L_B, of individual bright spiral galaxies (asterisk symbols), individual bright elliptical galaxies (solid dots), and groups, clusters and superclusters of galaxies (solid triangles) assuming a Hubble constant of H_o = 100 km $s^{-1}Mpc^{-1}$. The X marks a determination from the cosmic virial theorem and the large error bars denote the range of various results from the Virgo-centric infall. The slanted solid lines denote the best fit to M/L_B = constant × R (see Fig. 5.33). The dashed horizontal lines denote the M/L_B values for a density parameter Ω_o = 0.3 and 1.0; the observed data, bounded by the solid horizontal lines, indicate a low-density Universe with Ω_o = 0.2 to 0.3 which is insufficient to stop the expansion of the Universe in the future. (Adapted from Bahcall, Lubin and Dorman, 1995).

5.4.6 Dark Matter

Hidden matter is apparently immersed within the bleak darkness surrounding bright galaxies. The dark, unseen material does not emit any kind of electromagnetic radiation, and is mainly seen by its gravitational effects on visible objects. The fast-paced velocities of galaxies within clusters and the rapid spins of spiral galaxies indicate that there is approximately ten to twenty times more dark stuff than luminous matter (see the previous Sect. 5.4.5). Halos of dark matter surround spiral galaxies, and entire clusters of galaxies seem pervaded by the stuff.

The bulk of the matter in the Universe might therefore be invisible. As Agnes Clerke expressed it nearly a century ago (Clerke, 1903):

"Unseen bodies may, for ought we can tell, predominate in mass over the sum total of those that shine. They supply possibly the chief part of the motive power of the Universe."

The idea that most of the Universe may be made out of stuff we cannot see has sparked the public imagination, leading to at least eight popular, readable books on the subject (Bartusiak, 1993; Disney, 1984; Goldsmith, 1995; Gribbin, 1987; Krauss, 1989; Riordan and Schramm, 1991; Trefil, 1988; and Tucker and Tucker, 1988).

Observational evidence for dark matter in galaxies and clusters of galaxies is reviewed by Trimble (1987); hypothetical dark matter candidates are also mentioned. A number of review papers related to observations are given in Kormendy and Knapp (1987). Dark matter in the context of modern cosmology, including big-bang neutrinos that lead to ionization of intergalactic and interstellar matter, are discussed by Sciama (1993). The relationship between galaxy formation and dark matter is presented by Rees (1987, 1989). The elusive substance may be composed of ordinary neutrons and protons, called baryons, or more exotic nonbaryonic particles produced in the big-bang. Baryonic dark matter is reviewed by Carr (1994, 1996) and nonbaryonic dark matter by Turner (1991, 1993) and Sadoulet (1996).

In the cosmological scheme of things, it is the current mass density, ρ_o, of the Universe that measures the relative importance of dark matter. It may be inferred from the mass-to-light ratios, M/L, for spiral galaxies and clusters of galaxies

$$\rho_o = (M/L_B)\langle \mathscr{L}_B \rangle \ , \tag{5.421}$$

where the mean visual luminosity density $\langle \mathscr{L}_B \rangle$ in the blue band is:

$$\langle \mathscr{L}_B \rangle \approx 2.0 \times 10^8 h \quad L_\odot \mathrm{Mpc}^{-3} \tag{5.422}$$

and the Hubble constant $H_o = 100h$ km s^{-1}Mpc^{-1}. The numerical coefficient of ≈ 2.0 is given as 1.8 ± 0.2, 2.3, 2.4, and 1.98 ± 0.8 by Kirshner, Oemler and Schechter (1979), Kirshner et al. (1983), Felten (1985) and Efstathiou, Ellis and Peterson (1988).

The measured values of the mass density, ρ_o, are then compared to the critical mass density, ρ_c, that separates models of the homogeneous, isotropic Universe that expand forever ($\rho_o < \rho_c$) from those that ultimately recollapse ($\rho_o > \rho_c$), assuming that the cosmological constant $\Lambda = 0$.

$$\rho_c = 3H_o^2/(8\pi G) = 1.88 \times 10^{-29} h^2 \quad \mathrm{g\ cm^{-3}} = 2.8 \times 10^{11} h^2 \quad M_\odot \ \mathrm{Mpc}^{-3}$$

and

$$\rho_c = 1.05 \times 10^4 h^2 \ \mathrm{eV\ cm}^{-3} \ , \tag{5.423}$$

where the mass, m, is sometimes given as an energy, E, using $E = mc^2$ for a velocity of light $c = 2.9979 \times 10^{10}$cm s^{-1}, and an energy unit of $1\mathrm{eV} = 1.602 \times 10^{-12}$ ergs.

The density parameter, Ω_o, which has also been dubbed the omega factor, is specified by

$$\Omega_o = \rho_o/\rho_c \ . \tag{5.424}$$

It acts as a sort of weighing scale for the Universe. At equilibrium with $\Omega_o = 1$, the scales are in balance with the matter density just at the critical value; the scales are tipped at other values of omega. Note that estimates of Ω_o can be independent of the Hubble constant $H_o = 100h$ km s^{-1} Mpc^{-1} since $\rho_o = (M/L)\langle L_B \rangle$ and ρ_c are both proportional to the square of H_o and to h^2. Here the subscript o to the density factor and the Hubble constant denotes their present value.

The relative cosmological importance of different ingredients of the Universe is established by their density parameter or omega factor with a subscript that denotes the contribution. The current value of Ω_o will be the sum of all such omega factors. Using the mass-to-light ratios of the previous Sect. 5.4.5, we have (Carr, 1994, 1996):

For the dark halos of nearby spiral galaxies:

$$\Omega_H = 0.03 \text{ to } 0.10 \ , \tag{5.425}$$

increasing with radius, R, out to $\Omega_H \approx 0.10$ at $R_H \approx 100$ kpc, and as large as $\Omega_H \approx 0.2$ at $R_H \approx 1$ Mpc (Ostriker, Peebles and Yahil, 1974).

For rich clusters of galaxies:

$$\Omega_C = 0.10 \text{ to } 0.20 \ , \tag{5.426}$$

which is insufficient by at least a factor of five to close the Universe (Peebles, 1986). A value of $\Omega_C \approx 0.2$ is also obtained for those few cases where clusters of galaxies act as a gravitational lens.

For superclusters of galaxies,

$$\Omega_{SC} \approx 0.2 \text{ to } 0.3 \ , \tag{5.427}$$

with a lower value from estimates of the masses of second-order clusters (Abell, 1961; Ford et al., 1981) and a preferred value of 0.30 when the mass density of large-scale structures is inferred from peculiar motions superposed on the Hubble flow or expansion (see Sect. 5.7.6.2 and Strauss and Willick, 1995).

By way of comparison, stellar material tends to have a mass-to-light ratio of:

$$M/L_B = (2 \text{ to } 8)M_\odot/L_\odot \tag{5.428}$$

in blue light. The number is greater than 1 since the stellar mass is dominated by low mass stars and stellar light by high luminosity ones. Choosing $M/L_B = 4$, we obtain a stellar density or omega factor of:

$$\Omega_S \approx 0.003 \ . \tag{5.429}$$

Persic and Salucci (1992) obtain:

$$\Omega_S \approx 0.002 \ ,$$

and (5.430)

$$\Omega_G \approx 0.001 \ ,$$

for visible stars, Ω_S, in nearby galaxies and for the hot gas, Ω_G, in clusters and groups of galaxies. So, we can estimate the density parameter for luminous matter in galaxies as:

$$\Omega_{LUM} \approx 0.003 \ . \tag{5.431}$$

Thus, observations indicate that $\Omega \approx 0.1$ to 0.2, at least a factor of ten times that of luminous matter. This discrepancy is attributed to the elusive dark matter that may lie beyond the luminous boundaries of galaxies; it may comprise more than 90 percent of the mass of the Universe. Still, the observations indicate a low-density Universe of $\Omega \approx 0.2$ or 0.3, that falls well short of the amount required to balance the Universe at $\Omega_o = 1.0$. Indeed, we would need a mass-to-light ratio, M/L, in solar units of:

$$M/L \approx 1000(M_\odot/L_\odot) \text{ for } \Omega_o = 1 \ , \tag{5.432}$$

which is about five times greater than that inferred from the dynamical effects of the most massive collections of galaxies.

In subsequent discussion, we will distinguish between baryonic and nonbaryonic matter. The term baryon refers to heavy particles, such as neutrons or protons, that experience the strong force that holds the atomic nucleus together. In cosmological terms, baryonic matter has come to mean ordinary matter consisting of the familiar neutrons and protons, as well as the lighter electrons. Hypothesized, nonbaryonic forms of matter never interact with anything through the strong force and are completely unlike anything we have yet experienced.

Light elements, concocted in the pressure cooker of the big bang, place limits on the amount of baryons present in the Universe in both visible and invisible form. When primordial nucleosynthesis computations are combined with the observed abundances of the light elements, limits are obtained for the baryon density parameter, Ω_B, of (see Sect. 5.7.4):

$$0.09h^{-2} \leq \Omega_B \leq 0.25h^{-2} \ , \tag{5.433}$$

for a Hubble constant $H_o = 100h$ km s^{-1} Mpc^{-1} and $0.5 \leq h \leq 1.0$. The upper and lower limits come from the upper bounds on the abundances of ^{4}He and ^{2}D + ^{3}He, respectively.

Since we know that visible matter has a density factor of $\Omega_{LUM} \leq 0.003$, primordial nucleosynthesis indicates that the bulk of baryons in the Universe has to be dark for any plausible value of the Hubble constant. The allowed range in baryon density is also in agreement with the density implied from the dynamics of spiral galaxies, including all the dark matter in their halos at $\Omega_H \approx 0.10$, provided that the Hubble constant, H_o, is at the low end of the permissible range $h \approx 0.5$. Higher values of the Hubble constant, or larger density factors such as a galaxy cluster value of $\Omega_C = 0.2$, require nonbaryonic matter.

Since the baryon density factor is well below unity for any plausible value of the Hubble constant, or for $h \leq 1$, the Universe cannot be closed with baryonic matter, a conclusion suggested long ago by Reeves, Audouze, Fowler and Schramm (1973). If the Universe is truly at its critical density, in equilibrium between expansion and closure with $\Omega_o = 1$, then substantial amounts of nonbaryonic matter are required, but it has yet to be found.

More than two decades ago, Gott et al. (1974) summarized the available evidence bearing on the density parameter or omega factor, Ω_o, concluding that the dynamical evidence favored a value of 0.1 or 0.2, and noted that if matter were all baryonic, the lower end of this range was compatible with the value favored by primordial nucleosynthesis. From the astronomer's perspective, the situation has barely changed since then.

Rubin (1993) reviews the dynamical evidence for spiral galaxies, showing that it is consistent with a low-density Universe of $\Omega_o = 0.1$, in which the required dark matter could be solely baryonic. Coles and Ellis (1994) argue that a low-density Universe is favored by many observations, and that there is as yet no convincing evidence that the Universe has a density near the critical value required for closure, reaffirming a similar argument by Peebles (1986). The baryon content of galaxy clusters is also apparently less than that required for closure, in conflict with $\Omega_0 = 1$ (White et al., 1993; White and Fabian, 1995). Bahcall (1993) and Bahcall, Lubin and Dorman (1995) reason that we live in a low-density Universe with $\Omega_0 = 0.2$ to 0.3, and that there is much less dark matter than that required for closure. Ostriker and Steinhardt (1995) also favor a low-density Universe but with a non-zero cosmological constant.

Some theoretical physicists fervently believe that nonbaryonic matter fills the Universe to the brink of closure, with a density factor of unity or $\Omega_0 = 1$. The Universe would then have to be much denser than it appears, and it would need to be dominated by unseen dark matter of indeterminate nature. Why should some theoreticians so adamantly wish this to be the case, despite all of the contradictory evidence? High-energy particle physicists are no doubt pleased to find cosmological justifications for their speculations. There are also some perfectly good rational arguments in favor of this theoretical prejudice; they have been given by Turner (1991, 1993), including:

1. Observations indicate that the density parameter may be about $\Omega_0 = 0.2$, within a factor of five of closure. Since the deviation of Ω from unity grows as a power of the scale factor or size of the Universe, this can only be achieved if Ω fell within 1 part in 10^{14} of unity during the epoch of primordial nucleosynthesis (Dicke and Peebles, 1979). So, one wonders why our present time is so special in this regard, and why Ω is only now beginning to differ from unity.
2. An inflation theory hypothesizes an early period of inordinate expansion that irons the curvature of space away, whatever it might have been, resulting in a flat Einstein – De Sitter Universe with $\Omega = 1$. Such an inflationary era can explain the horizon problem, or why some widely-separated regions now have about the same temperature and density (see Sect. 5.7.2). However, this speculation involves the same kind of particle physics that yields a huge cosmological constant that conflicts with observations (see Sect. 5.7.1).
3. Galaxy formation is less difficult to explain, but not necessarily solved, with a density factor of unity since it provides more time for the growth of density perturbations. Smaller perturbations are required with $\Omega_0 = 1$ than with $\Omega_0 = 0.2$. Perturbations of abundant, hypothetical nonbaryonic dark particles can begin growing earlier than those containing ordinary baryonic matter, perhaps aiding the growth of galaxies and their clusters (see Sects. 5.7.6.4 and 5.7.6.5).

There is little doubt that the Universe contains far more invisible matter than either the matter that shines as stars or the interstellar and intercluster gas that respectively emits radio waves and X-rays. The main unresolved questions are how much dark matter is around, and what it is composed of. The models can be simplified into two extremes, one with a density factor of $\Omega_0 = 0.2$, derived from dynamical observations of visible matter, and one with $\Omega_0 = 1.0$ which is a theoretical speculation. If we live in a low-density Universe with

$\Omega_o = 0.2$, then most of the dark stuff could be similar to the ordinary baryonic matter we are familiar with, made of protons, neutrons and electrons. If $\Omega_o = 1.0$, some new, hypothetical nonbaryonic particles might dominate the Universe, most-likely uncharged relics of the big-bang. Whatever it is, dark matter must clump about spiral galaxies to form their halos, and be less concentrated toward the galactic centers than is the light. By definition, the elusive, unseen substance emits neither visible light nor any other form of electromagnetic radiation that we can detect, such as radio waves or X-rays, and it does not appreciably obscure sight of the visible Universe of galaxies.

Some astronomers suspect that all the dark stuff consists of ordinary matter, but in a form difficult to detect. Indeed, we know that some dark baryonic matter must exist, since the lower bound for the baryonic density factor, Ω_B, is greater than the upper limit to Ω_{LUM}, the amount of luminous matter in the Universe. Possible hiding places include a variety of low luminosity objects, such as failed stars (brown dwarfs), dead, collapsed stars (white dwarfs, neutron stars or black holes), or extremely dim galaxies, as well as intergalactic gas that does not shine by itself but does absorb the radiation of distant quasars. Such baryonic candidates have been reviewed by Carr (1990, 1994, 1996).

A brown dwarf has a low mass, M, of less than 0.065 solar masses, or $M \leq 0.065 M_{\odot}$. It can be up to 65 times more massive than Jupiter, whose mass is about 0.001 solar masses, but still below the mass required to ignite and sustain the nuclear hydrogen burning in its core. So, a brown dwarf cannot shine like a normal star. At birth, it might have a surface temperature of about 6,000 K, largely due to gravitational heat built up in its formation by collapse from an interstellar cloud, and it would then cool down like the embers of a dying fire. Some brown dwarfs have been detected by their low luminosities and temperatures, and the spectroscopic signature of lithium that is readily consumed by nuclear fusion in true stars (Rebolo et al., 1996; Thakrah, Jones and Hawkins, 1997).

If brown dwarfs are out in the halo of our Galaxy in appreciable numbers, then their collective presence might contribute significantly to its dark matter; but no one really knows how many are out there. Thousands of nearby, burned-out white dwarf remnants of normal intermediate mass stars have been observed in our Milky Way (Lang, 1992), so there might also be significant numbers of white dwarfs in the halo. Both brown and white dwarfs can also be observed indirectly by their gravitational effect on a stars light.

Dark-matter objects in our halo can be detected when they pass in front of a distant star. The starlight is briefly magnified by gravitational bending, like the lens of a magnifying glass. Gravitational lenses were suggested by Einstein (1936), and have been observed as arcs in galaxy clusters and as multiple quasar images (see Sect. 5.5.4). Refsdal (1964, 1966, 1970) has stressed the cosmological aspects of gravitational lenses, and Press and Gunn (1973) suggested that they might be used to detect a critical density of dark matter. Paczynski (1986, 1996) then suggested that dark bodies in the halo of our Galaxy could be detected when they act as gravitational microlenses, amplifying the light from stars in nearby galaxies and causing them to grow brighter for a few hours or days depending on the mass.

The probability of catching a glimpse of an unseen halo object when it passes directly between ourselves and an individual distant star is exceedingly low, about one in a million. Nevertheless, such rare gravitational microlensing events have been detected by monitoring about 10 million stars for at least a year (Alcock, 1997; Paczynski, 1996), in the direction of either the Large Magellanic Cloud (Alcock et al., 1993, 1995, 1996; Aubourg et al., 1993, 1995) or the bulge of our Galaxy (Udalski et al., 1993; Alcock et al., 1995). The dark micolensing objects are referred to generically as Massive Compact Halo Objects, or Machos for short. One international collaboration has adopted MACHO for its logo (Alcock et al., 1995), while a French one uses EROS for Experience de Recherche d'Objets Sombres (Aubourg et al., 1993).

As a Macho passes near the line of sight to a background star, the star appears to be magnified by a factor of (Refsdal, 1964, Alcock et al., 1996)

$$A = (u^2 + 2)/[u(u^2 + 4)^{1/2}] \ , \tag{5.434}$$

where $u = b/R_{\mathrm{E}}$ is the undeflected impact parameter of the light ray in units of the "Einstein radius", R_{E}, and b is the separation of the lens from the observer's line of sight to the star. In the case of perfect alignment, the star will appear as an "Einstein ring" with a radius in the lens plane of:

$$R_{\mathrm{E}} = [4GMDx(1 - x)/c^2]^{1/2} \ , \tag{5.435}$$

or

$$R_{\mathrm{E}} = 2.5\,[(M/M_{\odot})(Dx(1 - x)/1\mathrm{kpc}]^{1/2} AU \ ,$$

where M is the lens mass, D is the observer-star distance, Dx is the observer-lens distance, AU denotes the astronomical unit, $M_{\odot}$ is the Sun's mass, and the distance of the Large Magellanic Cloud is $\mathrm{D} \approx 55$ kpc. The time, t, for the halo lensing object to move through twice the Einstein radius at a velocity V perpendicular to the line of sight is (Alcock, et al., 1996):

$$t = 2R_{\mathrm{E}}/V \approx 130(M/M_{\odot})^{1/2} \text{ days} \ . \tag{5.436}$$

The experiments are sensitive to a broad range of mass between 10^{-7} and 10 solar masses, including invisible brown dwarfs, white dwarfs, neutron stars and black holes. Since the Machos are detected solely by their gravitational effects, one cannot tell exactly what they are, but models suggest that the white dwarfs account for less than 10 percent of the halo of our Galaxy (Charlot and Silk, 1995). Low-mass halo objects such as brown dwarfs or other objects of planetary mass, produce rapid microlensing and the tightest constraints are available for them. If the halo consisted entirely of objects with masses under about 10^{-4} solar masses, the average duration of microlensing would be less than 1.5 days, and the events would last only about 3 hours if the halo were made of $10^{-6} M_{\odot}$ objects. Analysis of two years of data toward the Large Magellanic Cloud indicate that the halo dark matter cannot be dominated by, or made entirely of, objects with low masses, M, in the range $2.5 \times 10^{-7} M_{\odot} < M < 8.1 \times 10^{-2} M_{\odot}$ (Alcock et al., 1996). The EROS and MACHO combined upper limits on the amount of planetary-mass dark matter in the Galactic halo indicate that less than 10% is made of Machos in the range 3.5×10^{-7} to 4.5×10^{-5} solar masses (Alcock et al., 1998), and the dozen microlensing events detected so far do not require a significant halo Macho population (Gates et al., 1998).

Another component of baryonic dark matter is the hydrogen and helium gas that is spread thinly throughout the seemingly empty space between the galaxies. As the light of remote quasars shines through a concentration of neutral, or unionized, intergalactic hydrogen at redshift z, a discrete Lyman α absorption line is produced at a wavelength of 1215.67 $(1+z)$Å. This effect, predicted by Bahcall and Salpeter (1965), produces a thicket of absorption lines, dubbed the Lyman α forest, due to numerous clumps of neutral hydrogen distributed along the line of sight to distant quasars (Bahcall, Greenstein and Sargent, 1968; Lynds, 1971; Sargent, Young, Boksenberg and Tytler, 1980; Weymann, Carswell and Smith, 1981; Blades, Turnshek and Norman, 1988; Lanzetta et al., 1991).

The Lyman α forest has been attributed to neutral hydrogen, or HI, clumped into galaxy-sized clouds, perhaps representing the progenitors of today's galaxies. In at least one case, radio astronomers have also observed an immense cloud of intergalactic neutral hydrogen, with a mass of $M \approx 10^{14}$ solar masses and redshift $z = 3.4$, from its redshifted 21-cm emission (Uson, Bagri and Cornwell, 1991).

As you look deeper into space, and go farther into the past, more and more of the Universe appears to be made up of gas rather than stars and galaxies. In a Friedmann Universe, the number density of Lyman α absorbers is expected to vary with redshift, z, according to the expression:

$$N(z) = N_o(1+z)(1+2q_o z)^{-1/2} , \tag{5.437}$$

for a deceleration parameter q_o, and

$$N(z) \propto (1+z)^{\gamma} , \tag{5.438}$$

where $\gamma = 1$ for $q_o = 0$ and $\gamma = 0.5$ for $q_o = 0.5$.

The density parameter, Ω_{HI}, due to neutral intergalactic hydrogen, HI, is roughly comparable to the density parameter of luminous matter in all the stars of nearby, present-day spiral galaxies, with (Sargent et al., 1980; Lanzetta et al., 1991):

$$\Omega_{HI} \approx 0.001\, h^{-1} , \tag{5.439}$$

for a Hubble constant of $H_o = 100h$ km s^{-1} Mpc^{-1}. The mass density of intergalactic neutral hydrogen gas is also comparable to that of the hot, ionized gas within clusters of galaxies, detected by its X-ray emission (Sarazin, 1986, 1988).

In fact, the intergalactic gas outside of galaxy clusters is also mainly ionized. Although the ionized hydrogen is invisible, intergalactic ionized helium can be detected in quasar light as a continuous absorption trough on the short-wavelength side of singly ionized helium at 304 $(1+z)$Å for a quasar of redshift z (Miralda-Escudé and Ostriker, 1992; Miralda-Escudé, 1993). This is analogous to the Gunn–Peterson effect for neutral hydrogen at wavelengths below redshifted Lyman α (Gunn and Peterson, 1965), which has never been detected because of the low density of unionized, neutral intergalactic hydrogen. In contrast, the ionized helium has been detected at ultraviolet wavelengths using space-borne telescopes (Jakobsen et al., 1994; Tytler et al.,

1996; Davidsen, Kriss and Zheng, 1996). These observations suggest that the hydrogen and helium gas in the intergalactic medium of the early Universe has a baryonic density parameter, Ω_B, given by (Bi and Davidsen, 1997):

$$\Omega_B = 0.015h^{-2} \text{ to } 0.025h^{-2} \ . \tag{5.440}$$

where the Hubble constant $H_o = 100\,h$ km s^{-1} Mpc^{-1}. This model-dependent result assumes that the Lyman α forest is due to fluctuations in a continuous intergalactic medium rather than discrete clouds. It shows that the remote intergalactic gas has a density comparable to the baryon density predicted by primordial nucleosynthesis, and that many of the invisible baryons might be in intergalactic space.

The limits on ordinary baryonic matter do not rule out the existence of hypothetical, extraordinary nonbaryonic matter. If the Universe is truly at its critical density and brought to the brink of closure, then most of the matter in the Universe would have to be nonbaryonic. A number of particle physicists have speculated about the forms it may take, and the methods by which it may be detected. They hope the unseen material consists of swarms of exotic elementary particles, not yet seen in the terrestrial laboratory, that were created in the earliest stages of the Universe and managed to stay around. Such hypothetical particles are reviewed in the context of possible detection mechanisms by Primack, Seckel and Sadoulet (1988).

Turner (1991, 1993) assumes that the weakly interacting relic particles comprise the mass density needed for $\Omega_o = 1$, and provides a summary of the unseen particles that might make this so. Three nonbaryonic candidates that may have been produced in the early Universe and could have the required mass, lifetime and cross section are: a very light axion of mass 10^{-6} eV to 10^{-4} eV; a light neutrino of mass 20 eV to 90 eV, and a heavy neutralino of mass 10 GeV to about 3 TeV. Particle physicists measure mass in energy units; to get the mass in grams convert the electron volts to c.g.s. energy units, using 1 eV $= 1.602 \times 10^{-12}$ erg, and divide by the square of the velocity of light, c^2, where $c = 2.9979 \times 10^{10}$ cm s^{-1}. Sadoulet (1996) and Spergel (1997) has also reviewed the evidence for nonbaryonic dark matter and searches for its particles – axions, neutrinos and weakly interacting massive particles, dubbed WIMPS. All of these possibilities can be tested by experiments that are either being planned or are underway.

We already know quite a bit about the neutrino, so it is somewhat more tangible than the other dark-matter particle candidates. The subatomic neutrino has no electric charge, a very small mass, and travels at or very near the velocity of light. It interacts so weakly with matter, that it streaks through nearly everything in its path, like a ghost that moves through walls. Each second about 70 billion solar neutrinos, resulting from nuclear reactions at the core of the Sun, fly through every square centimeter of the Earth, and massive, subterranean detectors have managed to catch just a few of them (Volume I). Many, many more neutrinos were produced by nuclear reactions in the big bang. Right now there are about 100 relic neutrinos from the big bang in every cubic centimeter of space, but the terrestrial detectors are not sensitive enough to see them. So there are plenty of neutrinos around; the Universe contains about a billion times more neutrinos than protons or electrons.

The cosmological significance of neutrinos hinges upon their mass. Gershtein and Zeldovich (1966) placed upper limits to the neutrino mass by considering its possible gravitational effect on the expanding Universe. Ramanath Cowsik and his colleagues pointed out that a neutrino rest mass of about ten electron volts could be sufficient for neutrinos to close the expanding Universe, conjecturing that they could also provide the unseen mass in galactic halos and clusters (Cowsik and Mc Clelland, 1973; Coswik and Ghosh, 1987). Although evidence for a muon neutrino rest mass has been reported (Fukuda et al., 1998), it would make little cosmological difference, having a mass (energy) of about 0.05 eV that is much smaller than the 30 eV needed for closure of the Universe. Moreover, interest in such fast-moving, hot, dark-matter particles has waned since they may not significantly affect the formation of the observed large-scale structure of the Universe (see Sect. 5.7.6.5).

Attention shifted toward the cold dark-matter particles that move very slowly – neutralinos and wimps because they are heavy and axions because they were born with very low momentum. Unfortunately, aside from theoretical conjecture, we know practically nothing about these candidates. They have never been detected on Earth, and they might be figments of our imagination.

The nonbaryonic, left-over particles of the big bang have to be very hard to detect. They don't emit any type of electromagnetic radiation very well, if at all. That is why they are dark, and since we can't see them we don't really know what they are or where they are located. And they hardly interact with ordinary matter, or they would have sunk into visible galaxies along with the ordinary gas and dust. The ongoing experiments are therefore very high-risk, with great importance if a relic dark-matter particle is discovered and with a high probability of detecting nothing.

If such nonbaryonic particles are discovered in substantial quantities, it will transform our view of the cosmos. Most of the Universe might then be made of something we know almost nothing about. The familiar natural world around us, our Earth and Sun, the stars and galaxies, and even our own bodies, may be minor constituents of the Universe. Their atomic ingredients are baryons, unlike the elusive nonbaryonic stuff. However, for the time being we have no observational clues about what the mysterious substance might be, and for all we know it might not even be there. Its sort of like the extraterrestrial life debate, fraught with intelligent speculation but essentially without solid observational evidence. Many astronomers therefore prefer a low-density Universe, perhaps composed primarily of familiar baryons.

5.5 Relativity and Its Consequences

5.5.1 Formulae of General Relativity

According to the general theory of relativity (Einstein, 1916, 1917, 1919) curvature in the geometry of space-time manifests itself as gravitation. Gravitation works on the separation of nearby world particle lines. In turn, particles and other sources of mass energy cause curvature in the geometry of space-time. Thus, gravity manifests itself in the geometry of space-time, which is described by the metric, ds, and is also sometimes called the space-time interval or the line element. In its most general and complex form, the metric, ds, is given by

$$ds^2 = \mathbf{g}_{ik} dx^i \, dx^k \quad , \tag{5.441}$$

where dx^i is a differential of the coordinate x^i, a space coordinate for $i = 1, 2, 3$, and the time coordinate, ct, for $i = 0$, the velocity of light is c, summation is implied when an index appears twice, and the metric tensor, $\mathbf{g}_{\mathrm{ik}}$, is itself a function of the space-time coordinates.

Without gravity there is no curvature of space-time, and in local, freely falling frames, called Lorentz frames, the local line element, ds, becomes the Minkowski metric given by (Minkowski, 1908; Hargreaves, 1908):

$$ds^2 = c^2 \, dt^2 - dx^2 - dy^2 - dz^2 \quad . \tag{5.442}$$

This metric applies to the special theory of relativity, whose tests are delineated in the next Sect. 5.5.2.

The line element, ds, for a non-rotating, mass in a vacuum is the Schwarzschild metric given by (Schwarzschild, 1916):

$$ds^2 = \left[1 - \frac{2GM}{c^2 r}\right] c^2 dt^2 - \frac{dr^2}{\left[1 - \frac{2GM}{c^2 r}\right]} - r^2 \, d\theta^2 - r^2 \sin^2 \theta \, d\varphi^2 \quad . \tag{5.443}$$

Here, r, θ, and φ are spherical coordinates whose origin is at the center of the massive object, and M is the mass which determines the Newtonian gravitational field GM/r, where G is the Newtonian gravitational constant. This metric is given in isotropic and post-Newtonian form in Sect. 5.5.2, where tests in the weak-field limit are described in detail. It is also extended to the rotating case, or Kerr metric, and applied to black holes in Sect. 5.6.2.

The line element, ds, for a homogeneous, isotropic expanding Universe, called the Robertson–Walker metric, is given by (Lanczos, 1922, 1923; Friedmann, 1922, 1924; Robertson 1935, 1936; and Walker, 1936):

$$ds^2 = c^2 \, dt^2 - R^2(t) \left[\frac{dr^2}{1 - kr^2} + r^2 \left(d\theta^2 + \sin^2 \theta \, d\varphi^2 \right) \right] \quad , \tag{5.444}$$

where the curvature index $k = -1, 0$ and $+1$, and $R(t)$ is called the radius of curvature or the scale factor of the Universe. The r coordinate has zero value for some arbitrary fundamental observer, the surface $r = \text{constant}$ has the geometry of the surface of a sphere, and θ, φ are polar coordinates. This metric is given in Sect. 5.7.1 and applied to classical tests of cosmological models in Sect. 5.7.4. It is used to provide definitions of cosmological distance in Sect. 5.2.9 and the age of the Universe in Sect. 5.3.12.

In the more general situation, the production of curvature by mass-energy is specified by Einstein's field equation (Einstein, 1916, 1919)

$$\mathbf{R}_{ik} - \frac{1}{2} \mathbf{R} \mathbf{g}_{ik} - \Lambda \mathbf{g}_{ik} = -\frac{\kappa}{c^2} \mathrm{T}_{ik} \quad , \tag{5.445}$$

where the cosmological constant, Λ, is often taken to be zero, $\mathbf{T}_{ik}$ is the energy-momentum tensor, Einstein's gravitational constant, $\kappa = 8\pi G/c^2 \approx 1.86 \times 10^{-27}$ dyn g^2 sec^2, where G is the Newtonian constant of gravitation and c is the velocity of light. Sometimes $\kappa_o = \kappa/c^2 \approx 2.07 \times 10^{-48} \mathrm{cm}^{-1} \mathrm{g}^{-1}$ sec^2 is also called the Einstein gravitational constant. The Robertson–Walker metric is, for example, derived from these equations by Islam (1992).

The Ricci tensor, $\mathbf{R}_{km}$, is given by

$$\mathbf{R}_{km} = \mathbf{R}^{i}_{klm}\mathbf{g}^{l}_{i} = \mathbf{R}^{i}_{kim} \quad , \tag{5.446}$$

where $\mathbf{R}^{i}_{klm}$ is the Riemann curvature tensor, and the Ricci scalar curvature of space-time, $\mathbf{R}$, is given by the contraction of the Ricci tensor

$$\mathbf{R} = \mathbf{R}_{ik}\mathbf{g}^{ik} = \mathbf{R}^{i}_{i} \quad . \tag{5.447}$$

Einstein's field equations follow from the assumptions that the ratio of gravitational and inertial mass is a universal constant, that the laws of nature are expressed in the simplest possible set of equations that are covariant for all systems of space-time coordinates, and that the laws of special relativity hold locally in a coordinate system with a vanishing gravitational field.

The Riemann curvature tensor, $\mathbf{R}^{i}_{klm}$, is the agent by which curves in space-time produce the relative acceleration of geodesics.

$$\mathbf{R}^{i}_{klm} = \frac{\partial \Gamma^{i}_{km}}{\partial x^{l}} - \frac{\partial \Gamma^{i}_{kl}}{\partial x^{m}} + \Gamma^{i}_{nl}\Gamma^{n}_{km} - \Gamma^{i}_{nm}\Gamma^{n}_{kl} \quad . \tag{5.448}$$

Here the Christoffel symbols Γ^{l}_{mn} are defined by the relations

$$\Gamma^{\mathrm{i}}_{\mathrm{kl}} = \frac{\mathbf{g}^{im}}{2}\left(\frac{\partial \mathbf{g}_{mk}}{\partial \mathbf{x}^{l}} + \frac{\partial \mathbf{g}_{ml}}{\partial x^{k}} - \frac{\partial \mathbf{g}_{kl}}{\partial x^{m}}\right) \tag{5.449}$$

where

$$\Gamma^{i}_{ik} = \frac{1}{2}\frac{\partial}{\partial x^{k}}(\ln -\mathbf{g}) = \frac{1}{\mathbf{g}^{1/2}}\frac{\partial}{\partial x^{k}}\left(-\mathbf{g}^{1/2}\right) \quad , \tag{5.450}$$

and $\mathbf{g}$ is the determinant $|\mathbf{g}_{ik}|$.

The amount of mass energy in a unit volume is determined by the stress-energy, or energy-momentum, tensor, $\mathbf{T}_{ik}$. For a gas or perfect fluid

$$\mathbf{T}^{ik} = (\epsilon + P)\mathrm{u}^{i}\mathrm{u}^{k} - P\mathrm{g}^{ik} \quad , \tag{5.451}$$

where $\epsilon = \rho c^2$ is the mass-energy density of the matter as measured in its rest frame, ρ is the mass density, P is the pressure, and u^i is the four-velocity of the matter, given by:

$$u^{i} = dx^{i}/ds \quad , \tag{5.452}$$

where $x^{i}(s)$ describes the world-line of matter in terms of the proper time $\tau = sc^{-1}$ along the world-line.

In a gravitational field, a particle moves along a geodesic line which is specified by the differential equation

$$\frac{d^2x^{i}}{ds^2} + \Gamma^{i}_{kl}\frac{dx^{k}}{ds}\frac{dx^{l}}{ds} = 0 \quad . \tag{5.453}$$

Light rays are represented by null geodesics for which $ds = 0$. An observer's proper time, τ, is governed by the metric along his world line

$$d\tau = \sqrt{ds^2} = \sqrt{\mathbf{g}_{ik} dx^i dx^k} \ , \tag{5.454}$$

where the world line is described by any time parameter, t,

$$x^i = x^i(t) \ . \tag{5.455}$$

The equation for the conservation of mass-energy and momentum is, in differential form:

$$\nabla \cdot \mathbf{T} = 0 \ , \tag{5.456}$$

where $\nabla \cdot$ denotes the covariant divergence, in components,

$$\frac{1}{\sqrt{-\mathbf{g}}} \frac{\partial \sqrt{-\mathbf{g}}\, \mathbf{T}_i^k}{\partial x^k} - \frac{\mathbf{T}^{km}}{2} \frac{\partial \mathbf{g}_{km}}{\partial x^i} = 0 \ , \tag{5.457}$$

where the fundamental determinant $\mathbf{g} = |\mathbf{g}_{ik}|$.

For a perfect fluid in flat space, this equation in Cartesian coordinates is

$$\frac{\partial \mathrm{T}^{ik}}{\partial x^k} = \frac{\partial P}{\partial x_i} + \frac{\partial}{\partial x^k} \left[(\rho + P) u^i u^k \right] = 0 \ , \tag{5.458}$$

and by substituting $u^i = v^i u^{\mathrm{o}}$ it becomes Euler's equation

$$\frac{\partial v}{\partial t} + (v \cdot \nabla) v = -\frac{(1 - v^2)}{(\rho + P)} \left(\nabla P + v \frac{\partial P}{\partial t} \right) \ . \tag{5.459}$$

Using the Christoffel symbols, the equation expressing the conservation of energy-momentum becomes

$$\mathbf{T}^{ik}_{;m} = \frac{\partial \mathbf{T}^{ik}}{\partial x^m} + \Gamma^i_{lm} \mathbf{T}^{lk} + \Gamma^k_{lm} \mathbf{T}^{il} \tag{5.460}$$

or

$$\mathbf{T}^{ik}_{;i} = \frac{\partial \mathbf{T}^{ik}}{\partial x^i} + \Gamma^i_{il} \mathbf{T}^{lk} + \Gamma^k_{il} \mathbf{T}^{il} = \frac{1}{-\mathbf{g}^{1/2}} \frac{\partial}{\partial x^i} \left(-\mathbf{g}^{1/2} \mathbf{T}^{ik} \right) + \Gamma^k_{il} \mathbf{T}^{il} \tag{5.461}$$

where the ; denotes the covariant divergence. For a contravariant vector, $\mathbf{V}^i$, and a covariant vector, $\mathbf{V}_i$, the covariant derivatives are

$$\mathbf{V}^i_{;k} = \frac{\partial \mathbf{V}^i}{\partial x^k} + \Gamma^i_{lk} \mathbf{V}^l \ , \tag{5.462}$$

and

$$\mathbf{V}_{i;k} = \frac{\partial \mathbf{V}_i}{\partial x^k} - \Gamma^l_{ik} \mathbf{V}_l \ , \tag{5.463}$$

where for any vector

$$\mathbf{V}^i_{;i} = \frac{1}{-\mathbf{g}^{1/2}} \frac{\partial}{\partial x^i} \left(-\mathbf{g}^{1/2} \mathbf{V}^i \right) \ . \tag{5.464}$$

The stress-energy tensor for the electromagnetic field is given by

$$\mathbf{T}^{ik} = -\frac{\mathbf{g}_{lm}}{4\pi}\mathbf{F}^{il}\mathbf{F}^{km} + \frac{\mathbf{g}^{ik}}{16\pi}\mathbf{F}_{lm}\mathbf{F}^{lm} , \quad (5.465)$$

where $\mathbf{F}_{lm}$ is the electromagnetic field tensor. In this case, $\mathbf{T}^{ik}$ gives a tension $(E^2 + B^2)/8\pi$ along the field lines, and a pressure $(E^2 + B^2)/8\pi$ perpendicular to the field lines, where $(E^2 + B^2)/8\pi$ is the energy density of the electromagnetic field.

5.5.2 Tests of the Special and General Theories of Relativity

The theory of relativity shows how our perception of physical phenomena depends upon our frame of reference and relative motion. The Special Theory of Relativity describes the laws of physics using inertial reference frames that are at rest or moving at constant velocity, and specifically excludes any consideration of gravity. The more General Theory of Relativity includes gravitation and accelerated motions. Both theories rest upon solid experimental foundations, are derived from fundamental postulates, and predict observable effects that have been precisely tested. In the following, we will first discuss these aspects of the special theory and then review their status for the general one.

The Special Theory of Relativity resolved an issue raised by Albert Michelson and Edward Morley's interferometer tests for the relative motion of the Earth and the luminiferous ether (Michelson and Morley, 1887). It was once thought that light might be transmitted in space by the vibrations of a hypothetical, invisible substance, called the ether, which permeated all space, but the Michelson–Morley interferometer failed to detect the Earth's motion through the ether. That is, they could measure no difference, Δc, in the speed of light, c, in two perpendicular paths of equal length, in the direction of the Earth's motion or transverse to it. Michelson and Morley found that light travels at the same velocity in all directions, with $\Delta c/c \leq 10^{-4}$, and that there is therefore no preferred inertial frame of reference for the relevant laws of physics.

The Newtonian notions of absolute space and time were abolished in Einstein's Special Theory of Relativity (Einstein, 1905), which rests upon the postulates:

1. The Principle of Relativity, which states that the laws of nature and the results of experiments performed in an inertial frame are independent of the uniform velocity of the system.
2. There exists in nature a limiting, invariant speed, the speed of light, c, a universal constant.

These postulates were anticipated by Poincaré (1904). They imply that no physical object can travel at a speed in excess of the velocity of light, $c = 2.99792458 \times 10^{10}$ cm s^{-1} in an inertial frame.

This meant that space and time are intertwined. As Minkowski (1908) expressed it:

> "The views of space and time which I wish to lay before you have sprung from the soil of experimental physics, and therein lies their strength. They are radical. Henceforth, space by itself, and time by itself, are doomed to fade away into mere shadows, and only a kind of union of the two will preserve an independent reality."

In the special theory, motions are described by the relative velocity, V, between two inertial frames of reference, and events are described by coordinates in both space (x, y, z) and time, t. The length of an object moving with the reference frame of an observer is called the proper length, and the time read on a clock in that frame is the proper time. Hendrik A. Lorentz derived the coordinate transformations from one inertial system to another, showing that Maxwell's equations are invariant when subjected to them (Lorentz, 1904). Such a Lorentz invariance was generalized by Einstein (1905) in the Principle of Relativity.

If an event has coordinates (x, y, z, t) as measured by an observer at rest in system K, and coordinates (x', y', z', t') as measured by an observer at rest in system K' that moves at velocity V relative to K along the x axis, then the Lorentz transformations are (Lorentz, 1904)

$$x = \frac{x' + Vt'}{\sqrt{1 - V^2/c^2}} = \gamma[x' + Vt'], \quad y = y', \quad z = z' , \tag{5.466}$$

where the x, y, z and x', y', z' axes are parallel, and

$$t = \frac{t' + V(x'/c^2)}{\sqrt{1 - V^2/c^2}} = \gamma[t' + (x'/c^2)V] , \tag{5.467}$$

where the Lorentz factor, γ, is given by

$$\gamma = \left[1 - \frac{V^2}{c^2}\right]^{-\frac{1}{2}} = [1 - \beta^2]^{-\frac{1}{2}} , \tag{5.468}$$

and $\beta = V/c$.

In local, freely falling frames, called local Lorentz frames, the nongravitational laws of physics are those of special relativity, and the local line element, ds, becomes the Minkowski metric given by (Minkowski, 1908; Hargreaves, 1908):

$$ds^2 = -c^2\, dt^2 + dx^2 + dy^2 + dz^2 \tag{5.469}$$

in rectangular coordinates,

$$ds^2 = -c^2\, dt^2 + dr^2 + r^2\, d\theta^2 + r^2 \sin^2\theta\, d\varphi^2 \tag{5.470}$$

in spherical coordinates and

$$ds^2 = -c^2\, dt^2 + dr^2 + r^2\, d\varphi^2 + dz^2 \tag{5.471}$$

in cylindrical coordinates, where in each case the coordinates are measured in an inertial frame of reference. In Minkowski space-time, ds is invariant under the Lorentz transformation, geometry is Euclidean, and space is flat.

If proper time, t, and time interval, Δt, between two events at one location, are measured in system K, then the time interval, $\Delta t'$, between the events as measured in system K$'$ moving with velocity V is

$$\Delta t' = \frac{\Delta t}{\sqrt{1 - V^2/c^2}} = \gamma \Delta \mathrm{t} . \tag{5.472}$$

This means that to the observer in system K', the moving clock appears to go slower. This is called time dilation. Such a time dilation can prolong the decay time of fast-moving unstable cosmic ray particles by several orders of magnitude, and noticeably lengthen the lifetime of elementary particles produced in man-made particle accelerators (Ayres et al., 1971).

Special relativity also predicts a relativistic Doppler shift in which a moving clock has its frequency, ν, modified by a factor $\left[1 - V^2/c^2\right]^{1/2} = 1/\gamma$ over the classical shift (see Volume I), where V is the velocity of the clock with respect to the observer. The presence of this time-dilation factor gives a redshift of order V^2/c^2 in the wavelength even if there is no first order term due to radial motion; such an effect is called the transverse Doppler shift.

If a detector moves at velocity, V, relative to a source emitting radiation at frequency, ν, then the detected frequency, ν', is given by (also see Volume I)

$$\nu' = \frac{\nu}{\gamma}(1 + \beta \cos \theta')^{-1} , \tag{5.473}$$

where $\beta = V/c$, the radiation is directed at angle θ' with respect to the X' axis, and the radiation is traveling opposite to the viewing direction. Thus, there is a relativistic frequency decrease, or time dilation, of ν/γ in addition to the classical Doppler shift to lower frequencies or longer wavelengths caused by radial motion, and a transverse Doppler effect is possible.

For proper spatial separation or length, Δx , in system K, there is a Lorentz contraction or shortening, $\Delta x'$, in the moving K' system given by

$$\Delta x' = \Delta x \sqrt{1 - \frac{V^2}{c^2}} = \frac{\Delta x}{\gamma} . \tag{5.474}$$

Thus, both space and time are relative in the Special Theory of Relativity.

If m_o denotes a particle's mass measured at rest, the moving particle's energy E', is given by

$$E' = \frac{m_o c^2}{\sqrt{1 - \frac{V^2}{c^2}}} = \gamma m_o c^2 . \tag{5.475}$$

This expression is verified in high-energy particle experiments that demonstrate that the energy of a particle such as an electron increases with its speed. It also shows that an infinite amount of work would be required to accelerate a particle to the velocity of light with $V = c$, implying that nothing can move faster than the speed of light in an inertial frame. Blokhintsev (1966) discussed the basis for special relativity provided by experiments in high energy physics.

Although Einstein's 1905 Special Theory of Relativity explained the Michelson–Morley 1887 experiment, it wasn't until the 1930s that two important confirming experiments were performed. Kennedy and Thorndike (1932) used an interferometer of unequal arm lengths to show that the speed of light in a moving system is independent of the velocity of the system. The transverse Doppler effect, or time dilation, was verified by Ives and Stillwell (1938, 1941). They measured the frequency of light emitted by fast-moving hydrogen ions along and opposite to the direction of the ion beam, showing

The status of experimental tests of the General Theory of Relativity was discussed by Will (1986), reviewed by Will (1990) on the occasion of its 75th anniversary, and additionally discussed in detail by Will (1993). An overview of the current limits to the relevant parameters is given in Table 5.16, indicating that Einstein's general theory has been confirmed with a precision of up to one part in a thousand. We will next discuss the predictions and observations of the gravitational redshift, the gravitational deflection of light, the gravitational time delay, the perihelion advance of Mercury, the gravitational precession of a

Table 5.16. Limits to post-Newtonian parameters that test metric theories of gravitation*

	What it measures relative to General Relativity (GR)	Value in GR	Experiment	Value or limit	Remarks
γ	How much space curvature is produced by unit rest mass?	1	Time delay	1.000 ± 0.002	Viking ranging
			Light deflection	1.0002 ± 0.0020	VLBI
β	How much "nonlinearity" is there in the superposition law for gravity?	1	Perihelion shift	1.000 ± 0.003	$J_2 = 10^{-7}$ assumed
			Nordtvedt effect	1.000 ± 0.001	$\eta = 4\beta - \gamma - 3$ assumed
ξ	Are there preferred-location effects?	0	Earth tides	$< 10^{-3}$	Gravimeter data
α_1	Are there preferred-frame effects?	0	Orbital preferred-frame effects	$< 4 \times 10^{-4}$	Combined solar system data
α_2		0	Earth tides	$< 4 \times 10^{-4}$	Gravimeter data
		0	Solar spin precession	$< 4 \times 10^{-7}$	Assumes alignment of solar equator and ecliptic are not coincidental
α_3		0	Perihelion shift	$< 2 \times 10^{-7}$	
			Acceleration of pulsars	$< 2 \times 10^{-10}$	Statistics of dP/dt
η**	Is WEP violated for self-gravitating bodies?	0	Nordtvedt effect	0.0005 ± 0.0011	Lunar laser ranging
ζ_1	Is there violation of conservation of total momentum?	0			
ζ_2		0	Self-acceleration	$< 4 \times 10^{-5}$	Binary pulsar
ζ_3		0	Newton's 3rd law	$< 10^{-8}$	Lunar acceleration
ζ_4		0			

* Adapted from Will (1990, 1993) and subsequent references in this text.
** The parameter $\eta = 4\beta - \gamma - 3$, and WEP stands for the Weak Equivalence Principle.

gyroscope within a satellite orbiting the Earth, the gravitational precession of the Earth–Moon orbit, the strong equivalence principle, and the binary pulsar.

If mass and energy are equivalent (Einstein, 1905), then radiation will have an effective mass and lose energy in overcoming gravity. Thus, the wavelengths of spectral lines will be shifted to the red when escaping a star or in moving to greater altitudes on Earth. The received wavelength at infinity, λ, of a spectral line emitted at a distance, R, from the center of spherical body of mass, M, is given by the equations

$$\lambda = \lambda_o + \Delta\lambda = \lambda_o \left[1 - \frac{2GM}{c^2 R}\right]^{-1/2} \tag{5.484}$$

and

$$z = \frac{\Delta\lambda}{\lambda_o} \approx \frac{GM}{c^2 R} \approx 1.477 \times 10^5 \left(\frac{M}{M_\odot}\right) R^{-1} \quad \text{if } GM < Rc^2 \ , \tag{5.485}$$

where z is the gravitational redshift, the numerical coefficient applies if the radius R is in centimeters, the solar mass $M_\odot = 1.989 \times 10^{33}$ grams, and λ_o is the wavelength emitted by the atom when the gravitational field is negligible. For light originating from the solar limb, $\Delta\lambda/\lambda_o = 2.17 \times 10^{-6}$, so an alternative expression is:

$$z = \Delta\lambda/\lambda_o \approx 2.17 \times 10^{-6} (M/M_\odot)(R_\odot/R) \ . \tag{5.486}$$

The gravitational redshift depends only on α which must be 1.0 in order for planetary orbits to agree with even the Newtonian theory. The redshift can be derived using the Newtonian theory of gravitation, assuming that Einstein's Equivalence Principle and the law of conservation of energy hold and that light with frequency ν has energy $h\nu$, where h is Planck's constant.

In considering the influence of gravitation on the propagation of light, Einstein (1907) derived the gravitational redshift, and subsequently viewed it as a test of the General Theory of Relativity (Einstein, 1916). According to Einstein, the rate of a clock will depend on the gravitational field or the curvature of space-time; the frequency of an atomic transition, and hence the wavelength of a spectral line, will appear changed if the observer has a different gravitational field than the atomic source. However, we now regard gravitational redshift experiments as tests for the existence of curved space-time, rather than a definitive test of general relativity. Such a redshift is consistent with any metric theory of gravity. Nevertheless, the gravitational redshift experiments are important tests of the Einstein Equivalence Principle, or Local Position Invariance.

Early attempts to measure the gravitational redshift of the Sun were inconclusive (St. John, 1917, 1928), primarily because thermal broadening of the solar lines is much greater than the gravitational redshift. Subsequent measurements of the solar redshift have been reported by Blamont and Roddier (1961), Brault (1963), Forbes (1970), and Snider (1972, 1974). The most precise measurement of the Sun's gravitational redshift is 99 ± 2 percent of that predicted by Einstein (Lo Presto, Schrader and Pierce, 1991).

White dwarf stars have masses roughly equal to that of the Sun, but radii comparable to the Earth's radius (Lang, 1992). The gravitational redshift for white dwarf stars should therefore be about 100 times larger than that of the Sun, with a solar radius of 6.96×10^{10} cm and an Earth radius of 6.37×10^{8} cm. Measurements of this effect are summarized by Lang (1992), Will (1993) and Bergeron, Liebert and Fulbright (1995). Early inconclusive attempts by Adams (1925) and Popper (1954) have, for example, been improved by Greenstein, Oke and Shipman (1971), Greenstein and Trimble (1972), Wegner (1979, 1980), Stauffer (1987) and Shipman, Provencal, Høg and Thejll (1997) – also see Volume I.

The most accurate tests of the gravitational redshift are made in the Earth's gravity, which is much weaker than that of the Sun or white dwarfs; the Earth mass is 5.97×10^{27} grams but the Sun's mass is 1.989×10^{33} grams. The greater terrestrial accuracy is possible because the wavelengths, or frequencies, of nuclear spectral lines and atomic clocks are much more precisely known than those of the spectral lines of any celestial object. The discovery of the narrow gamma ray spectral lines of nuclei (Mössbauer, 1958) led to the first detection of the terrestrial gravitational redshift. Using the Mössbauer effect, Pound and Rebka (1959, 1960) measured a value of $\Delta\lambda/\lambda_o = (2.57 \pm 0.26) \times 10^{-15}$ as opposed to the predicted value of 2.46×10^{-15}. The agreement between theory and experiment was subsequently improved to about one percent (Pound and Snider, 1964, 1965). Vessot et al. (1980) have compared the time of a hydrogen maser clock in a rocket with a similar clock on the ground, confirming the relativistic gravitational redshift at the 70×10^{-6} level.

When describing his new theory of gravitation, Einstein (1915) explained the anomalous advance in the perihelion of Mercury and predicted a bending or deflection of starlight by the Sun. The calculated deflection for a light ray grazing the Sun was 1.75 seconds of arc, or twice that previously obtained by Einstein (1911) using the Newtonian gravitational field and the weak principle of equivalence between gravitational and inertial mass. The smaller gravitational deflection of light, amounting to one-half the value predicted by general relativity, had also been inferred more than a century earlier using Newtonian theory with light treated as a corpuscle (Soldner, 1801), as well as by Henry Cavendish in unpublished work around 1784 (Will, 1988).

For the General Theory of Relativity, in which gravity curves space-time, a light ray passing a minimum distance, R_o, from the center of an object of mass, M, will be deflected from a straight line path by the angle (Einstein, 1911, 1915, 1916)

$$\varphi = 2(\alpha + \gamma)\frac{GM}{R_o c^2} \text{ radians} , \tag{5.487}$$

where G is the Newtonian gravitational constant. For the Schwarzschild metric, $\varphi = 4GM/(R_o c^2)$, which is twice that predicted from the Newtonian theory of gravitation. For the Sun, the light bending is

$$\varphi = 1.749\,(R_\odot/R)\,(1 + \gamma)/2 \text{ seconds of arc,} \tag{5.488}$$

where $R/R_\odot$ is the distance of closest approach of the ray to the Sun measured in units of the solar radius $R_\odot = 6.96 \times 10^{10}$ cm. In this expression, the first factor of 1/2 holds for any metric theory, and can be derived from the principle of equivalence and the corpuscular theory of light. The second $\gamma/2$ factor varies from theory to theory and measures the amount of curvature of space-time. At the limb of the Sun, $\varphi = 1.749$ arc seconds if $\alpha = \gamma = 1$.

A complete account of Einstein's three predictions – the gravitational redshift, Mercury's hitherto unexplained perihelion advance, and the deflection of light by the Sun – was provided by Eddington (1918), stimulating interest in a British expedition to measure the deflection of starlight by the Sun during the total solar eclipse of May 29, 1919. The eclipse measurement of a solar deflection of 1.98 ± 0.16 seconds of arc (Eddington, 1919; Dyson, Eddington and Davidson, 1920) brought Einstein international recognition.

Although the prediction of the bending of light by the Sun was a great success of general relativity, the eclipse measurements were not accurate enough to precisely test the theory. For example, in a review article Mikhailov (1959) concluded that the eclipse observations to date resulted in a weighted mean deflection at the solar limb of 1.93 ± 0.03 seconds of arc, about 10 percent over Einstein's prediction, but Merat, Pecker, Vigier and Yourgrau (1974) showed that the eclipse results have confirmed Einstein's prediction for the deflection of light by the Sun as a function of increasing distance from the Sun, at least within the uncertainties of the data. The difficulty is that inclement weather and terrestrial atmospheric distortions limit angular resolution to about 0.3 seconds of arc. Such limitations have persisted for optical eclipse results, leading, for example, to $(1 + \gamma)/2 = 0.95 \pm 0.11$ for the solar eclipse of June 30, 1973 (Jones, 1976).

In contrast to observations at optical or visible wavelengths, cosmic radio radiation is unaffected by stormy weather and angular resolutions as small as 0.0003 seconds of arc are possible using radio interferometers with transcontinental baselines. Moreover, the Sun is a relatively weak interferometric radio source, so the gravitational deflection can be observed whenever radio sources pass near the Sun, without a total solar eclipse. When close groups of radio sources are observed, one can measure their differential deflection relative to each other as they pass near the Sun, and simultaneous detection at different radio frequencies can be used to sort out small plasma effects in the solar corona and solar wind.

Since the initial suggestion by Shapiro (1967), the error in the measured radio deflection near the limb, or edge, of the Sun has been steadily reduced, from something like 10 percent to much less than 1 percent. Measurements of radio sources by Seielstad, Sramek, and Weiler (1970), Muhleman, Ekers and Fomalont (1970), Sramek (1971), Hill (1971) and Riley (1972) lead to respective values of $\varphi = 1.77 \pm 0.20$, $1.82^{+0.24}_{-0.17}$, 1.57 ± 0.08, 1.87 ± 0.3 and 1.82 ± 0.08 seconds of arc.

Fomalont and Sramek (1975) determined the relative bending of the microwave radiation from three radio sources as they passed near the Sun using a radio interferometer with a 35 kilometer baseline, obtaining

$$\varphi = 1.775 \pm 0.019 \text{ seconds of arc} \ , \tag{5.489}$$

where the uncertainty is the standard error. The mean gravitational deflection obtained by Fomalont and Sramek (1976) was 1.007 ± 0.009 (standard error) times the Einstein prediction of general relativity.

Even larger interferometric telescope separations are possible, over baselines of 10,000 kilometers that are comparable to the Earth in size. This Very Long Baseline Interferometry, or VLBI for short, is used for geodetic purposes, as well as for accurate radio gravitational deflections by the Sun (Robertson, 1991). Such bending is sometimes expressed by the solar deflection (Misner, Thorne and Wheeler, 1973)

$$\varphi = (1+\gamma)(M_{\mathrm{S}}/R_{\mathrm{E}})[(1+\cos\alpha)/(1-\cos\alpha)]^{1/2} , \quad (5.490)$$

for an object at an angle α from the center of the Sun, where $M_{\mathrm{S}} = GM_{\odot}/c^2 = 1.477 \times 10^5$ cm is the mass of the Sun in geometrized units and R_E is the distance from the Earth to the Sun in centimeters. An exact expression for the bending angle in a static, spherically symmetric space-time of arbitrarily strong curvature is derived by Cowling (1984), providing the apparent positional shift for a spacecraft close to superior conjunction with the Sun.

Early VLBI results (Counselman et al., 1974) led to a measurement of $\gamma = 0.98 \pm 0.06$ (standard error), while subsequent VLBI measurements of the radio bending by the Sun lead to precise estimates of (Robertson, Carter and Dillinger, 1991)

$$\gamma = 1.0002 \pm 0.0020 \,(\text{standard error}) , \quad (5.491)$$

and Lebach et al. (1995)

$$\gamma = 0.9996 \pm 0.0017 \,(\text{standard error}) . \quad (5.492)$$

So this prediction of general relativity has been confirmed to one part in a thousand, or to the third decimal place, by the VLBI results.

In 1964 Irwin Shapiro suggested another important test of the propagation of light through curved space-time, made possible through the development of modern radio technology. This test measures an excess propagation delay, now commonly called the Shapiro time delay, of radiation passing through a region of curved space near a massive body compared to the propagation time for the radiation passing far from the body. In other words, because of the presence of the gravitational field of a massive body, an electromagnetic signal will take a longer time to traverse a given distance than it would if the Newtonian theory were valid. The excess, round-trip time delay, $\Delta\tau$, is given by (Shapiro, 1964, 1966, 1967, 1968)

$$\Delta\tau = \frac{2GM}{c^3}(1+\gamma)\ln\left[\frac{r_{\mathrm{e}}+r_{\mathrm{p}}+R}{r_{\mathrm{e}}+r_{\mathrm{p}}-R}\right] , \quad (5.493)$$

where r_{e} and r_{p} are the respective radial distances of the Earth and the planet from the Sun, and R is the Earth-planet distance. This expression is sometimes given in terms of $r_{\odot} = 2GM_{\odot}/c^2 = 2.954 \times 10^5$ cm, the gravitational radius of the Sun, which is twice the Sun's mass in geometrized units. The coefficient of the natural logarithmic term then becomes $(2r_{\odot}/c)(1+\gamma)/2 = (4M_{\mathrm{S}}/c)$ $(1+\gamma)/2$.

The light deflection, φ, and the time delay, $\Delta\tau$, are two aspects of a single phenomenon, and both depend on the factor $(1+\gamma)$. One can obtain the deflection of light by appropriate differentiation of the time delay expression, essentially because rays passing closer to the deflector are slowed more.

The additional time delay is at a maximum when the planet is on the far side of the Sun from the Earth, at superior conjunction, when the excess delay is (Will, 1993)

$$\Delta\tau = (2r_{\odot}/c)\,[(1+\gamma)/2]\ln(4r_{\mathrm{e}}\,r_{\mathrm{p}}/d^2)\ , \tag{5.494}$$

or

$$\Delta\tau \approx [(1+\gamma)/2][240 - 20\ln(d/R_{\odot})^2(a/r_{\mathrm{p}})]\ \text{microseconds}\ , \tag{5.495}$$

where d is the distance of closest approach of the ray to the Sun, $R_{\odot}$ is the radius of the Sun, for a grazing ray $d = R_{\odot}$, and a is the astronomical unit. For example, from elongation to superior conjunction, $\Delta\tau$ increases from 2 to about 200 microseconds for Venus. Further discussion of the time delay is given by Muhleman and Reichley (1964) and Ross and Schiff (1966).

Like light bending, the time delay test places limits on the parameter γ that measures the amount of curvature of space-time. For example, Shapiro et al. (1971) measured the time delay for radar echoes from Mercury and Venus, obtaining

$$\gamma = 1.03 \pm 0.04\ . \tag{5.496}$$

By way of comparison, two-frequency radio range measurements generated during the solar conjunction of the Voyager spacecraft in 1985 resulted in (Krisher, Anderson and Taylor, 1991):

$$\gamma = 1.00 \pm 0.03\ . \tag{5.497}$$

In the meantime, in 1972 the Mariner 9 Mars orbiter and in 1977-78 the Viking Mars landers and orbiters were used as active retransmitters of radio signals. At superior conjunction, their radio signal suffers a delay given by (Will, 1990):

$$\Delta\tau \approx 250[\,(1+\gamma)/2\,][\,1 - 0.16\ \ln d]\ \text{microseconds}\ , \tag{5.498}$$

where d is the distance of closest approach of the ray to the Sun, in units of the solar radius. The two Viking landers were anchored on the solid surface of Mars, whose trajectory could be determined accurately, and dual-frequency transmission from the Viking orbiters permitted estimates of the time delay caused by the solar coronal plasma. The Viking measurements resulted in (Reasenberg et al., 1979):

$$\gamma = 1.000 \pm 0.002\ , \tag{5.499}$$

where the quoted uncertainty, twice the formal standard deviation, allows for possible systematic errors.

An anomalous precession of Mercury's perihelion provides another test of gravitational theories. Instead of always tracing out the same ellipse, the orbit of Mercury pivots around the focus occupied by the Sun. That is, the point of closest approach to the Sun, the perihelion of Mercury's orbit, advances in space at a rate of about 575 seconds of arc per century. The French celestial mechanician Urbain Jean Joseph Le Verrier showed that the observed rate of advance was about 38 seconds of arc per century greater than could be accounted for by the Newtonian gravitational pull of the planets (Leverrier,

time variation of G as the Universe expands and its mean density decreases. Lunar laser ranging provides experimental bounds on the time variation dG/dt in the strength of gravity with (Müller, Schneider, Soffel and Ruder, 1991):

$$(dG/dt)/G = (0.1 \pm 10.4) \times 10^{-12}\ \mathrm{yr}^{-1}\ , \tag{5.525}$$

which is comparable to that obtained from planetary radar ranging observations (Hellings et al., 1983)

$$(dG/dt)/G = (2 \pm 4) \times 10^{-12}\ \mathrm{yr}^{-1} \tag{5.526}$$

and the binary pulsar (Damour, Gibbons and Taylor, 1988; Damour and Taylor, 1991)

$$(dG/dt)/G = (11 \pm 11) \times 10^{-12}\ \mathrm{yr}^{-1}\ , \tag{5.527}$$

but see Nordtvedt (1990). These constraints are roughly a factor of ten smaller than the observed expansion rate of the Universe, which is about 10^{-10}.

The discovery of the pulsar PSR 1913+16, colloquially known as the binary pulsar, by Russell A. Hulse and Joseph H. Taylor (Hulse and Taylor, 1975; Hulse, 1994) provided a new laboratory for studying relativistic effects. The exceedingly stable pulsar clock, with a period, P, of only 59 milliseconds that is measured with an accuracy of 10^{-14} seconds, has made it possible to precisely monitor the dynamics of its orbital motion about an invisible companion star (Table 5.17). In addition to the Keplerian orbital elements, such as the orbital period, P_b, orbital eccentricity, e, and projected semimajor orbital axis $x = a_1 \sin i$, four "post-Keplerian" parameters have been established – the general relativistic periastron shift, $d\omega/dt$, the special relativistic time dilation parameter γ_p, the secular change of the orbital period, dP_b/dt, and the effect of the Shapiro time delay of pulses caused by their propagation through the gravitational field of the pulsar's companion (Taylor, Fowler and Mc Culloch, 1979; Taylor and Weisberg, 1989; Taylor, 1994).

Table 5.17. Orbital elements and relativistic parameters for the binary pulsar PSR 1913+16*

Pulsar period, P	59.029997929613(7) milliseconds
Rate of change of pulsar period, dP/dt	$8.62713(8) \times 10^{-18}$ s s^{-1}
Orbital period, P_b	27906.98089(12) seconds
Orbital eccentricity, e	0.6171314(10)
Orbital axis, $x/c = (a_1 \sin i)/c$	2.341754(9) seconds
Periastron shift, $d\omega/dt$	4.22659(4) degrees yr^{-1}
Time dilation parameter, γ_p	4.289(10) milliseconds
Rate of change of orbital period, dP_b/dt	$-2.427(26) \times 10^{-12}$ s s^{-1}
$s = \sin i$	0.734
Total mass**, $M = m_1 + m_2$	2.82837(4) solar masses
Pulsar mass**, m_1	1.442(3) solar masses
Companion mass**, m_2	1.386(3) solar masses

* After Taylor and Weisberg (1989); figures in parenthesis are uncertainties in the last digits quoted. The velocity of light $c = 2.99792458 \times 10^{10}$ cm s^{-1}.

** General relativity is assumed to be the correct theory of gravity in deriving the masses, which are given in units of the Sun's mass $M_\odot = 1.989 \times 10^{33}$ grams.

The binary pulsar PSR 1913+16 orbits once about its companion star every 7.752 hours. The orbit is very close, with a projected semimajor axis of $x = a_1 \sin i = 7.015 \times 10^{10}$ cm that is comparable to the size of the Sun (radius $R_\odot = 6.96 \times 10^{10}$ cm) and about 200 times less than the distance between the Earth and the Sun (one AU = 1.495×10^{13} cm). Furthermore, the system is relativistic with a pulsar orbit velocity of about one thousandth the speed of light.

The periastron shift is $d\omega/dt = 4.2266$ degrees per year, compared with Mercury's 43 seconds of arc per century. Although the pulsar's periastron shift cannot be used to test the General Theory of Relativity, because of the unknown total mass, M, one can infer a total mass of $M = 2.8284$ solar masses assuming that general relativity is the correct theory of gravity. The effects on pulse arrival times of the gravitational redshift, caused by the companion's gravitational field, and of the special relativistic time dilation caused by the pulsar's orbital motion, have been additionally used to infer separate masses of 1.44 solar masses for the pulsar and 1.39 solar masses for the companion (Taylor and Weisberg, 1989). Both stars are believed to be neutron stars with masses comparable to the Chandrasekhar limit of ≈ 1.4 solar masses compressed within a radius of about 10 kilometers. [High-precision timing of other pulsars in close binary systems have resulted in similar relativistic mass estimates, consistent with a pair of neutron stars with roughly the Sun's mass (Lyne and Bailes, 1990; Wolszczan, 1991; Ryba and Taylor, 1991), and the general relativistic Shapiro time delay has been additionally detected for another binary millisecond pulsar (Camilo, Foster and Wolszczan, 1994)].

The binary pulsar PSR 1913+16 contains strong gravitational fields, with surface fields of $GM/(c^2R) \approx 0.2$ as compared with $\approx 10^{-6}$ for the Sun. The post-Newtonian limit, used for the relatively weak gravitational field of the Sun, is therefore inadequate to discuss compact objects such as the binary pulsar. Damour and Taylor (1992) have provided a detailed account of a parameterized post-Keplerian formalism designed for strong-field tests of relativistic gravity using timing observations of binary pulsars. The equations for the five most significant parameters are (Taylor, 1994; also see Damour and Taylor, 1992, for more complex relations):

$$\begin{aligned}
\frac{d\omega}{dt} &= \dot{\omega} = 3\left[\frac{P_b}{2\pi}\right]^{-5/3} (T_o M)^{2/3}\left(1-e^2\right)^{-1} \;, \\
\gamma_p &= e\left[\frac{P_b}{2\pi}\right]^{1/3} T_o^{2/3} M^{-4/3} m_2 (m_1 + 2m_2) \;, \\
\frac{dP_b}{dt} &= \dot{P}_b = -\frac{192\pi}{5}\left[\frac{P_b}{2\pi}\right]^{-5/3}\left[1+\frac{73}{24}e^2+\frac{37}{96}e^4\right] \\
&\quad \times \left(1-e^2\right)^{-7/2} T_o^{5/3} m_1 m_2 M^{-1/3} \;, \\
r &= T_o m_2 \;, \\
s &= x\left[\frac{P_b}{2\pi}\right]^{-2/3} T_o^{-1/3} M^{2/3} m_2^{-1} \;.
\end{aligned} \tag{5.528}$$

where the total mass $M = m_1 + m_2$, the m_1 and m_2 respectively denote the masses of the pulsar and companion in units of the Sun's mass, and the parameter, x, is related to the mass function, f, by:

$$f_1(m_1, m_2, s) = \frac{(m_2 s)^3}{(m_1 + m_2)^2} = \frac{x^3}{T_o (P_b/2\pi)^2} \quad . \tag{5.529}$$

$$s = \sin i$$

and

$$T_o = GM_\odot/c^3 = 4.925490947 \times 10^{-6}\ \mathrm{s} \ , \tag{5.530}$$

where G is the Newtonian constant of gravitation, $M_\odot$ is the Sun's mass and c is the velocity of light.

Backer and Hellings (1986) use convenient units in the expressions:

$$\frac{d\omega}{dt} = \dot{\omega} = \left\{ \begin{matrix} 1.100351 \times 10^{-10}\ \mathrm{s}^{-1} \\ 0.198956\ \mathrm{deg\ yr}^{-1} \end{matrix} \right\} \left(\frac{P_b}{\mathrm{days}} \right)^{-5/3} \left(\frac{M}{M_\odot} \right)^{2/3} (1 - e^2)^{-1} \ , \tag{5.531}$$

and

$$\gamma_p = (0.006935498\ \mathrm{s}) \left(\frac{P_b}{\mathrm{days}} \right)^{1/3} \left(\frac{m_2}{M^{1/3}} \right) \left(1 + \frac{m_2}{M} \right) e \ . \tag{5.532}$$

According to Einstein's General Theory of Relativity, the orbital period of the binary pulsar should be gradually decreasing as the system loses energy in the form of gravitational waves. The rate of energy loss is given by a quadrupole formula, which leads to the previous expression for the orbital period change, dP_b/dt. Such an expression was derived by Peters and Mathews (1963). The quadrupole formula was applied by Wagoner (1975) to evaluate the evolution of the semimajor axis, orbital frequency and eccentricity of the binary pulsar (also see Esposito and Harrison, 1975 and Wagoner and Will, 1976). Will (1977) discussed gravitational radiation from binary systems in alternative metric theories of gravitation, and Iyer and Will (1993) described the gravitational radiation reaction in the equations of motion for binary systems to post-Newtonian order beyond the quadrupole approximation. Damour (1987) provides a review of the quadrupole approximation for the gravitational radiation reaction.

The measured inspiral, or decrease in the orbital period, of the binary pulsar provides strong indirect verification of gravitational radiation. Using the measured orbital elements and two masses determined for the binary pulsar, the General Theory of Relativity predicts that gravitational radiation will cause the orbital period to decrease at a rate $dP_b/dt = -2.403 \times 10^{-12}\ \mathrm{s\,s}^{-1}$. Observations confirm this quadrupole general relativistic prediction with an accuracy of 0.8 percent (Damour and Taylor, 1991; also see Table 5.17 and Taylor and Weisberg, 1989), indirectly verifying the existence of gravity waves. Although the gravitational radiation now emitted by PSR 1913+16 is far too weak to be directly observed, there are other cosmic objects whose gravity waves might be observed, directly verifying this prediction of general relativity (see the next Sect. 5.5.3).

5.5.3 Gravitational Radiation

It follows directly from Einstein's field equations that any nonspherical, dynamically changing system must produce gravitational waves. These waves create radiation-reaction forces in their source and these forces extract energy from that source at the same rate as the rate at which the gravitational waves carry off energy. The waves carry energy at the speed of light, c, so that the energy flux is the product of c and the energy density. As with electromagnetic waves, the amplitude falls off as the inverse of the distance, and the flux falls off as the inverse square of the distance travelled.

As shown in the previous Sect. 5.5.2, Einstein's General Theory of Relativity has a perfect record in describing how space-time is curved in the relatively weak gravitational field of the Sun, a comparatively small, slow-moving mass. It follows from his equations that a moving mass will also produce gravity waves, provided that the dynamics are nonspherical and that the source therefore changes shape.

The expected effects of gravitational radiation have been detected in the decreasing orbital period of the binary pulsar, which consists of a pair of compact neutron stars with strong gravitational fields but relatively slow motion. In this case, a varying quadrupole moment explained the gravitational radiation (also see Sect. 5.5.2). However, no instrument has directly observed gravity waves as they ripple through the fabric of space and time.

Gravitational-wave astronomy has been reviewed by Press and Thorne (1972) and Tyson and Giffard (1978). Thorne (1987) has reviewed all aspects of gravitational-wave research, including the physical and mathematical description, generation, propagation, astrophysical sources, and detection of gravitational waves. Laser-beam instruments that might lead to the detection of gravitational waves are reviewed by Abramovici et al. (1992). A possible source of detectable gravitational waves is the coalescence of binary neutron stars, that might also explain the enigmatic gamma ray bursts; these topics are presented in very readable form by Piran (1995). In this section we will describe the strain, energy flux and luminosity of gravity waves from possible cosmic sources, and then provide a description of detection sensitivity.

Gravitational radiation will be emitted most strongly by the coherent bulk motions of matter moving at nearly the speed of light within strong gravitational fields. The final infall and coalescence of binary neutron stars might, for example, produce gravity waves of sufficient strength to be detected at Earth. Other candidates for observation include massive gravitational collapse to form a black hole and the powerful stellar explosions called supernovae, with their associated core collapse to form a neutron star. All of these cataclysmic events might produce short-lived bursts of detectable gravitational waves, lasting just a few minutes or less. They could be superposed upon an all-pervasive continuum of primordial gravity waves, fossils from the first moments of the big bang.

Gravitational radiation from collapsing stars has been introduced by Zeldovich and Novikov (1964), Müller (1982) and Stark and Piran (1985). The use of gravitational-wave observations to test relativistic gravity theory was suggested by Eardley et al. (1973). Dyson (1963) considered the gravitational

Here the frequency $f = c/2\pi ƛ = c/\lambda$ for reduced wavelength ƛ, and the numbers correspond to a strongly emitting supernova in the Virgo cluster of galaxies where there are several supernovae per year.

For most astrophysical sources, the quadrupole formalism provides an order-of-magnitude estimate for the generation of gravitational waves. In this case,

$$h_{jk}^{\mathrm{TT}} = \frac{2}{D}\left[\ddot{I}_{JK}^{\mathrm{TT}}\right]_{\mathrm{ret}} , \tag{5.536}$$

where $\ddot{}$ denotes the second derivative with respect to time, D is the distance from the source's center, the proper time as measured at rest with respect to the observer is t, the ret denotes the retarded time $t - D$, in dimensionless units, and $I_{\mathrm{JK}}(t - D)$ is the source's mass quadrupole moment evaluated at the retarded time.

For rough calculations accurate to about a factor of ten, we can use

$$h \approx \frac{E_{\mathrm{kin}}^{\mathrm{ns}}}{D} \tag{5.537}$$

for the magnitude h of the gravity-wave field h_{jk}^{TT} at Earth, where $E_{\mathrm{kin}}^{\mathrm{ns}}$ is the portion of the sources internal kinetic energy associated with non-spherical motions, and dimensionless, geometrized units have been employed in which Newton's gravitational constant G and the speed of light, c, are set equal to unity. To convert to physical units, $E_{\mathrm{geom}} = GE_{\mathrm{phys}}/c^4 = 0.826 \times 10^{-49} E_{\mathrm{phys}}$ and length or distance remains unchanged.

From the quadrupole formula, for slow motion, weak-gravitational-field systems, one can compute the flux of energy, F, of gravitational radiation emitted in a given direction with unit vector $\mathbf{n}_j$ (Einstein 1916, 1918; Landau and Lifshitz, 1962; Peters, 1964; Zeldovich and Novikov, 1971; Press and Thorne, 1972)

$$F = \frac{1}{8\pi D^2}\frac{G}{c^5}\sum_{j,k}\left\langle\left(\dddot{\mathbf{I}}_{jk}^{\mathrm{TT}}\right)^2\right\rangle_{\mathrm{ret}} = \frac{c^3}{16\pi G}\left\langle\left[\dot{h}(t)\right]^2\right\rangle \text{ erg cm}^{-2}\text{ s}^{-1} , \tag{5.538}$$

where $\langle\rangle$ denotes an average over a few wavelengths, $\cdot$ denotes differentiation with respect to time, the third derivative with respect to time is denoted by $\cdots$, and the

$$\mathbf{I}_{jk}^{\mathrm{TT}} = \sum_{l,m}(\delta_{jl} - \mathbf{n}_j\mathbf{n}_l)\mathbf{I}_{lm}(\delta_{mk} - \mathbf{n}_m\mathbf{n}_k) , \tag{5.539}$$

where δ_{ij} is the Kronecker delta function, $\mathbf{n}_j$ is a unit vector in the j direction, and the reduced quadrupole moment tensor, $\mathbf{I}_{jk}$, is given by

$$\mathbf{I}_{jk} = \int \rho x_j x_k \, d^3x - \frac{1}{3}\delta_{ik}\int \rho D^2 \, d^3x , \tag{5.540}$$

where ρ is the mass density.

Slow-motion, weak-gravitational-field systems are those for which the radius, R, and internal velocity, v, satisfy the inequalities

$$R > \frac{GM}{c^2}$$

and

$$v \ll c \ , \tag{5.541}$$

where M is the system mass, $2GM/c^2$ is the Schwarzschild radius and c is the velocity of light. The total luminosity, L, radiated in quadrupole gravitational waves from a source of radius, R, effective mass, M_{eff}, and characteristic velocity $V = cR/\lambda$ is given by (Press and Thorne, 1972):

$$L = \frac{1}{5}\frac{G}{c^5}\sum_{j,k}\left\langle \left(\dddot{I}_{jk}\right)^2 \right\rangle \approx L_{\text{o}}\left(\frac{2GM_{\text{eff}}}{c^2R}\right)^2\left(\frac{V}{c}\right)^6 \ , \tag{5.542}$$

where M_{eff} is the effective mass for which amplitude changes in $\text{I}_{jk} = M_{\text{eff}}R^2$, the wavelength $\lambda = cR/V$, and the natural unit of luminosity, L_{o}, is given by

$$L_{\text{o}} = \frac{c^5}{G} = 3.63 \times 10^{59} \text{erg sec}^{-1} = 2.03 \times 10^5 M_{\odot}c^2 \text{ sec}^{-1} \ , \tag{5.543}$$

where $M_{\odot} \approx 2 \times 10^{33}$ grams is the solar mass. The wave amplitudes, h, are given by (Misner, Thorne and Wheeler, 1973; Tyson and Giffard, 1978):

$$h \approx (GM_{\text{eff}}/Dc^2)(V/c)^2 \approx 10^{-16}(M_{\text{eff}}/M_{\odot})(V/c)^2(1 \text{ kpc}/D) \ , \tag{5.544}$$

where $M_{\odot}$ is the Sun's mass and in the numerical approximation the source distance, D, is given in kiloparsecs or kpc.

For a binary star system of orbital period, P, semi-major axis, a, reduced mass $\mu = M_1M_2/(M_1 + M_2)$, where M_1 and M_2 are the respective masses of the two components, and total mass $M = M_1 + M_2$, the gravitational wave luminosity is given by (Peters and Mathews, 1963; Peters, 1964; Press and Thorne, 1972; also see previous Sect. 5.5.2 with the binary pulsar).

$$\begin{aligned} L &= \frac{32}{5}\frac{G^5}{c^{10}}\frac{\mu^2M^3}{a^5}L_{\text{o}}f(e) \\ &\approx 3.0 \times 10^{33} \text{erg sec}^{-1}\left(\frac{\mu}{M_{\text{o}}}\right)^2\left(\frac{M}{M_{\text{o}}}\right)^{4/3}\left(\frac{P}{1 \text{ hour}}\right)^{-10/3} f(e) \end{aligned} \tag{5.545}$$

where

$$f(e) = \frac{1 + \frac{73}{24}e^2 + \frac{37}{96}e^4}{(1 - e^2)^{7/2}} \ , \tag{5.546}$$

and e is the orbital eccentricity. The radiation is emitted at a fundamental frequency equal to twice the orbital frequency, and at increasing harmonics. If gravitational radiation is the dominant force in changing the orbital period, and if the orbit is nearly circular,

$$\frac{1}{P}\frac{dP}{dt} = -\frac{96}{5}\frac{G^3}{c^5}\frac{\mu M^2}{a^4} = \left(\frac{1}{2.8 \times 10^7 \text{ yr}}\right)\left(\frac{M}{M_{\odot}}\right)^{2/3}\left(\frac{\mu}{M_{\odot}}\right)\left(\frac{1 \text{ hr}}{P}\right)^{8/3} . \tag{5.547}$$

Wagoner and Will (1976) discuss post-Newtonian gravitational radiation from orbiting point masses. Iyer and Will (1993) provide expressions for the gravitational radiation reaction and the equations of motion for a binary system to post-Newtonian order beyond the quadrupole approximation.

Bondi et al. (1962) describe general relativistic gravitational waves from axisymmetric isolated systems. For a slightly deformed, nonaxisymmetric homogeneous neutron star with moment of inertia $I = 2MR^2/5$, mass, M, radius, R, rotation period, P, and nonaxial deformation, ε, the gravitational wave luminosity is (Ostriker and Gunn, 1969)

$$L = \frac{32}{5}\frac{G}{c^5} I^2 \varepsilon^2 \left(\frac{2\pi}{P}\right)^6$$

$$\approx 10^{38}\ \text{erg sec}^{-1} \left(\frac{I}{4 \times 10^{44}\ \text{g cm}^2}\right)^2 \left(\frac{P}{0.033\ \text{sec}}\right)^{-6} \left(\frac{\varepsilon}{10^{-3}}\right)^2 , \quad (5.548)$$

and at a time, t, after its birth (Press and Thorne, 1972)

$$L \approx 10^{45}\ \text{erg sec}^{-1} \left(\frac{4 \times 10^{44}\ \text{g cm}^2}{I}\right)^{1/2} \left(\frac{10^{-3}}{\varepsilon}\right) \left(\frac{10^6\ \text{sec}}{t + 10^4\ \text{sec}}\right)^{3/2} . \quad (5.549)$$

The quadrupole formalism that we have been using here is strictly applicable to slow-motion systems, but the most strongly emitting systems may violate the mathematical approximations that underlie these formulae. That is, the strongest emitters of gravitational radiation will be those with the largest internal masses and with internal velocities approaching the speed of light. Thorne (1987) provides a catalog of alternate wave-generation formulations that do not entail any slow-motion assumptions. He also reviews astrophysical sources of gravitational radiation within the context of their possible detection, including burst, periodic and stochastic background gravitational-wave sources.

According to Thorne (1987), the strength, rate of occurrence and even the very existence of sources of gravitational waves are very uncertain, with the single exception of binary stars and their coalescence. We will therefore provide a brief synopsis of possible sources, followed by more detail for merging binary neutron stars, noting that "it seems very likely that when gravitational waves are finally seen, they will come predominantly from sources we have not thought of or we have underestimated".

The gravitational radiation from a given astrophysical source has a characteristic signal, h_c, that can be compared with the noise amplitude, $h_n(f_c)$, of the detector at the characteristic frequency, f_c, to provide a signal, S, to noise, N, ratio:

$$\frac{S}{N} = \frac{h_c}{h_n(f_c)} . \quad (5.550)$$

Burst sources emit gravitational waves with complicated time dependencies, $h(t)$, and with durations short compared with the observation time. In most cases, such as supernovae, the wave form is so uncertain that one uses the approximation:

$$h_c = \left(\frac{3}{2\pi^2}\frac{\Delta E_{GW}/f_c}{D^2}\right)^{1/2} = 2.7 \times 10^{-20} \left(\frac{\Delta E_{GW}}{M_\odot c^2}\right)^{1/2} \left(\frac{1\ \text{kHz}}{f_c}\right)^{1/2} \left(\frac{10\ \text{Mpc}}{D}\right) \quad (5.551)$$

where ΔE_{GW} is the total energy radiated as gravitational waves, D is the distance to the source, $M_\odot$ is the mass of the Sun, and 10 Mpc is the approximate distance to the center of the Virgo cluster of galaxies.

For a supernova explosion with core collapse to a neutron star, various estimates suggest (Müller, 1982; Eardley, 1983):

$$\begin{aligned} f_c &\approx 1{,}000\ \mathrm{Hz} \\ \Delta E_{GW} &\approx (10^{-7}\ \text{to}\ 10^{-2}) M_\odot c^2 \\ h_c &\approx 10^{-23}\ \text{to}\ 10^{-21}\ (10\ \mathrm{Mpc}/D)\ . \end{aligned} \tag{5.552}$$

By comparing the arrival times of the first bursts of light and gravitational waves from a distant supernova, one could verify general relativity's prediction that electromagnetic and gravitational waves propagate with the same speed, the velocity of light, c.

When a star of mass, M, collapses to a black hole, radiating gravitational waves with an efficiency $\varepsilon = \Delta E_{GW}/(Mc^2)$, the characteristic frequency, f_c, and amplitude, h_c, are given by (Stark and Piran, 1985; Piran and Stark, 1986)

$$\begin{aligned} f_c &\approx \frac{1}{5\pi M} = (1.3 \times 10^4)\left(\frac{M_\odot}{M}\right)\ \mathrm{Hz}\ , \\ h_c &\approx \left(\frac{15}{2\pi}\varepsilon\right)^{1/2}\frac{M}{D} = 7 \times 10^{-22}\left(\frac{\varepsilon}{0.01}\right)^{1/2}\left(\frac{M}{M_\odot}\right)\left(\frac{10\ \mathrm{Mpc}}{D}\right) \\ &= 1.0 \times 10^{-20}\left(\frac{\varepsilon}{0.01}\right)^{1/2}\left(\frac{10^3\ \mathrm{Hz}}{f_c}\right)\left(\frac{10\ \mathrm{Mpc}}{D}\right)\ . \end{aligned} \tag{5.553}$$

More than half the stars in our Galaxy belong to binary systems, and perhaps one in 100 of the most massive pairs can remain bound together as orbiting neutron stars after they both undergo supernova explosions. The binary pulsar, discussed in the previous Sect. 5.5.2, is an example, apparently emitting gravitational radiation with a rate of energy loss of about one fifth the Sun's total luminosity in visible electromagnetic radiation. Although this radiation is still too weak to be directly measured by gravity-wave detectors on Earth, the neutron stars will eventually move closer together and coalesce in 3.5×10^8 years, producing much more intense gravitational radiation.

Coalescing neutron-star binaries and low-mass black hole binaries have (Thorne, 1987):

$$\begin{aligned} h_c &= 0.237\frac{\mu^{1/2}M^{1/3}}{D f_c^{1/6}} \\ &= 4.1 \times 10^{-22}\left(\frac{\mu}{M_\odot}\right)^{1/2}\left(\frac{M}{M_\odot}\right)^{1/3}\left(\frac{100\ \mathrm{Mpc}}{D}\right)\left(\frac{100\ \mathrm{Hz}}{f_c}\right)^{1/6} \end{aligned}$$

where the total mass, M, and reduced mass, μ, are given by:

$$M = M_1 + M_2\ \text{ and }\ \mu = M_1 M_2/(M_1 + M_2) \tag{5.554}$$

for component stars of mass M_1 and M_2, and the frequency, f, of the waves is given as a function of time, t, by:

$$f = \frac{1}{\pi}\left[\frac{5}{256}\frac{1}{\mu M^{2/3}}\frac{1}{(t_o - t)}\right]^{3/8} . \tag{5.555}$$

Energy conservation requires that the energy carried away by the gravitational waves of binary neutron stars must be supplied by their orbital energy, causing them to spiral together until they collide and merge. The closer they get, the faster they orbit each other and the more intense the gravitational wave emission. As the two bodies spiral together, the frequency sweeps upward to a maximum of $f_{\max} \approx$ 1,000 Hz for neutron stars. In the final minute, the merging stars will send out an energetic signal with a distinctive signature that might be detected by terrestrial gravitational-wave detectors (Piran, 1995). Because the two neutron stars each have a maximum mass between 1.4 and 2.4 solar masses, their combined mass should exceed the Chandrasekhar limit and collapse to a singularity is expected. LIGO might be sensitive enough to observe the gravitational waves emitted from a pair of colliding neutron stars during the last few seconds before coalescence. If so, the project could provide the first direct confirmation of the existence of gravitational waves.

Paczynski (1986) has suggested that the enigmatic gamma-ray bursts may be coming from the cataclysmic collisions of neutron stars in remote galaxies at cosmological distances, releasing about 10^{50} erg/sr on a timescale of a few seconds. The neutrinos released in binary neutron star mergers would produce electron-positron pairs and gamma rays with the required energy (Eichler et al., 1989). (Gamma rays from neutron star/black hole mergers are discussed by Mochkovitch et al., 1993). Such high-energy bursts can last anywhere from a few thousandths of a second to several minutes, and vary rapidly in intensity over that period, suggesting relatively compact sources. Their exact origin has nevertheless eluded astronomers for more than two decades since their discovery (Klebesadel, Strong and Olson, 1973; Cline, Desai, Klebesadel and Strong, 1973). Yet, the bursts are not concentrated in the plane of the Milky Way (Fishman and Meegan, 1995), as would be expected if they originated from collapsing or merging stars in the plane of our Galaxy. Gamma-ray bursts might originate in distant galaxies, and in this case the duration of the more remote, and presumably weaker bursts, would lengthen as the radiation makes its way across the expanding Universe (Piran, 1995). Such a correlation between the fainter and longer bursts has been found (Norris et al., 1995). Gamma ray bursts are also discussed in Volume I where the afterglows of their fireballs are included. The large redshift determinations of the optical counterparts of at least two gamma-ray bursts have demonstrated that they orginate at cosmological distances and have peak luminosities exceeding 10^{52} erg s^{-1} (Metzger et al., 1997; Kulkarni et al., 1998).

Calculations suggest that merging neutron stars will produce about 900 bursts per year, within the volume of space detected so far, and that is just the rate observed. Thus, colliding neutron stars offer a plausible scenario for the mysterious high-energy bursts. If a unique gravitational-wave signal is detected in coincidence with gamma-ray bursts, then it will confirm this theory for their origin (Kochanek and Piran, 1993).

For detection purposes, one evaluates the gravitational wave force, F, which acts on a mass, M, is given by

$$F = -M\nabla\varphi = -\sum_j \sum_k M\mathbf{R}_{\mathrm{joko}} x_k \ , \tag{5.556}$$

where φ is the equivalent Newtonian gravitational potential of the gravitational wave, and the components of the Riemann curvature tensor, $\mathbf{R}_{\mathrm{joko}}$ are given by

$$\mathbf{R}_{\mathrm{joko}} = \frac{\partial^2\varphi}{\partial x_j \partial x_k} = -\frac{G}{c^4 D}\left[\frac{d^4\mathbf{I}_{jk}^{\mathrm{TT}}}{dt^4}\right]_{\mathrm{ret}} \ , \tag{5.557}$$

where D is the distance to the source. When a gravitational wave acts on a pair of free, or almost free, masses separated by a distance, L, it will produce a change in distance, ΔL, given by

$$\frac{\Delta L}{L} = h(t) = \left[h_+^2(t) + h_x^2(t)\right]^{1/2} \ , \tag{5.558}$$

where $h(t)$ is the dimensionless field strength, and h_+ and h_{x} are the magnitudes of the perturbations in the metric tensor, $\mathbf{g}_{\mathrm{ik}}$. For a resonant type of gravitational wave detector, the displacement is between the two ends of an elastic solid, and

$$\frac{\Delta L}{L} \approx \pi\nu_{\mathrm{o}} h(\nu_{\mathrm{o}}) \ , \tag{5.559}$$

where ν_{o} is the resonant frequency, and $h(\nu_{\mathrm{o}})$ is the Fourier transform of $h(t)$ evaluated at ν_{o}.

Integrating the equation for gravitational-wave energy flux, F, over time and using Parseval's theorem, we obtain a total energy, E, per burst of

$$E = 4\pi\frac{c^3 D^2}{16\pi G}\int_{-\infty}^{+\infty}(2\pi\nu)^2 h^2(\nu)d\nu \ , \tag{5.560}$$

where D is the distance of the source, we assume that the source radiates isotropically, and $h(\nu)$ is the Fourier transform of $h(t)$. If the gravitational wave has a flat frequency spectrum out to the resonant frequency, ν_{o}, a resonant detector gives the limit

$$E \geq \frac{c^3}{4G}\left(\frac{2\Delta L}{L}\right)^2 \nu_{\mathrm{o}} D^2 \,\mathrm{erg} \geq 5.6\times 10^{-17}\left(\frac{2\Delta L}{L}\right)^2 \nu_{\mathrm{o}} D^2 (M_\odot c^2)\,\mathrm{erg} \ , \tag{5.561}$$

where the solar mass $M_\odot = 2\times 10^{33}$ gm.

Thorne (1987) measures the detector sensitivity by the parameter $h_{3/\mathrm{yr}} \approx (3$ to $5)h_{\mathrm{n}}(f_{\mathrm{c}})$. If a source of strength $h_{3/\mathrm{yr}}$ is seen three times per year by two identical detectors operating in coincidence, we can be 90% confident the detectors are not just seeing their own noise. Here $h_{\mathrm{n}}(f_{\mathrm{c}})$ is the noise amplitude at the characteristic frequency, f_{c}.

The sensitivity of a bar detector to short bursts is (Thorne, 1987)

$$h_{3/\mathrm{yr}} \approx 5h_{\mathrm{n}} \approx 11\left[\frac{G}{c^3}\frac{kT_{\mathrm{n}}}{\int \sigma_{\mathrm{o}} df}\right]^{1/2} = 2.0 \times 10^{-16}\left[\frac{T_{\mathrm{n}}/1\mathrm{K}}{\int \sigma_{\mathrm{o}}\, df/10^{-21}\,\mathrm{cm}^2\,\mathrm{Hz}}\right]^{1/2}, \tag{5.562}$$

where f_{o} is the resonant frequency of the bar detector, σ_{o} is the detector cross section, and T_{n} is its noise temperature. The integrated cross section for a cylindrical bar of mass, M, and length, L, somewhat greater than the radius, R, is

$$\begin{aligned}\int \sigma_{\mathrm{o}}\, df &= \frac{8}{\pi}\frac{GMv_{\mathrm{s}}^2}{c^3}\left[1+\frac{1}{2}\nu(1-2\nu)(\pi R/L)^2+\cdots\right] \\ &= 1.6\times 10^{-21}\left(\frac{M}{10^3\,\mathrm{kg}}\right)\left(\frac{v_{\mathrm{s}}}{5\,\mathrm{km\,s}^{-1}}\right)^2 \mathrm{cm}^2\,\mathrm{Hz}\ ,\end{aligned} \tag{5.563}$$

where ν is the Poisson ratio of the bar's material, and the internal sound velocity $v_{\mathrm{s}} = (E/\rho)^{1/2}$ with a Young's modulus, E, and density, ρ.

For LIGO-type detectors, laser beams are made to bounce back and forth in arms of length, L, for a large number, B, of round-trip times, with a rms shot noise amplitude of (Thorne, 1987):

$$\begin{aligned}h_{\mathrm{shot}} &\approx \left[\frac{2\hbar c\, \lambda\!\!\!^{-}_{\mathrm{e}}}{I_{\mathrm{o}}\eta}\frac{f}{(2BL)^2}\right]^{1/2} \\ &\approx 7.2\times 10^{-21}\frac{50}{B}\frac{1\,\mathrm{km}}{L}\left(\frac{\lambda\!\!\!^{-}_{\mathrm{e}}}{0.082\,\mu\mathrm{m}}\right)^{1/2}\left(\frac{10\ \mathrm{Watts}}{I_{\mathrm{o}}\eta}\right)^{1/2}\left(\frac{f}{1000\ \mathrm{Hz}}\right)^{1/2},\end{aligned} \tag{5.564}$$

where $\lambda\!\!\!^{-}_{\mathrm{e}} = \lambda_{\mathrm{e}}/2\pi$ is the reduced wavelength of the light, I_{o} is the laser output power, $\eta = 0.4$ to 0.9 is the photon counting efficiency of the photodetector, and the gravity-wave burst has a characteristic frequency f.

5.5.4 Gravitational Lenses

In the first edition of his Optiks, published in 1704, Sir Isaac Newton had already speculated that massive bodies might bend nearby light rays. The actual deflection, ϕ, computed under the assumption that light is composed of material corpuscles that are attracted by the Newtonian force of gravity, was published a century later (Soldner, 1801) but apparently calculated two decades earlier by Henry Cavendish as the result of his correspondence with John Michell (Will, 1988).

A light ray passing a minimum distance, ξ, from the center of an object of mass, M, will be deflected from a straight line path by the angle:

$$\alpha = \frac{2GM}{c^2\xi} = \frac{R_{\mathrm{s}}}{\xi}\ \text{radians} \quad \text{(Newton theory)}\ , \tag{5.565}$$

where one radian $= 2.06265 \times 10^5$ seconds of arc, and the Schwarzschild radius

$$R_{\mathrm{s}} = \frac{2GM}{c^2} \approx 2.95 \times 10^5\ \frac{M}{M_{\odot}}\ \mathrm{cm}\ ,$$

the gravitational constant $G = 6.6726 \times 10^{-8}\ \mathrm{cm}^3\,\mathrm{g}^{-1}\,\mathrm{s}^{-2}$, the velocity of light $c = 2.9979 \times 10^{10}\ \mathrm{cm\,s}^{-1}$ and the Sun's mass $M_{\odot} = 1.989 \times 10^{33}$ grams.

Michell (1784), and apparently independently Laplace (1796), noted that the attractive gravitational force of a body could be so large that no light could escape, introducing the modern idea of a black hole for which the radius $R < R_s$ (also see Sect. 5.6.2).

Einstein (1911), unaware of previous work, obtained the same value for the deflection angle α from the principle of equivalence and the assumption of flat, uncurved Euclidean space. Four years later he added the effect of space curvature near the gravitating body, obtaining the "Einstein angle" (Einstein, 1915)

$$\alpha = \frac{4GM}{c^2 \xi} = \frac{2R_s}{\xi} \text{ radians (Einstein theory)} \tag{5.566}$$

which was confirmed during the total eclipse of 1919 (Dyson, Eddington and Davidson, 1920) and numerous subsequent experiments (see Sect. 5.5.2).

The idea that an intervening stellar mass might act as a gravitational lens for the light of a distant star was introduced by Lodge (1919) and Eddington (1920), including the concepts of multiple images and the magnification of image flux (Eddington, 1920). Chwolson (1924) showed that perfect alignment of the background and foreground stars would result in a ring-shaped image, and Einstein (1936) noted that the apparent luminosity of a distant star could be significantly magnified if the intervening, light-deflecting star was suitably aligned. An examination of Einstein's research notes by Renn, Sauer and Stachel (1997) indicates that Einstein explored the possibility of gravitational lensing in 1912, three years before completing his General Theory of Relativity (Einstein 1915, 1916), deriving the basic features of the lensing effects; but he did not publish the results until 24 years later (Einstein, 1936) in response to prodding by an amateur scientist.

When a star is perfectly aligned with an intervening gravitational lens of mass, M, it will appear as an "Einstein ring" with a radius, R_E, in the lens plane of (Chwolson, 1924; Einstein, 1936; Refsdal, 1964; Alcock et al., 1996):

$$R_E = [4GMLx(1-x)/c^2]^{1/2} , \tag{5.567}$$

where L is the observer-star distance, and Lx is the observer-lens distance. The star appears magnified by a factor of:

$$A = (u^2+2)/[u(u^2+4)^{1/2}] , \tag{5.568}$$

where $u = b/R_E$ is the undeflected impact parameter of the light ray in units of the "Einstein radius" R_E, and b is the separation of the lens from the observer's line of sight to the star. This effect has been used to detect massive compact dark objects, called Machos, in the halo of our Galaxy, and to place limits on their abundance (Alcock et al., 1996, and Sect. 5.4.6).

Then Zwicky (1937) suggested that extragalactic nebulae, or galaxies, offered a much better chance than stars for observing the gravitational lens effect, and reasoned that the light deflection would be cosmologically important by establishing the mass of the intervening galaxy or clusters of galaxies, and by magnifying distant sources that might be otherwise unobservable. The formidable gravitational fields of an intervening galaxy or cluster of galaxies can divert the path of light from a remote object and focus it, bending and splitting the light like a stone in a stream.

The discovery of quasi-stellar objects in 1963–64, with their large redshifts, small angular extents, and unexpectedly remote distances, roughly coincided with renewed theoretical considerations of gravitational lenses, including the use of gravitational lenses for distance measurement (Klimov, 1963, Liebes, 1964, Refsdal, 1966) and testing cosmological theories (Refsdal, 1964, 1966).

The rapid discovery of the doubly-imaged quasar, Q0957 + 561AB (Walsh, Carswell and Weymann, 1979), its lensing galaxy (Stockton, 1980), and the triple-quadrupole quasar Q1115 + 0.80A_1, A_2, B, C (Weymann et al., 1980; Hege et al., 1980, 1981) put the concept of gravitational lenses on a firm observational footing (also see Young et al., 1980, 1981). This was followed by other examples, including the "Einstein cross" (Q2237 + 031, Huchra et al., 1985) and the "clover leaf" (H1413 + 117, Magain et al., 1988).

A second type of gravitational lens system is the luminous arcs, the distorted images of high-redshift galaxies seen through a compact cluster of galaxies. Thanks to their large masses, clusters of galaxies make especially good gravitational lenses, spreading the light of a distant object into an array of faint arcs (Lynds and Petrosian, 1986, 1989; Petrosian, 1989; Soucail et al., 1987, 1988). The entire cluster becomes one big gravitational lens, providing information on the dark or invisible mass within it (Tyson et al., 1986, 1990).

A third type is the radio rings (Hewitt et al., 1988; Langston et al., 1989). Examples of all three types of gravitational lenses are given in the accompanying Table 5.18; new candidates continue to be discovered (Luppino et al., 1993; Ratnatunga et al., 1995; Warren et al., 1996) and the old ones continue to be observed in new ways (Gorenstein et al., 1988; Kayser et al., 1990; Dahle, Maddox and Lilje, 1994; Pelt et al., 1996; Conti, Leitherer and Vacca, 1996).

As we have previously seen, the notion of distance for extragalactic objects depends on space-curvature and cosmological models, and the theory becomes additionally complicated by the spatial distribution of the imaging object. A comprehensive monograph on gravitational lenses is given by Schneider, Ehlers and Falco (1992), and the cosmological applications of gravitational lensing are reviewed by Blandford and Narayan (1992). Here we will limit our formulae to the simple case of lensing by a point mass in uncurved Euclidean space; followed by references to some of the more comprehensive papers. For further details see the above-mentioned monograph and review. Unsolved problems in gravitational lensing are discussed by Blandford (1997).

If a light ray from a source, S, at an angular diameter distance, D_{s}, passes a gravitational lens at a minimum distance or impact parameter, ξ, the observer sees an image of the source at the angular position

$$\theta = \frac{\xi}{D_{\mathrm{d}}} , \tag{5.569}$$

where the angular diameter distance from the observer to the gravitational lens, or deflector, is D_{d}. Assuming that the lens is thin compared to the total distance, D_{s}, its influence can be described by a deflection angle $\hat{\boldsymbol{\alpha}}(\boldsymbol{\xi})$ (a two vector) suffered by the ray on crossing the "lens plane". For a lens of mass, M,

$$\hat{\alpha}(\xi) = \frac{4GM\xi}{c^2\xi^2} = 2R_s\frac{\xi}{\xi^2} \quad . \tag{5.570}$$

where the Schwarzschild radius $R_s = 2GM/c^2$.

For a point mass, or Schwarzschild lens, we use the geometry shown in Fig. 5.35. A characteristic angle, α_o, and a characteristic length, ξ_o, in the lens plane, are given by

$$\alpha_o = \left(2R_s\frac{D_{ds}}{D_d D_s}\right)^{1/2} = \left(\frac{4GM}{c^2 D}\right)^{1/2} \tag{5.571}$$

$$\xi_o = \left(2R_s\frac{D_d D_{ds}}{D_s}\right)^{1/2} = \left(\frac{4GM}{c^2}\frac{D_d D_{ds}}{D_s}\right)^{1/2} = \alpha_o D_d \ , \tag{5.572}$$

where the effective lens distance is

$$D = \frac{D_d D_s}{D_{ds}} \quad . \tag{5.573}$$

The angular position of the source in the absence of the lens is β, its true direction in the sky.

Table 5.18. Multiple-imaging gravitational lenses together with the redshift of the source, z_s, the redshift of the intervening imaging deflector, z_d, and the angular extent, θ_{max}, of the image (adapted from Blandford and Narayan, 1992)

Source	z_s	z_d	θ_{max}	Source	z_s	z_d	θ_{max}
Multiple Quasars				*Arcs*			
Q0957 +561AB	1.41	0.36	6.1″	Abell 2218	?	0.17	
Q0142 −100AB	2.72	0.49	2.2″	Abell 2390	0.92	0.23	
Q2016 +112ABC	3.27	1.01	3.8″	Cl 0024 +17	?	0.39	
Q0414 +053ABCD	2.63	?	3″	Cl 0302 +17	?	0.42	
Q1115 +080A_1A_2BC	1.72	?	2.3″	Cl 0500 −24	0.91	0.32	
H1413 +117ABCD	2.55	?	1.1″	Cl 1409+52	?	0.46	
Q2237 +031ABCD	1.69	0.039	1.8″	Cl 2244−02	2.23	0.33	
Arcs				*Radio Rings*			
Abell 370	0.72	0.37		MG1131 +0456	?	?	2.2″
Abell 963	0.77	0.21		0218+357	?	?	0.3″
Abell 1352	?	0.28		MG1549 +3047	?	0.11	1.8″
Abell 1525	?	0.26		MG1634 +1346	1.75	0.25	2.1″
Abell 1689	?	0.18		1830−211	?	?	1.0″
Abell 2163	?	0.17					

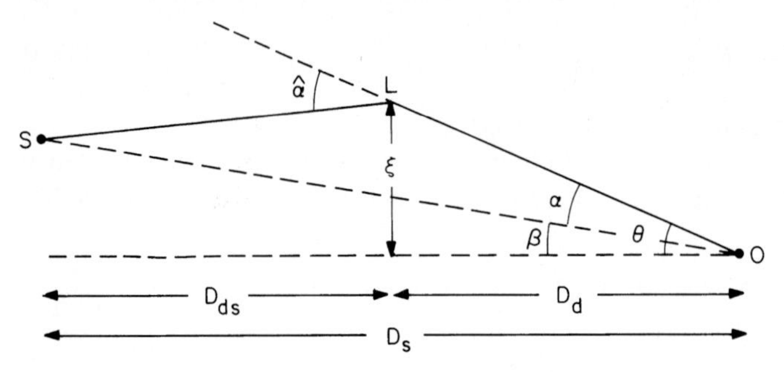

Fig. 5.35. Basic ray geometry of gravitational lensing. A light ray from a source S at redshift z_S is incident on a deflector or lens L at redshift z_d with impact parameter ξ relative to some fiducial lens "center". Assuming the lens is thin compared to the total path length, its influence can be described by a deflection angle $\hat{\boldsymbol{\alpha}}(\boldsymbol{\xi})$ (a two-vector) suffered by the ray on crossing the lens plane. The deflected ray reaches the observer O, who sees the image of the source at position θ on the sky. The true direction of the source, or its position on the sky in the absence of the lens, is indicated by β. Also shown are the angular diameter distances D_d, D_s, and D_{ds} separating the source, deflector, and observer. (Adapted from Blandford and Narayan, 1992).

$$\beta D_s = \frac{D_s}{D_d}\xi - \frac{2R_s}{\xi}D_{ds} \ , \tag{5.574}$$

where D_{ds} is the angular diameter distance of the source from the lens. In the simple case of noncurved, Euclidean space $D_{ds} = D_s - D_d$, but in general, $D_{ds} \neq D_s - D_d$.

The lens equation is

$$\beta = \theta - 2R_s \frac{D_{ds}}{D_d D_s}\frac{1}{\theta} = \theta - \frac{2R_s}{\theta D} \ , \tag{5.575}$$

where we allow β and θ to have either sign. Using the quadratic equation

$$\theta^2 - \beta\theta - \alpha_o^2 = 0 \ , \tag{5.576}$$

with two solutions θ_1 and θ_2, an off-axis source produces two images on opposite sides of the point-mass, or Schwarzschild lens, of angular separation, $\Delta\theta$, given by

$$\Delta\theta = \theta_1 - \theta_2 = \sqrt{4\alpha_o^2 + \beta^2} \geq 2\alpha_o \ , \tag{5.577}$$

$$\theta_{1,2} = \frac{1}{2}\left(\beta \pm \sqrt{4\alpha_o^2 + \beta^2}\right) \ , \tag{5.578}$$

and the "true" angular separation between the source and the deflector is related to the image positions by

$$\theta_1 + \theta_2 = \beta \ . \tag{5.579}$$

For source positions $\beta \neq \alpha_o$, the two images have roughly comparable magnifications, but with $\beta \gg \alpha_o$ the magnification of one image diminishes as $(\alpha_o/\beta)^4$. So, for the observable images of a point-mass lens we have $\Delta\theta$ close to its minimum value $2\alpha_o$ and a typical impact parameter of ξ_o. That is, the angular separation between the two images of a point-mass lens is given by

$$\Delta\theta = 2\alpha_o = 2\left(\frac{4GM}{c^2D}\right)^{\frac{1}{2}} = 6\left(\frac{M}{M_\odot}\right)^{\frac{1}{2}}\left(\frac{D}{10^9\ \mathrm{pc}}\right)^{-\frac{1}{2}} \times 10^{-6}\ \text{seconds of arc} \tag{5.580}$$

where the solar mass $M_\odot = 1.989 \times 10^{33}$ grams, and 10^9 pc = a billion parsec $= 3.18 \times 10^{27}$ cm. For an on-axis source, in which the source, lens and observer are precisely aligned, the image is a complete "Einstein ring" of angular diameter $\Delta\theta = 2\alpha_o$ (Chwolson, 1924; Einstein, 1936). The effective cross section for lensing of either on- or off-axis lensing is usually taken to be $\pi\alpha_o^2$.

The simple point-mass model is appropriate for computing deflections by individual stars and black holes, and for rough estimates of imaging by more extended lenses such as galaxies.

The time delay, Δt, for the two images is of the same order as the time it takes light to traverse a distance of about one Schwarzschild radius (Schneider, Ehlers and Falco, 1992), or

$$\Delta t \approx \frac{R_s}{c} = \frac{2GM}{c^3} \approx 0.98 \times 10^{-5}\frac{M}{M_\odot}\ \text{seconds}\ , \tag{5.581}$$

which for a galaxy mass of $M = 3 \times 10^{12} M_\odot$ is comparable to 1 year $= 3.15 \times 10^7$ seconds.

Light deflection in a gravitational field not only changes the direction of a light ray, but also the cross-section of a bundle of rays. The surface brightness, I, of a source is unchanged during light deflection, but the flux of the image is magnified, since the flux is the product of the surface brightness and the solid angle the source subtends in the sky (Schneider, Ehlers and Falco, 1992). When the lens mass distribution has a length scale that is much larger than the size of the image region, one can use the convergence parameter κ which measures the isotropic part of the magnification (Blandford and Narayan, 1992)

$$\kappa = \frac{\Sigma}{\Sigma_{\mathrm{cr}}}\ , \tag{5.582}$$

where Σ is the surface density,

$$\Sigma_{\mathrm{cr}} = \frac{c^2}{4\pi GD'} = 0.35\left(\frac{D'}{10^9\ \mathrm{pc}}\right)^{-1}\ \mathrm{g\ cm^{-2}} \tag{5.583}$$

and

$$D' = \frac{D_{\mathrm{ds}}D_{\mathrm{d}}}{D_{\mathrm{s}}}\ . \tag{5.584}$$

The Σ_{cr} is known as the critical density and D' is a second effective lens distance. Normally a lens will have $\Sigma > \Sigma_{\mathrm{cr}}$ somewhere in order to produce multiple images, but this is not strictly necessary.

The intermediate case, when the image separations are comparable with the size of the potential, is harder to treat. For galaxies and clusters, simple spherical models have been often used. These models can be described by the one-dimensional velocity dispersion $\sigma = 300\sigma_{300}$ km s^{-1}. For a singular isothermal sphere, the deflection angle is constant and given by $\hat{\boldsymbol{\alpha}} = 4\pi\sigma^2/c^2 = 2.6''\sigma_{300}^2$, so that

$$\alpha_o = \frac{4\pi\sigma^2 D_{ds}}{c^2 D_s} = 2.6''\sigma_{300}^2 \frac{D_{ds}}{D_s} \quad . \tag{5.585}$$

For source positions $\beta < \alpha_o$, there are two images at $\theta = \beta \pm \alpha_o$. Technically, there is also a third image at $\theta = 0$, but this has zero magnification because the surface density is singular at the center of the lens. The cross section for multiple-imaging is given by $\pi\alpha_o^2$.

In practice, two cases of lensing situations are of importance. First, if both lens and source are at cosmological distance, the factors $D_d D_{ds}/D_s$ and $D_d D_s/D_{ds}$ will be fair fractions η and η', respectively, of the Hubble length c/H_o (Schneider, Ehlers and Falco, 1992):

$$\frac{D_d D_{ds}}{D_s} = \eta \frac{c}{H_o} \quad , \tag{5.586}$$

$$\frac{D_d D_s}{D_{ds}} = \frac{\eta' c}{H_o} \quad , \tag{5.587}$$

where H_o is the Hubble constant. Measuring H_o in units of $50\,\mathrm{km\,s^{-1}\,Mpc^{-1}}, H_o = h_{50} \times 50\,\mathrm{km\,s^{-1}\,Mpc^{-1}}$, the Hubble length is

$$\frac{c}{H_o} \approx 1.8 \times 10^{28}\, h_{50}^{-1}\,\mathrm{cm} \approx 5.8 \times 10^9\, h_{50}^{-1}\,\mathrm{pc} \quad , \tag{5.588}$$

where 1 pc $= 3.086 \times 10^{18}$cm.

The angular scale α_o is then

$$\alpha_o \approx 1.2 \times 10^{-6} \sqrt{\frac{M}{M_\odot}} \sqrt{\frac{h_{50}}{\eta'}} \text{ seconds of arc} \tag{5.589}$$

where $M_\odot = 1.989 \times 10^{33}$ g is the Sun's mass, and $4GM_\odot/c^2 = 2\mathrm{R}_{s\odot} \approx 5.9$ km. The corresponding length scale is

$$\xi_o \approx 10^{17} \sqrt{\frac{M}{M_\odot}} \sqrt{\frac{\eta}{h_{50}}}\,\mathrm{cm} \approx 0.03 \sqrt{\frac{M}{M_\odot}} \sqrt{\frac{\eta}{h_{50}}}\,\mathrm{pc} \quad . \tag{5.590}$$

The second case that is frequently discussed is $D_d \ll D_{ds} \approx D_s$; then

$$\alpha_o \approx 3 \times 10^{-3} \sqrt{\frac{M}{M_\odot}} \left(\frac{D_d}{1\,\mathrm{kpc}}\right)^{-1/2} \text{ seconds of arc} \tag{5.591}$$

and

$$\xi_o \approx 4 \times 10^{13} \sqrt{\frac{M}{M_\odot}} \left(\frac{D_d}{1\,\mathrm{kpc}}\right)^{1/2} \mathrm{cm} \quad . \tag{5.592}$$

For the nearest observed galaxy gravitational lens system, the lens is a spiral galaxy at redshift $z_d \approx 0.039$; if we take $M = 10^{12} M_\odot$, we get $\xi_o \approx 7h_{50}^{-1/2}$ kpc.

Technical papers dealing with the theory of gravitational lenses include those by Klimov (1963), Liebes (1964), Refsdal (1964, 1966), Barnothy and

Barnothy (1968), Bourassa, Kantowski and Norton (1973), Silva (1974), Bourassa and Kantowski (1975, 1976), Young et al. (1980, 1981), Bray (1984), Schneider (1984), Saslaw, Narasimha and Chitre (1985), and Blandford et al. (1989). Those dealing with mass determinations include Refsdal (1964), Barnothy (1970), Bahcall and Salpeter (1970) and Dyer and Roeder (1982) and Borgeest (1986). The theoretical discussions of gravitational lensed arcs are found in Kovner (1987), Blandford and Kovner (1988), Grossman and Narayan (1988, 1989), Narasimha and Chitre (1988), and Petrosian (1989), Cosmological aspects are discussed in Zeldovich (1964), Dashevskii and Zeldovich (1965), Dashevskii and Slysh (1966), Refsdal (1966, 1970), Kantowski (1969), Dyer and Roeder (1972, 1973, 1974), Weinberg (1976), Zeldovich (1980) and Watanabe and Tomita (1990). A comprehensive treatment of all these topics is found in the monograph by Schneider, Ehlers and Falco (1992).

5.6 Compact Objects and Accretion Power

5.6.1 White Dwarf Stars, Neutron Stars, Pulsars, X-Ray Binaries, and Accretion

White dwarfs, neutron stars and black holes interest astrophysicists because they are the inevitable consequences of the deaths of normal stars, and tell us something about the last stages of stellar evolution. They are the corpselike remnants of ordinary stars, crushed into oblivion at the endpoints of their lives. Stars like the Sun shine as the result of thermonuclear reactions deep down within their central cores, replenishing the energy lost by radiation and keeping them hot inside. The hot, moving material creates a pressure that supports the star and holds it up against the inner pull of its gravity. When a star has exhausted its nuclear fuel, there is nothing left to support it, and the star collapses or implodes under its own gravity. After that, the fate of a star depends on its mass, ending up as a white dwarf, neutron star or black hole.

White dwarf stars were discovered more than eighty years ago (Adams, 1914, 1915); receiving their name from their white-hot temperatures and small size. Their low luminosities and high surface temperatures imply, from the Stefan-Boltzmann law, that they have sizes comparable to that of the Earth, which is about 100 times smaller than the Sun. White dwarfs typically have masses on the order of the solar mass, so their average mass densities are about a million times greater than that of the Sun.

Conservation of magnetic flux requires that the magnetic field of a collapsing star will increase as its surface area decreases (Ginzburg, 1964; Woltjer, 1964). Since an ordinary star like the Sun has surface magnetic field strengths of 10 to 1,000 Gauss, white dwarf stars would be expected to have surface magnetic fields amplified by collapse to 10^5 to 10^7 Gauss. This intense magnetism has been measured for more than a dozen white dwarfs, providing independent confirmation of their small size (Kemp et al., 1970; Angel, 1977; Chanmugam, 1992; Landstreet, 1992). – also see Volume I.

Since atoms are largely empty space, you can strip the electrons from the atoms and squeeze the free electrons and ions together into a relatively small volume. Even in the dense core of a white dwarf, the average distance between electrons is more than 100,000 times their size. The enormous self-gravity of white dwarf stars is supported by the high-speed motions of their densely-packed electrons, but these motions are not due to the star's internal temperature or heat. According to a stipulation of quantum mechanics, known as the Pauli exclusion principle (Pauli, 1927), no two electrons can have exactly the same quantum-mechanical state. When the electrons are squeezed into a small volume, they begin to invade one another's territory. The electrons then dart away at high speeds that can approach that of light, just to keep their own space. They thereby exert an outward push, called degeneracy pressure, that resists further compression, stabilizing the white dwarf star and preventing the electrons from being packed too closely together.

The resulting pressure persists when the gas is cooled. Even if the individual electrons are at absolute zero temperature, they have a kinetic energy and a combined pressure that can support a white dwarf star indefinitely without any need of nuclear reactions (Fowler, 1926). So, the pressure has nothing to do with heat, and it is also unrelated to the electrical repulsion between electrons. Neutrons, which have no charge, also create a degenerate pressure. Without these effects, predicted by quantum statistics, neither a white dwarf nor a neutron star could exist, and every star would eventually collapse to a black hole.

For nonrelativistic motions, when the electrons move much slower than the speed of light, the electron degeneracy pressure, P, varies as the 5/3 power of the mass density, ρ, or the equation of state is $P \propto \rho^{5/3}$ (Fowler, 1926; also see Volume I). When a self-gravitating sphere of mass, M, is supported by this pressure, its radius R decreases with increasing mass, with $R \propto M^{-1/3}$. If the mass was large enough, the radius would shrink to zero, which is preposterous, so there is a limit to the mass that can be held in equilibrium by a degenerate electron gas.

For masses just above that of the Sun, the central densities of white dwarfs exceed 10^6 g cm^{-3}, and the average electron energies are relativistic with electron velocities that approach the speed of light. The equation of state is then $P \propto \rho^{4/3}$ (Anderson, 1929; Stoner, 1930; also see Volume I). Because the degenerate, relativistic electron gas provides less pressure at high density than the nonrelativistic formula would indicate, the high-speed equation of state sets the upper limit to the white dwarf mass that can be supported by degenerate electron pressure. This makes common sense, since the electrons will move at greater speeds with increasing stellar mass, but the electrons cannot move faster than the speed of light.

Because of the extreme pressure, the matter inside white dwarfs is completely ionized. The more massive ions supply the mass and gravitational compression of a white dwarf star, while the electrons provide the pressure that holds gravity at bay. The mass density, ρ, is given by

$$\rho = (A/Z) m_{\mathrm{p}} n_{\mathrm{e}} \ , \tag{5.593}$$

for ions of atomic number Z and atomic weight A. Here m_{p} is the mass of the proton and n_{e} is the electron density. Since white dwarf stars are an end state of stellar evolution, we can expect the stellar hydrogen to have been converted into heavier elements, and take A/Z = 2.0. Assuming the ionization is complete and that the composition is limited to a single atomic species, the mean electron molecular weight is $\mu_e = A/Z$, and the relativistic expression for degenerate electron pressure leads to an upper mass limit of:

$$M_c = 0.20\left(\frac{Z}{A}\right)^2\left(\frac{hc}{Gm_p^2}\right)^{3/2} m_p \approx \frac{5.80}{\mu_e^2} M_\odot \approx 1.44\ M_\odot\ , \tag{5.594}$$

where the solar mass $M_\odot = 1.989 \times 10^{33}$ grams. Although the exact value of the limiting mass will depend to some extent on the chemical composition, we know that there can be no white dwarf stars that are more than about 40 percent more massive than the Sun. This is in accord with astrometric measurements of their masses, with an average mass of about 0.6 Suns, and from masses inferred from their gravitational redshifts, with an average of about 0.8 solar masses (Lang, 1992).

Only if the white dwarf's mass is less than about 1.4 solar masses can the squeeze of gravity be counterbalanced by degenerate electron pressure. Although both Wilhelm Anderson and Edmund Stoner had previously called attention to this upper mass limit, it is often called the Chandrasekhar limit, because Subrahmanyan Chandrasekhar was the first to derive the detailed equilibrium configurations in which degenerate, relativistic electron gases support their own gravity (Chandrasekhar, 1931, 1934, 1935).

A collapsing star that exceeds the critical mass of about 1.4 Suns cannot be held in equilibrium by degenerate electron pressure, no matter how much you compress it. Such a star is no longer able to support itself, and must implode. Both Lev Landau and Chandrasekhar thought that these stellar corpses would collapse without limit (Landau, 1932; Chandrasekar, 1934). However neutrons provide another quantum-mechanical resistance to compression, and can stabilize the star at nuclear densities provided that the stellar mass is not too high.

Just two years after the discovery of the neutron, by bombarding atomic nuclei with high-energy radiation (Chadwick, 1932), Walter Baade and Fritz Zwicky invented the concept of a neutron star, consisting of neutrons, and introduced the notion that supernovae are powered by the gravitational energy released during the transformation of an ordinary star into a neutron star (Baade and Zwicky, 1934). When a sufficiently massive star exhausts its central nuclear fuel, the core mass might exceed the limiting mass for a white dwarf star. The gravity is then strong enough to force negatively charged electrons to combine with the positively charged protons in the nuclei of the star's constituent ions, forming neutrons that have no charge. Since atoms are largely empty space, with electrons orbiting a tiny nucleus at very large distances, the elimination of the electrons means that the matter can be compressed to very high densities.

The core will implode until it reaches the density of an atomic nucleus, or about 10^{15} g cm^{-3}. At these densities, a star slightly more massive than the Sun would have collapsed into a neutron star with a radius of about 10 kilometers. Moreover, in just one second, the collapse releases about 10 times more energy than the star has emitted during its entire life time, creating a supernova explosion.

The details of the implosion and subsequent explosion have been given by Burbidge, Burbidge, Fowler, and Hoyle (1957), Hoyle and Fowler (1960), Colgate and White (1966), May and White (1966), Arnett (1966, 1967 and 1968), Arnett and Cameron (1967), Trimble (1982), Woosley and Weaver (1986), and Bethe (1990). In brief, the material of the collapsing star moves inward, at first slowly, then more and more rapidly, and the core accelerates faster than the outer stellar envelope. When a star of between 8 and 30 times the Sun's mass exhausts the nuclear fuel in its central regions, the core collapse is halted at nuclear densities to form a neutron star. At that point, the implosion stops, and the infalling outer layers bounce off the hard neutron core. The resulting shock wave blows the outer envelope out into space, creating a visible supernova and leaving a neutron star behind. For a short time, the expanding debris can become 10 billion (10^{10}) times more luminous than the Sun, and briefly outshine an entire galaxy.

As with white dwarfs, there is a maximum mass for a stable neutron star, the tiny cinders that remain after a supernova explosion. The first detailed treatment of the equilibrium conditions for neutron stars was made by J. Robert Oppenheimer and George Volkoff. They assumed that the degeneracy pressure of densely packed neutrons provides the force that opposes gravity, and realized that the gravity of a neutron star is so strong that it needs to be described by Einstein's General Theory of Relativity. This can be seen by evaluating the parameter $GM/(c^2R)$ for a star of mass, M, and radius, R. It has a value of $\approx 10^{-4}$ at the surface of a white dwarf star and ≈ 0.1 at the surface of a neutron star.

Using precise expressions for the internal gravity of a spherical, nonspinning star (Schwarzschild, 1916) together with the equation of state of a degenerate neutron gas up to relativistic velocities, Oppenheimer and Volkoff (1939) found that a stable neutron star can exist only if the mass lies between 0.1 and 0.7 solar masses, and confirmed their results with analytical solutions provided in a companion paper by Tolman (1939). However, the neutrons are influenced by nuclear forces whose specifics were not fully understood, so the maximum mass for a neutron star is sensitive to the details and turns out to be a few times higher than this initial approximation.

The upper limit to the mass of a stable neutron star, called the Tolman–Oppenheimer–Volkoff limit, has been raised to about three solar masses using improved computer technology and a better understanding of the nuclear force (Hartle, 1978; Baym and Pethick, 1979). The first detailed treatment required electronic calculations and physical insights resulting from studies of atomic bombs (Harrison, Wakano and Wheeler, 1958; Colgate and White, 1966; May and White, 1966).

As suggested by Oppenheimer and Volkoff (1939), the mass of a stable neutron star becomes maximum for the stiffest possible equation of state. A rigorous solution, based on this approach together with Einstein's relativity theory, led to a maximum mass, $M_{\max}$, for a neutron star of 3.2 solar masses, or $M_{\max} = 3.2M_{\odot}$ (Rhoades and Ruffini, 1974). More recently, Kalogera and Baym (1996) have derived a firm upper bound of $M_{\max} = 2.9M_{\odot}$, including a possible contribution due to rapid, uniform rotation (Friedman and Ipser, 1987). The final collapse cannot be halted for imploding stellar cores of larger mass.

Stars with initial masses above forty solar masses almost certainly have core masses above M_{max}. Since fewer than one star in ten million is this massive, dying stars that do not become white dwarfs or neutron stars may be quite rare.

The very existence of the hypothetical neutron stars remained conjectural until 1968 when Jocelyn Bell and her thesis advisor, Anthony Hewish, discovered radio pulsars (Hewish et al., 1968). They emit rapid bursts of radio emission with a clocklike periodicity of about one second. Thomas Gold explained the short duration and the regularity of the periodic bursts, with periods accurate to better than 1 part in 1 million, in terms of rotating neutron stars, and correctly predicted that the gradual loss of rotational energy would lead to a slow lengthening of the pulsar periods with time (Gold, 1968). Pulsar astronomy is discussed by Lyne and Graham-Smith (1990).

Due to the conservation of magnetic flux during collapse, neutron stars would have surface magnetic fields of up to 10^{12} Gauss. The observed luminosity of the Crab Nebula supernova remnant was explained by the electromagnetic radiation of such a powerful, spinning magnet shortly before the discovery of radio pulsars (Pacini, 1967). The detection of a pulsar with a short period, of 33 milliseconds, in the Crab Nebula soon clinched the case for identifying radio pulsars with isolated, rotating, magnetized neutron stars (Staelin and Reifenstein, 1968; Comella et al., 1969).

Radio pulsars are interpreted as spinning neutron stars with an intense dipole magnetic field misaligned with the rotation axis. Beams of radio radiation, emitted by electrons accelerated along the polar magnetic fields, sweep around the sky once per revolution, accounting for the periodic bursts. Radio pulsars are energized by the rapid rotation of neutron stars, a byproduct of the conservation of angular momentum during stellar collapse. For a neutron star of radius, R, angular velocity, $\omega = 2\pi/P$, period, P, and mass, M, the angular momentum, J, is $J = M\omega R^2$, and the rotational energy, E, is:

$$E = \frac{M\omega^2 R^2}{2} = 4 \times 10^{46} \left(\frac{M}{M_\odot}\right) \left(\frac{R}{10\,\mathrm{km}}\right)^2 \left(\frac{1\,\mathrm{sec}}{P}\right)^2 \mathrm{erg} \ . \tag{5.595}$$

If the neutron star has a magnetic field of surface strength, B_s, inclined at an angle α with respect to the rotation axis, then the luminosity, L, of the rotating magnet is (Pacini, 1967; Ostriker and Gunn, 1969; also see Volume I)

$$L = -\frac{2\omega^4}{3c^3} R^6 B_s^2 \sin^2 \alpha = I\omega\dot{\omega} \ , \tag{5.596}$$

where c is the velocity of light, the moment of inertia $I = MR^2$ for a neutron star of mass, M, and radius, R, and $\dot{\omega} = d\omega/dt$ is the rate of change of $\omega = 2\pi/P$.

A characteristic pulsar age, τ_c, is given by:

$$\tau_c = \frac{P}{2\dot{P}} \ . \tag{5.597}$$

where P is the pulsar period and the spin-down rate is $\dot{P} = dP/dt$. Lang (1992) provides P, $\dot{P}$, and τ_c for hundreds of pulsars, with typical values of $P = 1$ second, $\dot{P} = 10^{-15}$ sec/sec and $\tau_c \approx 10^7$ years.

Radio pulsars are observed to be slowing down, and this observation is consistent with a rotating object for which the source of free energy is the kinetic energy stored in rotation. A pulsar with a period of $P \approx 1$ second has a stored energy of $E \approx 10^{46}$ ergs, and the energy loss rates are $L \approx 10^{31}$ erg s^{-1} for $P \approx 1$ second and a typical period increase of $dP/dt = \dot{P} \approx 10^{-15}$ sec/sec. This is about 100,000 times the luminosity in radio waves. Most of the pulsar energy output is thought to be in the form of a relativistic magnetized "wind" flowing away from the pulsar. Such a wind apparently lights up the Crab Nebula supernova remnant, but is usually invisible.

If the loss of rotational energy from the pulsar equals the amount of magnetic dipole radiation emitted by the neutron star, then the surface magnetic field strength is approximated by:

$$B_{\mathrm{s}} = \left(\frac{3c^3 I}{8\pi^2 R^6} P\dot{P}\right)^{1/2} \approx 3.2 \times 10^{19} (P\dot{P})^{1/2} \text{ Gauss} , \tag{5.598}$$

assuming that the neutron star has a radius $R = 10^6$ cm, a moment of inertia $I = MR^2 = 10^{45}$ g cm^2, the period P is expressed in seconds, $\dot{\mathrm{P}}$ is dimensionless, and that the spin axis is perpendicular to the magnetic axis, or an orthogonal rotator with $\sin \alpha = 1.0$.

The discovery of radio pulsars with extremely short periods inspired a renaissance in pulsar astronomy. They include the millisecond pulsar PSR 1937 + 21 with a period of $P = 1.5$ milliseconds (Backer et al., 1982) and the eclipsing pulsar PSR 1957 + 20 with $P = 1.61$ milliseconds (Fruchter, Stinebring and Taylor, 1988). The binary pulsar 1913+16, with a pulsar period of 59 milliseconds was described in Sect. 5.5.2 and Table 5.17 in the context of testing relativity theory (also see Hulse and Taylor, 1975; Taylor and Weisberg, 1989; Damour and Taylor, 1992; Hulse, 1994; and Taylor, 1994). Shocks apparently driven into the interstellar medium by a relativistic wind from the millisecond pulsar PSR 1957+20 have been detected by their Hα emission (Kulkarni and Hester, 1988). The parameters of the fastest pulsars are given by Lang (1992). Fast older radio pulsars, such as those found within globular star clusters, are probably old pulsars that have been spun-up by the accretion of matter from a binary stellar companion (Smarr and Blandford, 1976; Alpar et al., 1982; Ruderman, Shaham and Tavani, 1989).

The vast majority of radio pulsars are isolated neutron stars with surface magnetic fields of 10^{11} to 10^{12} Gauss. The radio pulsars that are members of binary systems have magnetic fields that are small compared to those of single radio pulsars, typically about 10^8 to 10^9 Gauss. These members of binary systems also spin more rapidly on average with periods of about 10 milliseconds; which has a bearing on their formation and evolution. Only about 3 percent are known to be members of binary systems. Binary and millisecond pulsars are reviewed by Phinney and Kulkarni (1994), while their formation and evolution are discussed by Bhattacharya and Van Den Heuvel (1991).

Radio observations of four neutron star binary pulsar systems, each consisting of two neutron stars in orbit about a common center of mass, have constrained the masses of eight neutron stars (Finn, 1994). The lower limit, M_{L}, and upper bound, M_{U}, of these neutron stars lie in the range:

$$1.01 < \frac{M_L}{M_\odot} < 1.34 \tag{5.599}$$

and

$$1.43 < \frac{M_U}{M_\odot} < 1.64 \tag{5.600}$$

where $M_\odot = 1.986 \times 10^{33}$ grams is the mass of the Sun. Thorsett et al. (1993) obtain neutron star masses of $M_{NS} = 1.35 \pm 0.27$ solar masses. These limits give observational support to neutron star formation scenarios that fall predominantly in the range 1.3 to 1.6 solar masses.

Rapidly-rotating, relatively-young radio pulsars are still found within visible supernova remnants. They include the Crab pulsar, PSR 0531 + 21, and the Vela pulsar, PSR 0833 + 45. They have periods, P, period changes, $\dot{P}$, and characteristic ages, τ_c, of $P = 33.3$ milliseconds, $\dot{P} = 4.2 \times 10^{-13}$ s s^{-1}, and $\tau_c \approx 10^3$ years for the Crab pulsar and P $= 89.3$ milliseconds, $\dot{P} = 1.24 \times 10^{-13}$ s s^{-1}, and $\tau_c \approx 10^4$ years for the Vela pulsar (Lang, 1992). They are both amongst the three strongest high-energy (greater than tens of MeV) γ-ray sources.

Now, after nearly twenty years of mystery, the other strongest γ-ray source, known as Geminga or 2GC 195 + 04, has also been identified with a nearby, isolated rotating neutron star, but without the characteristic radio pulses. Geminga was identified with an X-ray source, 1E 0630 + 178 (Bignami, Caraveo and Lamb, 1983), and about a decade later found to emit pulsed radiation with a period of $P = 237$ milliseconds at both X-rays (Halpern and Holt, 1992) and γ-rays (Bertsch et al., 1992). The period and period derivative of $\dot{P} = (11.4 \pm 1.7) \times 10^{-15}$s s^{-1} suggest that Geminga is a nearby, isolated, spinning neutron star with a magnetic field of about 10^{12} Gauss and a characteristic age of $\tau_c \approx 3 \times 10^5$ years. Yet, this anomalous pulsar does not emit detectable radio pulses, as the other isolated neutron stars do, perhaps because the radio beam is very much narrower than the beam at high energies and thus does not intersect our line of sight.

In addition to radio pulsars, neutron stars are also found in interacting binary systems that emit intense X-rays (Giacconi et al., 1972; Schreier et al., 1972; Tananbaum et al., 1972). In such binaries, a neutron star closely orbits a normal, optically visible star, and draws gas away from it. The infalling, accreted gas is heated to millions of degrees and emits X-rays. The X-ray binary stars are discussed by Lewin, Van Paradijs and Van Den Heuvel (1995). The origin and evolution of both X-ray binaries and binary radio pulsars are reviewed by Verbunt (1993), and the origin of neutron stars in binary stars is reviewed by Canal, Isern and Labay (1990).

Unlike radio pulsars, the X-ray pulsars spin up with increasing time. The decrease, $\dot{P}$, in pulsar X-ray period, P, is given by (Joss and Rappapaport, 1984):

$$\frac{\dot{P}}{P} = -3 \times 10^{-5} f \left(\frac{P}{1\,\mathrm{s}}\right) \left(\frac{L_X}{10^{37}\,\mathrm{erg\,s^{-1}}}\right)^{6/7} \tag{5.601}$$

where L_X is the absolute X-ray luminosity, and the dimensionless function f is expected to be of order unity for a neutron star and contains parameters that are not yet measurable for most or all of the X-ray pulsars. The periods, period derivatives and X-ray luminosities of binary X-ray pulsars are given by Lang (1992); values range from $P = 0.7$ to 835 seconds, $\dot{P}/P = -10^{-6}$ to -10^{-2} year^{-1}, and $L_X = (0.1 \text{ to } 100) \times 10^{37}$ erg s^{-1} .

Measurements of the pulse arrival times from six binary X-ray pulsars have been combined with optical observations of their companions to infer the neutron star masses, obtaining values between 1 and 2 solar masses (Joss and Rappaport, 1984; Lang, 1992). Thus, observations of both radio pulsars and X-ray pulsars indicate that no known neutron star has a mass exceeding 2 to 3 solar masses, confirming theoretical expectations for their upper mass limit. We now know that stars more massive than that must collapse without limit to form a black hole.

The infall of gas onto a star or other gravitating system is called accretion. In X-ray binary systems matter can be accreted by the compact companion star. That is, if the neutron star is in sufficiently close orbit, its gravitational attraction may pull off the outer layers from the companion. A rotating neutron star in a close binary system can also form an X-ray pulsar. The strong magnetic field of a neutron star can disrupt the accreting material far from the stellar surface, deflecting the accretion flow toward the magnetic polar caps. The infalling gas is heated and emits X-rays, observed each time a pole rotates into view.

Such tidal mass transfer is called Roche lobe overflow. The Roche lobe is a critical surface such that matter exterior to it is not bound to the star, but is attracted by the nearby compact star. Also, when the visible star is massive, it produces a strong stellar wind and the neutron star may accrete part of the ejected wind material even while its companion is confined inside its Roche lobe.

Mass transfer between a pair of stars is not a new concept. When one stellar member of a close binary system evolves and expands, its outer material can be drawn toward the other component by its gravitational attraction. Gerard Kuiper has, for example, discussed the flow of matter between contact binaries that have a common envelope (Kuiper, 1941).

By studying old novae, when the intense light of the nova outburst has faded to a relatively weak level, Robert Kraft showed that they are members of short-period (hours) binary stars consisting of a blue white dwarf star and a cool red star, and that their short orbital periods meant that they are close together (Kraft, 1964). As the normal red companion evolves, its extended hydrogen-rich atmosphere can overflow onto the white dwarf companion, creating the explosive visible nova outburst (see Schatzman, 1949).

When the bright X-ray source Sco X-1 was identified with an optical object that resembled an old nova, the nova mechanism was extended by Iosif Shklovskii to suggest that the X-rays could be produced by a neutron star that accretes gas from a nearby visible companion (Shklovskii, 1967). Because the X-rays shine at much higher energies than conventional optically visible novae, the collapsed component of the binary system had to be much denser than a white dwarf, perhaps a neutron star or even a black hole. Prendergast and Burbidge (1968) then developed the concept, showing that rotation of the compact object will drive the infalling material into a whirling accretion disc that emits X-rays as its material spirals inward.

The most luminous galactic X-ray sources are close, interacting binary systems which contain an optically visible component and a neutron star or black hole that is completely invisible optically. Such cosmic X-ray sources have been discovered using X-ray telescopes aboard satellites, including the Uhuru, Ariel, High Energy Astrophysical Observatory, Einstein and Ginga satellites. Well known examples of intense galactic X-ray sources include Centaurus X-3, Cygnus X-1, Cygnus X-3 and Scorpius X-1, named after the constellation they appear in followed by an X for X-ray and a number ordered by decreasing brightness.

The neutron stars are sometimes identified by their periodic X-ray emission. At least twenty of the compact X-ray binaries consist of a bright, massive, giant visible star, of at least 10 solar masses, and an X-ray pulsar that is interpreted as a spinning neutron star (Lang, 1992). Centaurus X-3 is an example. The enormous magnetic field of the neutron star diverts the flow of infalling gas and channels it onto the magnetic polar caps, producing periodic X-ray emission as they rotate into view.

Other compact X-ray binaries, at least forty of them, have been identified with an optically-faint visible companion, with a mass comparable to that of the Sun (Lang, 1992). Examples include Cygnus X-3 and Scorpius X-1. Their high absolute X-ray luminosities, $L_X \approx 10^{38}$ erg s^{-1}, are comparable to the maximum X-ray emission of an accreting neutron star.

If all the kinetic energy of infalling matter is given up to radiation at the stellar surface, then the accretion luminosity, L_{acc}, of a star of mass, M, and radius, R, is given by:

$$L_{acc} \approx \frac{G\dot{M}M}{R} \approx 10^{41}\dot{M}\left(\frac{M}{M_\odot}\right)\left(\frac{R_\odot}{R}\right) \text{erg s}^{-1} \quad (5.602)$$

and

$$L_{acc} \approx 1.3 \times 10^{36}\,\dot{M}_{16}(M/M_\odot)(10^6\,\text{cm}/R)\,\text{erg s}^{-1} \ , \quad (5.603)$$

where the mass accretion rate is, $\dot{M}$, which is equivalent to $\dot{M} = 10^{16}\dot{M}_{16}$ g s^{-1}, the accretion rate of 10^{16} g s^{-1} $\approx 1.5 \times 10^{-10}$ solar masses per year, and for a neutron star $M/M_\odot \approx 1$ and (10^6 cm/R) ≈ 1. In the first numerical approximation, $M_\odot$ and $R_\odot$ are the solar values and $\dot{M}$ is the accretion rate in units of solar masses per year.

Eddington (1926) showed that there is a maximum luminosity, now called the Eddington limit, L_{Edd}, for any source of radiation before it blows away the surrounding matter. In effect, the radiation pressure pushes against the accreted material. When the radiation pressure balances the gravitational force, the radiation luminosity is:

$$L_{Edd} = \frac{4\pi G m_p c M}{\sigma_T} \approx 10^{38}\left(\frac{M}{M_\odot}\right)\text{erg s}^{-1} \approx 10^{4.5}\left(\frac{M}{M_\odot}\right)L_\odot \ , \quad (5.604)$$

which depends only on the mass, M, of the accreting object. Here $\sigma_T = 6.7 \times 10^{-25}$cm^2 is the Thomson scattering cross section for free electrons, since the scattering cross-section for protons is a factor of $(m_e/m_p)^2$ smaller, for an electron mass, m_e, and a proton mass of m_p. The proton mass enters the

Detailed properties of circular orbits around a Kerr black hole are given by Bardeen, Press and Teukolsky (1972). The coordinate radius of the innermost stable orbit is given by

$$r = r_{\rm ms} = \text{ outermost root of } r^2 - 6Mr \pm 8aM^{1/2}r^{1/2} - 3a^2 = 0 \ . \quad (5.632)$$

For $a = 0$, we have $r_{\rm ms} = 6M$, for $a = M$ we have $r_{\rm ms} = M$ or $9M$. The innermost stable orbit is the maximally bound orbit. The binding energy of the maximally bound orbit varies from

$$\left(1 - \frac{2\sqrt{2}}{3}\right)mc^2 \quad \text{for Schwarzschild } (a = 0) \quad (5.633)$$

to

$$\left(1 - \frac{\sqrt{3}}{3}\right)mc^2 \quad \text{for extreme Kerr } (a = M) \ , \quad (5.634)$$

where m is the rest mass of the orbiting particle.

To round out our discussion of the theory of black holes, we summarize the various theorems and laws that describe them. The rest of this section will then focus on the actual observations of black holes, and the theory behind the radiation that is used to detect them.

Birkhoff (1923) theorem: Any spherically symmetric geometry of a given region of space-time which is a solution to the Einstein field equations in vacuum is necessarily a piece of the Schwarzschild geometry. That is, a spherically symmetric gravitational field in empty space must be static (zero angular momentum) with a Schwarzschild metric.

Isreal (1967, 1968) theorem: Any static black hole with event horizon of spherical topology has external fields determined uniquely by its mass, M, and charge, Q, and moreover those external fields are the Schwarzschild solution if $Q = 0$ and the Reissner–Nordstrom solution if $Q \neq 0$. Hawking had already shown that a stationary black hole must have a horizon with spherical topology, and that it must either be static or axially symmetric, or both. The Reissner (1916)–Nordstrom (1918) metric is given by

$$ds^2 = -\left(1 - \frac{2M}{r} + \frac{Q^2}{r^2}\right)dt^2 + \left(1 - \frac{2M}{r} + \frac{Q^2}{r^2}\right)^{-1} dr^2 + r^2(d\theta^2 + \sin^2\theta \, d\varphi^2) \ , \quad (5.635)$$

and the electric field, E, is given by

$$E = \frac{Q}{r^2} \ , \quad (5.636)$$

where the charge is Q. Here geometrized units have been employed.

Penrose (1969) theorem: By injecting matter into a Kerr–Newman black hole in the proper way one can extract rotational energy from a spinning black hole. That is, when certain particles traverse the field of a black hole they may gain energy at the expense of the rotational energy of the black hole.

Carter (1971) theorem: All uncharged, stationary, axially symmetric black holes with event horizons of spherical topology fall into disjoint families not deformable into each other. The black holes in each family have external gravitational fields determined uniquely by two parameters – the mass, M, and the angular momentum, J, or spin $a = J/M$. The only such family presently known is the family of Kerr metrics.

No-Hair Theorem: All of the properties of a black hole are precisely predictable from its mass, angular momentum and electric charge. Whatever additional characteristics one tries to attach to the collapsing matter, they all vanish as far as the external space-time is concerned. In other words, a large amount of information is lost when a body collapses to form a black hole. All memory is apparently removed of any particle content or the types of matter that went into making the black hole. Such a conclusion has become known as the no-hair theorem, the hair being anything that might stick out of the hole to reveal the details from which it formed. That may be just a fancy way of saying that black holes have no identifiable external features. It essentially means that all the details of the infalling matter are washed out and removed, and that the final configuration is uniquely determined by three parameters which may be identified with the mass, the angular momentum, and the electric charge of the source. Any black hole is therefore uniquely described by the Kerr or Newman metrics (Ginzburg, 1964; Doroshkevich, Zeldovich and Novikov, 1965; Israel, 1967; Carter, 1971; Mazur, 1982). The question of what happens to the information in matter destroyed by a black hole is a current topic of intense speculation and research (Hawking and Penrose, 1996; Susskind, 1997).

Laws of Black Hole Mechanics: The similarity between the evolution of an accreting black hole and the laws of thermodynamics was noted by Christodoulou (1970). Subsequent comparisons with thermodynamics suggested analogous black-hole laws, with entropy and temperature respectively replaced by horizon area and surface gravity (Bekenstein, 1972, 1973; Hawking, 1972, 1974). Using geometrized units in which $G = c = 1$, the angular velocity Ω, the surface gravity κ, and the surface area A of a Kerr black hole are, in terms of the its mass, M, and angular momentum, J:

$$\Omega = J/\{2M(M^2 + (M^4 - J^2)^{1/2})\} \qquad (5.637)$$

$$\kappa = (M^4 - J^2)^{1/2}/\{2M(M^2 + (M^4 - J^2)^{1/2})\} \qquad (5.638)$$

$$A = 8\pi(M^2 + (M^4 - J^2)^{1/2}) \ . \qquad (5.639)$$

The surface area, A, of the event horizon of a Kerr–Newman black hole is given by

$$A = 8\pi[M^2 + M(M^2 - a^2 - Q^2)^{1/2} - Q^2/2] \ , \qquad (5.640)$$

Here, again, geometrized units have been employed for the mass, M, angular momentum, J, and for the spin $a = J/M$. The area of a nonspinning (Schwarzschild) black hole is proportional to the square of the mass, and a spinning (Kerr) hole can have a smaller surface area than a nonrotating hole of the same mass. A process in the ergosphere can therefore slow down the hole's spin and extract energy from it, without decreasing the area of the hole.

The analogy with thermodynamics has been consolidated into four laws of black hole mechanics (Bardeen, Carter and Hawking, 1973; Sciama, 1976):

The zeroth law. The surface gravity, κ, of a stationary black hole is constant over the horizon.

The first law. In a transformation from one state to a nearby state, the energy, E, of the system changes by:

$$\Delta E = \frac{c^2}{G}\frac{\kappa \Delta A}{8\pi} + W \ , \tag{5.641}$$

where A is the surface area of the horizon and W is the total of any work done in changing the rotation of the black hole and any work done on any matter that may be outside the black hole. We can also take this law in the form:

$$\delta M = \frac{\kappa \delta A}{4\pi} + \Omega \delta J \ , \tag{5.642}$$

which corresponds to the first law of thermodynamics, in which total energy is conserved, if the surface gravity κ is proportional to the temperature, T, and δA to the change in entropy δS. Here δM is the change in mass, δJ the change in momentum, and Ω is the angular velocity at the event horizon.

The second law. During any process in an isolated system, the total area of all horizons is non-decreasing. For any process involving black holes, the sum of the surface areas, A, of the event horizons of all black holes involved can never decrease. The area of the horizon is like entropy in thermodynamics and always grows. This means that if two black holes merge, the horizon of the final hole must have an area equal to or larger than the sum of the original two.

The third law. It is impossible by any finite sequence of steps to bring the surface gravity to the value zero. One form of the conventional third law of thermodynamics is that one cannot lower the temperature of a physical system to the absolute zero in a finite number of steps. The black hole version would simply replace temperature by surface gravity (Isreal, 1986). Absolute zero temperature would then correspond to a surface gravity of zero, or in the Kerr solution to the maximum rotation rate in which $J_{\max} = GM^2/c = M^2$ in geometrized units. If this could be achieved in a finite number of steps, then a naked singularity would be possible.

Evaporation of Black Holes. All black holes, spinning or nonspinning, radiate precisely as though they had a temperature that is proportional to their surface gravity and therefore evaporate by creation of particle–antiparticle pairs from the vacuum near the horizon (Hawking, 1974). The emitted particles have a thermal, or blackbody, spectrum with a precise temperature, T, given by (Raychaudhuri, Banerji and Banerjee, 1992):

$$T = \frac{\hbar c^3}{8\pi k G M} \approx 6.2 \times 10^{-8} \left(\frac{M_\odot}{M}\right) \text{degrees} \tag{5.643}$$

and an entropy, S, given by:

$$S = \frac{1kc^3}{4G\hbar} A = \frac{k}{4L_p^2} A \approx 10^{48} A \, \text{erg K}^{-1} \, \text{cm}^{-2} \ , \tag{5.644}$$

where $\hbar$ is Planck's constant, k is Boltzmann's constant, the area of the horizon is A, the surface gravity of the black hole is κ, and the Planck length, L_p, is given by

$$L_p = (\hbar G/c^3)^{1/2} \approx 1.6 \times 10^{-33} \text{cm} \ . \tag{5.645}$$

The time τ for complete evaporation is:

$$\tau = \frac{256 \, \pi^3 G^2 k^4}{3 \quad \sigma e^6 \hbar^4} M^3 \approx 10^{10} \left(\frac{M}{10^{15} \, \text{g}} \right)^3 \text{years} \ . \tag{5.646}$$

The temperature of a solar mass, $M_\odot \approx 10^{33}$ grams, black hole is less than 10^{-7} degrees Kelvin, and its lifetime against evaporation is about 10^{64} years. So, black holes formed by stellar collapse do not evaporate over the current age of the Universe, about 15 billion years. However, mini-black holes created in the big bang, with masses M of about 10^{15} grams, might be evaporating now, or about 15 billion years after the big bang.

How can you detect a black hole? Neither matter nor radiation can escape the strong gravitational pull within the event horizon, so such objects would be truly black and invisible if they were isolated in space. But we can expect X-ray radiation from a black hole in a binary system where it can capture gas from a nearby luminous companion. This gas, rushing into the hole, can be heated to temperatures high enough to result in X-ray emission (Zeldovich and Novikov, 1966, 1971). The X-rays come not from the compact star itself, but from gas that is drawn from the normal star to the black hole by its powerful gravitational attraction. As the hot, stolen material flows inward, it forms an intense X-ray whirlpool with blackness at its core. The X-rays are emitted as a final gurgle, as a sort of last hurrah, just before the swirling material disappears into the black hole, somewhat like the final brilliant flash of a light bulb just before it burns out.

As the hole pulls matter from its companion, the infalling material accelerates, moving faster and faster, and gains kinetic energy through the action of the ever-increasing gravitational forces. This increased energy of motion is changed into heat energy by collisions between the rapidly moving particles as they are compressed to fit into the hole, resulting in million-degree temperatures and X-ray radiation. The swirling, infalling material is so hot that it emits almost all of its radiation at X-ray wavelengths, while remaining undetected at visible wavelengths. The power source for the X-rays arises directly from the release of gravitational energy of the matter that has accreted from the companion star.

Because a black hole has no well-defined surface features, such as magnetic poles, the X-ray radiation from matter falling towards it will not show any regular or periodic behavior. So, a high-luminosity X-ray binary that does not emit periodic X-ray emission may qualify.

Both black holes and neutron stars have sufficiently high surface gravities to produce hot X-ray emitting gas near the limiting Eddington value, so the candidate black holes have to be distinguished from neutron stars by direct measurements of their greater mass. That is, since mass falling into a black hole

(see Sect. 5.6.3), the radio outbursts from GRO J1655-40 provide evidence for material emerging from the accretion disc at the velocity of light before matter disappears into the black hole.

Formidable black holes are found in the cores of ordinary galaxies in the nearby Universe. They resemble scaled-up versions of stellar black holes with millions if not billions of solar masses packed into a region a few light years across. Like their stellar counterparts, these massive black holes cannot be directly observed. Their presence is indirectly inferred from stars and gas orbiting the core, being drawn into the center at ever increasing velocities by the black hole's relentless gravity.

Strong circumstantial evidence for massive black holes is provided by measurements of the mass, M, and absolute luminosity, L, near the center of nearby galaxies. The mass, $M(r)$, within a radius, r, can be determined from the orbital velocity, V, obtained from Doppler-shift measurements of spectral lines. For observable matter, we have to a first approximation:

$$M(r) = \frac{V^2 r}{G} , \tag{5.648}$$

where G is the Newtonian gravitational constant. Thus, one signature of a black hole will be the presence of stars moving anomalously fast very near the nucleus of the galaxy. A supermassive black hole will produce ever-faster orbital motions as their centers are approached, and observations close to the nucleus will permit calculations of the mass contained within them (see Table 5.21). Exceptionally high mass-to-light ratios, M/L, within the innermost regions of a galaxy indicate the presence of a massive dark object that is most likely a supermassive black hole. A typical stellar system might, for example, have M/L of 3 to 5 $M_{\odot}/L_{\odot}$, whereas measurements of $M/L \geq 100\, M_{\odot}/L_{\odot}$ are found within the central 10 to 100 light years of some galaxies (also see Table 5.21), indicating that they emit very little light relative to their mass. The observations imply that the central object, whatever it may be, is a localized concentration of large amounts of non-luminous matter. Some of these dense concentrations of mass have been squeezed into regions as small as 0.6 light years across, or no larger than our solar system.

The most compelling and earliest evidence for a dark, supermassive object in the nucleus of a galaxy resulted from spectroscopic and photometric observations of M87, also called NGC 4486, located about 50 million light years away. It is one of the biggest and brightest elliptical galaxies in the nearby Virgo cluster of galaxies, and also associated with an intense radio source named 3C 274 or Virgo A. In seminal papers written about two decades ago, Wallace Sargent, Peter J. Young and their colleagues showed that the velocity dispersion and brightness of M87 increase strongly toward the center. The dynamical evidence, including Doppler broadening of up 1,500 km s^{-1} for lines of ionized oxygen, indicated that a mass equivalent to about 5×10^9 Suns is concentrated within the central 1.5 seconds of arc (Sargent et al., 1978). Thus, about 5 billion solar masses had to be packed within a very small volume, less than 100 parsecs across, or roughly 300 light years in diameter. The central mass-to-light ratio had to exceed 60 $M_{\odot}/L_{\odot}$, suggesting that this object

Table 5.21. Candidate supermassive black holes in the nuclei of nearby galaxies.*

Galaxy	Type	Distance (millions of light-years)	Galaxy Luminosity (10^9 Suns)	Black-hole Mass (Suns)	Mass-to-light-ratio (Solar Units)	Distance from galaxy's center (light-years)
Milky Way core	Sbc	0.028	1.9	2×10^6	–	1.0
NGC 221 M32	E2	2.3	0.25	3×10^6	>10	2
NGC 224 M31	Sb	2.3	5.2	3×10^7	350	3
NGC 4258 M106	Sbc	24	1.3	4×10^7	–	0.6
NGC 3115	S0	27	14.2	2×10^9	85	13
NGC 4594 M104	Sa	30	47	5×10^8	55	58
NGC 3377	E5	32	5.2	8×10^7	3	75
NGC 3379	E1	32	13	5×10^7	–	–
NGC 4486 M87	E0	50	56	3×10^9	500	60
NGC 4261	E2	90	33	1×10^9	2,100	50

* Adapted from Douglas O. Richstone (private communication) and Ford and Tsvetsanov (1996). The galaxy luminosity is in units of $10^9 L_{\odot} = 3.85 \times 10^{42}$ erg s^{-1}, the solar luminosity is $L_{\odot} = 3.85 \times 10^{33}$ erg s^{-1}, the black hole mass is in solar units $M_{\odot} = 1.989 \times 10^{33}$ grams, the mass-to-light ratio for the observed nuclear region is in solar units of $M_{\odot}/L_{\odot} = 0.52$ gm s erg^{-1}, and the distance from the galaxy center is for the innermost dynamical measurement.

included substantial dark mass, such as a supermassive black hole (Young et al., 1978).

By getting a clearer image of the light at the center of M87, the Hubble Space Telescope provided compelling dynamical evidence for a supermassive black hole, revealing a central disk of ionized gas. Observations at just 0.25 seconds of arc to either side of the center showed that the gas on one side is receding from us at about 500 km s^{-1}, and the other side is approaching at a similar velocity, strongly implying that the gas disk is orbiting the center of M87 at that velocity (Harms et al., 1994). These measurements indicate that $(2.4 \pm 0.7) \times 10^9$ solar masses are contained within 18 parsecs, or 59 light years, of the nucleus (Ford et al., 1994). Moreover, the mass-to-light ratio within that volume is 500 times that of the Sun. Since visible stars cannot account for so much nonluminous matter, the disk of ionized gas seems to be feeding a supermassive black hole in the center of M87.

Energy released from gas falling toward M87's center probably produces a jet of electrons moving nearly at the speed of light. Early in this century, Heber D. Curtis discovered a plume, or jet, of hot, optically luminous gas issuing from the galaxy's core (Curtis, 1918, also see Baade and Minkowski, 1954). The optical radiation of bright features, called knots, in the jets is strongly linearly polarized (Baade, 1956), suggesting the synchrotron radiation of electrons moving at relativistic velocities near the speed of light, probably ejected from the nucleus of the giant galaxy (Shklovskii, 1977). M87 is also a strong radio source, dubbed Virgo A or 3C 274 (Bolton, Stanley and Slee, 1949), and the jet provided early evidence for continuous collimated flow that apparently connects a central galactic nucleus to more extended radio emission (Rees, 1978). Its unusual jet of luminous matter, stretching 50,000 light years across, has been traced down to the galaxy core using radio interferometric techniques (Hardee, Owen and Cornwell, 1988; Biretta and Reid, 1988, Reid et al., 1989), and connects outward to at least one of the two giant radio lobes that extend beyond the visible extent of the galaxy. Hubble Space Telescope images show that an inner optical disk is oriented nearly perpendicular to the galaxy's one-sided jet, supporting the notion that the jet emerges from an accretion-fed compact object at the center of the galaxy.

High-resolution observations of the nuclei of several nearby galaxies also show a sharp rise in the velocities of stars near their center, accelerations that would be produced by the gravity of dark mass concentrations. Central masses of a million to a billion (10^6 to 10^9) Suns have been inferred, including the nucleus of our own Milky Way (also see Table 5.21).

Stellar- and gas-dynamical evidence for supermassive black holes in nearby galactic nuclei is reviewed by John Kormendy and Douglas Richstone (Kormendy and Richstone, 1995) – also see Tremaine (1997). The detailed measurements evaluate the mass, $M(r)$, given by:

$$M(r) = \frac{V^2 \mathrm{r}}{G} + \frac{\sigma_\mathrm{r}^2 r}{G}\left[-\frac{d \ln \nu}{d \ln r} - \frac{d \ln \sigma_\mathrm{r}^2}{d \ln r} - \left(1 - \frac{\sigma_\theta^2}{\sigma_\mathrm{r}^2}\right) - \left(1 - \frac{\sigma_\phi^2}{\sigma_\mathrm{r}^2}\right)\right], \tag{5.649}$$

where V is the rotation velocity; σ_r, σ_θ and σ_ϕ are the radial and azimuthal components of the velocity dispersion; and G is the gravitational constant. The density ν is not the total mass density ρ; it is the density of the tracer population whose kinematics are measured. Most of the total mass density is not seen, for the stars that emit most of the light contribute almost none of the mass. In practice, calculations are made assuming $\nu(r) \propto$ volume brightness, which means that M/L for the tracer population is independent of radius.

Both the nearest large spiral galaxy M31, also called Andromeda and NGC 224, and its small elliptical galaxy satellite, M32 or NGC 221, exhibit rapid rotation, high-velocity dispersion, and a high mass-to-light ratio in the central few parsecs, suggesting dark, central objects of roughly 10 million solar masses (Tonry, 1987; Dressler and Richstone, 1988; Kormendy, 1988; Van Der Marel et al., 1997). The edge-on spiral galaxy, NGC 3115, similarly weighs in with a probable black hole equivalent to a million solar masses (Kormendy and Richstone, 1992), while stellar dynamics of the nuclear bulge in the Sombrero galaxy, also called M104 or NGC 4594, suggest that it also harbors a supermassive black hole (Kormendy, 1988). There are no signs of substantial activity in the nucleus of Andromeda, M32 or the Sombrero galaxy, suggesting that their inner regions my be swept relatively clean of massive infalling material.

If black holes reside in many nearby galaxies, our own Galaxy, the Milky Way, might be expected to contain one. As suggested by Lynden-Bell and Rees (1971), it is likely that the Milky Way contains a massive black hole, probably associated with the compact, nonthermal radio source located at the dynamical center of our Galaxy (Riegler and Blandford, 1982; Lo, 1986; Brown and Liszt, 1984; Genzel and Townes, 1989; Liszt, 1988; Townes, 1989; Blitz et al., 1993 and references contained therein). This radio source, called Sagittarius A* or Sgr A* for short, is very compact, with a linear size less than 10 AU, requiring an object of stellar dimensions or less. (The radius of the Earth's orbit is 1 AU = 1.5×10^{13} cm = 0.0016 light years.) The observed radial velocities of ionized gas and stars near Sgr A*, obtained from Doppler shift measurements of their spectral lines, suggest that the innermost light-year or two of the galactic nucleus contains a mass of about 3 million solar masses (Blitz et al., 1993; Jayawardhana, 1995; Krabbe et al., 1995). Observations of stellar proper motions confirm this interpretation, with $(2.45 \pm 0.4) \times 10^6$ solar masses located within a fraction of a light year of Sagittarius A* (Eckart and Genzel, 1996). The apparent gravitational effect of Sgr A* on its surroundings, together with its small size, imply the presence of a supermassive black hole at the center of our Galaxy. Still, it has a rather feeble radio and X-ray output, suggesting a dormant black hole.

Radio and X-ray astronomers have greatly bolstered the evidence for black holes by observing two Seyfert galaxies, named after Carl Seyfert's discovery of extraordinary activity in the nuclei of these spiral galaxies (Seyfert, 1943). He used spectral observations of visible light to show that hot ionized gas in the small bright nuclei is moving at velocities of more than 1,000 km s^{-1}. Continent spanning radio telescopes, linked together in the Very Long Baseline Array, or VLBA, provide the astounding angular resolution of 0.0002 seconds of arc, or 0.2 milliarcseconds, thereby resolving the details of the rotating, nuclear gas of at least one Seyfert galaxy. Makoto Miyoshi, James Moran and their colleagues demonstrated that masers, produced by water molecules in the central disk of M106, or NGC 4258, move in orbits controlled by a mass of 36 million Suns, compressed in a region less than 0.13 parsec, or 0.42 light years, in radius (Miyoshi et al., 1995). The X-ray spectrum of another Seyfert galaxy, dubbed MCG 6-30-15, is of exactly the right shape to be coming from the innermost regions of a rotating accretion disk (Tanaka et al., 1995). Japan's Advanced Satellite for Cosmology and Astrophysics (ASCA) has probed the X-ray spectral line emitted by ionized iron, whose width corresponds to a velocity of 100,000 km s^{-1}, or about one third the velocity of light. The X-rays are emitted by heated gas, falling down into the black hole, and at these speeds the infalling gas must be just a few Schwarzschild radii outside the event horizon. Although such a small region cannot be resolved, and masses cannot be determined because of the high infall speed, the asymmetric, redshifted X-ray line resembles that expected from the innermost parts of an accretion disc.

Unfortunately, only relatively nearby galaxies are close enough to view their central regions with sufficient detail, or angular resolution, using conventional optical techniques. Solid evidence for nuclear black holes is therefore largely constrained to the older, more proximate galaxies that are relatively quiescent versions of younger, active ones. That is, because of their advanced age, the dormant cores of nearby galaxies are starving with a dwindling supply of fuel that once fed a higher rate of activity. More distant galaxies, whose light was generated at a younger stage of evolution, seem to be passing through a turbulent early phase with brilliant active centers that are powered by massive black holes.

Indeed, Jaffe et al. (1993) have used the Hubble Space Telescope to image a nuclear disc of gas and dust confined to the central 17 parsecs, or about 50 light years, of galaxy NGC 4261, which apparently energizes the radio galaxy 3C 270. The ionized gas is in Keplerian motion around a central mass of $(4.9 \pm 1.0) \times 10^8$ solar masses with a mass-to-light ratio of $M/L \approx 2,100\, M_{\odot}/L_{\odot}$, suggesting a dark, massive central black hole (Ferrarese, Ford and Jaffe, 1996). We therefore now turn to the more powerful radio galaxies and quasars.

5.6.3 Radio Galaxies, Quasars, Superluminal Motions, Active Galactic Nuclei, Supermassive Black Holes, and Galactic Micro-Quasars

The discovery of discrete, extragalactic radio sources, now called radio galaxies, led to a new picture of an active, violent Universe rather than a quiet, sedate one. These powerful extragalactic radio sources are characterized by nonthermal radiation and by the expulsion of energy in two oppositely directed beams on linear scales from parsecs to Megaparsecs. A radio galaxy often consists of two extended regions containing magnetic fields and synchrotron-emitting relativistic electrons moving at nearly the speed of light. Each of these radio lobes is linked by a jet to a central compact radio source located in the nucleus of the associated optically-visible galaxy. An example of such a double-lobed radio galaxy is Cygnus A, shown in Fig. 5.36. This nucleus is believed to power the jets and the extended nonthermal radio emission.

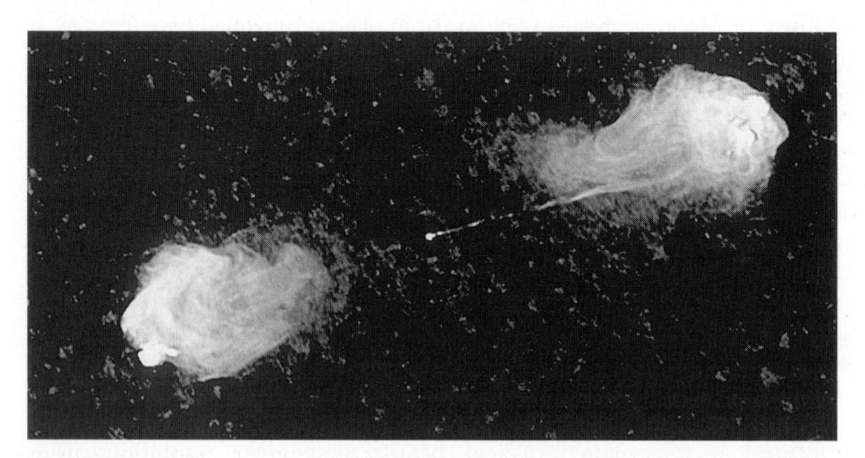

Fig. 5.36. Radio map of the source Cygnus A (3C405) produced by R. A. Perley and J. W. Dreher, Visible features are a) the compact core in the center of the galaxy, b) the jets emanating from the core and carrying energy and particles to the lobes, and c) the lobes themselves and the backflowing cocoons with wispy regions of enhanced emission. Barely visible in the overexposed lobes are the "hot spots" where the jets are terminated. (Courtesy of the National Radio Astronomy Observatory.)

To fully understand their nature, distances had to be determined from redshifts using optical spectroscopy, and this required accurate radio positions and angular sizes. Because angular resolution increases linearly with wavelength, a radio telescope has to be a million times as large as an optical telescope to provide the same angular resolution, and this was accomplished by connecting radio telescopes over large distances, creating interferometers that eventually provided far sharper images than the best optical instruments.

The very first identifications of the most intense radio sources, specified by the constellation in which they are found, included Virgo A, associated with the elliptical galaxy M87 (Bolton, Stanley and Slee, 1949), and Cygnus A, which coincided with a distant elliptical galaxy with forbidden emission lines similar to those found in the nuclei of Seyfert galaxies (Baade and Minkowski, 1954). Cygnus A, which has the highest apparent radio flux of any extragalactic source, exemplifies the characteristic features of strong radio galaxies (also see Fig. 5.36), with fairly symmetrical double structure, a power-law, non-thermal spectrum and substantial linear polarization. During the subsequent decade, the radio galaxies were resolved and were generally found to emit radio waves from giant double lobes on opposite sides of a central counterpart [see Jennison and Das Gupta (1953) for Cygnus A; Hogg et al. (1969) and Miley, Hogg and Basart (1970) for M87 or Virgo A, and De Young (1976) and Miley (1980) for reviews].

The radio emission of extragalactic radio sources is attributed to the synchrotron radiation of high-energy electrons accelerated to speeds approaching the velocity of light, spiraling about magnetic fields. This type of radiation was first observed as linearly polarized light emitted from electrons accelerated to high energies in the strong magnetic fields of terrestrial synchrotron particle accelerators, and then considered as a possible source for radio radiation from our Milky Way Galaxy (Alfvén and Herlofson, 1950; Kiepenheuer, 1950; Ginzburg, 1956). The synchrotron origin for the radio emission of our Milky Way was confirmed by its nonthermal spectrum and linear polarization. These attributes also fingerprint the synchrotron radiation mechanism for extragalactic radio sources.

The observed emission of radio galaxies is produced by a nonthermal population of relativistic electrons gyrating in a magnetic field. The radio radiation has an intensity that increases at longer wavelengths or lower frequencies. That is, the observed flux density, $S(\nu)$, at frequency, ν, obeys $S(\nu) \propto \nu^{\alpha}$, where the negative spectral index α lies in the range -0.5 to -1.0 for the brighter regions of the extended radio sources. The radio emission is also linearly polarized, indicating high-energy electrons radiating in the presence of a magnetic field (Gardner and Whiteoak, 1966; Saikia and Salter, 1988). The theory for the synchrotron radio radiation of high-speed, energetic electrons (100 MeV to 10 GeV, where 1 MeV $= 1.6 \times 10^{-6}$ erg, and 1 GeV $= 10^{3}$ MeV) gyrating in magnetic fields (10^{-6} to 10^{-1} Gauss) is given in Volume I, and also provided by Pacholczyk (1970), Moffet (1975), and Verschuur and Kellermann (1988).

As noted by Geoffrey Burbidge, the total energy content of the radio lobes of an extended radio galaxy such as Cygnus A is enormous (Burbidge, 1959). When the observed intensities and spectra of the radio emission are interpreted

streams, or jets, of hot, magnetized, fast-moving gas, created in the nucleus of the central galaxy and squirting outward in diametrically opposite directions at about the speed of light into the radio lobes, in some cases generating more power than all the stars in our Galaxy. As we shall next see, additional evidence for jet material flowing at relativistic speeds close to the velocity of light is provided by unexpectedly rapid radio variations, superluminal radio expansion and the one-sidedness of radio jets.

The rapid time variability of active galactic nuclei suggests a compact source, such as a black hole. If the luminosity changes significantly during a time interval, t_v, then the source cannot be much larger than the distance that light can travel in the same time, or the radius, R, of the source is usually limited to:

$$R \leq ct_v \ . \tag{5.650}$$

For a variation on the time scale of one day, or 86,240 seconds, we have $R \leq 2.6 \times 10^{15}$ centimeters. Very-Long-Baseline Interferometry, or VLBI, has shown that the radio cores of some active galactic nuclei are unresolved with angular sizes, $\varphi = 2R/D$, smaller than 0.001 seconds of arc, corresponding to a radius of $R < 2.5 \times 10^{-7}\ D = 8.3 \times 10^{18}$ centimeters for a distance of one Megaparsec, or 3.08×10^{24} cm. So, both rapid brightness changes and VLBI observations show that active galactic nuclei are small; only a black hole can be so massive and so small that changes can occur over such short time scales.

Intraday variability of quasars at various electromagnetic bands is reviewed by Wagner and Witzel (1995), and the theory for the variability of galactic nuclei and quasars, caused by instabilities of disk accretion onto black holes, is given by Shakura and Sunyaev (1976). The X-ray spectra and time variability of active galactic nuclei are reviewed by Mushotzky, Done and Pounds (1993).

When the radiating source is moving toward the observer at nearly the velocity of light, then there is a correction to the limiting radius, R, given by the light-travel-time argument. In the presence of relativistic motion, the causality argument that variability cannot be faster than the light-crossing time for the emitting volume, gives an upper limit to the size (Cohen and Unwin, 1984; Urry, 1988)

$$R \leq \delta ct_v \ , \tag{5.651}$$

where the time variation t_v is an observer frame quantity, and the ratio of an emitted time interval, Δt_{em}, or emitted frequency, ν_{em}, and the observed time interval, Δt_{obs}, or observed frequency, ν_{obs}, is given by the Doppler shift factor:

$$\delta = \Delta t_{em}/\Delta t_{obs} = \nu_{obs}/\nu_{em} \ . \tag{5.652}$$

The Doppler factor, δ, for a jet of material moving at velocity, v, at a small angle, θ, with respect to the line of sight is:

$$\delta = \gamma^{-1}(1 - \beta \cos\theta)^{-1} \ , \tag{5.653}$$

where $\beta = v/c$ and the Lorentz factor γ is given by

$$\gamma = (1 - \beta^2)^{-1/2} \ . \tag{5.654}$$

For $\gamma \gg 1$, the radiation is beamed into a small angle of halfwidth $1/\gamma$, the angle $\theta = \sin^{-1}(1/\gamma)$, and $\delta \approx \gamma$.

The observed brightness cannot vary too rapidly. Since you cannot see anything within the Schwarzschild radius $R_s = 2GM/c^2$, the shortest observable time-scale, t_s, is the light-crossing time (Urry, 1988):

$$t_s = R_s/(\delta c) \approx 10^3 \delta^{-1} M_8 \text{ seconds} \tag{5.655}$$

for an object of mass, $M = 10^8$ solar masses $= 10^8 M_\odot$.

The idea that compact radio sources might expand with relativistic velocities, approaching that of light, was advanced by Rees (1966, 1967) and Rees and Simon (1968) to explain the rapid time variation in the flux density of some quasars at radio wavelengths. They suggested tiny sizes that were incompatible with the theory for their synchrotron radio emission. When the suggested dimensions were used to compute the brightness temperatures, T_B, within the sources, they exceeded by factors up to 1,000 the restrictions of $T_B \leq 10^{12}$ K imposed by inverse Compton scattering of the relativistic electrons with the synchrotron radiation they produce (Hoyle, Burbidge and Sargent, 1966; Rees, 1967; Kellermann and Pauliny-Toth, 1968, 1981; Jones and Burbidge, 1973; Jones, O'Dell and Stein, 1974). A source moving at relativistic speed toward the observer can avoid the inverse Compton catastrophe implied by the rapid flux variations, causing the observer to underestimate the actual size of the source (Rees, 1966, 1967; Blandford and Mc Kee, 1977; Marsher and Broderick, 1981). Although expanding sources cannot by themselves explain rapid time variations (Terrell, 1977; Jones and Tobin, 1977; Vitello and Pacini, 1978), relativistic linear motion oriented nearly along the line of sight does explain the rapid flux density outbursts, as well as the apparent superluminal motions that we next describe.

VLBI observations have demonstrated that the apparent transverse velocity, v_{Tapp}, of some quasar radio cores apparently exceeds the velocity of light, c, moving at superluminal velocities. The observed phenomenon, called superluminal expansion, is interpreted as an illusion caused by jets, or beams, of radio emitting material moving toward the observer at very high, relativistic speeds that are close to the speed of light but do not exceed it.

As indicated in Table 5.22, VLBI observations of the differential proper motions of the quasars 3C 273, 3C 279 and 3C 345 indicate apparent transverse speeds of about ten times that of light, or $v_{\text{Tapp}} \approx 10c$ (Whitney et al., 1971; Cohen et al., 1971, 1976, 1979; Pearson et al., 1981; Unwin et al., 1983). The radio galaxy 3C 120 also shows evidence for superluminal motions over large scales (Benson et al., 1988). Reviews of these superluminal motions are given by Cohen and Unwin (1984), Kellermann (1985), Pearson and Zensus (1987), Porcas (1987), Readhead, Pearson and Barthel (1988), and Zensus and Pearson (1987, 1988). Parsec-scale jets in extragalactic radio sources are reviewed by Zensus (1977)

The apparent transverse velocity, v_{Tapp}, projected onto the sky is inferred from observations of an angular separation, φ, and its rate of change, or angular proper motion, $\mu = d\varphi/dt$, with time, t, by the relation:

$$v_{\rm Tapp} = \mu(1+z)D_\theta = \mu D_{\rm L}/(1+z) \ , \tag{5.656}$$

where the first factor of $(1+z)$ corrects for the relativistic time dilation due to cosmological redshift z, the angular size distance, D_θ, is used to derive the projected linear size, l, from the observed angular size, θ, and the luminosity distance, $D_{\rm L}$, is related to D_θ through a factor $(1+z)^2$, or:

$$D_\theta = D_{\rm L}/(1+z)^2 \ . \tag{5.657}$$

The luminosity distance depends on Hubble's constant, $H_{\rm o}$, and the deceleration parameter, $q_{\rm o}$, by the expressions (also see Sect. 5.2.9):

$$D_{\rm L} = \frac{cz}{H_{\rm o}} \frac{(1+\sqrt{1+2q_{\rm o}z}+z)}{(1+\sqrt{1+2q_{\rm o}z}+q_{\rm o}z)} \ , \tag{5.658}$$

where c is the speed of light and the cosmological constant Λ is assumed to be zero.

The angular proper motion, μ, is often measured in milliarcseconds per year, or mas yr^{-1}, and the apparent transverse velocity is then given by:

$$v_{\rm Tapp} = \beta_{\rm Tapp}c = 47.4\mu c\left[\frac{q_{\rm o}z + (q_{\rm o}-1)(\sqrt{1+2q_{\rm o}z}-1)}{hq_{\rm o}^2(1+z)}\right] \ , \tag{5.659}$$

where $H_{\rm o} = 100h$ km s^{-1} Mpc^{-1}. Values of $h\beta_{\rm Tapp} = hv_{\rm Tapp}/c$ are given in Table 5.22.

Theoretical models which are capable of explaining the appearance of superluminal expansion are reviewed by Blandford, Mc Kee and Rees (1977), Marscher (1979), Marscher and Scott (1980), and Scheuer (1984). The favored interpretation is in terms of bulk relativistic motions which are nearly oriented along the line of sight. According to this model, "blobs" or "knots" of material travel at a true velocity $v \approx c$ toward the observer in a beam that lies at a small angle, θ , with respect to the line of sight. The apparent transverse velocity is then given by

$$v_{\rm Tapp} = v\sin\theta/(1-\beta\cos\theta) \tag{5.660}$$

or

$$v_{\rm Tapp} \approx \gamma c \quad \text{for}\ \gamma \gg 1 \ , \tag{5.661}$$

where $\beta = v/c$, the Lorentz factor $\gamma = (1-\beta^2)^{-1/2}$ and $\theta \approx \sin^{-1}(1/\gamma)$ for $\gamma \gg 1$.

When two sources move in opposite directions from a common origin, the apparent perpendicular velocity of separation, $v_{\rm TSapp}$, is given by:

$$v_{\rm TSapp} = 2v\sin\theta/(1-\beta^2\cos^2\theta) \ , \tag{5.662}$$

which also has a maximum value of $v_{\rm TSapp} \approx \gamma c$ for $\beta \approx 1$ and $\sin\theta \approx 1/\gamma$.

Observations of the one-sidedness of radio jets, in which they are seen on only one side of a galaxy, may also be explained by relativistic beaming effects. This is because the radiation is no longer isotropic, but beamed along the directions of bulk relativistic motion within an angle of half width $\approx 1/\gamma$. If the radiating material is moving at speeds close to the velocity of light, then the

Table 5.22. Superluminal radio sources*

Source	z	μ mas yr^{-1}	$h\beta_{\mathrm{Tapp}}$ $q_o = 0.05$	$h\beta_{\mathrm{Tapp}}$ $q_o = 0.5$
1253–055 3C 279	0.538	0.5	10.4	9.2
1226+023 3C 273	0.158	1.20	8.3	8.0
0430+052 3C 120	0.033	2.66	4.1	4.1
1641+399 3C 345	0.595	0.48	10.8	9.5
0723+679 3C 179	0.846	0.19	5.7	4.8
0333+321 NRAO 140	1.258	0.15	6.2	4.8
2200+420 BL Lac	0.070	0.76	2.4	2.4
0923+392 4C 39.25	0.699	0.16	4.1	3.5
1901+319 3C 395	0.635	0.64	~15	~13
2251+158 3C 454.3	0.859	0.35	10.7	8.8
1845+797 3C 390.3	0.0569	0.74	1.9	1.9
1928+738	0.302	0.6	7.5	7.0
1642+690	0.751	0.34	9.3	7.9
0850+581	1.322	0.12	5.1	3.9
0212 +735	2.367	0.09	6.0	3.9
1150+812	1.25	0.13	5.3	4.1
0906+430 3C 216	0.669	0.11	2.7	2.4
1137+660 3C 263	0.652	0.06	1.5	1.3
1951 +498	0.466	~0.07	~1.3	~1.2
1040+123 3C 245	1.029	0.11	3.9	3.1
1721+343 4C 34.47	0.206	0.36	3.2	3.1
0735+178	0.424	0.18	3.0	2.8
2230+114 CTA 102	1.037	~0.65	~23	~18
0851+202 OJ 287	0.306	2.28	3.6	3.3

* Adapted from Porcas (1987); when more than one component were given, only the one with the largest apparent transverse motion is reproduced here.

apparent surface brightness of the source will be affected by relativistic aberration. In the simplest case of an optically thin blob moving at an angle, θ, towards the observer with a speed $v = \beta c$, and emitting an isotropic flux density $S_o(\nu)$ at frequency ν, the observer will measure an enhanced apparent flux density, $S_{\mathrm{app}}(\nu)$ given by (Ryle and Longair, 1967; Rees, 1978; Scheuer and Readhead, 1979; Blandford and Königl, 1979; Pearson and Zensus, 1987; Kellermann and Owen, 1988)

$$S_{\mathrm{app}}(\nu) = \delta^3 S_o(\nu/\delta) = \delta^{3-\alpha} S_o(\nu) \; , \tag{5.663}$$

for a source with a power-law spectrum of spectral index α, so that the flux density varies with frequency as $S_{\mathrm{app}}(\nu) \propto \nu^{\alpha}$ and because α is a negative number there is less flux at higher frequencies and shorter wavelengths as is the case for a nonthermal radiator.

The observed radiation from a relativistically moving body is therefore enhanced by a factor, called Doppler boosting, given by

$$S_{\mathrm{app}}(\nu)/S_o(\nu) = \gamma^{-3}(1 - \beta\cos\theta)^{\alpha-3} \tag{5.664}$$

for a radio source of negative spectral index α. For $\gamma \gg 1$, the $\theta \approx \sin^{-1}(1/\gamma)$ and for $\alpha \approx 0$ and $\delta \approx \gamma$, we have:

$$S_{\rm app}(\nu) \approx \gamma^3 S_{\rm o}(\nu) \ , \tag{5.665}$$

so the observed emission is enhanced by about γ^3. If the radio emission consists of an approaching and receding jet, and the approaching component is seen head on at small θ, then the approaching beam is enhanced by about $8\gamma^3$ and the receding one is diminished by $1/(8\gamma^3)$ and is essentially invisible.

So, the different images of radio galaxies seem to depend on the observing perspective, or how they are oriented relative to our line of sight. If a pair of jets moves at relativistic velocities from a galactic nucleus, we will see both jets if they are traveling at a large angle with respect to our line of sight. When the jets are aligned at a small angle with respect to the line of sight, we detect the jet pointed toward us and the other one traveling away is invisible. In this situation, a bright one-sided jet is seen, foreshortened by the small viewing angle. The statistical properties of the radio emission from quasars are also consistent with the simple relativistic-beam model (Orr and Browne, 1982).

Beams or jets transport energy from an active galactic nucleus to more extended structures. High resolution radio images of radio galaxies and quasars indeed show very long and narrow connections between a compact central radio component and the outer lobes (also see Fig. 5.36; Perley, Willis and Scott, 1979; and Bridle and Perley, 1984 for a review). In order for jets to produce the observed radio emission, relativistic electrons must be continuously accelerated within them, and the high degrees of observed linear polarization imply an ordered magnetic field. The radio jets apparently mark physical conduits along which mass, momentum, energy and magnetic flux are supplied from the nucleus to the outer components. The detailed physics and production mechanisms for such jets are reviewed by Begelman, Blandford and Rees (1984), including such topics as continual flow, collimation, particle acceleration, magnetic field amplification, nozzles, funnels, and winds. This comprehensive article considers the underlying theory of relativity, gravitation, fluid mechanics and shocks, respectively found in the fundamental books by Pauli (1958), Misner, Thorne and Wheeler (1973), Landau and Lifshitz (1959), and Zeldovich and Raizer (1966), and also includes more recent investigations of relativistic thermal plasmas (Araki and Lightman, 1983; Gould, 1982; Lightman and Band, 1981; Lightman, 1982; and Svensson, 1982, 1984) and accretion (Callahan, 1977; Begelman, 1978, 1979).

The observed jets are extraordinarily straight and suprizingly stable. They can extend with a rock steady orientation for hundreds of kiloparsecs. The central source is emitting charged particles in the same direction without disruption for hundreds of millions of years, enough time to build the radio lobes into powerful sources. This elephantine memory and long temporal stability are made possible if the radio jets are firmly anchored in a gigantic, spinning black hole that acts like a gyroscope, holding its spin axis steady over long periods of time. As shown by James Bardeen and Jacobus Petterson, the inner parts of an accretion disk are held fixed within the equatorial plane of the black hole (Bardeen and Petterson, 1975). As a result, jets that are shot along the spin axis of the black hole are held in a fixed and unchanging direction.

In technical terms (Rees, 1978, 1984), a Lense–Thirring precession enforces axisymmetry on any inward-spiraling accretion pattern near a black hole. An orbit around a spinning (Kerr) hole that does not lie in the equatorial plane precesses around the hole's spin axis with an angular velocity of (Bardeen and Petterson (1975)

$$\omega_{\mathrm{BP}} \approx 2\left(\frac{R}{R_{\mathrm{s}}}\right)^{-3}\left(\frac{c}{R_{\mathrm{s}}}\right)\left(\frac{J}{J_{\max}}\right) . \tag{5.666}$$

Here the radius is R, the Schwarzschild radius is R_{s}, the angular momentum is J, and the maximum angular momentum is $J_{\max} = GM/c$. This precession has a time scale longer than the orbital period by a factor of $\approx (R/R_{\mathrm{s}})^{3/2}(J/J_{\max})^{-1}$. However, if material spirals slowly inward (at a rate controlled by viscosity) in a time much exceeding the orbital time, then the effects of this precession can mount up. The flow pattern near the black hole, within the radius where $2\pi/\omega_{\mathrm{BP}}$ is less than the inflow time, can then be axisymmetric with respect to the hole irrespective of the infalling material's original angular momentum. As shown by Rees (1978), any directed outflow initiated in the relativistic domain will therefore be aligned with the hole's spin axis and will squirt in a constant direction. Thus, an accreting black hole naturally explains the long-lived directionality and stability of the radio jets.

The central optical galaxy NGC 4261 of the radio galaxy 3C 270 contains, for example, a nuclear disk of dust that is oriented roughly perpendicular to the radio jet (Jaffe, et al., 1993, 1996). Spectroscopic observations of ionized gas, concentrated within 17 parsecs, are interpreted in terms of Keplerian orbital motion around a central mass of $(4.9 \pm 1.0) \times 10^8$ solar masses with a mass-to-light ratio of 2,100 times that of the Sun, suggesting a massive black hole at the center (Ferrarese, Ford, and Jaffe, 1996).

Quasars are believed to be very luminous versions of the same blue nuclei that Seyfert observed in the center of nearby spiral galaxies (Seyfert, 1943), and the optical counterparts of radio galaxies also exhibit strong forbidden emission lines similar to those found in the nuclei of Seyfert galaxies. Todays astronomers distinguish between Seyfert 1 galaxies, which are bright and rich in the X-rays and ultraviolet light that bespeak violent events, and Seyfert 2 galaxies, which are dimmer and give off light at infrared wavelengths suggesting cooler conditions. Another category of variable extragalactic objects is named after BL Lacertae, a powerful radio source and an optical galaxy overshadowed by its brilliant core that does not exhibit emission lines (Schmitt, 1968; Blake, 1970). The BL Lac objects vary in radiation intensity across the entire electromagnetic spectrum, from radio waves to gamma rays, and many of them are low-redshift elliptical galaxies (Wolfe, 1978).

All of these types of variable extragalactic objects belong to a common class, known collectively as active galactic nuclei. They all radiate so powerfully over the entire range of the electromagnetic spectrum that they cannot possibly consist of ordinary stars, which emit most of their luminous output in a narrow band of wavelengths grouped around visible light. In contrast, supermassive black holes can account for the prodigious energy output, violent activity, and rapid brightness variations of active galactic nuclei, as well as their columns of gas or jets that move at extremely high, relativistic speeds approaching the speed of light.

Rees (1990) provides a popular review of the idea that the hubs of both nearby galaxies and quasars are dominated by black holes. An excellent popular account of both the observational and theoretical aspects of active galactic nuclei, including evidence for their central black holes, is provided by Begelman and Rees (1996). The relationship of radio galaxies and quasars to black holes is also enthusiastically presented in a non-technical way by Longair (1996).

Observational spectroscopy and continuum radiation of active galactic nuclei has been reviewed by Davidson and Netzer (1979), Angel and Stockman (1980), Weymann, Carswell and Smith (1981), Miller (1985, 1989), Osterbrock (1984, 1991), Osterbrock and Mathews (1986), Miller and Wiita (1988), Osterbrock and Miller (1989), Sanders (1989), and Turner and Pounds (1989). Seyfert galaxies and quasars have been respectively described by Weedman (1977) and Weedman (1986). BL Lac objects are discussed in Maraschi, Maccacaro and Ulrich (1989). Variability of active galactic nuclei is reviewed by Ulrich, Maraschi and Urry (1997). Compact extragalactic radio sources, including those coincident with galactic nuclei and quasars, are discussed by Kellermann and Pauliny-Toth (1981), and extragalactic radio jets are reviewed by Bridle and Perley (1984) and Zensus (1997). Extranuclear clues to the origin and evolution of activity in galaxies were reviewed by Balick and Heckman (1982).

How the central black hole and its oppositely-directed jets are perceived from the Earth depends largely on the angle from which they are observed, and this can explain many of the differences in character of what seem to be very disparate objects. Seyfert 1 cores may be jets and nuclei observed nearly head on, while the nuclei of Seyfert 2 galaxies might be viewed sideways through an obscuring accretion disc (Miller and Antonucci, 1983). A strong radio galaxy like Cygnus A is also seen sideways, revealing outward-moving jets more than the obscured central source. When the jets are pointed toward the Earth, and observed almost head on, one detects the superluminal, rapidly-varying quasars and the one-sided jets of some radio galaxies.

The tremendous luminosity and energy output of active galactic nuclei can also be explained by the black hole model. Gas swirling around and down toward the hole accelerates almost to the speed of light, providing a tremendous source of energy that can power radio galaxies and quasars. However, the pressure of the intense radiation that is produced would expel all matter in the vicinity unless the gravity of the central mass is a sufficiently strong countervailing force. A characteristic luminosity, the Eddington limit, at which gravity balances radiation is (Eddington, 1926)

$$L_{\mathrm{Edd}} = \frac{4\pi G M m_{\mathrm{p}} c}{\sigma_{\mathrm{T}}} \approx 1.3 \times 10^{46} M_8 \ \mathrm{erg\ s^{-1}} \ , \tag{5.667}$$

where $G = 6.67 \times 10^{-8}$ cm^3 g^{-1} s^{-2} is the Newtonian gravitation constant, $m_{\mathrm{p}} = 1.76 \times 10^{-24}$ grams is the proton mass, $c = 2.9979 \times 10^{10}$ cm s^{-1} is the velocity of light, $\sigma_{\mathrm{T}} = 0.665 \times 10^{-24}$ cm^2 is the Thomson scattering cross section for electrons, M is the mass of the black hole, and $M_8 = 10^8 M_{\odot} = 100$

million solar masses, where the Suns mass $M_\odot = 1.989 \times 10^{33}$ grams. If supplied by accretion, then the observed luminosity, L, must be less than the Eddington limit, L_{Edd}. For quasars this implies a minimum possible mass of at least 100 million Suns, and for the most powerful ones at least one billion Suns.

Accretion onto a black hole is also much more efficient than other methods of producing energy. Up to 40 percent of the rest mass energy, or $0.4Mc^2$, can be extracted from material of mass, M, spiraling slowly inward toward the event horizon of a black hole (Pringle, 1981). In contrast, nuclear fusion within stars transforms only about 0.7 percent of the rest-mass energy, or $0.007Mc^2$, of nuclei into radiation and neutrinos.

Related to the Eddington limit is the time scale (Salpeter, 1964)

$$t_{\mathrm{Edd}} = \frac{\sigma_{\mathrm{T}} c}{4\pi G m_{\mathrm{p}}} \approx 4 \times 10^8 \text{ years} \ . \tag{5.668}$$

This is the time it would take an object to radiate its entire rest mass if its luminosity were L_{Edd}.

Constraints on the most rapid time variations, t_{v}, obtained from the light-crossing time for the Schwarzschild radius, $R_{\mathrm{s}} = 2GM/c^2$, and the condition that the observed luminosity, L, cannot exceed the Eddington limit, L_{Edd}, or $L < L_{\mathrm{Edd}}$, results in:

$$t_{\mathrm{v}} \geq 2GM/c^3 = 3 \times 10^{-10}(L/L_\odot) \text{ seconds} \ , \tag{5.669}$$

where the solar luminosity $L_\odot = 3.85 \times 10^{33}$ erg sec^{-1}.

The characteristic blackbody temperature, T_{Edd}, if luminosity at the Eddington limit, L_{Edd}, is emitted from the Schwarzschild radius, R_{s}, is

$$T_{\mathrm{Edd}} \approx 3 \times 10^5 M_8^{-1/4} \text{ K} \ . \tag{5.670}$$

Some other fundamental quantities involving accretion onto supermassive black holes are given by Rees (1984) and Begelman, Blandford and Rees (1984). The critical accretion rate, $\dot{M}_{\mathrm{Edd}}$, associated with the Eddington limiting luminosity, L_{Edd}, is

$$\dot{M}_{\mathrm{Edd}} = L_{\mathrm{Edd}}/c^2 \ . \tag{5.671}$$

If accretion with efficiency, ε, provides the power, the mass inflow rate, $\dot{M}$, needed to supply a luminosity, L, is

$$\dot{M} = \frac{1}{\varepsilon \dot{M}_{\mathrm{E}}}\left(\frac{L}{L_{\mathrm{E}}}\right) \approx 1.5\left[\frac{0.1}{\varepsilon}\right]\left[\frac{L}{10^{46}\text{ erg s}^{-1}}\right] \text{ solar masses per year} \ , \tag{5.672}$$

where ε is the fraction of the rest mass radiated. The particle density, n, at radius, R, corresponding to an inflow rate $\dot{M}$ is

$$n \approx 10^{11} \dot{M} M_8^{-1} (R/R_{\mathrm{s}})^{-3/2} \text{ cm}^{-3} \ , \tag{5.673}$$

where the Schwarzschild radius of the black hole of mass M is $R_{\mathrm{s}} = GM/c^2$. At radius $R = R_{\mathrm{s}}$, the fiducial particle density becomes

$$n_{\mathrm{Edd}} = c^2/(\sigma_{\mathrm{T}} GM) \approx 10^{11} M_8^{-1} \text{ cm}^{-3} \ . \tag{5.674}$$

A characteristic magnetic field strength, $B_{\rm Edd}$, obtained from $(B^2/8\pi) = n_{\rm Edd} m_{\rm p} c^2$, is:

$$B_{\rm Edd} = \left[\frac{8\pi c^2 m_{\rm p}}{\sigma_{\rm T} G M}\right]^{1/2} \approx 4 \times 10^4 M_8^{-1/2} \text{ Gauss} . \tag{5.675}$$

If a field $B_{\rm Edd}$ were applied to a black hole with maximum angular momentum, $J = J_{\rm max} = GM/c$, the electromagnetic power extraction would be about the Eddington limit $L_{\rm Edd}$. The expected field strengths induced by accretion flows can be of this order. The corresponding cyclotron frequency is:

$$\nu_{\rm cEdd} \approx 10^{11} M_8^{-1/2} \text{ Hz} . \tag{5.676}$$

Most astronomers now think that the machine, engine or prime mover, imbedded in the nucleus of an active galaxy, is a rotating, supermassive black hole that derives its energy from the accretion of surrounding matter and channels it along the rotation axis in long, thin, collimated relativistic jets (Blandford and Rees, 1974; Rees, 1978; Lynden-Bell, 1978; Rees, 1984). As it spins and pulls in matter, a black hole builds up an enormous store of rotational energy in a swirling accretion disk. Part of that spin energy can be used to power jets that emerge from the very center of active galactic nuclei. Roger Penrose first suggested, in general terms, that such energy extraction was possible (Penrose, 1969), and Roger Blandford proposed that electromagnetic processes might convert the rotational energy into a powerful outflow at relativistic speeds (Blandford, 1976). According to a mechanism proposed by Blandford and Roman Znajek, the magnetic field embedded in, and produced by, the accretion disc gets twisted up by its rotation, creating a magnetic brake that slows the rotation of the gas and extracts energy from it (Blandford and Znajek, 1977; Mac Donald and Thorne, 1982). The slowing gas falls toward the black hole and is ultimately shaped into two jets along the hole's spin axis (Blandford and Königl, 1979; Blandford and Payne, 1982; Lovelace, Wang and Sulkanen, 1987; Königl, 1989).

Some of the details of the electromagnetic properties of black hole models for active galactic nuclei are given by Rees (1984) and Begelman, Blandford and Rees (1984). Frank, King and Raine (1992) scale the thin-disc accretion equations (5.616) to typical parameters for active nuclei such as the mass density $\rho = \Sigma/H$, disc thickness H, central temperature, $T_{\rm c}$, optical depth, τ, viscosity, υ, and radial velocity, $V_{\rm R}$, given by:

$$\begin{aligned}
\Sigma &= 5.2 \times 10^6 \alpha^{-4/5} \dot{M}_{26}^{7/10} M_8^{1/4} R_{14}^{-3/4} f^{14/5} \text{ g cm}^{-2} \\
H &= 1.7 \times 10^{11} \alpha^{-1/10} \dot{M}_{26}^{3/10} M_8^{-3/8} R_{14}^{9/8} f^{7/5} \text{ cm} \\
\rho &= 3.1 \times 10^{-5} \alpha^{-7/10} \dot{M}_{26}^{11/20} M_{14}^{5/8} R_{14}^{-15/8} f^{11/5} \text{ g cm}^{-3} \\
T_{\rm c} &= 1.4 \times 10^6 \alpha^{-1/5} \dot{M}_{26}^{3/10} M_8^{1/4} R_{14}^{-3/4} f^{6/5} \text{ K} \\
\tau &= 3.3 \times 10^3 \alpha^{-4/5} \dot{M}_{26}^{1/5} f^{4/5} \\
\upsilon &= 1.8 \times 10^{18} \alpha^{4/5} \dot{M}_{26}^{3/10} M_8^{-1/4} R_{14}^{3/4} f^{6/5} \text{ cm}^2 \text{ s}^{-1} \\
V_{\rm R} &= 2.7 \times 10^4 \alpha^{4/5} \dot{M}_{26}^{3/10} M_8^{-1/4} R_{14}^{-1/4} f^{-14/5} \text{ cm s}^{-1}
\end{aligned} \tag{5.677}$$

where $f = 1 - (6GM/Rc^2)^{1/2}$, and the inner radius R_* is taken to be the last stable orbit for a Schwarzschild black hole.

The effective temperature of the active-nuclei accretion disc photosphere is:

$$T = 2.2 \times 10^5 \dot{M}_{26}^{1/4} M_8^{1/4} R_{14}^{-3/4} \text{ K} . \quad (5.678)$$

Gas pressure will be greater than radiation pressure at radii

$$R \geq 5.2 \times 10^{14} \alpha^{8/30} \dot{M}_{26}^{14/15} M_8^{1/3} f^{56/15} \text{ centimeters} . \quad (5.679)$$

Miniature versions of the powerful quasars have been found in our own stellar backyard, as relativistic jets emitted from the compact companions of X-ray binaries. These micro-quasars include the intense X-ray source Cygnus X-3 and SS 433, numbered 433 in the catalogue of hydrogen emission line stars by C. Bruce Stephenson and Nicholas Sanduleak (Stephenson and Sanduleak, 1977). Both stars exhibit synchrotron radio emission of relativistic electrons with jets moving at nearly the velocity of light (Lang, 1992).

The combination of high X-ray luminosity $L_X = 10^{38}$ erg s^{-1} (at 2 to 20 keV with an assumed distance of 8.5 kiloparsecs), short orbital period, $P = 4.79$ hours, giant radio flares with flux densities reaching 20 Jy (Hjellming, 1973), and reports of energetic gamma-ray emission make Cygnus X-3 a unique object amongst X-ray binaries (see Bonnet-Bidaud and Chardin, 1988, for a review). Its position has been accurately determined from radio interferometry measurements. The flaring radio emissions have been interpreted as the synchrotron radiation of an expanding double-sided jet, aligned with the rotation axis of an accretion disc and moving at bulk velocities between 16 and 31 percent of the velocity of light, or $0.16c$ to $0.31c$ (Molnar, Reid and Grindlay, 1988).

One of the most intriguing phenomena is the star SS 433, or V1343 Aquilae, the optical counterpart of an X-ray and nonthermal radio source whose mysterious emission lines have wavelengths that change with time, seemingly wandering back and forth on time scales of days. Optical spectroscopic observations indicate Doppler shifts of up to $50,000$ km s^{-1} toward the red and up to $30,000$ km s^{-1} toward the blue (Margon et al., 1979; Margon, 1981, 1982, 1984). They are explained by a pair of identical, narrow collimated jets of matter that are ejected from a compact star at velocities of 26 percent of the speed of light, or $0.26c$, which precesses around a fixed axis in space every 164 days (Abell and Margon, 1979; Fabian and Rees, 1979; Margon, 1982, 1984). The discovery and eventual explanation of SS 433, using techniques of visible light, radio and X-ray astronomy, are recounted by Clark (1985) – see for example Clark and Murdin (1978) and Ryle et al. (1978). Its overloaded accretion disc apparently converts accretional energy into the kinetic energy of relativistic jets with only weak X-ray emission of absolute luminosity $L_X \approx 10^{35}$ erg s^{-1}.

SS 433 is close enough that the motions of the precessing jet can be imaged with a radio interferometric telescope such as the Very Large Array, or VLA. When dense blobs of ejected gas move outward along the two jets, emitting intense radio radiation, they trace out a widening corkscrew pattern due to the precession of the jet's axis around a cone every 164 days. Robert Hjellming and

Kenneth Johnston have used the VLA to map the outward spiral of the radio-emitting plasma (Hjellming and Johnston, 1981; Hjellming, 1988).

Astronomers have also now used radio images to obtain direct evidence for superluminal motions in two other nearby micro-quasars named GRS 1915 + 05 and GRO J1655–40 (Mirabel and Rodriquez, 1994; Tingay et al., 1995; Hjellming and Rupen, 1995). These two strong Galactic X-ray transient sources exhibit high-speed motions that are probably associated with relativistic jets emanating from a black hole in an X-ray binary. We have previously mentioned evidence for a black hole in the eclipsing X-ray nova GRO J1655–40, also see Bailyn et al. (1995) and Section 5.6.2.

Very Long Baseline Interferometry, VLBI, radio observations of GRO J1655–40 reveal an unprecedented proper motion of 65 ± 5 milliarcseconds per day (Tingay et al., 1995). The subcomponents of GRS 1915 + 105 have proper motions of 17.53 and 9.04 milliarcseconds per day and SS 433 is at 8.77 milliarcseconds per day. The extremely high proper motions of all three stellar sources are attributed to relativistic jets, which magnify the transverse velocity, and to the relative nearness of these particular sources. The Very Large Array, VLA, has been used to follow the evolution of the radio jets of GRO J165540, and the Very Large Baseline Array, VLBA, employed to study the details, indicating episodic ejection of two highly collimated relativistic jets, one on each side of the star, which expand and decay over a few days and are ejected at 92 percent the speed of light (Hjellming and Rupen, 1995). Levinson and Blandford (1996) examine the physical conditions within these Galactic sources, discuss their interaction with their environment and their possible formation, and contrast them with their extragalactic counterparts. The observed properties of the superluminal jet sources GRO J1655–40 and GRS 1915 + 105 strongly suggest that each contains a spinning black hole (Zhang, Cui and Chen, 1997).

5.7 Big Bang Cosmology

5.7.1 The Homogeneous and Isotropic Universe

Hubble (1926), Hubble and Humason (1931), and Shapley and Ames (1932) observed that the total number of galaxies to various limits of total magnitude vary directly with the volumes of space represented by the limits, and concluded that the density of observable matter is constant. Their additional observation that the distribution of nebulae is spatially isotropic across the sky led to the conclusion that the Universe is filled with a spatially homogeneous and isotropic distribution of matter. In other words, the large-scale features of the Universe would appear the same to an observer in any galaxy no matter in which direction he or she looked. The idea that the Universe presents the same aspect from every point in space, except for local irregularities, has become known as the cosmological principle, a term coined by Milne (1935). Although there is also considerable structure in the large-scale distribution of galaxies (Sect. 5.7.6.1), the overall uniformity of the Universe on the largest scales is confirmed by the extraordinary smoothness or isotropy of the cosmic microwave background radiation (Sect. 5.7.5).

Weyl (1923, 1930) first postulated that the galaxies are on a bundle of geodesics in space-time which converges toward the past, and showed that the line element, ds, is given by

$$ds^2 = c^2\, dt^2 + \mathbf{g}_{ik}\, dx^i\, dx^k \ , \tag{5.680}$$

where c is the velocity of light, the cosmic time, t, measures the proper time of an observer following the geodesic, and $i, k = 1, 2, 3$. Following the first derivation of the line element of a nonstationary Universe (Lanczos, 1922, 1923), it was shown that the general form for the line element of a spatially homogeneous and isotropic Universe is given by (Friedmann, 1922, 1924)

$$ds^2 = c^2\, dt^2 - R^2(t)\left[\frac{dr^2}{1 - kr^2} + r^2(d\theta^2 + \sin^2\theta\, d\varphi^2)\right] \ , \tag{5.681}$$

where the curvature index $k = -1, 0$, and $+1$, and $R(t)$ is called the radius of curvature or the scale factor of the Universe. The r coordinate has zero value for some arbitrary fundamental observer, the surface r = constant has the geometry of the surface of a sphere, and θ, φ are polar coordinates. Although this metric was first used by Friedmann (1922, 1924) for $k = \pm 1$ it is often called the Robertson–Walker metric after its rigorous deduction by Robertson (1935, 1936) and Walker (1936) (cf. also Robertson, 1928, 1933).

Einstein (1917) initially applied his General Theory of Relativity, with the equations given in Sect. 5.5.1, to a static, or non-expanding Universe by introducing the cosmological constant, Λ, that opposed gravity at large distances and balanced a static Universe of zero pressure with a mass density $\rho = c^2\Lambda/(8\pi G)$. [Some authors use either the symbol Λ or λ for the quantity Λc^2 and designate that symbol as the cosmological constant (see for example Weinberg, 1972).] However, astronomical limits to the cosmological constant are exceedingly small. If we use the Hubble constant, H_o, to define a reduced cosmological constant, λ_o, by:

$$\lambda_\mathrm{o} = c^2\Lambda/(3H_\mathrm{o}^2) \ , \tag{5.682}$$

where the subscript o denotes the present-day value, we have the following limits (Review of Particle Properties, 1994):

$$-1 < \lambda_\mathrm{o} < 2 \ . \tag{5.683}$$

In mass density units, the astronomical limits derived from the minimum age of the Universe and the mere existence of high-redshift objects is (Carroll, Press and Turner, 1992):

$$-2 \times 10^{29}\ \mathrm{g\ cm^{-3}} \le c^2\Lambda/(8\pi G) \le 4 \times 10^{-29}\ \mathrm{g\ cm^{-3}} \ . \tag{5.684}$$

For example, the existence of quasars with redshift $z_{\mathrm{max}} = 4$ requires that $(2\lambda_\mathrm{o}/\Omega_\mathrm{o})^{1/3} < z_{\mathrm{max}}$, or that (Combes, Boissé, Mazure and Blanchard, 1995):

$$\Lambda < 10^{-54}\ \mathrm{cm^{-2}} \ , \tag{5.685}$$

where the density parameter Ω_o, defined below, is $\Omega_\mathrm{o} < 1.0$. These limits to the cosmological constant Λ are in rough accord with those established more than two decades earlier from the redshift-magnitude relation for galaxies (Peach, 1970):

$$-2 \times 10^{55}\ \mathrm{cm^{-2}} \le \Lambda \le 2 \times 10^{-55}\ \mathrm{cm^{-2}} \ . \tag{5.686}$$

Thus, astronomical observations indicate that the cosmological constant is very small. It is many orders of magnitude smaller than estimated in modern theories of elementary particles; and Weinberg (1989) has reviewed various theoretical approaches to this paradox.

For many purposes in astrophysics it is convenient to assume that the cosmological constant $\Lambda = 0$. We have, for example, provided expressions for the luminosity distance (Sect. 5.2.9) and the age of the Universe (Sect. 5.3.12) for various cosmological models assuming $\Lambda = 0$, but for completeness we also gave the formulae for that age for the case of reduced cosmological constant $\lambda_o = 1 - \Omega_o$ with a density parameter Ω_o (also see Sect. 5.3.12).

Isotropic, homogeneous Universes with zero cosmological constant $\Lambda = 0$ are often called Friedmann Universes after the first solution for this case by Friedmann (1922, 1924). Here we will provide expressions for the $\Lambda = 0$ case, and give more complete, and complicated, expressions for nonzero Λ at the end of this section. For zero cosmological constant, Einstein's field equations lead to the following expressions involving the scale factor, $R(t)$, in a Robertson–Walker, or homogeneous and isotropic, Universe at time t.

$$\left(\frac{\dot{R}}{R}\right)^2 = -\frac{kc^2}{R^2} + \frac{8\pi G}{3}\rho \quad \text{for } \Lambda = 0 \tag{5.687}$$

$$\frac{2\ddot{R}}{R} + \left(\frac{\dot{R}}{R}\right)^2 = -\frac{kc^2}{R^2} - \frac{8\pi G}{c^2}P \quad \text{for } \Lambda = 0 \ , \tag{5.688}$$

and therefore

$$\frac{\ddot{R}}{R} = \frac{-4\pi G}{3}\left(\rho + \frac{3P}{c^2}\right) \quad \text{for } \Lambda = 0 \ . \tag{5.689}$$

Here the $\dot{}$ denotes differentiation with respect to cosmic time, t, $k = -1, 0$ or $+1$ is the space curvature constant found in the Robertson–Walker metric, the mass-energy density of the Universe is ρ, the pressure is P, and G is the Newtonian constant of gravitation. The quantity $\kappa = 8\pi G/c^2 \approx 1.86 \times 10^{-27}$ dyn g^{-2} s^{-2} is called Einstein's gravitational constant.

If light was emitted from an extragalactic object at time t_e, then its redshift, z, is given by:

$$1 + z = R(t_o)/R(t_e) = R_o/R(t_e) \ , \tag{5.690}$$

where the subscript zero is used to denote the present epoch, and $R_o = R(t_o)$ is the present value of $R(t)$ at time t_o. For instance, a quasar of redshift 3.0 emitted the light we now see when the Universe was one fourth its present size.

At any time, t, we can define a Hubble expansion parameter, $H(t)$, by:

$$H(t) = \frac{\dot{R}}{R} \tag{5.691}$$

a deceleration parameter, $q(t)$, by

$$q(t) = -\frac{\ddot{R}R}{\dot{R}^2} \tag{5.692}$$

and a density parameter, $\Omega(t)$, by:

$$\Omega(t) = \frac{8\pi G\rho}{3H^2} \quad . \tag{5.693}$$

Following Robertson (1955) and Hoyle and Sandage (1956), it is customary to define different cosmological models in terms of the present value of the Hubble expansion parameter, H_o, also called the Hubble constant, H_o, and the present value of the deceleration parameter, q_o.

$$H_o = \frac{\dot{R}_o}{R_0} \quad , \tag{5.694}$$

and

$$q_o = -\frac{\ddot{R}_o R_o}{\dot{R}_o^2} = \frac{\ddot{R}_o}{R_o H_o^2} = \frac{\Omega_o}{2} - \frac{\Lambda c^2}{3H_o^2} \tag{5.695}$$

or

$$q_o = -\frac{\ddot{R}_o R_o}{\dot{R}_o^2} = \frac{4\pi G}{3H_o^2}\rho_o \quad \text{for } \Lambda = 0 \quad , \tag{5.696}$$

for vanishing pressure. Nowadays the density parameter, or omega factor, $\Omega_o = 2q_o$ has become the favored expression in cosmological lore. Here the subscript zero is used to denote the present epoch, and R_o is the present value of $R(t)$. For zero cosmological constant $\Lambda = 0$, and in the present matter-dominated era for which the pressure term $P_o = 0$, and the radiation density can be omitted, the present mean density of matter is given by

$$\rho_{mo} = \frac{3q_o H_o^2}{4\pi G} = 3.8 \times 10^{-33} q_o H_o^2 \ g\,\text{cm}^{-3} \quad , \tag{5.697}$$

where in the numerical approximation H_o is in km sec^{-1} Mpc^{-1} and 1 Mpc $= 3.18 \times 10^{24}$ cm, and

$$\frac{kc^2}{R_o^2} = H_o^2(2q_o - 1) \quad \text{for } \Lambda = 0 \quad . \tag{5.698}$$

The present value of the density parameter, Ω_o, is given by:

$$\Omega_o = 2q_o = \rho_{mo}/\rho_c \quad \text{for } \Lambda = 0 \quad , \tag{5.699}$$

where the critical mass density, ρ_c is given by

$$\rho_c = \frac{3H_o^2}{8\pi G} = 1.879 \times 10^{-29} h^2 \ \text{g cm}^{-3} \quad , \tag{5.700}$$

and $H_o = 100h$ km s^{-1} Mpc^{-1} and h is a constant. Big-bang nucleosynthesis (Section 5.7.4) constrains the baryon density parameter, Ω_B, to:

$$0.009h^{-2} \leq \Omega_B \leq 0.025h^{-2} \quad , \tag{5.701}$$

or for any realistic $h = 0.4$ to 1.0,

$$0.009 \leq \Omega_B \leq 0.15 \quad , \tag{5.702}$$

for visible or invisible baryonic matter. Neutrons and protons are baryons, so ordinary matter is composed of baryons. However, there may be extraordinary, invisible, nonbaryonic matter present that could boost the density parameter up beyond the baryon value (Sect. 5.4.6).

For model Universes with negligible cosmic pressure, $P_o = 0$, and zero cosmological constant, $\Lambda = 0$, we have

$$\dot{R}^2 = \left(\tfrac{8}{3}\pi G\rho\right) - kc^2 \ , \tag{5.703}$$

where $\dot{R} = dR(t)/dt$ is the time differential of the scale factor $R(t)$, which is now increasing with time, t, since the Universe is observed to be expanding. It therefore follows that:

- If k $= 1$ then $q_o > 0.5$, $\rho_o > \rho_c$, and $\Omega_o > 1$ for elliptical closed space and an oscillating Universe, in which R(t) reaches a maximum in the future, when the two terms on the right side of the equation cancel, and then decreases. For $\Omega_o > 1$ the Universe will eventually curl back upon itself and reform the dense fireball of its youth.
- If $k = 0$ then $q_o = 0.5$, $\rho_o = \rho_c$, and $\Omega_o = 1$ for a flat, Euclidean space and an ever-expanding Einstein – De Sitter Universe (Einstein and De Sitter, 1932), in which $R(t)$ continues to forever increase with time, t. At $\Omega_o = 1$ the Universe stands poised between open and closed, right on the dividing line.
- If k $= -1$ then $0.0 < q_o < 0.5$, $\rho_o < \rho_c$, and $\Omega_o < 1$ for a hyperbolic open space and an ever-expanding Milne Universe (Milne, 1935), in which R(t) also continues to forever increase with time, t. For $\Omega_o < 1$ the inward pull of gravity is too weak to ever quell the outward expansion of the Universe.

The matter density, $\rho_m(t)$, radiation energy density, $\rho_r(t)$, and radiation temperature, $T_r(t)$, in a Universe dominated by matter go as (Einstein, 1917; Tolman, 1934)

$$\begin{aligned} \rho_m(t) &\propto [R(t)]^{-3} \\ \rho_r(t) &\propto [R(t)]^{-4} \\ T_r(t) &\propto [R(t)]^{-1} \ . \end{aligned} \tag{5.704}$$

Here t is the time since the big bang, and

$$\frac{\rho_m(t)T_r(t)}{\rho_r(t)} = \text{constant for all time} \ . \tag{5.705}$$

We also have:

$$\rho_r(t) = aT_r^4(t) \propto [R(t)]^{-4} \tag{5.706}$$

where the radiation constant

$$a = \frac{8\pi^5 k^4}{15c^3h^3} = 7.5641 \times 10^{-15} \text{ erg cm}^{-3}\ {}^\circ\text{K}^{-4} \ .$$

The radiation temperature, $T_r(t)$, at the current time, t_o, after the big bang is $T_r(t_o) = 2.726 \pm 0.010\ ^\circ$K, the temperature of the cosmic microwave background radiation (Mather et al., 1994; see the next Sect. 5.7.5). The uniformity of this radiation provides the best evidence for the large-scale homogeneous and isotropic structure of the Universe (also see Sect. 5.7.5).

At a time, t_{eq}, when the matter density $\rho_m(t_{eq})$ equals the radiation energy density $\rho_r(t_{eq})$, the radiation temperature, $T_r(t_{eq})$ is given by:

$$T_{\mathrm{r}}(t_{\mathrm{eq}}) = \frac{\rho_{\mathrm{m}}(t_{\mathrm{o}})T_{\mathrm{r}}(t_{\mathrm{o}})}{\rho_{\mathrm{r}}(t_{\mathrm{o}})} \tag{5.707}$$

$$\begin{aligned} T_{\mathrm{r}}(\mathrm{t}_{\mathrm{eq}}) &= \frac{3\Omega_{\mathrm{o}}H_{\mathrm{o}}^2}{8\pi \mathrm{G}a\mathrm{T}_{\mathrm{r}}^3(\mathrm{t}_{\mathrm{o}})} \\ &\approx 10^5 \Omega_{\mathrm{o}} h^2 \ {}^{\circ}\mathrm{K} \ , \end{aligned} \tag{5.708}$$

where the density parameter $\Omega_{\mathrm{o}} = 2q_{\mathrm{o}}$, the deceleration parameter is q_{o}, Hubble's constant $H_{\mathrm{o}} = 100h$ km s^{-1} Mpc^{-1}, and we have assumed a cosmological constant $\Lambda = 0$.

In the early Universe in the radiation-dominated era when $\rho_{\mathrm{r}}(t) > \rho_{\mathrm{m}}(t)$, the formulae for the adiabatic expansion of the radiation in a homogeneous, isotropic Universe (Robertson–Walker metric) can be used to obtain the radiation temperature, $T_{\mathrm{r}}(t)$, at time, t (Gamow, 1948, 1956):

$$T_{\mathrm{r}}(t) = \left[\frac{3c^2}{32\pi aG}\right]^{1/4} \frac{1}{t^{1/2}} = \frac{1.5 \times 10^{10}}{t^{1/2}} \mathrm{K} \ , \tag{5.709}$$

In the numerical approximation the time, t , is in seconds.

The time, t_{eq}, at which the radiation energy density is equal to the mass density is therefore given by:

$$t_{\mathrm{eq}} \approx 10^{20}/T_{\mathrm{r}}^2(t_{\mathrm{eq}}) \approx 10^{10}\Omega_{\mathrm{o}}^{-2}h^{-4} \ \text{seconds} \ , \tag{5.710}$$

or

$$t_{\mathrm{eq}} \approx 10^3 \Omega_{\mathrm{o}}^{-2} h^{-4} \ \text{years} \ . \tag{5.711}$$

For those models for which the cosmological constant is zero, the present value of the density of matter and radiation, ρ_{o}, and the present value of the spatial curvature, kc^2/R_{o}^2, are related by the equations

$$\rho_{\mathrm{o}} + \frac{3P_{\mathrm{o}}}{c^2} = \frac{3H_{\mathrm{o}}^2 q_{\mathrm{o}}}{4\pi G} \quad \text{for } \Lambda = 0 \ , \tag{5.712}$$

$$\frac{kc^2}{R_{\mathrm{o}}^2} = \frac{4\pi G}{3q_{\mathrm{o}}}\left[\rho_{\mathrm{o}}(2q_{\mathrm{o}} - 1) - \frac{3P_{\mathrm{o}}}{c^2}\right] \quad \text{for } \Lambda = 0 \ . \tag{5.713}$$

and for nonzero cosmological constant

$$q_{\mathrm{o}} = -\frac{\ddot{R}_{\mathrm{o}}}{R_{\mathrm{o}}H_{\mathrm{o}}^2} = -\frac{\ddot{R}_{\mathrm{o}}R_{\mathrm{o}}}{\dot{R}_{\mathrm{o}}^2} = \frac{-1}{H_{\mathrm{o}}^2}\left[\frac{\Lambda c^2}{3} - 4\pi G\left(\frac{\rho_{\mathrm{o}}}{3} + \frac{P_{\mathrm{o}}}{c^2}\right)\right] \ . \tag{5.714}$$

For a Universe with negligible present cosmological pressure, $P_{\mathrm{o}} \approx 0$,

$$q_{\mathrm{o}} = \frac{\Omega_{\mathrm{o}}}{2} - \frac{\Lambda c^2}{3H_{\mathrm{o}}^2} \ , \tag{5.715}$$

where the density parameter $\Omega_{\mathrm{o}} = 8\pi G\rho_{\mathrm{mo}}/(3H_{\mathrm{o}}^2)$ and the present mass density is $\rho_{\mathrm{mo}} = \rho_{\mathrm{o}}$. The cosmological constant is therefore given by:

$$\Lambda = \frac{3H_{\mathrm{o}}^2}{c^2}\left(\frac{\Omega_{\mathrm{o}}}{2} - q_{\mathrm{o}}\right) \ . \tag{5.716}$$

When the Robertson–Walker metric is used with Einstein's field equations, the following equations are obtained for the scale factor $R(t)$ for non-zero cosmological constant (Lemaitre, 1927; Raychaudhuri, 1955, 1957)

$$\frac{3\dot{R}^2}{R^2} + \frac{3kc^2}{R^2} = \kappa\rho c^2 + \Lambda c^2 \ , \tag{5.717}$$

or

$$\left(\frac{\dot{R}}{R}\right)^2 = \frac{8\pi G\rho}{3} - \frac{kc^2}{R^2} + \frac{\Lambda c^2}{3} \tag{5.718}$$

and

$$\frac{2\ddot{R}}{R} + \frac{\dot{R}^2}{R^2} + \frac{kc^2}{R^2} = -\kappa P + \Lambda c^2 \tag{5.719}$$

or

$$\frac{\ddot{R}}{R} = \frac{\Lambda c^2}{3} - \frac{4\pi G}{3}\left[\rho + \frac{3P}{c^2}\right] \ , \tag{5.720}$$

where $R = R(t)$ is the radius of curvature or the scale factor of the Universe, the $\dot{}$ denotes differentiation with respect to the cosmic time, t, Einstein's gravitational constant $\kappa = 8\pi G/c^2 = 1.86 \times 10^{-27}$ dyn g^{-2} sec^2, the mean density of matter and energy in the Universe is $\rho(t)$ (the energy density is ρc^2), the isotropic hydrodynamic pressure of matter and radiation is $P(t)$, the cosmological constant is Λ and k/R^2 is the Riemannian space curvature. The index k takes values of $+1$, -1, or 0 according to whether the space is closed, open, or Euclidean.

The reduced curvature constant, α , is defined by:

$$\alpha = \frac{kc^2}{H^2R^2} = \Omega + \lambda - 1 \ . \tag{5.721}$$

where the reduced cosmological constant $\lambda = \Lambda c^2/(3\ H_0^2)$. Inflationary models of the primordial Universes (Sect. 5.7.2) require that:

$$\lambda_0 = 1 - \Omega_0 \ , \tag{5.722}$$

or $\alpha_0 = 0.0$ and no curvature. Using $\Omega_0 \approx 0.1$ to 0.3, the prediction of inflation scenarios would then be

$$\lambda_0 = \Lambda\, c^2/(3H_0^2) \approx 0.7 \text{ to } 1.0 \ . \tag{5.723}$$

5.7.2 The Inflation Scenario

A relatively recent conjecture, called inflation, imagines that the infant Universe underwent a brief spasm of explosive expansion, when space-time ballooned outward at an exceptional rate, long before settling down to its present, more leisurely expansion. The swift, fierce acceleration, for just a fleeting instant at the dawn of time, ironed the Universe out flat, pushing its curvature to the very brink of closure.

If this speculation is true, then the mass density of the Universe, ρ_0, is poised at the critical value, ρ_c, between open and closed. That is, the inflationary hypothesis requires that the omega factor or density parameter $\Omega_0 = \rho_0/\rho_c = 1$ and that the space curvature constant $k = 0$. As seen in Sects. 5.4.5 and 5.4.6, that would mean that dark, invisible matter is hundreds of times more plentiful than the luminous stars and galaxies. The Universe would then be filled with exotic, weakly interacting particles left over from the big bang; and although physicists are hunting for them, no one has yet caught a glimpse of the imaginary stuff. In a recent twist to the theory, the Universe is supposed to have burped twice, undergoing two rounds of inflation (Bucher and Turok, 1995; Sasaki, Tanaka and Yamamoto, 1995). The first round might have created the uniformity of the Universe, whereas the second round of inflation could have left the Universe curved. Another approach imagines that open inflation and a smooth Universe arose out of emptiness (Turok and Hawking, 1998), but this also increases the complexity of the theory and thereby makes it less attractive.

Descriptions of this inflationary model, suitable for the general reader, are given by Guth and Steinhardt (1984, 1989) and Guth (1997). Narlikar and Padmanabhan (1991) have explained inflation for astronomers. The old, new and chaotic varieties of the inflation theory are given in technical papers by Starobinsky (1980), Guth (1981), Linde (1982), Albrecht and Steinhardt (1982), and Linde (1983). A short updated overview of inflation has been given by Guth (1993). A good technical account is given in the book by Kolb and Turner (1990).

The fluctuations from which galaxies formed may have been created during the inflationary expansion of the very early Universe. The generation of perturbations during inflation has been discussed by Bardeen, Steinhardt and Turner (1983), Guth and Pi (1982), Hawking (1982), and Starobinsky (1982), obtaining results similar to those of Harrison (1970) and Zeldovich (1972) on other grounds. The scale-invariant spectrum of the microwave background fluctuations observed by the COsmic Background Explorer (COBE) are consistent with these predictions (Guth, 1993; Bennett et al., 1996; also see Sects. 5.7.5 and 5.7.6.3)

When first introduced by Guth (1981) and Linde (1982), the inflationary Universe was proposed as a possible solution to the horizon and flatness problems, reviewed by Dicke and Peebles (1979). The first difficulty is that regions cannot communicate beyond the horizon distance, the maximum distance that a light signal could have traveled since the regions originated (Rindler, 1956). The observed Universe is remarkably uniform over distances that are large compared with the horizon distance, resulting in the horizon problem. Since the widely-separated regions could not have communicated, it is difficult to see how they could have evolved to be so similar, since otherwise they cannot "know" that they have to be the same.

As an example, the microwave background radiation observed in opposite directions on the sky has the same intensity and spectrum to a precision of one part in 100,000. This radiation originates from a last scattering surface at the epoch of recombination when the Universe was about 1,000 times smaller than its present size. The distance that light can travel along this last scattering surface since the origin of the big bang corresponds to an angle of only 5 degrees in the sky, which means that regions in opposite directions could not have been in causal communication.

Inflation gets around the horizon problem by supposing that all the regions of the Universe were once very close together, so they could then communicate with each other and become uniform and homogeneous. Then inflation caused the Universe to expand exponentially, driving regions apart to distances where they could no longer communicate with each other by light signals.

The second flatness problem is related to the near flatness of the present-day Universe. It is too close to flat for comfort, and inflation removes this problem by supposing that the Universe has been precisely that way ever since the early inflation epoch. That is, inflation requires a condition of precise equilibrium in which the density parameter $\Omega_o = 1$, so the present mass density equals the critical value, and the space curvature constant $k = 0$ and the Universe has a flat Euclidean geometry, an idea first proposed by Einstein and De Sitter (1932). The Universe does now seem to be expanding at a velocity close to its own escape velocity, with the dynamical evidence for both visible and invisible matter amounting to $\Omega_o \approx 0.1$ in spiral galaxies and $\Omega_o \approx 0.2$ in clusters of galaxies. To obtain this rough accord at the present time, the mass density must have been much closer to the critical value in the remote past. Just one second after the big bang, the kinetic energy of expansion would have to balance the gravitational potential energy with an accuracy of 1 part in 10^{14} (Dicke and Peebles, 1979). This fine-tuned, delicate balance at early epochs can be explained by inflation; but it also requires that the density parameter Ω_o continues to be exactly unity and the space curvature constant k continues to be precisely zero. However, the observational evidence suggests that the mass density of the Universe, including invisible dark matter, is now less than the critical mass density (see Sects. 5.4.5 and 5.4.6). So inflation remains a "vague idea" in need of substantiation.

To put the inflation idea in perspective, the widely-accepted big-bang model supposed that the Universe began between ten and twenty billion years ago as a primeval fireball of extreme density and temperature, and it has been expanding and cooling ever since. This picture of the evolving Universe has stood the test of time and observation, successfully predicting or explaining the expansion of the Universe, the cosmic microwave background, and the relic abundances of the lightest elements (Peebles, Schramm, Turner and Kron, 1991). Astronomers have therefore confidently extrapolated this model back to the early stages of the expanding Universe, including the first few seconds of primordial nucleosynthesis.

The big-bang model cannot, however, be completely traced backward into time to stipulate the initial conditions from which the observable Universe evolved. In the very early stages, all matter was so hot that it was decomposed into its elementary particle constituents, and quantum mechanical forces applied, that have no counterpart in classical physics. The conditions of the very early Universe are described by the theory of high-energy particles, at energies unattainable in terrestrial laboratories.

One theoretical hypothesis is the "grand" unified theory, or GUT for short, that supposes that the basic forces of elementary particle interaction become unified at high enough energy. The strong electromagnetic force and the weak force are then described by a single force due to an underlying symmetry and order. Such a unification could take place about 10^{-34} seconds after the big bang, when the energies $E_{\rm GUT} \approx 10^{14}$ GeV and the temperature $T \approx 10^{27}$ degrees Kelvin (1 GeV = one billion electron volts $= 1.6 \times 10^{-3}$ erg).

When the Universe was about 100 times older, or at about 10^{-32} seconds after the big bang, the symmetry might have been broken, due to the lower energy and temperature, giving rise to distinct electromagnetic and weak forces. This would explain currently observable particles, such as the electron and neutrino, that behave very differently.

The temporary epoch at times, t, between a beginning time $t_{\rm b} \approx 10^{-34}$ seconds and stopping time $t_{\rm s} \approx 10^{-32}$ seconds has become known as the inflationary era, since the Universe may have then inflated and expanded exponentially (Kolb and Turner, 1990). Before and immediately after inflation, the Universe is radiation-dominated, with a cosmic scale factor, $R(t)$, given by

$$R(t) \propto t^{1/2} \quad \text{for } t \leq t_{\rm b} \quad \text{and } t \geq t_{\rm s} \text{ (up to decoupling)} \ . \tag{5.724}$$

During inflation there is a dramatic, in-flight correction in which the scale factor $R(t)$ expands exponentially, or

$$R(t) \propto \exp\left(\frac{t}{\tau}\right) \quad \text{for } t_{\rm b} \leq t \leq t_{\rm s} \ . \tag{5.725}$$

Over the period from 10^{-34} seconds to 10^{-32} seconds, the scale factor R(t) increases exponentially by an enormous factor of $\exp(100) \approx 10^{43}$ (see Fig. 5.37). The temperature drops drastically during inflation, but is expected to rise to the initial temperature of 10^{27} degrees, or an energy of 10^{14} GeV, at the end of inflation (also see Fig. 5.37).

The mechanism of inflation is described in terms of scalar fields that can cause a symmetry in the theory to be spontaneously broken. These scalar fields have a negative pressure, P, given by

$$P = -u_{\rm f} = -c^2 \rho_{\rm f} \ , \tag{5.726}$$

where $u_{\rm f}$ and $\rho_{\rm f}$ respectively denote the energy density and the density of the false vacuum, and c is the velocity of light. The energy density $u_{\rm f} \approx E^4_{\rm GUT}/(hc)^3 \approx 10^{93}$ erg cm^{-3}, where $E_{\rm GUT} \approx 10^{14}$ GeV $= 1.6 \times 10^{11}$ erg, Planck's constant $h = 6.6 \times 10^{-27}$ erg s, and the velocity of light $c = 2.9979 \times 10^{10}$ cm s^{-1}. Particle physicists use the word "vacuum" to mean the state of lowest energy density, and describe a temporary vacuum as a "false vacuum" (also see Kazanas, 1980; Sato, 1981).

The negative pressure produces a sort of tension that drives the expansion, and it can be described in terms of a cosmological constant Λ that temporarily opposes gravitational attraction. The Λ becomes enormously large during inflation, but then reverts to an exceedingly small value. The scale factor $R(t)$ in a Robertson–Walker Universe obeys the equation of motion

$$\ddot{R}(t) = -\frac{4\pi G}{3c^2} u(t) R(t) \ , \tag{5.727}$$

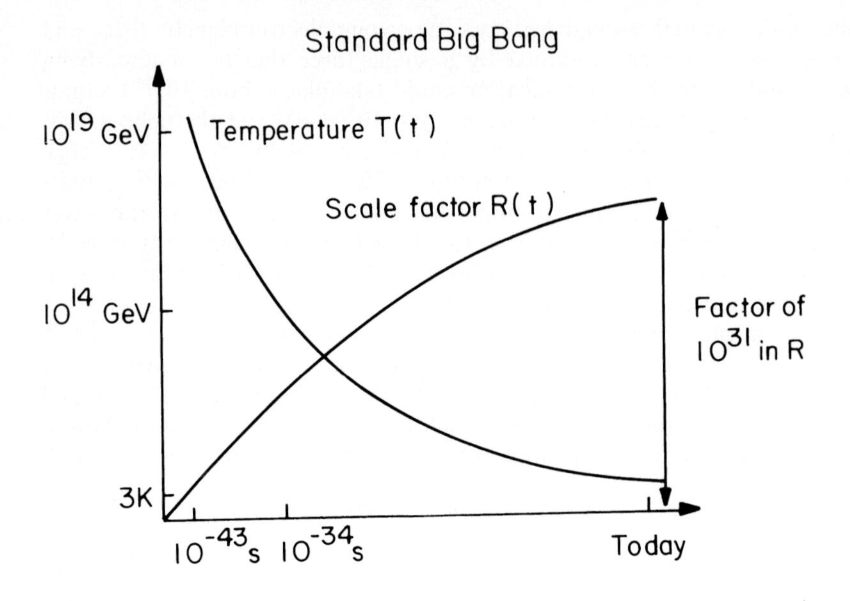

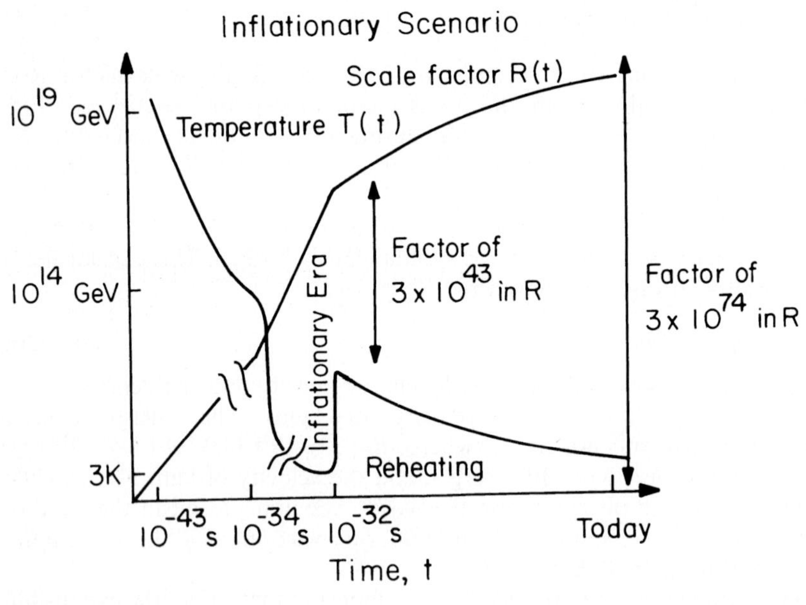

Fig. 5.37. The evolution of the temperature, T(t), and scale factor, $R(t)$, of the Universe with increasing time, t, in the standard big-bang model (top) and the inflationary model (bottom). The scale factor is a measure of the distance between any two points that partake in the expansion of the Universe. In the inflation model, the scale factor increases by a factor of 3×10^{43} between 10^{-34} and 10^{-32}seconds after the big bang. After $t = 10^{-32}$ seconds, the expansion is adiabatic, with the product $R(t)\ T(t)$ constant, as it is during much of the big-bang model. The temperature dropped to nearly zero during inflation, but was reheated at the end of inflation to the temperature before inflation began. (Adapted from Kolb and Turner (1990) and Partridge (1995)).

where $u(t)$ denotes the energy density at time t and

$$[\dot{R}(t)]^2 = \frac{8}{3}\frac{\pi G}{c^2} u_f [R(t)]^2 \ . \tag{5.728}$$

If the Universe is dominated by the large negative pressure of a false vacuum during the inflationary era, then

$$\ddot{R}(t) = \frac{4\pi G}{3c^2} u_f R(t) = \frac{\Lambda}{3} R(t) \ , \tag{5.729}$$

and

$$[\dot{R}(t)]^2 = \frac{\Lambda}{3} [R(t)]^2 \ , \tag{5.730}$$

where the cosmological constant Λ is in units of s^{-2} (sometimes $\Lambda = \lambda c^2$ is used with λ in units of cm^{-2}).

These equations have the solution

$$R(t) \propto \exp\left(\frac{t}{\tau}\right) \quad \text{for } t_b \leq t \leq t_s \tag{5.731}$$

where

$$\tau = \left(\frac{3c^2}{4\pi G u_f}\right)^{1/2} = \left(\frac{3}{\Lambda}\right)^{1/2} \approx 10^{-33} \text{ seconds} \ , \tag{5.732}$$

and $u_f = (10^{14}\ \mathrm{GeV})^4/(\hbar c)^3 \approx 10^{93}\ \mathrm{erg\ cm^{-3}}$.

Even if the inflation scenario is correct, it does not mean that inflation has solved all the problems of initial conditions for the Universe. As we probe further back into time, the energies become higher and higher, until we reach a time when a quantum theory of gravity is required. This is known as the Planck era.

Although the relevant theory is still unknown, we can estimate the conditions under which quantized gravity effects become important. If a mass was so compact that it was contained in its own Schwarzschild radius, then on quantum mechanical grounds that radius should be comparable to the Compton radius, or Harwit (1988):

$$\frac{2mG}{c^2} = \frac{h}{mc} \tag{5.733}$$

which permits us to solve for the mass, m, in terms of the other natural constants. We define the Planck mass

$$m_p \equiv \left(\frac{hc}{2\pi G}\right)^{1/2} = 2.18 \times 10^{-5} \text{ grams} \ , \tag{5.734}$$

where a factor of π has been included, as a matter of convention. If this mass is inserted in the expression for the Compton wavelength, we obtain the Planck length

$$l_p \equiv \left(\frac{hG}{2\pi c^3}\right)^{1/2} = 1.61 \times 10^{-33} \text{ cm} \tag{5.735}$$

where again, by convention, Planck's constant is divided by 2π. The shortest time during which it makes sense to talk about such a mass is the length of time light would take to traverse the Planck length. This is the Planck time,

$$t_{\mathrm{P}} \equiv \left(\frac{hG}{2\pi c^5}\right)^{1/2} = 5.38 \times 10^{-44} \text{ seconds} \tag{5.736}$$

Over shorter intervals than that, one end of the Planck mass distribution would cease to be aware of the presence of the other end, so that the laws of causality would no longer apply. The Planck time must therefore be considered the earliest time in the existence of the Universe for which the Einstein field equations could apply.

Finally, dividing the Planck mass by the cube of the Planck length gives us a measure of the density the early Universe could have attained. Again, the laws of relativity could not be expected to apply at higher densities than this Planck density

$$\rho_{\mathrm{P}} \equiv \frac{2\pi c^5}{hG^2} = 5.18 \times 10^{93} \text{ g cm}^{-3} \ . \tag{5.737}$$

It makes sense to also ask about the temperature that might have existed at the Planck time. When the number of available species is of order unity, the temperature becomes

$$T_{\mathrm{P}} \sim \left(\frac{c^2 \rho_{\mathrm{P}}}{a}\right)^{1/4} \sim 10^{32}\,^{\circ}\mathrm{K} \ , \tag{5.738}$$

where a is the radiation constant.

As shown by Penrose (1965) and Hawking (1966), the Einstein field equations contain at least one singularity at the big bang, and this precludes any meaningful extrapolation to before the start of the expansion. Einstein (1945) realized this dilemma, reasoning that his equations may not be valid for very high densities, and that they may not be extrapolated past the beginning of the expansion. So, in the end, even if inflation is true it has just pushed our uncertainties about the beginning of things back a little further in time.

5.7.3 Classical Tests of Cosmological Models

Astrophysical cosmology deals with the physical aspects of the origin, structure and evolution of the Universe. It is firmly rooted in observational evidence that supports a big bang model of a homogeneous, isotropic expanding Universe (Peebles, Schramm, Turner and Kron, 1991). The observed redshifts of extragalactic objects indicate that the Universe is expanding with recession velocities that increase with distance according to Hubble's law. This linear relation between redshift and distance indicates that the expansion is uniform, except for certain peculiar motions described in Sect. 5.7.6.2, and that the galaxies are being homogeneously swept apart, carried by a uniformly enlarging fabric of space-time. That expansion started from a big bang, when all matter and energy were compressed to infinite density. The cosmic microwave background radiation is the relic of the primeval heat that accompanied the big bang.

The observations of the cosmic microwave background radiation, that bathes the cosmos uniformly (see Sect. 5.7.5), indicate that the Universe is the same at every point and in every direction when averaged over large enough areas of the sky, as expected from the cosmological principle. Any observer in any galaxy would therefore observe the same linear relationship between distance and redshift caused by the uniform expansion of the Universe. This means that we are not in a privileged position in the Universe, and that the Universe has no preferred shape or center.

The mathematical model used to describe the expanding Universe is Einstein's General Theory of Relativity, simplified to the case of an isotropic, homogenous, expanding Universe. The Robertson–Walker metric, and the scale factor, $R(t)$, at time, t, for this model were provided in Sect. 5.7.1. They depend on three parameters, the present rate of expansion, or Hubble's constant, H_o, the deceleration parameter, q_o, and the cosmological constant, Λ, which is usually taken to be zero, $\Lambda \approx 0$, because of its extremely small value. If $R_o = R(t_o)$ denotes the scale factor at the present time, t_o, after the big bang, then we have:

$$H_o = \frac{\dot{R}_o}{R_o} \ , \tag{5.739}$$

and

$$q_o = -\frac{\ddot{R}_o R_o}{\dot{R}_o^2} = \frac{\Omega_o}{2} - \frac{\Lambda c^2}{3H_o^2} \tag{5.740}$$

or

$$q_o = \frac{\Omega_o}{2} \quad \text{for } \Lambda = 0 \ , \tag{5.741}$$

and the density parameter, Ω_o, at the present time is given by:

$$\Omega_o = \frac{8\pi G \rho_o}{3H_o^2} \tag{5.742}$$

for a present mass density, $\rho_{mo} = \rho_o$. Provided that the cosmological constant $\Lambda = 0$, the value of Ω_o or q_o, or equivalently H_o and ρ_o, determine the fate of the Universe.

The Universe is changing with time. There is a boundary, or edge, to time when the expansion started, the contents of the Universe evolve with time, and there is a predictable outcome for it all. There is no way of knowing what happened before the big bang, and whatever events occurred then have no consequence for the present. But by determining the current value of Ω_o or q_o we can forecast the fate of the Universe. In one possibility, for $\Omega_o > 1$ or $q_o > 0.5$, the rate of expansion is sufficiently slow, and the amount of matter sufficiently large, to eventually halt the expansion in the future. If $\Omega_o = 1$ or $q_o = 0.5$, the galaxies are moving apart at just the critical rate to avoid recollapse, and for $\Omega_o < 1$ or $0 < q_o < 0.5$, the expansion continues forever.

Thus, the big-bang cosmological models for a homogeneous, isotropic expanding Universe depend on how much matter is presently contained in the Universe, as well as its rate of expansion. This matter curves space-time and therefore plays a role in interpreting the luminosity, distance, age, angular size and number density information for extragalactic objects. Classical tests of cosmological models examine these observable aspects of the models in an effort to refine measurements of H_o and q_o. They have been reviewed by Sandage (1988) and Yoshii and Takahara (1988). Here we will provide the basic formulae and outline the results for these tests, noting at the outset that they suggest source evolution effects and generally fail to discriminate between cosmological models.

Observations of the redshift, z, of galaxies support the linear relation

$$V_r = cz = H_o D_L \quad \text{for } V_r \ll c \quad \text{and } z \ll 1 \ , \tag{5.743}$$

where the recession velocity, V_r, increases with the luminosity distance, D_L, the Hubble constant is H_o , the velocity of light is c, and the redshifts are small with $z \ll 1$. The observational evidence for this linear relation between distance and redshift is reviewed by Peebles (1993). The redshift, z, is given by:

$$z = \frac{\lambda_o - \lambda_e}{\lambda_e} \ ,$$

where λ_e is the emitted wavelength of the spectral line and λ_o is the wavelength at which it is observed. For large recession velocities,

$$V_r = c \times \left[\frac{(z+1)^2 - 1}{z+1)^2 + 1}\right] \quad \text{for } V_r \approx c \quad \text{and } z \geq 1 \ . \tag{5.744}$$

For more than three decades there has been an unresolved controversy over the exact value of the Hubble constant, H_o, which lies somewhere between 50 and 100 km s^{-1} Mpc^{-1} (see Sect. 5.2.8). The luminosity distance, D_L, depends on H_o, the deceleration parameter, q_o, and the redshift, z, and the relevant formulae for the luminosity distance in homogeneous, isotropic expanding Universes were given in Sect. 5.2.9. The expansion age of the Universe, or the time t_o since the big bang, depends on the same parameters, with a similar uncertainty. The ages of the oldest known astronomical objects provide a lower limit to t_o, and for a given value of H_o they can be used to infer q_o (the relevant formulae are given in Sect. 5.3.12). Such a test may be devoid of direct evolutionary effects, but it depends upon the enormous current uncertainty in the value of H_o.

To get back to Hubble's law, Slipher (1917) showed that bright nearby spiral galaxies are moving away from us at high velocities, sometimes exceeding 1,000 km s^{-1}. Wirtz (1921, 1924), demonstrated an approximate linear dependence of the recession velocity, V_r, and apparent magnitudes, m, of spiral nebulae, and Hubble (1929) showed that their V_r increases with distance, D. The relation $V_r = H_o \times D_L$ is now known as Hubble's law, although he did not fully state it.

Although some galaxies are aggregated into groups or clusters, Hubble (1926, 1936) found that they are uniformly distributed when viewed on very large scales. As he probed deeper into space to fainter and fainter galaxies, their numbers increased by amounts that agreed with the assumption of a uniform distribution with constant intrinsic brightness or absolute magnitude.

As shown by Lemaitre (1927), the observed recessional velocities of the extragalactic nebulae are a cosmical consequence of the expansion of a homogeneous, isotropic Universe. Indeed, the models for such an expanding Universe, based on Einstein's relativity theory, were anticipated by Friedmann (1922, 1924) even before the discovery of Hubble's law. The linear dependence of velocity on distance was first predicted by Weyl (1923) using the static De Sitter model; discussed by Lanczos (1923) for the nonstationary $k = +1$ model; given as a function of luminosity distance by Tolman (1930); and shown by Milne (1935) to be an immediate consequence of the assumed homogeneity and isotropy of the Universe.

For such a model, the ratio of the scale factor, $R(t_e)$, at the time, t_e, that the light was emitted to the scale factor, $R(t_o)$, at the present time, t_o, is given by:

$$\frac{R(t_e)}{R(t_o)} = \frac{1}{1+z} = \frac{\lambda_e}{\lambda_o} \quad . \tag{5.745}$$

Thus, for example, the light we now see from a quasar of redshift $z = 4.0$ was emitted when the Universe was one fifth its present size.

Since distances can only be directly measured for relatively nearby extragalactic objects (see Sect. 5.2.7), Hubble's law $V_r = cz = H_o D_L$ was reformulated in the form:

$$\log(cz) = 0.2 \text{ m} + \text{B} \quad , \tag{5.746}$$

where m is the apparent magnitude of the object and the constant B depends on the Hubble constant and the intrinsic brightness of the object. The slope of 0.2 in the redshift-magnitude plot, also called the Hubble diagram, is what one would expect for a homogeneous, isotropic expanding Universe provided that all extragalactic objects have the same absolute magnitude and there are no evolutionary effects on a cosmic scale. Hubble and Humason (1934) and Humason, Mayall and Sandage (1956) examined the Hubble diagram for galaxies, showing that the observed redshift-magnitude relation is, within the accuracy of the data, linear with a slope of value 0.2.

When radio galaxies and quasars were subsequently used to extend the Hubble diagram to greater distances, or larger redshifts, it was found that it is also consistent, within the data dispersion, with that expected for a homogeneous, isotropic expanding Universe (Lang et al., 1975). Unfortunately, the large scatter in the observed data makes it impossible to determine the details of the expansion, and to discriminate between different cosmological models. They instead suggest that extragalactic objects evolve in time. Because the light with which we view distant objects left them some time ago, the observed data suggest that younger extragalactic objects are brighter and more active than older, nearby ones.

For detailed comparisons with cosmological models and the theoretical evolution of cosmic objects, one recasts the redshift-magnitude equation in the form

$$m_{\text{bol}} = 5 \log D_L + M_{\text{bol}} + 25 \quad , \tag{5.747}$$

for an apparent bolometric magnitude, m_{bol}, an absolute magnitude, M_{bol}, and a luminosity distance D_{L}, in Megaparsecs or Mpc for short. For a homogeneous, isotropic, expanding Universe with zero cosmological constant $\Lambda = 0$, this relation becomes (Mattig, 1958, 1959)

$$m_{\mathrm{bol}} = 5 \log \frac{c}{H_{\mathrm{o}} q_{\mathrm{o}}^2} \left\{ q_{\mathrm{o}} z + (q_{\mathrm{o}} - 1) \left[(1 + 2q_{\mathrm{o}} z)^{1/2} - 1 \right] \right\} + M_{\mathrm{bol}} + 25 \quad \text{for } \Lambda = 0 \tag{5.748}$$

and $q_{\mathrm{o}} > 0$. Using the more precise, equivalent expression for luminosity distance, we have (Terrell, 1977)

$$m_{\mathrm{bol}} = 5 \log \frac{cz}{H_{\mathrm{o}}} \left\{ 1 + \frac{z(1 - q_{\mathrm{o}})}{(1 + 2q_{\mathrm{o}} z)^{1/2} + 1 + q_{\mathrm{o}} z} \right\} + M_{\mathrm{bol}} + 25 \quad \text{for } \Lambda = 0 \ . \tag{5.749}$$

When $z \ll 1$, both of these expressions reduce to the redshift-magnitude relation

$$m_{\mathrm{bol}} = 5 \log \frac{cz}{H_{\mathrm{o}}} + M_{\mathrm{bol}} + 25 \quad \text{for } z \ll 1 \quad \text{and } \Lambda = 0 \ ,$$

which is obtained using Hubble's law for the luminosity distance.

As with the radio galaxy and quasar data, the observed data for galaxies have an extremely large scatter, presumably caused by a large spread in the absolute bolometric magnitudes due to evolutionary effects. It has even been suggested that the dispersion in the observed redshift-magnitude relation is so great that it fails to confirm the predictions of the Friedmann-Lemaitre cosmology (Segal, 1993).

It was thought that one method of reducing the uncertainty was to choose the brightest members of clusters of galaxies, with presumably similar absolute magnitudes, and to thereby determine the deceleration parameter q_{o}. However, the redshift-magnitude data for such first-member cluster data has not determined whether q_{o} is greater or less than 0.5. For example, the Hubble diagram of first-ranked cluster galaxies has been used to obtain $q_{\mathrm{o}} = 2.6 \pm 0.8$ (Humason, Mayall and Sandage, 1956), $q_{\mathrm{o}} = -1.27 \pm 0.7$ (Gunn and Oke, 1975), $q_{\mathrm{o}} = 0.33 \pm 0.7$ (Sandage et al., 1976), $q_{\mathrm{o}} = 1.6 \pm 0.4$ (Kristian et al., 1978). This uncertainty is further compounded when evolutionary effects are taken into account (Sandage, 1961).

The magnitude change, $K(z)$, due to the redshift of the energy curve and the change in magnitude, $E(z)$, due to evolution are included in comparisons with the observed Hubble diagram through the relation (Sandage, 1988)

$$m_{\mathrm{bol}} = 5 \log D_{\mathrm{L}} - K(z) - E(z) + M_{\mathrm{bol}} + 25 \ .$$

Because of the fixed detector effective wavelength and finite detector bandwidth, any shift in the spectrum towards the red, described by the redshift z, causes the observed apparent magnitudes to differ from those that would have been observed at zero redshift. This K correction is given by (Humason, Mayall and Sandage, 1956; Oke and Sandage, 1968):

$$K = 2.5\ \log(1+z) + 2.5\ \log \frac{\int_{\mathrm{o}}^{\infty} I(\lambda) s(\lambda) d\lambda}{\int_{\mathrm{o}}^{\infty} I\left(\frac{\lambda}{1+z}\right) s(\lambda) d\lambda}\ \text{magnitudes}\ , \tag{5.750}$$

where the first term arises from the narrowing of the photometer pass-band in the rest frame of the galaxy by the factor $(1+z)$, and the second term is due to the fact that the radiation received by the observer at wavelength, λ, is emitted by the galaxy at wavelength $\lambda/(1+z)$. Here $I(\lambda)$ is the incident energy flux per unit wavelength observed at wavelength, λ, and corrected for absorption, and $s(\lambda)$ is the photometer response function. A summary of the rest-frame spectral energy distributions that have been used to calculate K is given by Yoshii and Takahara (1988).

Since galaxies are composed of stars, and stars are now being born, evolving and dying, one can expect stellar evolution effects to alter the overall brightness of a galaxy. For small redshifts, $z \ll 1$, the correction due to evolution is approximately 0.07 magnitude per 10^9 years (Sandage, 1988). The changes, $E(t)$, over time from 10^7 to 1.5×10^{10} years are given by Yoshii and Takahara (1988) for E/S0 and Sdm galaxies. They should be converted into a redshift term E(z) using the relationship between the look-back time and the redshift, z, and deceleration parameter, q_{o}, given in Sect. 5.3.12. For example,

$$t_{\mathrm{o}} - t_{\mathrm{e}} = H_{\mathrm{o}}^{-1}\left(\frac{z}{1+z}\right) \quad \text{for } q_{\mathrm{o}} = 0\ , \tag{5.751}$$

and

$$t_{\mathrm{o}} - t_{\mathrm{e}} = \frac{2}{3} H_{\mathrm{o}}^{-1}\left[1 - \frac{z}{(1+z)^{3/2}}\right] \quad \text{for } q_{\mathrm{o}} = 0.5\ , \tag{5.752}$$

where t_{o} is the observed time and t_{e} is the time the radiation was emitted. Very complex calculations are required to determine $E(z)$ at very large redshifts when the look-back times are of the order of the age of the Universe.

The general conclusion from the observed Hubble diagrams, or redshift-magnitude relations, is that evolution must be invoked at very large redshifts if the standard model for homogeneous, isotropic expanding Universes is applicable (Spinrad, 1986; Sandage, 1988); and that such investigations have not been able to distinguish between cosmological models of various deceleration parameters q_{o}.

Another classical test for cosmological models is based on the variation of angular size, θ, with redshift, z. A spherical source of linear diameter, l, and redshift, z, in a homogeneous, isotropic expanding Universe with zero cosmological constant $\Lambda = 0$ will have the apparent angular diameter, θ, given by

$$\theta = \frac{l(1+z)^2}{D_{\mathrm{L}}} = \frac{lH_{\mathrm{o}}}{c} \frac{q_{\mathrm{o}}^2 (1+z)^2}{\{q_{\mathrm{o}} z + (q_{\mathrm{o}} - 1)[(1+2q_{\mathrm{o}} z)^{1/2} - 1]\}}\ , \tag{5.753}$$

and

$$\theta = \frac{lH_{\mathrm{o}}}{c}\frac{(1+z)^2}{(z+z^2/2)} \quad \text{for } q_{\mathrm{o}} = 0 \ . \tag{5.754}$$

Here we have assumed that all extragalactic objects have the same linear size, l, that does not evolve or change with time, and the luminosity distance is $D_{\mathrm{L}} = D_{\mathrm{A}}(1+z)^2$ for angular size distance D_{A} so in a Friedmann Universe the angular size $\theta = l/D_{\mathrm{A}}$ (Whittacker, 1931; Etherington, 1933).

In a static, or non-expanding, Universe, the angular size, θ, for a source of fixed linear size would decrease with increasing redshift, z, as $1/z$. However, for an isotropic, homogeneous expanding Universe with any deceleration parameter $q_{\mathrm{o}} > 0$, the angular size decreases to a minimum at some finite z and then increases with increasing z (Hoyle, 1959). As an example, for $q_{\mathrm{o}} = 0.5$, or density parameter $\Omega_{\mathrm{o}} = 1$, the angular size has a minimum at redshift $z_{\mathrm{min}} = 1.25$. Thus, if extragalactic objects have some standard linear size, then observation of the angular diameter and redshift could help to determine the correct cosmological model, and the determination of z_{min} might establish q_{o}.

Miley (1971) applied this test to double-lobe radio sources, showing that the largest angular separation, θ, of the radio components of quasi-stellar sources and radio galaxies decreases with increasing redshift, z, as $\theta \propto 1/z$ up to $z \approx 2$, without a detectable flattening or minimum at lower redshift. In an expanding Friedmann model, this result indicates either size evolution or an inverse correlation between luminosity and size. The observed data do not, by themselves, suggest that the linear dimensions of the sources evolve with cosmological epoch (Hooley, Longair and Riley, 1978; Saikia and Kulkarni, 1979), but might be attributed to a selection effect in which smaller, intrinsically more luminous objects are preferentially observed at larger distances. Yet, Kapahi (1987) and Sandage (1988) indicate the need for size evolution of the sources. Either selection effects or size evolution almost certainly obscure the cosmological information contained within the angular size-redshift data. However, cosmological parameters may be constrained if the active lifetime of the collimated outflows are shorter at higher redshifts (Guerra and Daly, 1998).

The Hubble Space Telescope has been used to derive the relationship between apparent brightness and angular size for faint galaxies in a deep field down to a visual magnitude of about 30, equivalent to one photon per week hitting the eye. These optical data also show a steady decrease in linear size at large redshifts $z > 2$ with no detected increase in angular size. The small angular sizes at faint magnitudes imply that, on average, the faint galaxies are physically more compact than the Milky Way (Ferguson, Williams and Cowie, 1997).

A final, inconclusive cosmological test is the number counts of galaxies at successively larger apparent magnitudes, or the source counts of extragalactic radio sources of varying flux density. If galaxies are uniformly distributed and have the same absolute luminosity, L, then the number $N(>f)$ of galaxies with apparent energy flux greater than f, is given by

$$N(>f) = \frac{4\pi}{3} D_{\rm L}^3 n = \frac{4\pi}{3} n \left(\frac{L}{4\pi f}\right)^{3/2} \propto f^{-3/2} \tag{5.755}$$

where n is the number density of galaxies, and the observed energy flux, f, is given by

$$f = \frac{L}{4\pi D_{\rm L}^2} \text{ erg cm}^{-2} \text{ sec}^{-1} , \tag{5.756}$$

where $D_{\rm L}$ is the luminosity distance. Shapley and Ames (1932) and Hubble (1934, 1936) found rough agreement with this relation for counts of optical galaxies down to the limiting magnitude $m_{\rm pg} = 20.7$ or $m_{\rm v} = 19.8$.

In addition to showing that the distribution of galaxies is uniform on a large scale, number counts might lead to tests of cosmological models. For homogeneous, isotropic expanding Universes with a cosmological constant of zero, the number of galaxies, $N(m)$, brighter than apparent magnitude, m, is given as a function of deceleration parameter, $q_{\rm o}$, by Mattig (1959) and Sandage (1962). Unhappily, the difference in the observed $N(m)$ for small ranges of $q_{\rm o}$ between 0 and 1 is less than the probable observation errors when observing as low as $m_{\rm R} = 22$. The Hubble Space Telescope has been used to extend the number counts in a deep field to galaxies with visual magnitudes of about 30. These number counts also do not, by themselves, provide a definitive test of cosmological models, although they tend to favor cosmologies that are either open or dominated by a cosmological constant (Ferguson, Williams and Cowie, 1997).

Radio source counts offer sampling to much greater redshifts than are available for some optical data, but they have also not led to definitive tests of cosmological models. The radio counts have instead shown strong evolutionary effects with cosmological epoch.

What is actually observed in radio source counts is the number $N(>S)$ of radio sources per steradian with flux densities greater than S. Because the volume of a sphere of radius, r, is $4\pi r^3/3$, the number of sources within such a sphere is proportional to r^3 provided that the sources are uniformly distributed. If all the sources have the same intrinsic luminosity, then the faintest source visible from the center of the sphere will have a flux, S, proportional to r^{-2}. It follows that the number $N_{\rm o}(>S)$ of sources per steradian with flux densities greater than S in an isotropic, static, Euclidean universe in which no evolution takes place is

$$N_{\rm o}(>S) = \frac{n_{\rm o} P_{\rm o}^{3/2}}{3} S^{-3/2} \propto S^{-3/2} , \tag{5.757}$$

where $n_{\rm o}$ is the mean number density of sources, and $P_{\rm o}$ is their mean brightness at the frequency under consideration. Number count data are then plotted logarithmically and the observed slope is compared to the -1.5 slope expected from a static Euclidean Universe. Departures from this slope might discriminate among cosmological models, but the actual counts agreed with no plausible cosmological model. That is, a significant excess in the number of faint extragalactic radio sources was found as compared with what would be expected for all the standard cosmological models (Ryle and Clarke, 1961; Ryle, 1968). In an expanding, Friedmann Universe, there should be relatively fewer of the fainter sources than in a static Euclidean Universe. The observed increase in N as S decreased could not be explained by space curvature effects, and could only be explained if radio sources were either more luminous or more numerous in the past (Longair, 1966).

This conclusion has been confirmed and enriched with the extension of radio source counts to very low flux densities (Bridle et al., 1972, Donnelly, Partridge and Windhorst, 1987; Condon, 1984, 1988; Windhorst, Van Heerde and Katgert, 1984; and Windhorst et al., 1995). The major factor controlling the counts of faint galaxies is the evolution of the galaxies themselves not cosmological factors describing the geometry of the Universe. Number counts of faint optical and radio galaxies contain much more information about the evolution of the galaxy than about the cosmological geometry. This is because the nuclear activity responsible for the radio emission and the massive stars responsible for the blue light are transient events by cosmological standards, producing enormous evolutionary changes in galaxy properties with redshift (Cowie and Songaila, 1993).

The radio source counts have nevertheless ruled out the steady state cosmology (Bondi and Gold, 1948; Hoyle, 1948). That cosmology is based on the perfect cosmological principle that assumes that the Universe is unchanging in both space and time, presenting the same aspect from any point in space or time. The radio data showed that the Universe evolves and changes with time, violating the perfect cosmological principle. The steady state cosmology is also inconsistent with the cosmic microwave background radiation and the observed helium abundance. However, no convincing detection of anisotropy in the sky distribution of extragalactic radio sources has ever been made, so the radio data is consistent with the cosmological principle that applies to space alone.

Although the details are outside the scope of the classical tests of cosmological models, alternatives to the big bang have been reviewed by Ellis (1984). Astrophysical formulas for Friedmann models with cosmological constant and radiation were given by Dabrowski and Stelmach (1987). The apparent luminosity-redshift and distance-redshift relations for a locally inhomogeneous Universe are given by Dyer and Roeder (1974), Weinberg (1976) and Watanabe and Tomita (1990).

Altogether different cosmological models involve the possibility that the Newtonian gravitational constant, G, might vary with time. As pointed out by Dirac (1937, 1938), the approximate age of the Universe, H_0^{-1} for Hubble constant H_0, in terms of the atomic unit of time, $e^2/(m_e c^3)$ for electron charge e and mass m_e, is of order 10^{39}, as is the ratio, $e^2/(G m_p m_e)$, of the electrical force between the electron and proton to the gravitational force between the same particles. This, according to Dirac, suggested that both ratios are functions of the age of the Universe, and that the gravitational constant, G, might vary with time. For example, Dirac chose

$$\left(\frac{\dot{G}}{G}\right)_0 = -3H_0 \ , \tag{5.758}$$

where $\dot{}$ denotes the first derivative with respect to time, the subscript o denotes the present value, and H_o is the present value of the Hubble constant. The cosmological model of Brans and Dicke (1961), Dicke (1962), gives (Weinberg, 1972)

$$\begin{aligned}\left(\frac{\dot{G}}{G}\right)_o &= \frac{-3q_o H_o}{\omega + 2}, \quad \text{for } q_o < 1 \\ &= \frac{-H_o}{\omega + 2} \quad \text{for } q_o = 0.5 \\ &= \frac{-1.71 H_o}{\omega + 2} \quad \text{for } q_o = 1\end{aligned} \tag{5.759}$$

and

$$= \frac{-3.34 H_o q_o^{1/2}}{\omega + 2} \quad \text{for } q_o > 1 \ ,$$

where ω is a dimensionless constant of the cosmological model, q_o denotes the present value of the deceleration parameter, and

$$H_o = [0.98 \times 10^{10} h^{-1}]^{-1} \text{ years}^{-1} \ ,$$

if $H_o = 100h$ km sec^{-1} Mpc^{-1}. Planetary radar ranging observations (Hellings et al., 1983), lunar laser ranging data (Müller, Schneider, Soffel and Ruder, 1991), and the binary pulsar (Damour and Taylor, 1991) all indicate that

$$\left(\frac{\dot{G}}{G}\right)_o \leq 10^{-11} \text{ years}^{-1} \ . \tag{5.760}$$

This brings us to a much more definitive test of cosmological models, the comparison of the abundances of the lightest elements with predictions of big bang nucleosynthesis, which provides important constraints to the baryon density of the Universe.

5.7.4 Big Bang Nucleosynthesis and the Baryon Density of the Universe

Light elements were formed at the dawn of time before the stars even existed, in the immediate aftermath of the big-bang explosion that produced the expanding Universe. These primordial elements include hydrogen, the most abundant element in the cosmos, and helium, the second-most abundant one, as well as traces of deuterium, a heavier form of hydrogen, and lithium. They were created at very high temperatures, where nuclear reactions can proceed, at about 10^9 K, and in a brief moment of time during the first few hundred seconds of the big bang.

Such a hot, explosive origin for the Universe is inferred from a backward extrapolation of the present expansion of the galaxies, which must have been propelled from a common origin in the big-bang explosion, and confirmed by the discovery of the cosmic microwave background radiation, a relic of the initial fireball. The overwhelming evidence for the relativistic, hot big-bang model for the expanding Universe is presented by Peebles, Schramm, Turner and Kron (1991).

Observations of the light element abundances agree with the predictions of big-bang nucleosynthesis, showing that the majority of atoms we see today are left-over remnants of the earliest times, created in the infant Universe, a sort of primordial nuclear reactor. Indeed, calculations based on homogeneous models of the early phases of the big-bang explosion have been remarkably accurate in explaining or predicting the primordial abundances relative to hydrogen of the light elements, D, ^{3}He, ^{4}He and ^{7}Li, deduced from observations over nine orders of magnitude in abundance. In addition, these results have been very successfully used to predict upper bounds on the number of light neutrino families.

Perhaps more importantly, the cosmological implications of big-bang nucleosynthesis are profound! The agreement of light-element abundances and predictions from primordial nuclear reactions works only if the density of ordinary matter in the Universe, in both visible and invisible forms, is less than 10 percent of the critical mass density, ρ_c, required to eventually stop the expansion of the Universe. If only presently known forms of matter exist, the cosmological constant is zero, and the big-bang nucleosynthesis models are correct, then they require an open, hyperbolic Universe that will continue to expand forever. In other words, if the Universe is at its critical density, then some kind of exotic, "nonbaryonic" matter is required. In fact, since estimates of cosmic mass densities from clusters of galaxies exceed the 10 percent limit from primordial nucleosynthesis, it appears that independent of whether the Universe is at its critical density, nonbaryonic matter is required.

The number of baryons is equal to the number of neutrons and protons. And since the present-day cosmos consists primarily of hydrogen, the current baryon density, ρ_{Bo}, is roughly equal to that of hydrogen atoms. It is usually given in terms of the critical mass density, ρ_c, between an open and closed Universe by using the density parameter

$$\Omega_B = \rho_{Bo}/\rho_c \ , \tag{5.761}$$

where

$$\rho_c = 3H_o^2/(8\pi G) = 1.88 \times 10^{-29}\, h^2 \text{ g cm}^{-3} \ ,$$

and the Hubble constant $H_o = 100h$ km s^{-1} Mpc^{-1} .

The idea that the early, dense stages of the Universe were hot enough to generate thermonuclear reactions and synthesize elements was introduced by George Gamow and his colleagues in the late 1940s and early 1950s. In discussing the relative importance of matter and radiation at different stages of the expanding Universe, Gamow (1946, 1956) and Alpher and Herman (1949) pointed out that the radiation energy density, $\rho_r(t)$, would greatly exceed the mass density, $\rho_m(t)$, in the early stages, when radiation would dominate the expansion.

For a homogeneous, isotropic Universe, described by the Robertson–Walker metric, the scale factor, $R(t)$, and radiation temperature, $T_r(t)$, at time, t, after the big bang are (Tolman, 1934):

$$\rho_{\mathrm{r}}(t) = aT_{\mathrm{r}}^4(t) \propto [R(t)]^{-4}$$
$$\rho_{\mathrm{m}}(t) \propto [R(t)]^{-3} \ , \tag{5.762}$$

where the radiation constant

$$a = \frac{8\pi^5 k^4}{15c^3h^3} = 7.5641 \times 10^{-15} \ \mathrm{erg\ cm^{-3}\ ^\circ K^{-4}} \ .$$

The baryon mass density, $\rho_{\mathrm{B}}(t)$, at time, t, decreases as the Universe expands, and the radius of the Universe, $R(t)$, increases. However, the total mass of baryons, $\rho_{\mathrm{B}}(t)[R(t)]^3$, will remain constant, and since the radiation temperature, $T_{\mathrm{r}}(t)$, decreases linearly with the radius, at any time, t.

$$\rho_{\mathrm{B}}(t)[T_{\mathrm{r}}(t)]^{-3} = \mathrm{constant} \ , \tag{5.763}$$

and the baryon mass density at the present time, t_{o}, can be inferred from that in the distant past from

$$\rho_{\mathrm{Bo}} = \rho_{\mathrm{B}}(t_{\mathrm{o}}) = \rho_{\mathrm{B}}(t)[T_{\mathrm{r}}(t_{\mathrm{o}})/T_{\mathrm{r}}(t)]^3 \ . \tag{5.764}$$

The temperature of the cosmic microwave background radiation is $T_{\mathrm{r}}(t_{\mathrm{o}}) = 2.726 \pm 0.010$ K (Mather et al., 1990, 1994). The observed abundances of the light elements provide constraints on the primordial abundances synthesized during the early stages of the big-bang, and permit a determination of the baryon density back then. Since the calculations depend on the mass density then, and can be extrapolated to now, big-bang nucloeosyntheis predicts $\rho_{\mathrm{B}}(t_{\mathrm{o}})$ and has therefore provided one of the cornerstones of modern cosmology.

According to theory, the relative abundances of the elements created during big-bang nucleosynthesis in a spatially homogeneous Universe depends on a single free parameter, the universal baryon-to-photon number ratio:

$$\eta = 10^{-10}\eta_{10} = n_{\mathrm{B}}/n_{\gamma} \approx 3 \times 10^{-10} \ , \tag{5.765}$$

which has remained unchanged since the epoch of electron-positron annihilation a few seconds after the big-bang. Here n_{B} is the number of baryons and n_{γ} is the number of photons, so the Universe contains fewer than one baryon (neutrons and protons) per billion photons. During the time of big-bang nucleosynthesis, the Universe was a dilute gas of photons, contaminated by only trace amounts of baryons.

The temperature of the cosmic background radiation tells us the number of photons that are now left over from the big bang. At present:

$$n_{\gamma\mathrm{o}} = 410.89(T_{\mathrm{r}}(t_{\mathrm{o}})/2.726)^3 \approx 411 \ . \tag{5.766}$$

Through the known temperature of the microwave background, we can infer that the baryon mass density, $\rho_{\mathrm{Bo}} = \rho_{\mathrm{B}}(t_{\mathrm{o}})$, of the Universe today is:

$$\rho_{\mathrm{Bo}} = m_{\mathrm{p}} n_{\gamma\mathrm{o}} \eta \approx 6.87 \times 10^{-32} \eta_{10} \ \mathrm{g\ cm^{-3}} \ , \tag{5.767}$$

where the proton mass $m_{\mathrm{p}} = 1.673 \times 10^{-24}$ grams. The baryon density parameter is therefore given by:

$$\Omega_{\mathrm{B}} = \rho_{\mathrm{Bo}}/\rho_{\mathrm{c}} \approx 0.00365 h^{-2} \ \eta_{10} \ . \tag{5.768}$$

For more than two decades, big bang nucleosynthesis has shown that:

$$10^{-9} \leq \eta \leq 10^{-10} \ , \tag{5.769}$$

or

$$0.1 \leq \eta_{10} \leq 1.0 \ , \tag{5.770}$$

so the total amount of visible and invisible (dark) baryons (the protons and neutrons of ordinary matter), fail by a wide margin to close the Universe ($\Omega = 1$) for any reasonable value of the Hubble constant.

During the early radiation-dominated era, when $\rho_r(t) > \rho_m(t)$, the formulae for the adiabatic expansion of the radiation in a homogeneous, isotropic Universe (Robertson–Walker metric) can be used to obtain the radiation temperature, $T_r(t)$, at time, t (Gamow, 1948, 1956):

$$T_r(t) = \left[\frac{3c^2}{32\pi aG}\right]^{1/4} \frac{1}{t^{1/2}} = \frac{1.5 \times 10^{10}}{t^{1/2}} \ \mathrm{K} \ , \tag{5.771}$$

or equivalently

$$t = \left(\frac{c^2}{32\pi GaT^4}\right)^{1/2} \approx 10^{20} \ T^{-2} \ \text{seconds} \ . \tag{5.772}$$

Gamow (1948) noted that the nuclear elements had to be synthesized as the Universe cooled down. Nuclear build up could not take place in the hottest, most condensed, state of the early Universe when $T_r(t) \geq 10^{10}$ K, or $t \leq 1$ second, because the radiation, or thermal photons, would be hot and energetic enough to break up bound particle groups. Throughout the radiation-dominated era, the radiation would cool during the expansion as if matter were not present, quickly reaching $T_r(t) \approx 10^9$ K at $t \approx 10$ seconds when nucleosynthesis can begin. Any element buildup, however, must be completed during the first few hundred seconds before all the free neutrons, n, decay into protons, p, according to the reaction

$$\mathrm{n} \rightarrow \mathrm{e}^- + \mathrm{p} + \overline{\nu}_\mathrm{e} \ , \tag{5.773}$$

where e^- denotes the electron, and $\overline{\nu}_e$ is the anti-electron neutrino. The neutron lifetime for such a beta decay is $\tau_n = 887 \pm 2.0$ seconds (Review of Particle Properties, 1994; also see Mampe et al., 1989, 1993, and Byrne, et al., 1990). Thus, any synthesis of the elements during the early stages of the expanding Universe is determined by events occurring in the epoch $t \approx 1$ to $1,000$ seconds.

Early ideas concerning primordial element formation have been reviewed by Alpher and Herman (1950) and Penzias (1979), including the ideas of Weizsäcker (1938) on synthesizing elements during the primeval "fireball" explosion as well as inside stars. Gamow and his colleagues considered the possibility that all chemical elements might be generated in the cooling primordial Universe, predicting the observed abundances using only one free parameter, the density of matter during the formation process. They assumed that the cosmic ylem, or primordial substance, initially consisted solely of neutrons at high temperatures of 10^{10} K. This hypothesis was based upon

Gamow's assumption that all the electrons and protons would be mashed together at high densities to make neutrons. The attractive feature of the neutron capture theory included the excellent inverse correlation between neutron capture cross sections and the abundance of both light and heavy elements. Protons were supposed to be formed by neutron decay, and successive captures of neutrons and protons led to the formation of the elements during the first few minutes of the big bang (Alpher, Bethe and Gamow, 1948, Gamow, 1948).

Gamow's hypothesis for the big bang nucleosynthesis of the lightest elements, such as hydrogen, helium and deuterium, was in error because his assumption that matter was first composed solely of neutrons was wrong. Moreover, since there is no stable nucleus of mass 5 or 8, the simple scheme of successive neutron capture will stop at helium 4, assuring that helium would be the second most abundant element (after hydrogen) and that heavier elements such as lithium 7 would be produced in only trace amounts. That is, by the time that the temperature of the expanding Universe had cooled enough for primordial helium to form, the density of the Universe was much too low to permit further fusion to form heavier elements in the time available by three-body collisions of helium nuclei. Heavier elements therefore have to be synthesized inside stars where the densities are high enough for triple collisions of helium to form carbon 12 (Öpik, 1951; Salpeter, 1952; Burbidge, Burbidge, Fowler and Hoyle, 1957; Cameron, 1957), rather than the big bang where only two-body reactions were important.

In the first few seconds of the expansion of the Universe, the temperature must have exceeded 10^{10} K or 1 MeV in energy units, and sufficient energy was available to form electron neutrinos, ν_e, and pairs of electrons, e^- and positrons, e^+. As Hayashi (1950) showed, the neutrons will interact with thermally excited positrons to form protons, p, and antineutrinos, $\overline{\nu}_e$, according to the reaction

$$e^+ + n \leftrightarrow p + \overline{\nu}_e \ , \tag{5.774}$$

and the neutrons also interact with the electron neutrinos by

$$\nu_e + n \leftrightarrow p + e^- \ . \tag{5.775}$$

As indicated by the double arrow in these reactions, the reverse processes happen just as rapidly, and the baryons (protons and neutrons) are maintained in their equilibrium abundance by weak interactions involving neutrinos.

At temperatures above 10 billion degrees, or above 1 MeV in energy units, equilibrium applies because the weak interaction time scale is much shorter than the age of the Universe. As the Universe cools, the time-scales, τ, for these weak interactions increase rapidly according to

$$\tau \approx 10^{50}\ T^{-5} \ \text{seconds} \ . \tag{5.776}$$

When the Universe is about one second old, or at a temperature $T \approx 10^{10}$ K, the weak interaction time scale becomes greater than the age of the Universe. The weak interactions then drop out of equilibrium, and the ratio of neutrons to protons changes slowly as the free neutrons decay to protons.

During the first few seconds after the big bang, the radiation was so powerful that it could be transformed into electrons and positrons and back again. That is, the electron, e^-, and positron, e^+, pairs persisted in equilibrium with high-energy, γ-ray photons by the electron–positron annihilation and production reactions

$$e^+ + e^- \leftrightarrow \gamma + \gamma \ . \qquad (5.777)$$

The high-energy, γ-ray radiation continues to produce matter and anti-matter until its thermal energy falls below the rest mass energy of the electron, $m_e c^2$, or until

$$T(t) \leq m_e c^2/\mathrm{k} \approx 5.9 \times 10^9 \ \mathrm{K} \qquad (5.778)$$

and

$$t \geq 3 \ \text{seconds} \ , \qquad (5.779)$$

where Boltzmann's constant $k = 1.38 \times 10^{-16}$ erg K^{-1} and $m_e c^2 = 8.187 \times 10^{-7}$ erg $= 0.511$ MeV for an electron mass of $m_e = 9.1094 \times 10^{-28}$ grams and the velocity of light $c = 2.9979 \times 10^{10}$ cm s^{-1}. At this stage, no more electron–positron pairs are produced, since the photon energies are too low.

Primordial nucleosynthesis begins when neutrons, n, and protons, p, fuse into the simplest compound nucleus, deuterium, D, sometimes designated as ^{2}H and also called a deuteron. The relevant thermonuclear reaction

$$\mathrm{n} + \mathrm{p} \leftrightarrow \mathrm{D} + \gamma \qquad (5.780)$$

can work in both directions. During the initial seconds, and high temperatures, $T \geq 10^9$ K or 0.1 MeV in energy units, the deuterium is torn apart by the energetic radiation, or γ rays. This is because baryons (protons and neutrons) are a trace constituent in a bath of high-temperature radiation, as indicated by the baryon-to-photon ratio $\eta = \mathrm{n_B}/\mathrm{n}_\gamma \approx 3 \times 10^{-10}$. The tail of that radiation distribution still contains some photons with energies above the deuterium binding energy per nucleon, about 1 MeV, down to temperatures of 10^9 K. At higher temperatures, deuterium does not survive and is continuously photodissociated into neutrons and protons.

As the Universe expands and cools, the photon (radiation) energy declines and the photodissociation of deuterium becomes less frequent. When the temperature has cooled to $\mathrm{T} \approx 10^9$ K, or about 0.1 MeV in energy units, deuterium can survive and nuclear reactions begin. The deuterons then combine with protons, neutrons and other deuterons, to form heavier elements. Since the baryon density is low (about 10 percent the density of terrestrial air), and the time scale is short (about 100 seconds), only reactions involving two-particle collisions occur, producing the stable isotope of helium, ^{4}He, by the following reactions:

$$\begin{aligned} \mathrm{D} + \mathrm{n} &\leftrightarrow {}^3\mathrm{H} + \gamma \\ \mathrm{D} + \mathrm{D} &\leftrightarrow {}^3\mathrm{H} + \mathrm{p} \\ {}^3\mathrm{H} + \mathrm{p} &\leftrightarrow {}^4\mathrm{He} + \gamma \\ {}^3\mathrm{H} + \mathrm{D} &\leftrightarrow {}^4\mathrm{He} + \mathrm{n} \ . \end{aligned} \qquad (5.781)$$

As well as:

$$\begin{aligned} D + p &\leftrightarrow {}^3He + \gamma \\ D + D &\leftrightarrow {}^3He + n \\ {}^3He + n &\leftrightarrow {}^4He + \gamma \\ {}^3He + D &\leftrightarrow {}^4He + p \\ {}^3He + {}^3He &\leftrightarrow {}^4He + 2p \ . \end{aligned} \tag{5.782}$$

Cross sections for the relevant nuclear reactions have been measured in the nuclear physics laboratories at energies of 0.1 to 1 MeV appropriate for primordial nucleosynthesis.

Since it is very hot, the reactions occur rapidly. Indeed, the synthesis of helium must be finished within the first few hundred seconds before all the free neutrons decay into protons. Moreover, for times $t \geq 1{,}000$ seconds and temperatures $T_r(t) \leq 4 \times 10^8$ K, the electrostatic repulsion of nuclei prevents nuclear reactions from proceeding, the cosmological expansion separates the particles, and big-bang nucleosyntheis is terminated.

Not all of the deuterium formed from the neutrons and protons will be further reacted to synthesize helium. Because of the rapid expansion of the Universe in its early phases, there is not enough time to destroy all the deuterium by energetic radiation or to convert it all into helium by further nuclear reactions. A slight residue will be left, with an abundance relative to hydrogen of about 10^{-5}. The exact amount of surviving deuterium depends sensitively on the density of baryons during the epoch of helium production. A higher baryon density at that time means a greater capture probability for nuclear reactions, and, therefore, a smaller fraction of surviving deuterium nuclei. The implications for the baryon mass density of the Universe are provided after we review estimates for the deuterium abundance.

Two-body reactions during big-bang nucleosynthesis can also produce trace quantities of lithium-7, with an abundance relative to hydrogen of about 10^{-10}. For $\eta \leq 3 \times 10^{-10}$, the lithium, ^{7}Li, is produced mainly by:

$$^4He + {}^3H \rightarrow {}^7Li + \gamma \ . \tag{5.783}$$

For higher η, most of the ^{7}Li comes from the decay of beryllium, ^{7}Be, produced by

$$^4He + {}^3He \rightarrow {}^7Be + \gamma \ . \tag{5.784}$$

Within the first 30 minutes after the big bang, this process of primordial nucleosynthesis is over, resulting in hydrogen nuclei, ^{1}H, or protons, p, and helium nuclei, ^{4}He, with traces of other lighter elements such as the heavy isotope of hydrogen, or deuterium, D = ^{2}H, the lighter isotope of helium, ^{3}He, and a little lithium, ^{7}Li. The big bang model predicts, for example, that about one quarter of the normal matter now in the Universe consists of helium synthesized during the first few minutes of the big-bang, and that the other three quarters consists of hydrogen that was also made in the very early stages of the expanding Universe.

Wherever we look, no celestial object is known to have a helium abundance of less than about 25 percent by mass, suggesting that helium formation operated on a cosmic scale in the early stages of the Universe. A year before the discovery of the three-degree relic background radiation of the big bang, Hoyle and Tayler (1964) pointed out that the observed luminosity of our Galaxy implies that the amount of helium synthesized inside stars over the lifetime of the Galaxy is no more than about ten percent of the amount observed in cosmic objects (O'Dell, 1963), which meant that roughly ninety percent of the helium now found in stars must have been made before the birth of our Galaxy. They showed that a helium abundance of about 25 percent by mass can be synthesized in the big bang, reaffirming a similar conclusion obtained by Alpher, Follin and Herman (1953).

Soon after the discovery (Penzias and Wilson, 1965) of the cosmic microwave background radiation, Peebles (1966) computed the big-bang nucleosynthesis of ^{3}He,^{4}He and D using the present radiation temperature of $T_r(t_o) \approx 3$ K as a boundary condition for extrapolation into the past. Like Alpher, Follin and Herman (1953), Peebles showed that a large amount of helium and smaller amounts of the other light elements could be produced in the big bang. Detailed calculations of element production in the early stages of a homogenous and isotropic expanding Universe were carried out by Wagoner, Fowler and Hoyle (1967), showing that D, ^{3}He, ^{4}He and ^{7}Li are produced in amounts comparable to those observed now if the present radiation temperature $T_r(t_o) \approx 3$ K. However in the 1960s it was thought that deuterium, D, and lithium, ^{7}Li, were primarily made by the T-Tauri phase of stars, so baryon density arguments were viewed as irrelevant at that time. Only after Reeves, Audouze, Fowler and Schramm (1973) argued that deuterium could not be made in stars, but must be made in the big bang, did primordial big-bang nucleosynthesis calculations enable an accurate measure of the baryon density at the level of about 5 percent of that required to close the Universe, with a present baryon density $\rho_B \approx 0.05\rho_c$. This meant that the amount of normal matter (baryons) in both invisible and visible form is insufficient to stop the expansion of the Universe in the future (also see Wagoner, 1973 and Gott, Gunn, Schramm and Tinsley, 1974).

Since then, the standard (homogeneous) big bang nucleosynthesis has been repeated, over and over again, with remarkable success in predicting the primordial abundances of deuterium, helium and lithium derived from observations, always concluding that baryons alone fail to close the Universe by a factor of ten or more (Yang et al., 1984; Walker et al., 1991; Copi, Schramm and Turner, 1995). Moreover, observations of the cosmic microwave background radiation have shown that on the largest scales the Universe is remarkably homogeneous and isotropic (see Sect. 5.7.5), confirming that the geometry described by the Robertson–Walker metric is appropriate for the epoch of primordial nucleosynthesis.

The big-bang nucleosynthesis of light elements has been reviewed by Schramm (1993), Reeves (1994) and Copi, Schramm and Turner (1995); Boesgaard and Steigman (1985) provided an earlier review of the relevant theory and observations. Important papers on primordial nucleosynthesis, including its cosmological consequences include, in chronological order, those of Peebles (1966), Wagoner, Fowler and Hoyle (1967), Reeves, Audouze, Fowler and Schramm (1973), Wagoner (1973), Schramm and Wagoner (1977), Yang et al. (1984), Olive, Schramm, Steigman and Walker (1990), Walker et al. (1991), Smith, Kawano and Malaney (1993), Schramm (1993) and Copi, Schramm and Turner (1995). These investigations have repeated essentially the same calculations for about three decades, with better values of reaction rates and better neutron lifetimes. However, the main improvements have resulted from better light element abundance measurements and an understanding of their evolution (Schramm, 1993), so we now review current observational abundance data combined with their detailed cosmological and physical consequences.

With the possible exception of remote, intergalactic deuterium, the abundances attributed to big bang nucleosynthesis, called primordial abundances, are not measured. They are inferred from the presently observed abundances of material that has undergone up to 10 billion years of chemical evolution (Reeves, Audouze, Fowler and Schramm, 1973; Audouze and Tinsley, 1974; Reeves, 1994). That is, the primordial abundances, which provide important constraints to the baryon density of the Universe, have changed as the result of nuclear reactions that take place in stars and by cosmic ray collisions. The observed abundances can nevertheless be used to provide important limits to the primordial amounts – see Pagel (1982, 1993) and Reeves (1994) for reviews.

Astronomers theorize that nuclear reactions during the first few minutes after the big bang would have produced roughly one helium atom for every 12 of hydrogen, so the primordial helium abundance relative to hydrogen is about 0.08 by number. Most cosmic objects contain additional helium, of up to about 20 percent of the big-bang amount, that was produced inside stars and ejected into the interstellar medium from which these objects formed. The observed abundance of helium therefore provides an important upper limit to its primordial amount. Deuterium, D, is produced in much smaller quantities during the big bang, with an abundance of about 10^{-5} relative to hydrogen, but it is readily destroyed inside stars and cannot be created by nucleosynthesis within them. Indeed, because deuterium is the most weakly bound, stable nucleus, it cannot survive at temperatures greater than half a million degrees, and no realistic astrophysical site for the production of significant amounts of deuterium exists other than the big bang (Epstein, Lattimer and Schramm, 1976). The deuterium abundance observed today therefore provides a lower bound to the primordial amount. Moreover, because ^{3}He is the main product resulting from the destruction of deuterium when interstellar material is cycled through stars, the composite abundance of D + ^{3}He observed today provides an upper limit to the primordial value of their combined abundance.

Primordial abundances can be closely approximated by observing chemically unevolved galaxies, recognized by a low abundance of heavy elements that are synthesized inside stars. They can also be inferred by looking back in time, when the Universe was young and composed of relatively pristine material. This has been accomplished by observing remote intergalactic gas illuminated by very distant (high redshift) quasars, those brightly shining objects that were fairly common in the early Universe.

The primordial helium abundance has been inferred from the optical recombination spectra of emission nebulae, or H II regions, in irregular galaxies and in the so-called extragalactic H II regions (compact blue galaxies). They have an underabundance of heavy elements, implying that a small fraction of their gas has been processed through stars and contaminated by stellar synthesis. The primordial mass fraction, Y_P, of helium has been obtained from the observed value, Y, by extrapolating the following linear relation to zero metallicity (Peimbert and Torres-Peimbert, 1976)

$$Y = Y_P + Z(dY/dZ) \ , \tag{5.785}$$

or

$$Y = Y_P + (O/H)[dY/d(O/H)] \ . \tag{5.786}$$

where Z denotes the mass fraction of heavy elements and O/H is the oxygen to hydrogen abundance ratio.

Most of the weight of an atom is in its nucleus, and a helium ion, consisting of two neutrons and two protons, is about four times as massive as a hydrogen ion, or single proton. The mass fraction of helium, Y, is therefore related to the ratio of the number, N, abundance of helium, $N(^4He)$, and hydrogen, $N(H)$, by:

$$\begin{aligned} Y &\approx \frac{4N(^4He)}{N(H) + 4N(^4He)} \\ &\approx \frac{[4N(^4He)/N(H)]}{1 + 4[N(^4He)/N(H)]} \ . \end{aligned} \tag{5.787}$$

Thus, for a helium to hydrogen number abundance of 1 in 12, or $N(^4He)/N(H) = 0.0833$, we obtain $Y \approx 0.25$.

Lequeux et al. (1979) obtained the primordial helium mass fraction of:

$$Y_P = 0.233 \pm 0.005 \ , \tag{5.788}$$

with $dY/dZ = 3.0 \pm 1.6$ inferred from the O/H abundance ratio, which corresponds to a relative abundance by number of:

$$[N(^4He)/N(H)]_P = 0.074 \pm 0.006 \ . \tag{5.789}$$

Kunth and Sargent (1983) did not find any detectable correlation between Y and O/H, but used optical emission lines from 12 extragalactic H II regions of low metallicity to obtain:

$$Y_P = 0.245 \pm 0.003 \ . \tag{5.790}$$

Pagel et al. (1992) and Pagel (1993) similarly find:

$$Y_P = 0.228 \pm 0.005 \text{ (standard error)} \ , \tag{5.791}$$

with $dY/dZ = 6.1 \pm 2.1$ and $Z = 20(O/H)$, while Campbell (1992) used spectra of 53 H II galaxies to obtain

$$[N(^4He)/N(H)]_P = 0.0759 \pm 0.0014 \ . \tag{5.792}$$

Balbes, Boyd and Mathews (1993) obtain

$$Y_P = 0.227 \pm 0.006(1\sigma) \ , \tag{5.793}$$

and Izotov, Thuan and Lipovetsky (1994) obtain

$$Y_P = 0.240 \pm 0.005 \ . \tag{5.794}$$

for low metallicity blue compact galaxies. Olive and Steigman (1995) use:

$$Y_P = 0.223 \pm 0.004 \ , \tag{5.795}$$

and Sasselov and Goldwirth (1995) examine the uncertainties caused by systematic errors in determining $N(^4\mathrm{He})/N(\mathrm{H})$ from the relative intensities of helium and hydrogen and theoretical recombination coefficients to obtain the upper bound of

$$Y_P \leq 0.255 \ . \tag{5.796}$$

Upper and lower bounds to the primordial helium abundance include:

$$\begin{aligned} &0.226 \leq Y_P \leq 0.248 \,(\text{Kunth and Sargent, 1983}) \\ &0.224 \leq Y_P \leq 0.254 \,(\text{Boesgaard and Steigman, 1985}) \\ &0.228 \leq Y_P \leq 0.238 \,(\text{Campbell, 1992}) \\ &0.223 \leq Y_P \leq 0.232 \,(\text{Pagel et al., 1992}) \\ &0.221 \leq Y_P \leq 0.243 \,(\text{Copi, Schramm and Turner, 1995}) \ , \end{aligned} \tag{5.797}$$

The ranges are dominated by assumptions on systematic errors and not on the statistical errors quoted above. A reasonable consensus of the prevailing view is (Walker et al., 1991; Reeves, 1994; Copi, Schramm, and Turner, 1995):

$$0.22 \leq Y_P \leq 0.25 \ . \tag{5.798}$$

By way of comparison, the solar system abundance, Y_{SS}, inferred from meteorites and the solar photosphere and corona, is (Anders and Grevesse, 1989; Lang, 1992)

$$Y_{SS} = 0.274 \pm 0.016 \ , \tag{5.799}$$

indicating that the Sun and planets formed from material that was enriched in helium by prior synthesis inside stars. The solar system is thought to be 4.6 billion years old, or less than half the age of our Galaxy. The observed helium abundances for other galactic objects range from 23 percent to 30 percent in fractional mass, showing that main-sequence stars have transformed up to 7 percent of the interstellar hydrogen into helium over the lifetime of our Galaxy.

More than 30 years ago, James P. Gunn and Bruce Peterson postulated that we might detect primordial hydrogen gas in intergalactic space by analyzing the light from quasars (Gunn and Peterson, 1965). They noticed that the Lyman alpha resonance line of neutral (unionized) hydrogen (Lyman α at 1215.67°A , where $1°\mathrm{A} = 10^{-8}$ cm), would be brought into the optical band at the high redshifts, z, of quasars at a wavelength of $1215.67(1+z)°\mathrm{A}$. This would give rise to a continuous absorption trough in the short-wavelength side of the emitted Lyman α. Such a broad absorption, now called the Gunn–Peterson effect, has never been detected.

The absence of the absorption of the continuum radiation of quasars below the redshifted Lyman α line means that the number density, N_{HI}, of neutral (unionized) intergalactic hydrogen must be (Shklovskii, 1964; Scheuer, 1965; Gunn and Peterson, 1965):

$$N_{HI} \leq 4.7 \times 10^{-12} h \text{ cm}^{-3} \quad \text{for } z \geq 2 \ , \tag{5.800}$$

where the Hubble constant $H_o = 100h$ km s^{-1} Mpc^{-1}. Such a low density for neutral hydrogen, the most abundant element in the Universe, suggests that the intergalactic medium is likely to be nearly one hundred percent ionized.

Also in 1965, John N. Bahcall and Edwin E. Salpeter suggested that the neutral intergalactic gas might not be uniformly distributed, and would therefore produce discrete Lyman α absorption lines at the redshifts of its concentrations along the line of sight to more distant quasars (Bahcall and Salpeter, 1965). The subsequent observations of large numbers, or a "forest", of absorption features seen on the short-wavelength side of the redshifted Lyman alpha radiation of quasars (Bahcall, Greenstein and Sargent, 1968; Lynds, 1971; Weymann, Carswell and Smith, 1981; Lanzetta et al., 1991) has been attributed to neutral hydrogen clumped into galaxy-sized clouds billions of light year away, perhaps coming together in the early stages of galaxy formation (Sargent et al., 1980; Blades, Turnshek and Norman, 1988). The properties of the Lyman forest systems suggest that these clouds are highly photoionized by metagalactic ultraviolet radiation. For example, the observed hydrogen line widths indicate temperatures of about 50,000 degrees for the forest clouds.

Calculations suggest that there is about 100,000 times more ionized hydrogen out there, but unlike neutral hydrogen the ionized form is invisible. Ultraviolet radiation is apparently stripping most of the primordial hydrogen atoms of their single electrons, ionizing each atom into a single proton that does not absorb light and hence cannot be seen.

But helium atoms have two electrons. Radiation might strip them of one electron and not always remove both. In this case the Gunn-Peterson effect might be detected for the redshifted line of singly ionized helium (He II) at wavelengths shorter than 304 $(1 + z)$ °A for a quasar of redshift z (Miralda-Escudé and Ostriker, 1990; Miralda-Escudé, 1993). An He II absorption effect has been found for quasars bright enough to have their ultraviolet light analyzed with the Hubble Space Telescope (Jakobsen et al., 1994 for quasar *Q*0302–003 at $z_q = 3.286$, Tytler et al., 1996 for quasar *Q*1935–67 at $z_q = 3.185$) and with the Hopkins Ultraviolet Telescope on the Space Shuttle (Davidsen, Kriss and Zheng, 1996 for quasar HS 1700+64 at $z_q = 2.730$). The absorption signature is found in ultraviolet light that has been traveling to Earth for roughly 10 billion years, indicating that ionized helium was present at or before the early stages of galaxy formation. This confirms that the young Universe was a harsh environment, hot enough to strip neutral helium atoms of one of their electrons and to ionize most of the intergalactic hydrogen.

More importantly, these detections show that helium really is primordial, and that substantial amounts of helium were synthesized in the big bang. They also suggest that both the intergalactic medium and the Lyman α clouds may be significant repositories for dark (nonluminous) baryons (Giroux, Fardal and Shull, 1995; Davidsen, Kriss and Zheng, 1996). The primordial matter in the intergalactic medium probably equals or might even exceed all the ordinary matter thought to be present in galaxies, but the ultraviolet observations of quasars do not yet tell us exactly how much primordial helium was manufactured. It nevertheless appears that the adolescent Universe's intergalactic medium contained at least half of the baryonic (normal) matter that cosmologists expect the big bang to have made (Bi and Davidsen, 1997).

The big-bang models are used to calculate just how much helium was synthesized back then. The free neutrons are, for example, almost entirely used up in the first few minutes of the big bang in the synthesis of helium, ^{4}He. If N_n and N_p respectively denote the number densities of neutrons and protons before the beginning of big bang element synthesis, then $N_n/2$ helium nuclei are formed since each helium nucleus contains two neutrons and two protons. The number of protons left over is $N_p - N_n$, so the relative number abundance of helium, $N(^4\mathrm{He})$, and hydrogen, $N(\mathrm{H})$, is:

$$\frac{N(^4\mathrm{He})}{N(\mathrm{H})} = \frac{N_n}{2(N_p - N_n)} \quad . \tag{5.801}$$

In terms of the mass fraction of helium, Y, relative to that of hydrogen, X,

$$\frac{Y}{X} = \frac{2N_n}{N_p - N_n} \tag{5.802}$$

where $X + Y \approx 1.0$, or

$$Y \approx \frac{2N_n}{N_n + N_p} = \frac{2(N_n/N_p)}{1 + (N_n/N_p)} \quad , \tag{5.803}$$

and N_n/N_p is the neutron–proton ratio at the time corresponding to the temperature at which deuterium can first survive photodissociation. When the generation of electron–positron pairs is no longer able to maintain the equilibrium between neutrons and protons, their abundance ratio becomes frozen in at the value:

$$\frac{N_n}{N_p} = \exp\left(-\frac{\Delta mc^2}{kT}\right) = 0.22 \quad , \tag{5.804}$$

at time

$$t \approx 1 \text{ second} \quad ,$$

where $\Delta m = m_n - m_p = 2.3055 \times 10^{-27}$ grams is the mass difference between neutrons and protons, the energy $\Delta m\, c^2 \approx 2.07 \times 10^{-6}$ erg ≈ 1.29 MeV, and N_n and N_p denote, respectively, the number density of neutrons and protons. Thereafter the ratio changes because the neutrons decay into protons with a mean life of roughly 887 seconds or about 14.8 minutes. This ratio was estimated by Hayashi (1950) to be $N_n/N_p \approx 0.25$. Alpher, Follin and Herman (1953) redid his calculations in greater detail, obtaining a ratio of between 0.22 and 0.17 depending on what they took as the neutron lifetime. They concluded that roughly 25 percent of the matter would be converted into helium, with the exact amount dependent only slightly upon the mass density, and that much smaller amounts of deuterium would be produced with an abundance that is a sensitive function of the mass density of the Universe.

The amount of helium produced in the big bang is not very sensitive to changes in the baryon density, but it is noticeably dependent on two physical constants that determine the neutron-to-proton ratio at the onset of big-bang nucleosynthesis. They are the number of "light" degrees of freedom, expressed in terms of the equivalent number of light neutrino species, N_ν, and the mean lifetime of the neutron, $\tau_n = \tau_{1/2}/0.693$ for a neutron half-life of $\tau_{1/2}$. Boesgaard and Steigman (1985) used computer calculations to evaluate the dependence of the primordial helium abundance, Y_P, on these parameters over a plausible range of $1.5 \leq \eta_{10} \leq 10$ for a baryon-to-photon ratio of $\eta = 10^{-10}\eta_{10}$, obtaining:

$$Y_P = 0.2320 + \Delta Y_P \ , \tag{5.805}$$

or

$$Y_P = 0.230 + 0.011 \ln(\eta_{10}) + 0.013(N_\nu - 3) + 0.014(\tau_{1/2} - 10.6) \ , \tag{5.806}$$

where $\tau_{1/2}$ is the neutron half-life in minutes, then taken to be $\tau_{1/2} = 10.6 \pm 0.2$ minutes. Bernstein, Brown and Feinberg (1989) used semianalytical methods, that do not involve complex calculations, to obtain a dependence:

$$\Delta Y_P = 0.009 \ln(\eta_{10}/5) + 0.014(N_\nu - 3) + 0.18[(\tau_n - 896)/\tau_n] \ , \tag{5.807}$$

where τ_n is the neutron's mean lifetime in seconds. More recently, Walker et al. (1991) use computer model calculations over the range $3 \leq \eta_{10} \leq 10$ to provide:

$$Y_P = 0.228 + 0.010 \ln(\eta_{10}) + 0.012(N_\nu - 3) + 0.185[(\tau_n - 889)/889] \ , \tag{5.808}$$

where the last term becomes $0.017(\tau_{1/2} - 10.27)$ for a neutron half-life $\tau_{1/2}$ in minutes.

Thus, the mass fraction of helium, Y_P, produced during the immediate aftermath of the big bang is but a slowly increasing function of the baryon-to-photon ratio, $\eta_{10} = 10^{10}\eta$. On the other hand, Y_P increases linearly with the number of neutrino species, N_ν, and the neutron lifetime, τ_n.

Steigman, Schramm and Gunn (1977) showed that the number of light, left-handed neutrino types, N_ν, had to be small to avoid overproduction of helium, and were able to use observational limits on the helium abundance to set an upper limit of $N_\nu \leq 5$ (also see Szalay, 1981). A larger number of neutrino families would have speeded up the Universe's expansion, fewer neutrons would have had time to decay to protons, and more helium would be produced than is observed. Over the years, this limit has improved with better helium abundance measurements, neutron lifetime measurements and limits on the lower bound to the baryon density. Yang et al. (1984) obtained

$$N_\nu = 3 \pm 1 \ . \tag{5.809}$$

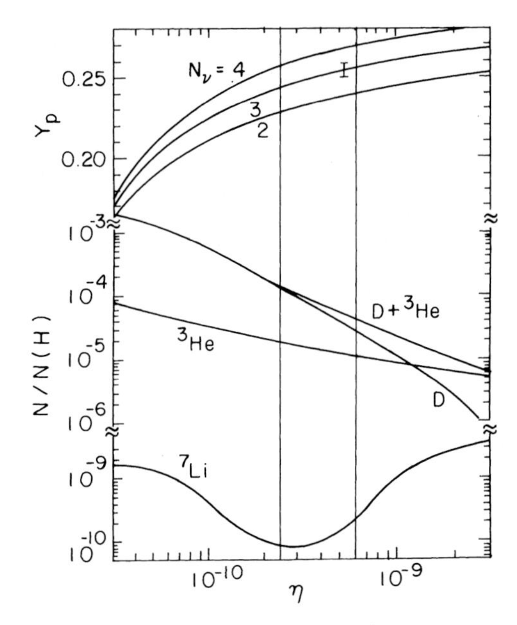

Fig. 5.38. Predicted abundance of ^{4}He (or Y_P by mass), D, ^{3}He, D + ^{3}He, and ^{7}Li (by number relative to H) as a function of $\eta = n_B/n_\gamma$, the nucleon-to-photon ratio (bottom). The three curves for the mass fraction of helium. Y_P, correspond to different choices for the number, N_ν, of neutrino species. The error bar shows the range of Y_P, with the neutron half-life, $\tau_{1/2}$, then thought to be 10.6 minutes. This figure was adopted from Yang et al. (1984), but see Walker et al. (1991) for a very similar one computed for $N_\nu = 3$ and $\tau_{1/2} = 10.27$ minutes, the currently accepted values. The vertical bars denote the bounds corresponding to $2.5 \times 10^{-10} \leq \eta \leq 6 \times 10^{-10}$ given by Copi, Schramm and Turner (1995). These limits also provide bounds to $\Omega_B = \rho_B/\rho_c$, the baryonic fraction of the critical density, of $0.006 \leq 0.009\,h^2 \leq \Omega_B \leq 0.02\,h^2 \leq 0.21$, with $0.4 \leq h \leq 1$ and a Hubble constant of $H_o = 100$ h km s^{-1} Mpc^{-1}.

or

$$N_\nu \leq 4 \tag{5.810}$$

for $\tau_{1/2} = 10.6 \pm 0.2$ minutes (also see Fig. 5.38).

For an inferred upper limit to the primordial helium abundance Y_P, various authors obtain the following limits to the number of light neutrino species:

$$
\begin{aligned}
&N_\nu \leq 3.2 \quad \text{for } Y_P \leq 0.242 \text{ (Olive et al., 1990)} \\
&N_\nu \leq 3.3 \quad \text{for } Y_P \leq 0.240 \text{ (Walker et al., 1991)} \\
&N_\nu \leq 3.25 \quad \text{for } Y_P \leq 0.243 \text{ (Steigman and Tosi, 1992)} \\
&N_\nu \leq 3.3 \quad \text{for } Y_P \leq 0.243 \text{ (Copi, Schramm and Turner, 1995)} .
\end{aligned} \tag{5.811}
$$

The neutron lifetime sets the rate for all reactions that interconnect neutrons and protons, and therefore affects calculations of the primordial helium abundance. For a larger lifetime, there will be more neutrons relative to protons at the freeze-out temperature, producing more helium. Laboratory measurements of the neutron lifetime, τ_n, include:

$$\begin{aligned} \tau_{\rm n} &= 887.6 \pm 3.0 \text{ seconds (Mampe et al.,1989)} \\ &\quad (\text{or } \tau_{1/2} = 10.25 \text{ minutes}) \\ \tau_{\rm n} &= 893.6 \pm 5.3 \text{ seconds (Byrne et al., 1990)} \\ \tau_{\rm n} &= 882.6 \pm 2.7 \text{ seconds (Mampe et al., 1993)} \ . \end{aligned} \tag{5.812}$$

However, notice that if $Y_{\rm P}$ and η_{10} are fixed, then an uncertainty in $\tau_{\rm n}$ of ± 7 seconds contributes only ± 0.12 or ≤ 4 percent to the uncertainty in N_ν.

The predictions of primordial nucleosynthesis for the number of light neutrino species have been confirmed by high energy particle experiments. They include:

$$\begin{aligned} N_\nu &= 3.27 \pm 0.30 \text{ (Aleph collaboration, 1989)} \\ N_\nu &= 3.24 \pm 0.46 \text{ (statistical) (L3 collaboration, 1992).} \\ &\quad \pm 0.22 \text{ (systematic)} \end{aligned} \tag{5.813}$$

Expressed another way, the particle physics experiments, with their measurements of N_ν and $\tau_{\rm n}$, provide a lower limit to the primordial helium abundance of (Izotov, Thuan and Lipovetsky, 1994)

$$Y_{\rm P} \geq 0.236 \pm 0.001 \ . \tag{5.814}$$

Because the amount of ^{4}He produced in the big bang is very insensitive to the cosmic baryon density, it has not been possible to make decisive conclusions from the observed helium abundance regarding the mean density of ordinary matter. Nevertheless, precise observations of the abundance of deuterium, D, as well as ^{3}He and ^{7}Li, have enabled an accurate determination of the average density of baryons in the Universe.

Since deuterium, D or ^{2}H, is so fragile, post big bang astrophysical processes destroy it, and its cosmic abundance at any one time is a lower limit to the primordial amount (Reeves, Audouze, Fowler and Schramm, 1973; Epstein, Lattimer and Schramm, 1976). Such limits have been provided from measurements of the number abundance, D/H, of deuterium to hydrogen atoms in the solar system, local interstellar space, and distant intergalactic clouds. All of these atoms of deuterium, wherever they are now found, are cosmic leftovers from the first minutes of creation.

Direct measurements of the abundance in the solar wind, $({\rm D/H})_{\rm SW}$, made during the Apollo 11 mission, indicate (Geiss and Reeves, 1972):

$$({\rm D/H})_{\rm SW} = 2.5 \times 10^{-5} \ , \tag{5.815}$$

while spectral measurements of Jupiter, J, provide (Encrenaz and Combes, 1982; Smith, Schempp and Baines, 1989):

$$({\rm D/H})_{\rm J} = (1.0 \text{ to } 3.2) \times 10^{-5} \ . \tag{5.816}$$

In contrast, about one molecule in 10,000 of seawater contains a deuterium atom in place of hydrogen. Both the terrestrial, T, and carbonaceous meteoritic, M, abundances of deuterium are (Black, 1971; Schramm and Wagoner, 1974):

$$(D/H)_{T,M} \approx 1.5 \times 10^{-4} \ . \tag{5.817}$$

The higher D/H ratio can be understood by chemical fractionation (Geiss and Reeves, 1972; Boesgaard and Steigman, 1985). The larger mass of the hydrogen isotope, D or ^{2}H, when compared with normal hydrogen, ^{1}H or H, allows it to have different chemical as well as nuclear properties. The creation of deuterated molecules is favored in the formation of water and at the low temperatures in the cool protosolar nebula from which primitive meteorites were formed. Chemical fractionation also favors the formation of molecules that contain deuterium in cold interstellar clouds, where an anomalous, high deuterium-to-hydrogen abundance is also inferred (Jefferts et al., 1973; Wilson et al., 1973). Still every deuterium atom that we now find in seawater, meteorites or interstellar molecular clouds is a remnant of the big bang.

The presolar deuterium abundance, $(D/H)_{PS}$, at the time of formation of the solar system is therefore taken to be (Geiss, 1993; Reeves, 1994; Vangioni-Flam, Olive and Prantzos, 1994; Copi, Schramm and Turner, 1995):

$$(D/H)_{PS} = (2.6 \pm 1.0) \times 10^{-5} \ . \tag{5.818}$$

The deuterium abundance in the local interstellar medium, $(D/H)_{LISM}$, within about 50 parsecs of the Sun, is determined by using spacecraft to observe the ultraviolet radiation of bright nearby stars. They illuminate the interstellar medium which produces ultraviolet absorption lines (Lyman series) of deuterium and neutral (unionized) hydrogen. The local interstellar deuterium abundance was first obtained using the Copernicus satellite, observing along the line of sight to Beta Centauri (Rogerson and York, 1973):

$$(D/H)_{LISM} = (1.4 \pm 0.2) \times 10^{-5} \text{ toward Beta Centauri} \ , \tag{5.819}$$

then inferred from International Ultraviolet Explorer (IUE) satellite observations of interstellar lines along the line of sight to other bright stars (Boesgaard and Steigman, 1985):

$$(D/H)_{LISM} = (0.8 \text{ to } 2.0) \times 10^{-5} \ ,$$

and determined more precisely using the Hubble Space Telescope to observe Capella (Linsky et al., 1993):

$$(D/H)_{LISM} = (1.65 \pm 0.15) \times 10^{-5} \text{ toward Capella} \ . \tag{5.820}$$

Since deuterium is steadily destroyed by processing in stellar interiors, the local deuterium abundance needs to be corrected for this depletion over the aeons, using models of galactic chemical evolution to arrive at a higher primordial value characteristic of the early Universe. That is, some of the local interstellar medium that we see today must have been "astrated", or processed inside stars, and ejected back into space with the complete destruction of its deuterium. The present abundance in the interstellar medium therefore provides a lower limit to the primordial abundance, $(D/H)_P$, that can be ascribed to the big bang. Some chemical evolution models that include the complete burning of some deuterium inside stars with subsequent return of the gas to the interstellar medium, suggest that (Clayton, 1985; Steigman and Tosi, 1992):

$$(\mathrm{D/H})_{\mathrm{P}} = (1.5 \text{ to } 3.0)(\mathrm{D/H})_{\mathrm{LISM}} \ . \tag{5.821}$$

A more direct approach to determining the primordial deuterium abundance is to observe the absorption spectra of quasars, the most luminous objects to illuminate the early Universe. In this case, the absorption lines of deuterium and hydrogen are observed at high redshift, where they fall within the terrestrial atmosphere's optical window. At a redshift, z, the Lyman alpha line of neutral hydrogen falls at a wavelength of $1215(1+z)°\mathrm{A}$, where $1°\mathrm{A} = 10^{-8}$ cm, and the same transition for deuterium is just 3 °A shorter, equivalent to a motion toward the observer of 82 km s^{-1}. Observations of these two spectral features enable one to determine the deuterium abundance, relative to hydrogen, in distant clouds of hydrogen gas, seen when the Universe was much younger and still close to its initial, chemically unevolved composition.

Measurements of the deuterium abundance in intergalactic clouds, $(\mathrm{D/H})_{\mathrm{IGC}}$, have been made by observing at least two quasars (QSO0014 + 813 with $z_{\mathrm{q}} = 3.42$ and QSO 1937–009 with $z_{\mathrm{q}} = 3.78$) with the Keck telescope, the world's largest at optical wavelengths. Songalia et al. (1994) and Rugers and Hogan (1996) obtain

$$(\mathrm{D/H})_{\mathrm{IGC}} = (1.9 \pm 0.5) \times 10^{-4} \text{ at redshift } z = 3.32 \text{ toward } 0014 + 813 \tag{5.822}$$

(also see Carswell et al., 1994; Hogan, 1996). This was a startling high value, implying that the Universe was much more rarefied than expected, but Rugers and Hogan on longer believe they were measuring the deuterium abundance (Glanz, 1997). An abundance that is about an order of magnitude lower is found by Tytler et al. (1996) with

$$(\mathrm{D/H})_{\mathrm{IGC}} = (2.3 \pm 0.3) \times 10^{-5} \text{ at redshift } z = 3.572 \text{ toward } 1937-009 \ , \tag{5.823}$$

but this value has subsequently been revised upward to (Songaila, Wampler and Cowie, 1997) about 4×10^{-5}. Levshakov, Kegel and Takahara (1998) argue that the D/H ratio at large redshifts toward three quasars, including Q1937–009, is

$$(\mathrm{D/H})_{\mathrm{IGC}} = (4.1 \text{ to } 4.6) \times 10^{-5} \ , \tag{5.824}$$

which is consistent with standard big bang nucleosynthesis if the baryon-to-photon ratio, η_{10}, is in the range 4.2 to 4.6 and $0.0155 \leq \Omega_{\mathrm{B}} h^2 \leq 0.0167$.

Since all of the deuterium that we see today is primordial, resulting from big-bang nucleosynthesis, its abundance serves as a "baryometer", establishing the density of ordinary matter (baryons). If the baryon density was high back then, numerous collisions between deuterium and other nuclei would convert more of it into helium. If the density was low, there would not be enough time for all of the deuterium to be converted into helium, leaving more deuterium behind. Calculations of the primordial abundance therefore depend sensitively on the baryon-to-photon ratio $\eta = 10^{-10}\eta_{10}$. Moreover, since η remains unchanged during the expansion of the Universe, it can be used to provide important limits to the density of baryons in the Universe today.

The observed deuterium abundance provides a lower limit to the primordial amount, resulting in a firm upper bound to η, and therefore to the baryon density parameter $\Omega_B = \rho_{Bo}/\rho_c = 0.00365h^{-2}\eta_{10}$, the ratio of the baryon mass density today, ρ_{Bo}, and the critical mass density, ρ_c, required to stop the expansion of the Universe in the future. Early measurements of the deuterium abundance of a few parts in 10^5 relative to hydrogen established that baryons could not contribute more than 20 percent of the closure density, and this conclusion holds today. The lower abundance limit of $D/H \geq 2 \times 10^{-5}$ gives an upper limit to $\eta_{10} \leq 7$, and therefore an upper limit to the baryon density of (Gott, Gunn, Schramm and Tinsley, 1974, Yang et al., 1984; Copi, Schramm and Turner, 1995)

$$\Omega_B \leq 0.025h^{-2} \ , \tag{5.825}$$

where the Hubble constant $H_o = 100h$ km s^{-1} Mpc^{-1}. Thus, for more than two decades it has been known that the baryon mass density cannot approach the critical density required to close the Universe if big-bang nucleosynthesis is correct. The new quasar abundances of deuterium reaffirm that conclusion.

To round out our discussion of big-bang nucleosynthesis, there are two other elements, helium-3 or ^{3}He and lithium-7 or ^{7}Li, produced in trace amounts during the early stages of the expansion of the Universe, and they provide additional constraints to the baryon density parameter. Stars like the Sun should have converted essentially all of the deuterium in their outer regions into helium-3 by the reaction

$$D + p \rightarrow {}^3He + \gamma \ , \tag{5.826}$$

so the abundance sum of D + ^{3}He provides an upper limit to their combined big-bang production. This, in turn, places a lower limit on the baryon density parameter, Ω_B.

Measurements of the solar wind helium abundance $(^3He/^4He)_{SW}$ using foil collectors on the Moon (Geiss and Reeves, 1972) and the instruments on the Ulysses spacecraft (Bodmer et al., 1995) indicate:

$$(^3He/^4He)_{SW} = (4.4 \pm 0.4) \times 10^{-4} \ , \tag{5.827}$$

which agrees with the value measured in a solar prominence (Hall, 1975). Measurements of primitive carbonaceous meteorites (Black, 1972; Wieler et al., 1991) give

$$(^3He/^4He)_M = (1.5 \pm 0.4) \times 10^{-4} \ , \tag{5.828}$$

which is taken to be the protosolar value before the solar system formed (Reeves, 1994).

When combined with deuterium abundance observations, these results provide an upper limit to the combined presolar abundance, $[(D + {}^3He)/H]_{ps}$ given by (Yang et al., 1984; Walker et al., 1991; Geiss, 1993; Copi, Schramm and Turner, 1995)

$$[(\mathrm{D} + {}^3\mathrm{He})/\mathrm{H}]_{\mathrm{PS}} \leq 1.1 \times 10^{-4} \ , \tag{5.829}$$

and hence a lower limit to the baryon-to-photon ratio

$$\eta \geq 2.5 \times 10^{-10} \tag{5.830}$$

and a corresponding lower limit to the baryon density parameter

$$\Omega_{\mathrm{B}} \geq 0.0091 h^{-2} \ . \tag{5.831}$$

Lithium-7, or ^{7}Li, is a fragile nucleus, easily destroyed inside stars by proton fusion at relatively low temperatures of $\approx 2 \times 10^6$ K. It nevertheless survives in the cool outer atmospheres of stars whose convective zones do not reach to such hot depths. Our knowledge of the primordial lithium abundance relative to hydrogen, $(^7\mathrm{Li/H})_{\mathrm{P}}$, is therefore based upon observations of the oldest, unevolved stars of our Galaxy, the metal-poor, population II halo stars.

The lithium abundance of very old stars levels off to a value that is independent of stellar metallicity, the so-called Spite plateau, that argues against modification by stellar depletion or cosmic ray nucleosynthesis. (Lithium can be produced by cosmic ray spallation mechanisms reviewed by Reeves, 1994.) The population II stellar data suggest a primordial abundance of (Spite and Spite, 1982; Spite, Maillard and Spite, 1984; Hobbs and Pilachowski, 1988; Rebolo, Molaro and Beckman, 1988; Thorburn, 1994):

$$(^7\mathrm{Li/H})_{\mathrm{P}} = (1.4 \pm 0.4) \times 10^{-10} \ . \tag{5.832}$$

Upper bounds to the interstellar lithium abundance, derived from interstellar absorption lines, are consistent with this estimate, and suggest that it represents the primordial value (Steigman, 1996).

Because the predicted primordial abundance of lithium is not a monotonic function of the baryon-to-photon ratio, η, comparison with the observed abundance can provide both upper and lower limits to the baryon density in the Universe today. Rebolo, Molaro, and Beckman (1988) and Deliyannis et al. (1989) use the lithium abundance of metal-deficient stars to infer

$$1.2 \leq \eta_{10} \leq 8 \ , \tag{5.833}$$

where $\eta_{10} = 10^{10}\eta$, and a present-day baryon density parameter, $\Omega_{\mathrm{B}} = 0.00365 h^{-2} \eta_{10}$ of:

$$0.0044 h^{-2} \leq \Omega_{\mathrm{B}} \leq 0.025 h^{-2} \ , \tag{5.834}$$

for a Hubble constant $H_{\mathrm{o}} = 100h$ km s^{-1} Mpc^{-1}.

Putting together the big-bang nucleosynthesis constraints on all four key cosmological nuclei, ^{4}He, D, ^{3}He, and ^{7}Li, the baryon-to-photon ration, η, is within the bounds (Yang et al., 1984; Walker et al., 1991; Smith, Kawano and Malaney, 1993; Krauss and Kernan, 1995; Copi, Schramm and Turner, 1995)

$$2.5 \times 10^{-10} \leq \eta \leq 6 \times 10^{-10} \ . \tag{5.835}$$

If η is within these limits, there is agreement between the predicted and measured abundance of all four light elements. The lower limit arises mainly from the upper limit on the $(\mathrm{D} + {}^3\mathrm{He})$ abundance, while the upper limit arises from the upper abundance limit to ^{4}He, with $Y_P \leq 0.245$, and/or the lower limit $D/H \geq 1.6 \times 10^{-5}$ to the deuterium abundance.

This leads to the decade-old limits to the present-day baryon density parameter, Ω_B,

$$0.009h^{-2} \leq \Omega_B \leq 0.025h^{-2} , \tag{5.836}$$

where $\Omega_B = \rho_{Bo}/\rho_c$, the present baryon mass density is ρ_{Bo}, and ρ_c is the critical mass density required to stop the expansion of the Universe in the future and eventually close it.

The baryon mass density today is therefore

$$1.72 \times 10^{-31} \text{ g cm}^{-3} \leq \rho_{Bo} \leq 5.1 \times 10^{-31} \text{g cm}^{-3} . \tag{5.837}$$

For a very generous range of $h = 0.4$ to 1.0, the baryon mass density contributes between 0.9 and 15 percent of the critical closure density, ρ_c. The allowed range in the baryon density today, ρ_{Bo}, is in good agreement with the mass density of luminous matter (Sect. 5.4.4, Persic and Salucci, 1992), with $\Omega_{\text{visible}} \approx 0.003$, as well as the dynamics of galaxies that suggest greater amounts of matter in invisible (dark) halos (see Section 5.4.5). Indeed, the big-bang nucleosynthesis constraints suggest that most of the baryons in the Universe could be nonluminous, or dark.

Moreover, if the Universe now has a density parameter $\Omega = 1$, as required by the inflation scenario, then it is dominated by nonbaryonic matter. That is, more than 80 percent of the mass of the Universe would have to be nonbaryonic. The search for dark, invisible matter, in both baryonic and nonbaryonic form was described in Sect. 5.4.6.

5.7.5 The Cosmic Microwave Background Radiation

The discovery of the cosmic microwave background radiation, now considered a relic of the big-bang explosion that gave rise to the expanding Universe, came about quite serendipitously when Arno A. Penzias and Robert W. Wilson were testing microwave receiving systems for the Bell Telephone Laboratories. They found an unexpected noise coming from all directions in the sky, with a temperature of 3.5 ± 1.0 degrees Kelvin at a wavelength of 7.35 cm (Penzias and Wilson, 1965). They were careful to exclude any discussion of the cosmological implications of their discovery, since they had not been involved in that kind of work and thought that their measurements were independent of theory and might outlive it. Nevertheless, in a companion letter Dicke et al. (1965) reasoned that the "excess noise" was that of a thermal (black-body) cosmic background, a relic of the primeval fireball or hot big bang. Penzias and Wilson received the 1978 Nobel Prize for Physics for their measurement (Penzias, 1979; Wilson, 1979).

This radiation had been both observed and predicted before 1965. Alpher and Herman (1948, 1949) examined the evolution and physics of a Universe expanding from the big bang, concluding that the present Universe would have a "background temperature" of 5 degrees Kelvin. Radiation with a temperature of a few degrees had been inferred from the rotational excitation of interstellar molecules observed at optical wavelengths (Adams, 1941; Mc Kellar, 1941), but the cosmological implications were not realized until after

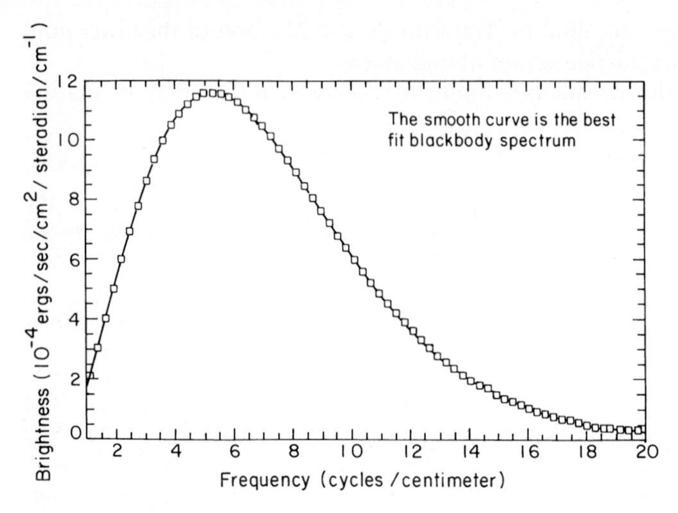

Fig. 5.39. Spectrum of the cosmic microwave background radiation over the range from 2 to 20 cm^{-1} compared to a black body of temperature 2.726 ± 0.010 degrees Kelvin. The units of the brightness on the vertical axis are 10^{-4} erg s^{-1} cm^{-2} sr^{-1} cm. (Adapted from Mather et al., 1990).

the discovery of the microwave background. Doroshkevich and Novikov (1964) noted that the remnant radiation would have a thermal black-body spectrum detectable at microwave wavelengths.

If the cosmic noise discovered by Penzias and Wilson is the relic fireball radiation, then it should have a thermal black body spectrum. Such a spectrum would be established in the early stages of the expanding Universe, when its material and radiation components were in thermal equilibrium. The resulting thermal spectrum will be preserved for all time as the Universe expands and cools. The present relic brightness, $B_\nu(T)$ at temperature, T, and frequency, ν, will therefore be given by (Planck 1901, 1913)

$$B_\nu(T) = \frac{2h\nu^3}{c^2} \frac{1}{[\exp(h\nu/kT) - 1]} \tag{5.838}$$

where c is the velocity of light, h is Planck's constant and k is Boltzmann's constant.

The thermal spectrum of the three-degree radiation was confirmed by measurements at 3 cm, 21 cm and 49–74 cm, respectively by Roll and Wilkinson (1966), Wilson and Penzias (1967), and Howell and Shakeshaft (1966, 1967), as well as by the expected thermal excitation of interstellar molecules at the shorter wavelength of 2.6 mm (Field and Hitchcock, 1966; Thaddeus and Clauser, 1966) During the ensuing decades, measurements of the spectrum of the cosmic background radiation have been shown to be consistent with a thermal spectrum with a present temperature $T_o = 2.726 \pm 0.010$ K (degrees Kelvin) over a wavelength, λ, range of 0.1 cm $\leq \lambda \leq$ 75 cm. The results are given in the excellent monograph by Partridge (1995) together with methods of measurement and references to the relevant technical papers.

More recently, the COsmic Background Explorer satellite, or COBE for short, has shown that the cosmic microwave background exhibits no detectable deviation from a thermal spectrum (see Fig. 5.39), and that its temperature is (Mather et al., 1990, 1994)

$$T_o = 2.726 \pm 0.010 \text{ K} \tag{5.839}$$

at the 95% confidence level systematic. Turner (1993) has explained, in review, why the temperature has a value it has, and how its thermal spectrum was preserved. Any nonthermal distortions in the spectrum are less than 0.03% of the peak brightness, with an rms value of 0.01% (Mather et al., 1994). The COBE results confirm that 99.97% of the background radiation was released within a year of the big bang.

Detailed calculations show that rapid production and destruction of photons (radiation) within the first year of the big bang, at redshifts greater than $z \approx 3 \times 10^6$, forced the radiation to have a black body spectrum (Burigana et al., 1991), and that its thermal form should be maintained through the recombination of electrons and nuclei to make neutral atoms at about 300,000 years after the big bang (Weinberg, 1972). Tolman (1934) showed that the radiation would preserve its black body spectrum after recombination, propagating freely and unimpeded in the expanding Universe to the present time.

A second important aspect of the cosmic microwave background radiation is that it is remarkably isotropic, or equally intense in all directions. The observed isotropy is strong evidence in support of the cosmological principle – the conjecture that the Universe is isotropic on a large scale. Such an isotropy was noted at the time of the discovery of the background radiation (Penzias and Wilson, 1965), placing an upper limit of roughly 20% on any anisotropy at 7.35 cm. Within two years of its discovery the background radiation had been found to be isotropic to 0.1%, or $\Delta T/T \leq 10^{-3}$, on angular scales of 15 to 180 degrees (Partridge and Wilkinson, 1967; Wilkinson and Partridge, 1967). Subsequent searches for anisotropy in the cosmic microwave background radiation have been reviewed by Partridge (1995).

A first-order anisotropy at large angular scales, termed the dipole anisotropy, was apparently detected from a high altitude site (Conklin, 1969) and by a balloon borne radiometer (Henry, 1971), and convincingly demonstrated with measurements from balloons (Corey and Wilkinson, 1976; Cheng et al., 1979; Boughn, Cheng and Wilkinson, 1981; Lubin et al., 1985), and U-2 aircraft (Smoot, Gorenstein and Muller, 1977; Smoot and Lubin, 1979; Gorenstein and Smoot, 1981). They show a striking cosine anisotropy with an amplitude of about 0.003 K, indicating that the background radiation has a maximum temperature in one direction and a minimum in the opposite direction. This result has been confirmed by COBE, with a dipole moment, T_1 of (Smoot, et al., 1991; Fixsen et al., 1994; Lineweaver, et al., 1996)

$$T_1 = 0.003358 \pm 0.000023 \text{ K} \tag{5.840}$$

or $T_1/T_o = 0.001232$ in the direction

$$l = 264.31\,^\circ \pm 0.20\,^\circ$$
$$b = +48.05\,^\circ \pm 0.10\,^\circ \quad , \tag{5.841}$$

where l, b are galactic coordinates. The direction in the celestial coordinates of right ascension, α, and declination, δ, at the epoch J2000, is (Lineweaver et al., 1996):

$$\alpha = 11^{\mathrm{h}}11^{\mathrm{m}}57^{\mathrm{s}} \pm 23^{\mathrm{s}}$$

and (5.842)

$$\delta = -7.22\,^\circ \pm 0.08\,^\circ \quad .$$

This is the heliocentric component of motion with respect to the cosmic microwave background radiation. Corrections to the centroid of the Local Group of galaxies are given in Sect. 5.7.6.2.

The dipole anisotropy is attributed to the Doppler effect caused by the motion of the observer with respect to the background radiation. An observer moving relative to this frame will see the radiation shifted to higher energy, shorter wavelength and higher frequency in the direction of motion; in the opposite direction the radiation is shifted to lower energy, longer wavelength and shorter frequency with a smooth variation in between given by $\cos\theta$, where θ is the angle between the line of sight and the direction of motion of the observer. At a velocity $v \ll c$, where c is the velocity of light, the change in the observed frequency, ν_{obs}, is given by (Lorentz, 1904)

$$\nu_{\mathrm{obs}} = \frac{\nu}{\left(1-\beta^2\right)^{1/2}}(1-\beta\cos\theta) \quad , \tag{5.843}$$

where the radiation is emitted at frequency ν, the parameter $\beta = v/c$, and $c = 2.9979 \times 10^{10}$ cm s^{-1}. The angular distribution of the temperature, $T(\theta)$, of the background radiation is thus given by (Peebles and Wilkinson, 1968)

$$T(\theta) = T_{\mathrm{o}}\frac{\left[1-(v/c)^2\right]^{1/2}}{1-(v/c)\cos\theta} \quad , \tag{5.844}$$

where $T_{\mathrm{o}} = 2.726$ K is the mean cosmic radiation temperature. For $v \ll c$, the angular distribution of the radiation temperature approximates a first-order spherical harmonic, or dipole anisotropy, given by

$$T(\theta) = T_{\mathrm{o}}[1+(v/c)\cos\theta] \quad . \tag{5.845}$$

The observed value of $T_1/T_{\mathrm{o}} = 0.001232 \approx \beta$ indicates a velocity $v = 369 \pm 3$ km s^{-1} . This motion can be compared with the Sun's velocity of 300 km s^{-1} $\pm$ 20 km s^{-1} toward the Local Group of galaxies (Yahil, Tammann and Sandage, 1977), and a motion of the Local Group at roughly twice this speed toward $l = 248^\circ \pm 9^\circ$ and $b = 40^\circ \pm 8^\circ$ (Yahil, Walker and Rowan-Robinson, 1986). The Sun's orbital velocity with respect to the center of our Galaxy is not important, since it lies roughly in the opposite direction to the apex of the dipole. The detected dipole anisotropy implies a motion of the entire Local Group of galaxies with a speed of about 600 km s^{-1} (Smoot et al., 1991; Partridge, 1995). Additional details of peculiar motions with respect to the cosmic microwave background radiation are given in Sect. 5.7.6.2.

When the dipole anisotropy is removed, the cosmic microwave background radiation is remarkably isotropic to one part in 100,000. Other anisotropies have been assiduously searched for, since they might reveal conditions prior to recombination (Rees and Sciama, 1967; Sachs and Wolfe, 1967; also see Sect. 5.7.6.3). Despite 25 years of effort, no anisotropy was found on any scale, except the dipole (Readhead et al., 1989; Readhead and Lawrence, 1992; Partridge, 1995), with a limit to the temperature difference, ΔT, of

$$\Delta T/T_{\rm o} \leq 1.7 \times 10^{-5} \tag{5.846}$$

on arcminute scales with the dipole anisotropy removed.

Then COBE discovered a statistically significant cosmic microwave background anisotropy with (Smoot et al., 1992)

$$\Delta T/T_{\rm o} \approx (1.1 \pm 0.1) \times 10^{-5} \ . \tag{5.847}$$

The fluctuations determined by COBE are parameterized by specifying the r.m.s. amplitude of the quadrupole moment, T_2, with (Smoot et al., 1992; Gould, 1993)

$$T_2 = (16 \pm 4) \times 10^{-6} \text{ K} \ . \tag{5.848}$$

This statistical detection has been confirmed by ground-based, multi-frequency radio observations that have detected individual features in the microwave background (Hancock et al., 1994; Scott et al., 1996). Indeed, an anisotropy of $\Delta T/T_{\rm o} \approx 10^{-5}$ has been detected by more than 10 experiments on angular sizes from 0.5 to 10 degrees (Scott, Silk and White, 1995).

These temperature fluctuations delineate primordial density perturbations, imprinted on the cosmic microwave background radiation at recombination some 300,000 years after the big bang. They emanate from a redshift of about 1000, and are the high-redshift precursors of the fluctuations that generated the large-scale structures that we see in the Universe today. The implications of anisotropies in the cosmic microwave background for cosmology, including the formation of large-scale structure, are included in Sect. 5.7.6.3 and also reviewed by White, Scott and Silk (1994), Scott, Silk and White (1995) and Gawiser and Silk (1998).

The COBE results are consistent with a scale-invariant primordial density fluctuation spectrum (Bennett et al., 1993, 1996; Smoot et al., 1992) with

$$\frac{\Delta T}{T} \propto \theta^{(1-n)/2} \tag{5.849}$$

that is independent of angular scale, θ, on the sky so that $n \approx 1$ (to be precise $n = 1.2 \pm 0.3$). This so-called Harrison (1970)–Zeldovich (1972) spectrum, with $n = 1$, is also predicted by models of the inflationary cosmology described in Sect. 5.7.2.

Current anisotropy measurements can be expressed in terms of a multipole expansion and a correlation function. The multipole expansion is

$$T(\theta, \phi) = \sum_l \sum_{m=-l}^{l} a_{l{\rm m}} Y_{l{\rm m}}(\theta, \phi) \ , \tag{5.850}$$

where $Y_{lm}(\theta, \phi)$ are spherical harmonic functions. Since galactic emission dominates the signal, galactic coordinates (l, b) are chosen, and we can rewrite the expansion as:

$$\begin{aligned} T(l,b) &= A_{00} + A_{11}\cos b\cos l + B_{11}\cos b\sin l \\ &\quad + A_{10}\sin b + A_{20}\left(\frac{3}{2}\sin^2 b - \frac{1}{2}\right) + \frac{3}{2}A_{21}\sin 2b\cos l \\ &\quad + \frac{3}{2}B_{21}\sin 2b\sin l + 3A_{22}\cos^2 b\cos 2l + 3B_{22}\cos^2 b\sin 2l \ . \end{aligned} \tag{5.851}$$

The amplitude of the dipole component, T_1, is given by

$$T_1 = \left(A_{11}^2 + B_{11}^2 + A_{10}^2\right)^{1/2}. \tag{5.852}$$

Since many anisotropy experiments are differential, the monopole term, T_0, is not observed. When the dipole term, T_1, is also removed, we can express the expansion as

$$\begin{aligned} Q(l,b) &= Q_1\left(3\sin^2 b - 1\right)/2 + Q_2\sin 2b\cos l \\ &\quad + Q_3\sin 2b\sin l + Q_4\cos^2 b\cos 2l + Q_5\cos^2 b\sin 2l \ , \end{aligned} \tag{5.853}$$

and obtain the r.m.s. amplitude of the quadrupole moment from

$$T_2 = Q_{\mathrm{rms}} = \frac{4}{15}\left[\left(\frac{3}{4}Q_1^2 + Q_2^2 + Q_3^2 + Q_4^2 + Q_5^2\right)\right]^{1/2} . \tag{5.854}$$

The measured correlation function determines the parameters of the fluctuation power spectrum. The correlation function is

$$C(\alpha) = \sum_{l>2} \Delta T_l^2 W(l)^2 P_l(\cos\alpha) \ , \tag{5.855}$$

where P_l are Legendre polynomials, and COBE's 3.2° rms Gaussian beam gives a weighting $W(l) = \exp\left[-1/2\left(l(l+1)/17.8^2\right)\right]$, and

$$\Delta T_l^2 = \frac{1}{4\pi}\sum_{\mathrm{m}} |a_{lm}|^2 \tag{5.856}$$

are the rotationally-invariant rms multipole moments. As with the spherical harmonic expansion, the $l = 0$ term is excluded from the correlation function since it is not measured by differential instruments, and the $l = 1$ term is excluded because it is contaminated by the kinematic dipole. The $l = 2$ quadrupole term is sometimes excluded since the quadupole has only $2l + 1 = 5$ degrees of freedom and thus has an intrinsically high statistical or "cosmic" variance, independent of the measurement. The $l = 2$ term is also significantly affected by galactic emission. For a power law primordial fluctuation spectrum the predicted moments, as a function of spectral index $n < 3$, are given by Bond and Efstathiou (1987)

$$\langle \Delta T_l^2 \rangle = (Q_{\mathrm{rms}})^2 \frac{(2l+1)\Gamma(l+(n-1)/2)\Gamma((9-n)/2)}{5\Gamma(l+(5-n)/2)\Gamma((3+n)/2)} \ . \tag{5.857}$$

The best-fitted COBE data correspond to (Bennett et al., 1996; Gorski et al., 1996; Hinshaw et al. , 1996; Kogut et al., 1996).

$$n = 1.2 \pm 0.3,$$
$$Q_{\rm rms} = (15.3 \pm 3.8) \times 10^{-6}\ {\rm K\ for\ best\ fit}\ ,$$
$$Q_{\rm rms} = (18 \pm 1.6) \times 10^{-6}\ {\rm K\ for}\ n\ {\rm held\ constant\ at\ 1}\ . \tag{5.858}$$

For $n = 1$, the expression simplifies to

$$\langle \Delta T_l^2 \rangle = (Q_{\rm rms})^2 \frac{6}{5} \frac{2l+1}{l(l+1)}\ . \tag{5.859}$$

Distortions from an exact blackbody spectrum could only occur if some energy-release process added heat and radiation after $z \approx 3 \times 10^6$ (Weymann, 1966; Zeldovich and Sunyaev, 1969; Sunyaev and Zeldovich, 1980; Partridge, 1995). Energy release or conversion between a year and 2,000 years after the big-bang, corresponding to the redshift range $3 \times 10^6 > z > 10^5$, would give rise to a radiation spectrum characterized by a non-zero chemical potential, μ. Thus there would be a Bose–Einstein spectral distortion with the photon occupation number

$$N(\varepsilon) \sim \frac{1}{{\rm e}^{(\varepsilon - \mu)/kT} - 1} \tag{5.860}$$

where ε is the photon energy, k is the Boltzmann constant, and T is the temperature.

A Compton distortion of the spectrum can become important after 2,000 years, but before recombination at 300,000 years. After recombination it becomes nearly impossible to distort the cosmic microwave background spectrum short of reionizing the Universe. The "Comptonized" spectrum is a mixture of blackbodies at a range of temperatures parameterized in terms of a Compton y parameter first introduced by Zeldovich and Sunyaev (1969) in considering the interaction of matter and radiation in a hot Universe.

$$y = \int_0^{z_h} \alpha \frac{dt}{dz}\, dz\ . \tag{5.861}$$

where the integral is along the line of sight, or back in time, to the redshift, z_h, where electron heating occurred, and $\alpha(z)$ is the rate at which microwave photons gain energy by inverse Compton scattering from hot electrons

$$\alpha(z) = \sigma_{\rm T} n_{\rm e}(z) c \frac{kT_{\rm e}(z)}{m_{\rm e}c^2}\ , \tag{5.862}$$

where the Thomson cross section $\sigma_{\rm T} = 0.665 \times 10^{-24} {\rm cm}^2$, the electron density at redshift, z, is $n_{\rm e}(z)$, c is the velocity of light, k is Boltzmann's constant, $m_{\rm e}$ is the mass of the electron, and $T_{\rm e}(z)$ is the electron temperature at redshift, z. Thus, the Compton y-parameter is given by

$$y = \frac{\sigma_{\rm T}}{m_{\rm e}c^2} \int n_{\rm e} k (T_{\rm e} - T_{\rm CMB}) c\, dt \tag{5.863}$$

where the integral is the electron pressure along the line of sight and T_{CMB} is the temperature of the cosmic background radiation released in the first 2,000 years after the big bang. A Compton distortion of the spectrum can become important when $(1+z)dy/dz > 1$, which occurs ~ 2000 years after the big bang.

COBE places stringent limits on both of the dimensionless cosmological distortion parameters (Mather et al., 1994)

$$\left|\frac{\mu}{kT}\right| < 3.3 \times 10^{-4} \tag{5.864}$$

and

$$|y| < 2.5 \times 10^{-5} \ , \tag{5.865}$$

both at the 95% confidence level.

Collisions between photons of the microwave background and fast-moving electrons in clusters of galaxies distort the microwave spectrum along the line of sight to the clusters. This creates a small cooling of the brightness temperature of the microwave background radiation toward a dense cluster. This has become known as the Sunyaev–Zeldovich effect after their prediction of it (Sunyaev and Zeldovich, 1972, 1980). The redistribution of microwave energy is caused by inverse Compton scattering in which photons from the cosmic microwave background obtain energy at the expense of the kinetic energy of hot, million degree electrons in the intergalactic gas of the clusters.

The Sunyaev–Zeldovich effect produces a change in brightness temperature, ΔT, expressed as a function of the observing frequency, ν, given by (Zeldovich and Sunyaev, 1969; Sunyaev and Zeldovich, 1980):

$$\frac{\Delta T}{T_0} = y\left(x\frac{\mathrm{e}^x+1}{\mathrm{e}^x-1} - 4\right) \ , \tag{5.866}$$

where $x = h\nu/(kT_{\mathrm{o}})$ and h and k respectively denote Planck's and Boltzmann's constant. In the Rayleigh–Jeans, or RJ, part of the cosmic blackbody spectrum, where $h\nu \ll kT_{\mathrm{o}}$, this becomes

$$\frac{\Delta T_{\mathrm{RJ}}}{T_0} = -2y \quad \text{for } h\nu \ll kT_{\mathrm{o}} \ . \tag{5.867}$$

For a cluster of galaxies of extent L, the Compton y-factor is given by

$$y = \frac{k\sigma_{\mathrm{T}}}{m_{\mathrm{e}}c^2}\int_0^L n_{\mathrm{e}}T_{\mathrm{e}}dl \ . \tag{5.868}$$

For rich clusters, $y \approx 10^{-4}$ along the line of sight toward the cluster center.

Detections of the Sunyaev–Zeldovich effect towards clusters of galaxies have been reported by Birkinshaw, Gull and Northover (1978) and Birkinshaw, Gull and Hardebeck (1984). Theoretical and observational work on the Comptonization of the cosmic microwave background radiation by hot gas in clusters of galaxies is reviewed by Rephaeli (1995). For an isothermal cluster, the effect can be specified as a function of the angular distance, θ, from the cluster center

$$\Delta T_{\mathrm{RJ}}(\theta) = \Delta T_{\mathrm{CRJ}}\left(1 + \theta^2/\theta_{\mathrm{CX}}^2\right)^{-1/4} , \tag{5.869}$$

where the central microwave decrement is ΔT_{CRJ} and θ_{CX} is the angular core radius inferred from X-ray data.

X-ray observations of clusters of galaxies can be combined with microwave observations of the Sunyaev–Zeldovich effect for the same clusters to infer cosmological information such as the Hubble parameter H_{o} (Gunn, 1978; Silk and White, 1978). Since the X-ray flux depends on $\int n_{\mathrm{e}}^2 T_{\mathrm{e}}^{1/2} d\ell/(D_{\mathrm{L}}^2)$, the luminosity distance D_{L} can be obtained if the microwave decrement ΔT_{RJ} is used to measure $\int n_{\mathrm{e}} T_{\mathrm{e}} d\ell$. The results depend upon models of the X-ray emission and X-ray spectra to infer T_{e}. For the Coma cluster of galaxies, Herbig et al. (1995) use

$$\Delta T_{\mathrm{CRJ}} = -5 \times 10^{-4}\ \mathrm{K} , \tag{5.870}$$

with X-ray data to obtain

$$H_{\mathrm{o}} = 71 \pm 25\,\mathrm{km\ s^{-1} Mpc^{-1}} . \tag{5.871}$$

Similar comparisons of the Sunyaev–Zeldovich effect and X-ray measurements for Abell 665 (Birkinshaw, Hughes and Arnaud, 1991) and Abell 2218 (Birkinshaw and Hughes, 1994) result in a combined estimate of

$$H_{\mathrm{o}} = 55 \pm 17\,\mathrm{km\ s^{-1} Mpc^{-1}} . \tag{5.872}$$

Cen (1998) describes systematic effects that cause uncertainties in the determination of the Hubble constant through the Sunyaev–Zeldovich effect of clusters of galaxies, suggesting a true value of $H_{\mathrm{o}} = 60$ to 80 km $\mathrm{s^{-1}}$ $\mathrm{Mpc^{-1}}$.

5.7.6 The Shape, Structure, Content and Formation of the Universe

5.7.6.1 Distribution of Galaxies in Space and Time

To survey the Universe and probe its unknown depths, to fix its shape and size, to identify its contents and establish our place in it; these are amongst the most noble objectives of astronomy and astrophysics. As George Ellery Hale expressed it, in his stirring call for a 200-inch optical telescope (Hale, 1928):

> "Like buried treasures, the outposts of the Universe have beckoned to the adventurous from immemorial times. Princes and potentates, political or industrial, equally with men of science, have felt the lure of the uncharted seas of space, and through their provision of instrumental means the sphere of exploration has rapidly widened.... Each expedition into remoter space has made new discoveries and brought back permanent additions to our knowledge of the heavens. The latest explorers have worked beyond the boundaries of the Milky Way in the realm of spiral "island universes", the first of which lies a million light-years from the earth while the farthest is unmeasurably remote."

Edwin Hubble had recently shown that the nearest spiral nebula, M31 or Andromeda, is an extragalactic system residing outside the boundaries of our Galaxy, the Milky Way (Hubble, 1925). Nowadays we use such beacons, called galaxies, to gauge the overall content and extent of the Universe, to indicate how its material content is distributed and how it is moving, and to tell of its origin, its texture and its fate. They are places where the brightest, most luminous matter has become more concentrated than average.

So the basic visible units of cosmological inquiry are the galaxies, either spirals, like the Milky Way and Andromeda, or ellipticals. Each large galaxy is an isolated concentration or roughly 100 billion stars, spanning about 100 thousand light years. In this section we describe the unexpected realization that the Universe of galaxies is highly organized on large scales vastly surpassing the distances between them.

The space-time of the expanding Universe is swiftly ballooning outward, carrying the galaxies with it, like birds in currents of air, all in accordance with Hubble's law (Hubble, 1929; see Sect. 5.2.8)

$$V_r = cz = H_o \times \mathrm{D} \ . \tag{5.873}$$

Here V_r is a galaxy's recession velocity, $c = 2.9979 \times 10^{10}$ cm s^{-1} is the velocity of light, z is the galaxy's redshift, D is its distance, and the Hubble constant, H_o, is often expressed as:

$$H_o = 100h \ \mathrm{km\ s^{-1}\,Mpc^{-1}} \ , \tag{5.874}$$

where h is the value of H_o in units of 100 km s^{-1} Mpc^{-1}. Cosmological or extragalactic distances are often quoted in Megaparsecs, abbreviated Mpc, where 1 Mpc $= 3.086 \times 10^{24}$ cm; in this section we will also use the more picturesque light year, the distance light travels in one year, for conversion 1 Mpc is equivalent to 3.26 million light years. The nearest large galaxy, Andromeda, is located about 2 million light years away, but galaxies are organized into structures that span hundreds of millions of light years.

Since the Hubble constant is uncertain by about a factor of two, with h lying between 0.5 and 1.0 (see Sect. 5.2.8), extragalactic distances are also expressed in velocity units. As an example, a velocity of 10,000 kilometers per second corresponds to a distance between 326 and 652 million light years for $h = 1$ and $h = 0.5$, respectively.

Instead of being randomly placed throughout expanding space, like dust scattered in the wind, thousands of galaxies swarm and knot together into great clusters that gather together into superclusters. This clumpy pattern in the distribution of galaxies, apparent in the first surveys, indicates that gravity triumphs over expansion on scales that are as large as we can see. Although space-time is continually stretching out, pulling most galaxies away from one another, gravity is still strong enough to pull them together and hold them close (see Fig. 5.40).

The fluctuations in galaxy density are large, but the average value corresponds to a low density, one that would by itself expand forever. On average, the space density of galaxies, n_G, is (Loveday, Peterson, Efstathiou and Maddox, 1992)

$$n_G = 5.52 \times 10^{-2} h^3 \,\mathrm{Mpc^{-3}} \ , \tag{5.875}$$

and galaxies are typically separated by a few Mpc, depending on the exact value of h. So the galaxies are typically separated by about 10 million light

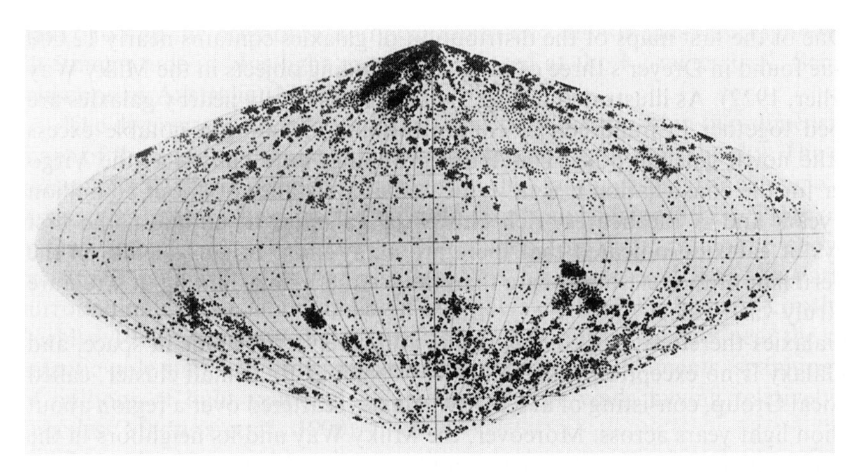

Fig. 5.40. C. V. L. Charlier's map of the distribution of 11,475 nebulae, now known to be mostly galaxies, in galactic coordinates. As illustrated in this diagram, bright galaxies group and cluster together in space, forming numerous concentrations. The grouping toward the north galactic pole is now known as the Virgo cluster of galaxies. (Adapted from Charlier, 1922).

years, or 100 galaxy diameters. Thus, the Universe appears to be largely empty space as far as the galaxies are concerned, like the room you are sitting in, the interior of the atom, the solar system, or the Milky Way.

The first step in surveying the Universe and establishing its structure is the construction of a catalogue of the positions of galaxies in the sky. Well before their extragalactic nature was realized, astronomers meticulously recorded the celestial coordinates and apparent magnitudes of the brightest nebulae, as galaxies and some other objects were then called. Just over 100 nebulae and star clusters are found in Charles Messier's famous list (Messier, 1771, 1784). William Herschel dramatically increased the number to 2,500, during twenty years of systematically sweeping the heavens, from 1783 to 1802. Herschel fixed his telescope on the meridian at a particular elevation, so that, as the Earth turned, all the objects at a particular declination, or elevation, could be noted down. The sidereal time of transit gave the right ascension, to complete the celestial coordinates. This sweep of the sky's northern hemisphere was continued by William's son, Sir John Herschel, who also extended it to the southern hemisphere, publishing data for 5,079 objects in his General Catalogue, the combined result of more than half a century of painstaking observations (Herschel, 1864).

Using the Herschel catalogue as a basis, J. L. E. Dreyer published his New General Catalogue, or NGC for short, of nebulae and star clusters, followed by two Index Catalogues, designated IC, with additional discoveries (Dreyer, 1888, 1895, 1908). Many galaxies, as well as emission nebulae and star clusters, are still known by their NGC and IC numbers. Dreyer's list of positions and magnitudes was updated with corrections, using the photographic prints from the Palomar Sky Survey, by Sulentic and Tifft (1973).

can even outweigh them (see Sarazin, 1986, 1988 for reviews) and 80 to 90 percent of the mass of some clusters is in nonluminous form that can only be detected by its gravitational effect (see Sects. 5.4.5 and 5.4.6).

Abell noticed that clusters of galaxies found on the survey photographs tended to occur in small groups, averaging five or six clusters, which he called second-order clusters (Abell, 1961, 1965), and that are now known as superclusters. A typical supercluster has a mass of 10^{16} to 10^{17} solar masses in a radius of about 50 Mpc, or about 150 million light years (Abell, 1961). Guido Chincarini and Herbert J. Rood showed that the well-known Coma cluster, located at a distance of about 100 Mpc or roughly 330 million light years, is much more extensive than was previously thought, with a radius of at least 100 million light years and a mass of about 10^{16} solar masses (Chincarini and Rood, 1975). Stephen Gregory and Laird Thompson then showed how Coma is joined to another cluster A1367, speculating that all rich clusters are located in superclusters (Gregory and Thompson, 1978).

Superclusters of galaxies are indeed quite common (Gregory and Thompson, 1982; Oort, 1983). They form elongated unsymmetric structures without a central concentration. Examples include the Local (Virgo), Coma, Perseus and Hercules superclusters, with axial ratios of about 1 to 5, or 0.2, lengths of between 100 and 500 million light years, and typical masses of 10^{16} solar masses corresponding to the mass of 100 thousand galaxies (Oort, 1983). Moreover, it is sometimes hard to tell where one supercluster ends and another begins, so they could be connected and linked together in filamentary chains forming a vast web of galaxies with large empty places, or voids, between them (Einasto, Jôeveer and Saar, 1980; Einasto et al., 1997). As an example, our Galaxy appears to join four superclusters containing 48 Abell clusters of galaxies that together form a flattened structure containing up to 10^{18} solar masses with a length of up to 500 h^{-1} Mpc, or about 1,500 million light years, and extending to velocity scales of up to a tenth the velocity of light, or 0.1c (Tully, 1986, 1987). Indeed, the density of bright optical galaxies is dominated by superclusters, such as Virgo, Hydra–Centaurus, Pisces–Perseus, Coma–A1367 and Telescopium–Pavo–Indus, out to a distance corresponding to a recession velocity of 8,000 km s^{-1} (Santiago et al., 1995).

Statistical analysis of the spatial distribution of Abell's rich clusters confirms that they are themselves clustered (Hauser and Peebles, 1973). The rich cluster spatial correlation function has the same shape and slope as that of the galaxy correlation functions, but it seems to be considerably stronger at any given scale than the correlation function for galaxies (Bahcall, 1988). As shown by Neta Bahcall and Ray Soneira, galaxy clusters seem more likely to come together and lie near each other than individual galaxies (Bahcall and Soneira, 1983). Because the force of gravity falls off with increasing distance, you would expect gravity to be less effective in organizing large structures than small ones, so the result seems strange. The cluster correlation also extends to greater separations than the scales of galaxies in general. The spatial correlation function becomes unity at a scale length, r_C, of (Bahcall and Soneira, 1983; Bahcall and West, 1992)

$$r_{\mathrm{C}} = (21 \text{ to } 22)h^{-1} \quad \text{Mpc for separations } r \leq 100\,h^{-1} \ , \tag{5.886}$$

with a spatial correlation function (Klypin and Kopylov, 1983)

$$\xi_{\mathrm{C}}(r) = (r_{\mathrm{C}}/r)^{1.8} \quad \text{for separations } r \text{ out to } 100\,h^{-1} \ . \tag{5.887}$$

However, some authors have argued that artificial clustering in Abell's catalogue may have enhanced the amplitude of $\xi_{\mathrm{C}}(r)$ above its true value (Efstathiou, 1993).

Two-dimensional maps, derived from the catalogued positions of galaxies on the celestial sphere, provide a blurred picture of the galaxy distribution in space. They portray galaxies near and far, piled upon one another at any given location in the sky. Chance superpositions of distant and nearby groups of galaxies could therefore be mistaken for real physical structures.

Redshift surveys use the radial velocity, or redshift, to gauge the approximate distance to galaxies, providing the crucial third dimension. They have transformed our understanding of galaxy clustering, bringing the galaxy distribution into shaper focus and greater relief. As is always the case in astronomy, what you see depends on how you look at it. In the new perspective, nearby galaxies are found in thin, enormously wide, sheet-like walls of hundreds of thousands of galaxies and at the junctions and surfaces of immense bubbles, tens of millions of light years across, whose insides appear to be largely devoid of matter.

One cosmic abyss, in the vicinity of the Böotes constellation, spans nearly 300 million light years from end to end. It was discovered by Robert Kirshner and his colleagues early in the course of a redshift survey (Kirshner et al., 1981, 1987). As far as they could determine, the Böotes void is an immense, empty volume containing virtually no bright galaxies; it is a million cubic Megaparsecs, or more than 10^{25} cubic light years, of practically nothing that the eye can see. The Böotes void contains less than one fifth, or below 20 percent, of the average density of galaxies.

At about the same time, redshifts for hundreds of galaxies in specific regions of the sky provided evidence for filamentary superstructures connected across equally vast distances (Gregory, Thompson and Tifft, 1981; Giovanelli, Haynes and Chincarini, 1986; Haynes and Giovanelli, 1986; Giovanelli and Haynes, 1991).

Then a decade-long, Center for Astrophysics, or CfA, redshift survey, startled the world, showing that the vacant places detected in the number-count maps and as redshift voids are immense bubbles with galaxies distributed on their surfaces. Marc Davis and his collaborators did the first redshift survey at the CfA; later Margaret Geller and John Huchra extended the survey to greater depths.

The CfA survey determined redshifts, z, for galaxies in the nearby Universe, out to a redshift of $z = 0.04$, or a radial velocity of $V_{\mathrm{r}} = cz = 12,000$ km s^{-1}. This extent corresponds to about 650 million light years, but the exact value is uncertain because we do not yet have precise values for either Hubble's constant or for extragalactic distances. (A redshift of $z = 0.04$ corresponds to

How can you create gigantic hollow spheres in space? Even before this intriguing discovery, Jeremiah Ostricker and Lennox Cowie, and independently Satoru Ikeuchi, imagined that explosions of massive collapsing stars might have been blowing bubbles when the Universe was young (Ostriker and Cowie, 1981; Ikeuchi, 1981). Only exceptionally powerful explosions, equal to the combined explosive punch of billions of supernovae, are capable of cleaning out the seemingly empty voids and blasting the galaxy material into thin spherical shells. Estimates of the energy output of the hypothetical explosions indicate that they cannot generate the largest voids. Explosions large enough to create huge clusters and voids should also have left a detectable imprint on the cosmic microwave background; such signatures have not yet been found.

Something else is needed, and gravity seems to be the missing ingredient. But again, the problem is one of sheer size. Structures like the Great Wall seem far too large and too massive to have resulted from the mutual gravitational attraction of their constituent galaxies even over the age of the Universe. That would require moving material over vast distances, as large as a billion light years, during the Universe's lifetime of roughly 10 billion light years. The matter would have to move at enormous velocities to travel that distance, at speeds of about a tenth the velocity of light or at $0.1\,c$. That is much faster than the typical velocity of galaxies today, some 300 kilometers per second or $0.001\,c$. In other words, the Universe is not old enough to have become as clumpy as it now appears by the gravity of its visible components.

Such enormous structures seem not to have been able to form as the result of gravitational clumping unless there were already significant lumps present in the Universe at its birth. Given enough time, gravity could amplify slight irregularities into distinct formations, magnifying density irregularities that were present at the big-bang beginning of time into the large-scale structures we see today. The observed structures would always retain something of their past, so today's clustering provides a fossil record of ancient formation processes.

Astronomers are therefore extending their redshift surveys to the far away realms of long ago, gaining an improved understanding of how galaxies condensed early in the history of time. The bigger the regions of the sky mapped, the bigger the structures they have found, like topographers exploring distant regions of the Earth's mountains and valleys. The largest features detected so far are comparable to the dimensions of the survey, suggesting that there are even larger ones. And as astronomers probe deeper into space, puncturing it with needle-like redshift probes, they keep on hitting walls of galaxies separated by vast volumes of space.

Observations of high-redshift objects suggest that they cluster together to form structures similar in size to the walls and voids found more locally (Cohen et al., 1996). The large-scale structures approximately $120\,h^{-1}$ Mpc in extent, or a few hundred million light years across, are common features (Landy et al., 1996). They apparently persist out to quasar redshifts of $z = 3$ (Quashnock, Vanden Berk and York, 1996), and might even recur with a spatial periodicity on that scale (Szalay et al., 1993). The network of rich superclusters and voids seems to form a three-dimensional lattice with regions of high density regularly separated by about $120\,h^{-1}$ Mpc (Einasto et al., 1997). Still, the Universe may eventually smooth out on the biggest scales.

In fact, the galaxy distribution does seem to make a transition from substantial angular fluctuations, seen on small scales, to a nearly smooth large-scale distribution at a size, l, of (Peebles, 1993):

$$l \approx 30\,h^{-1}\,\text{Mpc} \quad \text{or} \quad H_o l \approx 3,000\,\text{km s}^{-1} \ , \tag{5.888}$$

where the angular fluctuations, δN, in the number counts, N, become

$$\delta N / \langle N \rangle = 0.5 \pm 0.1 \ , \tag{5.889}$$

for a mean count value of $\langle N \rangle$. Here

$$(\delta N)^2 = \langle (N - \langle N \rangle)^2 \rangle = \langle N^2 \rangle - \langle N \rangle^2 \ .$$

This break in the number count fluctuations is present in the Lick catalogue and the ATM survey of millions of galaxies, as well as in the sparser IRAS (InfraRed Astronomical Satellite) redshift survey of about 5,000 bright galaxies.

Somewhere out in space, the lumpiness in the distribution of galaxies ought to fade into uniformity. We know this because the cosmic microwave background, the pervasive, glowing remnant radiation of the big bang, is as smooth as silk. As we shall see, it does contain important, miniscule ripples, the probable seeds of future galaxies. Yet, it suggests that all the places in the cosmos will look alike on the largest scales, with no center and no edge. That would confirm Hubble's initial survey of faint galaxies (Hubble, 1934), and verify Milne's cosmologial principle that asserts there are no preferred locations in the Universe (Milne, 1935).

But first, our attention will be focused on another departure from uniformity, the localized peculiar motions of the galaxies. As it turns out, the vast concentrations of matter detected in galaxy sky maps and in redshift surveys, pull surrounding galaxies toward them, imposing localized currents of motion within the expanding Universe.

5.7.6.2 Large-Scale Peculiar Motions of the Galaxies

The Universe is mostly empty space, and every material object can only hold itself up in that space by moving around very fast. Everything we see is moving and rushing about, from atoms and people to stars and galaxies, usually on well defined trajectories. It seems that you can't exist without motion, otherwise gravity will pull you down.

Propelled by the big bang explosion, any galaxy moves away from us and every other galaxy, at a speed that increases with its distance or separation. This is the normal, smooth and regular expansion of the Universe described by Hubble's law and designated the Hubble flow.

Yet, you can occasionally catch galaxies colliding, merging or passing through each other, when they ought to be continuously moving further apart. So we know that individual galaxies dart about, like gnats or mosquitoes in a swarm. In addition, an entire swarm is moving off on its own, independent of the expansion, as if blown by powerful, capricious winds.

The galaxies are not expanding outward in a smooth and regular fashion; but move around on their own as well, responding to the gravitational pull of mass. Astronomers measure this departure from uniformity by dividing the observed radial velocity, V_r, of each galaxy into two components, the sum of the smooth expansion, called the Hubble flow, and the departure from that flow, designated a peculiar velocity. That is

$$V_r = cz = H_o D + V_p \ , \tag{5.890}$$

where c is the velocity of light, z is the redshift of the galaxy, H_o is the Hubble constant, D is the distance to the galaxy, and V_p is the galaxy's peculiar velocity independent of the Universe's general expansion.

Because the galaxies are moving in consort over vast distances of hundreds of millions of light years, the streaming motions are called large-scale peculiar motions. They are large in space, but not so big in velocity when compared with the expansion speed of distant objects. Most of the radial velocity and redshift is usually due to the expansion of space between us and the distant galaxy, and not due to the peculiar streaming motion of the galaxy itself. This means that a redshift survey can provide a reasonably exact map of the third distance dimension; for example voids that are 5,000 km s^{-1} across surpass the localized peculiar motions that generally amount to no more than 700 km s^{-1}.

Astronomers sometimes refer to the cosmic Mach number, M, the ratio of either the Hubble flow velocity, $H_o D$, or the streaming velocity, V_p, in some patch, or scale R, to the small-scale random velocity dispersion, σ, of galaxies within the same patch of the Universe. After correction for local streaming motion, the Hubble flow has a low Mach number, with $\sigma < 200$ km s^{-1}, for recession velocities that are much larger (Rubin et al., 1976), something like water directed from a hose. When referred to the velocity of streaming motion, the Mach numbers have values of about unity, constraining various models for the origin of structure in the Universe (Ostriker and Suto, 1990; Suto, Cen and Ostriker, 1992; Strauss, Cen and Ostriker, 1993).

To calculate peculiar motion, astronomers need some way of measuring distance independent of redshift, which is not an easy task. Techniques include estimating the intrinsic brightness of a galaxy, and hence its distance, from some additional observable such as its speed of rotation or its internal velocity dispersion (see Sect. 5.2.7). Once the distance of a galaxy is known, its observed radial velocity can be compared with that predicted for its distance and a given value of the Hubble constant. Assuming a uniform Hubble flow, any deviation is attributed to a peculiar motion.

Early evidence for local departures from the Hubble expansion was summarized by Davis and Peebles (1983) and Burstein (1990), and a readable popular discussion of the large-scale streaming of galaxies is given by Dressler (1987). A collection of seminal review articles on these motions is provided by Rubin and Coyne (1988); it includes the flow of the local group (Faber and Burstein, 1988; Lynden-Bell and Lahav, 1988) and the peculiar velocity field indicated by the InfraRed Astronomical Satellite, or IRAS, galaxy survey. Evidence for large-scale motions out to a cosmological distance corresponding to a recession velocity of 5,000 km s^{-1} is reviewed by Burstein (1990), and discussed by Peebles (1993); the dynamics of such cosmic flows are reviewed by Dekel (1994). The density and peculiar velocity fields of nearby galaxies are reviewed by Strauss and Willick (1995). Bahcall (1997) reviews observational studies of the large-scale structure of the Universe and describes the peculiar-velocity fields on large scales.

All motion is relative, depending on the frame of reference used to measure it. From one perspective, our Galaxy appears to be rushing toward a distant point, but from that place it looks like the Galaxy is being pulled through space. And if the reference point is itself moving, we've got all the measured velocities wrong.

The smooth, faint glow of microwave radiation, left over from the big bang, solves this perspective dilemma. It fills the sky in all directions, and provides the backdrop, or celestial rest frame, of the early, and hence distant, Universe. Our motion toward this cosmic frame of reference distorts its uniformity, producing an increase in temperature in the direction of motion and a temperature decrease in the opposite direction. This asymmetry in the angular distribution of the radiation is called the dipole anisotropy. The latest measurements of the background radiation anisotropy indicate that the Sun is moving at a peculiar velocity of (Lineweaver et al., 1996)

$$V_p = 369 \pm 3 \text{ km s}^{-1} \ , \tag{5.891}$$

toward galactic coordinates

$$l = 264.31° \pm 0.17° \text{ and } b = +48.05° \pm 0.1° \tag{5.892}$$

and celestial coordinates

$$\alpha = 11^h 11^m 57^s \pm 23^s \text{ and } \delta = -7.22° \pm 0.08° \ . \tag{5.893}$$

Our place in space is not so terribly unique, and astronomers often prefer to use the centroid of the Local Group of galaxies as the reference point for galaxy motion. Indeed, if we just observed the apparent motions of the nearby galaxies, we wouldn't even know the Universe is expanding (Hubble, 1939). Their motion is largely determined by the rotation of our Galaxy. The Sun and nearby stars are rotating at a velocity of 220 km s^{-1} about the Galactic center located 8.5 kiloparsecs away (see Sect. 5.2.6). The nearest small galaxy, the Large Magellanic Cloud, at about 50 kiloparsecs away, is so close that it orbits our Galaxy; its radial velocity has nothing to do with the expansion of the Universe. And the nearest large spiral galaxy, Andromeda or M31 at about 725 kiloparsecs, has a negative heliocentric radial velocity of about -300 km s^{-1}, or -120 km s^{-1} referred to the center of our Galaxy; its blueshift suggests that M31 is approaching us rather than expanding away. The two galaxies are orbiting in a gravitationally bound system, called the Local Group, that consists more or less of our Galaxy, Andromeda and their dwarf galaxies (also see Table 5.15 of Sect. 5.4.5).

Current investigations therefore reference all cosmic motions, including galaxy redshifts, to the rest frame, or barycenter, of the Local Group. The observed heliocentric radial velocities of non-Local Group galaxies are corrected for the velocity, V_{LG}, of the Sun's motion toward the centroid of the Local Group given by (Yahil, Tammann and Sandage, 1977)

$$V_{LG} = -79 \cos l \cos b + 296 \sin l \cos b - 36 \sin b \ \text{km s}^{-1} , \tag{5.894}$$

for galactic coordinates l, b. In approximate form, the correction to the centroid of the Local Group is:

$$V_{LG} \approx +300 \sin l \cos b \ \text{km s}^{-1} . \tag{5.895}$$

The maximum deviation between the full and approximate correction is $\pm 87 \text{ km s}^{-1}$, which may be important for nearby galaxies with small redshifts.

When corrected for the Sun's motion with respect to the Local Group, the dipole anisotropy in the background radiation indicates that the peculiar velocity of the Local Group is (Smoot et al., 1991):

$$V_p \text{ (Local Group)} = 622 \pm 20 \text{ km s}^{-1} \tag{5.896}$$

toward galactic coordinates of

$$l = 277 \pm 2 \text{ degrees}, b = +30 \pm 2 \text{ degrees} \tag{5.897}$$

or celestial coordinates of

$$\alpha = 10.5 \pm 0.2 \text{ hours}, \delta = -26 \pm 3 \text{ degrees} . \tag{5.898}$$

When the motion of the Local Group is inferred from the relative distances of remote clusters of galaxies (Aaronson et al., 1986; Jerjen and Tammann, 1993), it is in statistical agreement with both the direction and magnitude of the motion indicated by the dipole of the cosmic microwave background, indicating a peculiar velocity of the Local Group of (Jerjen and Tammann, 1993)

$$V_p = 745 \pm 106 \text{ km s}^{-1} , \tag{5.899}$$

in the same direction as that inferred from the background radiation, and showing that it is caused by a local streaming motion. Full-sky surveys indicate that the dipole anisotropy is indeed due to the motion of the Local Group, and that this motion is gravitationally induced (Strauss et al., 1992).

Over vast intergalactic distances, the dominant movement of galaxies ought to be outward, at a speed proportional to distance. However, all of the galaxies in our region of space are not just going along for the ride with the pervasive Hubble flow. They, too, are being swept away in vast currents of matter, hurtling together in a forced, mass migration from one place to another. This was suggested some time ago by Rubin et al. (1976) who showed that some spiral galaxies, at distances corresponding to recessional velocities of up to $6,500 \text{ km s}^{-1}$, are participating in a bulk, directed motion with speeds of hundreds of kilometers per second, and that the Milky Way is moving edge-on in the same direction.

Nearby galaxies are rushing *en masse*, like a high-speed celestial convoy, toward the same remote point in space. This streaming motion is superimposed on the Universe's smooth expansion, like strong local currents running in a larger cosmic sea. Astronomers now detect such bulk flows, superimposed on top of the Universe's smooth expansion, out to distances of hundreds of millions of light years. This means that all of these galaxies are moving toward, or being pulled by, a distant powerful mass. A large mass outside the largest region of streaming motion is affecting the local expansion.

The first suggested explanation was the Virgo cluster of galaxies, the nearest strong excess of gravitating matter. The gravitational forces of such a cluster, or supercluster, can act as a brake on, or accelerator of, the otherwise unimpeded expansion of the Universe, modifying the speed of flight for galaxies in its vicinity. From the large-scale, bird's eye view of cosmology, the Virgo cluster will produce an apparent local retardation of the cosmological expansion. The infall of the Local Group will reduce the observed radial velocities of remote galaxies below that expected from the general Hubble flow, an effect first noted by De Vaucouleurs (1958) who reasoned that our local motion might be additionally affected by the rotation of the Local Supercluster.

The motion of the Local Group toward the microwave background is not toward the Virgo cluster ($\alpha = 12.5\,\text{hours}, \delta = +12.7\,\text{degrees}$), so the Virgo cluster of galaxies cannot account for all of the observed large-scale motions. The contribution is nevertheless important. Huchra (1988) has reviewed the history of studies of the Virgo infall, including the work of Tonry and Davis (1981), Hoffman and Salpeter (1982) and Aaronson et al. (1982), concluding that the Local Group of galaxies is falling toward the Virgo cluster at a speed of

$$V_{\text{INFALL}}\ (\text{Local Group}) = 250 \pm 64\ \text{km s}^{-1} \ . \tag{5.900}$$

This does not mean that we are actually moving toward the Virgo cluster, but only that we are moving away less rapidly than in a Hubble flow unperturbed by a large mass concentration in Virgo. Peebles (1976) also obtained a value of about 250 km s^{-1}. The cosmic expansion velocity of the cluster is Huchra (1988):

$$V_{\text{COSMIC}}\ (\text{Virgo}) = 1{,}404 \pm 50\,\text{km s}^{-1} \ , \tag{5.901}$$

with a range in estimates of between 1,200 and $1,600\,\text{km s}^{-1}$ due to uncertain corrections in the observed heliocentric velocity for the motion of our Galaxy toward the Local Group, the Virgo-centric infall of the Local Group and the uncertain membership and random velocities of the cluster itself. The Hubble constant inferred from these motions falls in the range H_0 (Virgo) = 53 to 88 km s^{-1} Mpc^{-1}, for distances of D (Virgo) = 14 to 23 Mpc. Sandage and Tammann (1990) use the observed motions of remote clusters of galaxies to refine their estimates of the cosmic expansion velocity for the Virgo cluster, obtaining an infall of the Local Group toward Virgo (the retarded expansion effect) of:

$$V_{\text{INFALL}}\ (\text{Virgo}) = 168 \pm 50\,\text{km s}^{-1} \ , \tag{5.902}$$

and a cluster Hubble flow velocity of:

$$V_{\text{COSMIC}}\ (\text{Virgo}) = 1{,}144 \pm 18\,\text{km s}^{-1} \ , \tag{5.903}$$

with H_o (Virgo) $= 52 \pm 2\,\mathrm{km\,s^{-1}\,Mpc^{-1}}$ and a distance of D (Virgo) $= 21.9 \pm 0.9$ Mpc.

Silk (1974, 1977) and Peebles (1976) calculated simple models for the distortion of the local Hubble flow caused by the Virgo cluster, pointing out that the infall velocity depends on the mass overdensity. Since the overdensity is the amount by which the mass density of the cluster galaxies exceed the mean density of galaxies, the velocity measurements could be used to estimate the density parameter, Ω_o, the ratio of the mean mass density of the Universe to the closure density. That is the same thing as determining the deceleration parameter $q_o = \Omega_o/2$.

A material concentration of mass, ΔM, located at a distance, D, from the Local Group will produce a peculiar velocity, V_p, given by (Peebles, 1976, Gunn, 1977)

$$V_p \approx \frac{G\Delta M}{H_o D^2} = \delta V_H f(\Omega_o) \ , \tag{5.904}$$

where G is the gravitational constant, H_o is the Hubble constant, the density contrast $\delta = \Delta\rho/\rho_o = (\rho - \rho_o)/\rho_o$ is the overdensity with respect to the mean density, ρ_o, of the Universe, the expansion velocity due to the Hubble flow is $V_H = H_o D$, and the function $f(\Omega_o) = (R/\delta)(d\delta/dt)$, at scale factor R and time t after the big bang, depends on the density parameter $\Omega_o = \rho_c/\rho_o$ for a critical mass density ρ_c, required to keep the Universe in equilibrium between closure and open expansion.

The peculiar velocities of the galaxies around the Virgo cluster have been calculated on the basis of simple spherical models introduced by Silk (1974, 1977) and Peebles (1976) who showed how these distortions of the local Hubble flow can be used to infer the mass density of the Universe. For a spherically symmetric mass concentration of density contrast, δ, at distance, D, the peculiar velocity, V_p, is given by (Silk, 1974; Peebles, 1976; Gunn, 1977; Hoffman and Salpeter, 1982):

$$V_p \approx \frac{1}{3}\delta D\, H_o \Omega_o^{0.66} \ , \tag{5.905}$$

or

$$\Omega_o^{0.66} \approx 3(V_p/V_H)(\Delta\rho/\rho_o)^{-1} \ . \tag{5.906}$$

With an infall velocity to the Virgo cluster of $V_p = 250\,\mathrm{km\,s^{-1}}$, and expansion velocity of $V_H = 1,404\,\mathrm{km\,s^{-1}}$, and an overdensity of $\Delta\rho/\rho_o = 2$ to 3, Huchra (1988) obtains a density parameter of:

$$\Omega_o = 0.25 \text{ to } 0.35 \ . \tag{5.907}$$

Davis, Tonry, Huchra and Latham (1980) similarly obtained $\Omega_o = 0.4 \pm 0.1$ from the motion of the Local Group toward the Virgo cluster.

Large-scale deviations from the Hubble flow have now been mapped out to distances in excess of 100 Megaparsecs, and used to characterize the mass density inhomogeneities that are thought to cause the motions (Strauss and Willick,

1995). One can, for example, infer the value of the density parameter Ω_0 from the pull of all the matter that one can see. The greater the mass concentration, the greater will be the flow velocity determined from the analysis of the peculiar motions of hundreds and thousands of individual galaxies. The result depends on the survey used, and whether or not there are substantial quantities of dark invisible matter. A somewhat higher value of $\Omega_0 = 0.7 \pm 0.1$ is, for example, suggested by the InfraRed Astronomical Satellite (IRAS) redshift survey (Rowan–Robinson et al., 1990; Rowan-Robinson, 1993), but with considerable uncertainty (Strauss et al., 1992). Nevertheless, in their comprehensive survey of such research, Strauss and Willick (1995) conclude that the density and velocity surveys favor $\Omega_0 \approx 0.3$, but do not rule out a larger value.

The best estimates for the infall velocity of the Local Group toward Virgo are, however, too small and in the wrong direction to explain the observed motion of the cosmic microwave background seen from the centroid of the Local Group. This is because the Virgo cluster is itself moving at high velocity in another direction. Both the Virgo cluster and the Local Group are falling toward a distant, massive Great Attractor (Dressler et al., 1987; Lynden-Bell et al., 1988; Bertschinger et al., 1990; Dressler, 1991).

If the Virgo cluster of galaxies accounts for all the local streaming motions, then the associated peculiar velocities should eventually disappear when looking far beyond the cluster. As galaxies get closer to whatever is producing the streams, their outward speed will increase, since the gravitational pull increases. Then, at the center of attraction there should be no peculiar motion at all, like the calm eye of a hurricane. Galaxies far beyond that still point will be pulled backward toward the center and the Earth, with reduced outward speed in comparison to those nearby.

Yet, galaxies hundreds of millions of light years away, up to four times the distance of the Virgo cluster and in every direction, are all moving toward the same point in the sky with velocities higher than predicted by the expansion of the Universe. The distances and velocities of hundreds of elliptical galaxies, with recession velocities up to $6,000\,\mathrm{km\,s^{-1}}$, indicate a common peculiar velocity of about $600\,\mathrm{km\,s^{-1}}$ in the general direction of the Hydra–Centaurus supercluster of galaxies (Dressler et al., 1987). The data suggest a flow induced by a Great Attractor centered at roughly $4,000\,\mathrm{km\,s^{-1}}$ away, or at a distance of $40\,h^{-1}$ Mpc, where the Hubble constant $H_0 = 100h\,\mathrm{km\ s^{-1}\ Mpc^{-1}}$ (Lynden-Bell et al., 1988; Dressler, 1988). That is about a hundred times further than the nearest galaxy, and between 150 and 300 million light years away, depending on the value of h.

The Great Attractor generates more infall velocity at the Local Group than the Virgo cluster, although the Great Attractor is more than two times further away. Altogether, the Local Group is being pulled through space at a net velocity of about $600\,\mathrm{km\,s^{-1}}$. The overall motion apparently consists of at least two components having a comparable size of roughly $300\,\mathrm{km\,s^{-1}}$; these are the Local Group motion within the Local Supercluster, primarily toward Virgo, and bulk Supercluster motion as a whole, in a direction lying close to that of the Great Attractor in our next nearest neighbor supercluster, Hydra–Centaurus (Aaronson et al., 1986; Lynden-Bell et al., 1988).

The implication is that galaxies in a region hundreds of millions of light years across are being pulled by the gravitational force of some huge mass, out at the limits of vision. The mass required to produce the observed peculiar motions is about 5×10^{16} solar masses, which is comparable to that inferred for the observed portion of the Great Wall. So, the Great Attractor has a mass equivalent to about 50 million billion stars like the Sun, and at least 50,000 galaxies like the Milky Way.

Although the Great Attractor is partly hidden behind the dusty plane of the Milky Way, much of its gravitational pull can be attributed to known superclusters of galaxies (Scaramella et al., 1989; Rowan-Robinson et al., 1990). Indeed, the Local Supercluster appears to be an extended ridge on the near flank of the Great Attractor, proceeding through the Virgo and Ursa Major superclusters and on beyond the Hydra-Centaurus one. The center of the Great Attractor probably coincides with a rich, massive galaxy cluster, Abell 3627, with a mass of about 5×10^{16} solar masses, located at a distance of about $49\,h^{-1}$ Megaparsecs (Kraan–Korteweg et al., 1996).

All those awesome concentrations of mass out there seem to be jerking us around in a cosmic tug of war. In the most distant regions, its hard to tell what is dominating the flow. Some investigations seem to show that very distant galaxies, are falling back toward the Great Attractor (Dressler and Faber, 1990); while other studies contradict this claim (Mathewson, Ford and Buchhorn, 1992). Detailed scrutiny of the various streaming motions indicates that they cannot arise solely from the Great Attractor; at least two comparable mass concentrations play a role, one in Perseus-Pisces and one in the Great Attractor complex (Willick, 1990; Han and Mould, 1992; Courteau, Faber, Dressler and Willick, 1993; Da Costa et al., 1996). Thus, as massive as the Great Attractor is, it does not account for all of the $600\,\mathrm{km\,s^{-1}}$ motion of the Local Group. Other masses must join in the tug of war on the local Universe.

The first tentative steps into more distant realms suggest even vaster rivers of matter. When Tod Lauer and Marc Postman used the brightest galaxies in clusters of galaxies to probe motions beyond $100\,\mathrm{h}^{-1}$ Mpc, the collective peculiar motions still did not go away. The entire group of galaxies, as far as they could measure, appears to be fleeing at about $700\,\mathrm{km\,s^{-1}}$ in yet another direction that does not coincide with either the Great Attractor or the background radiation dipole (Lauer and Postman, 1994; Postman, 1995). If they are right, then a tremendous mass located beyond the edge of their survey, at recessional velocities of more than $10,000\,\mathrm{km\,s^{-1}}$, may be pulling on all these galaxies, and mass concentrations beyond $100\,h^{-1}$ Mpc, or outside the nearest 300 to 600 million light years, may be affecting the motion of the Local Group.

So far we've just taken a myopic glimpse at our astronomical backyard. Only the nearest million cubic Megaparsecs, or about 10^{25} cubic light years, have been surveyed. That sounds like a lot, but its a small part of the observable Universe, less than 0.1 percent. The exploration of unknown, distant territory is therefore continuing. After all, we perceive the Earth as flat locally, but know its round from the vantage point of space. When astronomers extend their horizons back far enough, the Universe ought to appear smoother, like the ocean with its waves and foam at the shore and a flat appearance from afar.

That is what the microwave background radiation suggests.

5.7.6.3 Ripples in the Background Radiation

The cosmic microwave radiation, that emanated from the big bang and now fills the Universe, provides a strong constraint on theories for the formation of galaxies and larger material structures. The faint afterglow that we now detect carries information about the state of the Universe at the time that the radiation last interacted with matter, roughly 300,000 years ago. Any angular temperature fluctuations present at that time have been retained in the background radiation; and these slight temperature differences preserve a fossil imprint of the first material structures to have arisen in the infant Universe, detected by a variation in the radiation temperature as a function of position on the sky. Relationships between such anisotropies in the cosmic microwave background radiation and the formation of contemporary structure in the Universe were reviewed by Sunyaev and Zeldovich (1980) and more recently by White, Scott and Silk (1994), Scott, Silk and White (1995), Bennett, Turner and White (1997) and Gawiser and Silk (1998).

In the early stages of the expanding Universe, no structures such as galaxies or stars could have existed. All space would have been filled with radiation and hot gas, consisting of the nuclei of hydrogen and helium and free electrons. Radiation was too powerful back then, and the subatomic particles too hot to settle down into anything you would recognize today. That is, for at least the first 100,000 years in the history of time, the electrons were too hot, and moving too quickly to be captured by the nuclei and retained in bound orbits. The hot radiation associated with the big bang kept the matter ionized and prevented the creation of neutral, unionized hydrogen atoms until its temperature was lowered enough, by the Universe's expansion, for electrons to combine with nuclei and the Universe was said to have "recombined".

After the fireball has cooled to about 3,000 degrees, the electrons stay bound within atoms. In this condition, they can no longer scatter radiation and the Universe becomes transparent. Today we see the cooled remnant of the radiation as the uniform 3 degree background, and any angular fluctuations that we now detect in that radiation were created when it last scattered at the recombination time. Since the temperature cools as the Universe expands, in inverse proportion to the size and redshift, we can conclude that the Universe has expanded by a redshift factor of about 1,000 since the recombination epoch when the temperature was 3,000 degrees.

To be precise about the timing, the process of recombination proceeds through the Saha equation (Volume I), and was first applied to conditions in the expanding Universe by Peebles (1968) and about the same time by Zeldovich, Kurt and Sunyaev (1969). Recent summaries are found in Kolb and Turner (1990) and Peebles (1993), indicating that the primeval plasma is free to combine to atomic hydrogen when the temperature, $T_{\rm rec}$, has cooled to

$$T_{\rm rec} = T_{\rm o}(1 + z_{\rm rec}) \approx 3500\,{\rm K} \approx 0.31\,{\rm eV} \ , \qquad (5.908)$$

where the current temperature of the background radiation is $T_{\rm o} = 2.726 \pm 0.010\,{\rm K}$, and the energy associated with 1 degree Kelvin is 1.38×10^{-16} erg and 9.617×10^{-5} eV, and the redshift, z_{rec}, of recombination is given by:

$$(1 + z_{rec}) \approx 1380(\Omega_B h^2)^{0.023} \approx 1240 \text{ to } 1380 \ , \tag{5.909}$$

where the range corresponds to $\Omega_B h^2 = 0.1$ to 1.0, the baryon density parameter $\Omega_B = 8\pi G\rho_B/H_o^2$ for a baryon density ρ_B, Hubble's constant $H_o = 100\,h$ km s^{-1} Mpc^{-1}, and h lies between 0.5 and 1.0.

Assuming that the Universe was matter dominated at recombination, then the age of the Universe at recombination is:

$$\begin{aligned} t_{rec} &= (2/3)H_o^{-1}\,\Omega_o^{-1/2}(1 + z_{rec})^{-3/2} \\ t_{rec} &\approx 4.4 \times 10^{12}\,\Omega_o^{-1/2}h^{-1} \quad \text{seconds} \\ t_{rec} &\approx 1.4 \times 10^{5}\,\Omega_o^{-1/2}h^{-1} \quad \text{years} \ , \end{aligned} \tag{5.910}$$

for a density parameter $\Omega_o = 8\pi G\rho_o/(3H_o^2)$ and a total mass density ρ_o.

The recombination time at a few hundred thousand years after the big bang is not significantly different from the decoupling time when matter becomes free to move through the radiation to form the first gravitationally bound systems. Before that time, the cosmic background radiation keeps the matter fully ionized and the radiation drag on the free electrons prevents the formation of such material structures (Peebles, 1965). We can therefore assume that the recombination time marks the epoch at which the density fluctuations of ordinary baryonic matter, consisting of atoms, could begin to grow.

This time is noticeably different from the time, t_{eq}, of about a thousand years at which the radiation energy density is equal to the mass density

$$t_{eq} \approx 10^3\Omega_o^{-2}h^{-4} \ \text{years,} \tag{5.911}$$

when the radiation temperature was

$$T_{eq} \approx 5.5\Omega_o h^2 \ \text{eV}. \tag{5.912}$$

Before that time the radiation controlled the action and dominated the expansion of the Universe, and after it matter took over. Density inhomogeneities of hypothetical, but not yet observed, non-baryonic matter might have started growing at this earlier time, when the matter first dominated radiation.

In any event, we can assume that any angular temperature fluctuations, ΔT, detected in the cosmic background radiation today, provide information about the surface of last scattering at a redshift of $z \approx 1000$ and at a time between 140,000 and 360,000 years after the big bang. This conclusion is independent of cosmological model, and does not significantly depend on the density parameters, Ω_o or Ω_B, or on Hubble's constant, H_o. Places of slightly higher or lower temperature mark off regions that had slightly greater or lesser concentrations of ordinary matter back then, like waves rippling through the new-born Universe.

A major problem lies in the uniformity of the background radiation in all directions, or in its isotropy. Any anisotropy, or ripple, is exceedingly small. So the background radiation, and by extension the matter once embedded in that radiation, is exceptionally smooth and unperturbed all around the sky. This provides strong constraints to any theory for the formation of structure in the Universe.

Still, from another perspective, the surprising uniformity is a good thing. It provides strong evidence in support of the cosmological principle – the conjecture that the Universe is isotropic on a large scale, and it confirms that the background radiation is indeed a relic of the big bang. The high intrinsic isotropy of the cosmic microwave background radiation also meant that the Robertson-Walker metric was an unexpectedly good approximation for the real Universe.

The only deviation from smoothness detected in 25 years of effort was the dipole one attributed to the Milky Way's motion with respect to the background radiation, which has a slightly higher temperature in the direction of the motion, and a slightly lower temperature in the opposite direction (see Section 5.7.5). Once this dipole term was removed from the data, any deviations from smoothness, specified by a quadrupole term, are below one part in ten thousand on angular scales from arcminutes to 180 degrees (Readhead and Lawrence, 1992; Partridge, 1995), or:

$$\Delta T/T_o < 10^{-4} \text{ for angular sizes } \theta \geq \text{a few arcminutes} , \tag{5.913}$$

where the current temperature of the radiation is $T_o = 2.756 \pm 0.010\,\mathrm{K}$. Theoretical predictions of these temperature fluctuations were therefore adjusted, refined and reduced to meet successively lower observational limits to $\Delta T/T_o$ (White, Scott and Silk, 1994).

There was one process, the adiabatic one, that could not be reconciled with the observations. In an adiabatic process, radiation and matter are perturbed so that their ratio is constant or conserved in space and time. As shown by Joseph Silk decades ago, adiabatic fluctuations in temperature and radiation at the recombination time are related by (Silk, 1967)

$$\left(\frac{\Delta T}{T_o}\right)_{\text{rec}} \approx \frac{1}{3}\left(\frac{\Delta \rho}{\rho_o}\right)_{\text{rec}} \tag{5.914}$$

for fluctuation $\Delta\rho$ in the mass density ρ at the recombination time.

Silk (1968) showed how such adiabatic, primordial density fluctuations could survive until the epoch of galaxy formation, implying an angular anisotropy, ΔT, in the background radiation on angular scales between 10 arcseconds and 30 arcseconds, depending on the mean value taken for the present mass density of the Universe. In the purely baryonic case, the growth of fluctuations begins at recombination, and the presently observed matter density contrast $\delta = \Delta\rho/\rho_o \approx 1$ must have grown from (Zeldovich, Einasto and Shandarin, 1982):

$$(\Delta\rho/\rho_o)_{\text{rec}} \approx 10^{-3}\,\Omega_o^{-1} \approx (1/z_{\text{rec}})\Omega_o^{-1} . \tag{5.915}$$

If the density parameter $\Omega_o \leq 1$ then at recombination we need $\Delta\rho/\rho_o \leq 10^{-3}$, and for the adiabatic case these density fluctuations correspond to angular fluctuations of the temperature of the cosmic microwave background radiation given by:

$$\Delta T/T_o \approx 3 \times 10^{-4}\,\Omega_o^{-1} \tag{5.916}$$

at recombination and now, which disagrees with the observational upper limits and eventual detection.

Thus, a pure baryonic Universe with adiabatic primordial fluctuations is inconsistent with observations of the anisotropies in the cosmic microwave background radiation and the present mass density variations for any value of the density parameter at or below the critical value required to close the Universe or for $\Omega_o \leq 1$. The upper limit to anisotropy may also rule out adiabatic primordial fluctuations in any low-density, $\Omega_o \leq 0.2$, Universe dominated by non-baryonic matter (Bond and Efstathiou, 1984).

The primeval ripples in the microwave radiation were finally found by George Smoot and his colleagues in data collected by the COsmic Background Explorer satellite, or COBE for short. They indicate an angular temperature fluctuation, ΔT, given by (Smoot et al., 1992; Bennett et al., 1993, 1996)

$$\Delta T/T_o \approx (1.1 \pm 0.1) \times 10^{-5} \text{ for angular sizes } \theta \geq 7 \text{ degrees} \tag{5.917}$$

where $T_o = 2.726 \pm 0.010$ K. Such a temperature ripple is incredibly significant to formation theories, establishing evidence for the primeval density fluctuations that seeded all the structure in the Universe. The discovery of the long-sought "wrinkles in the fabric of space-time" certainly exited the public imagination; no less than five popular books exploring its implications and significance were published within a year of the discovery, including Smoot's own volume (Gribbin, 1993; Parker, 1993; Rowan-Robinson, 1993; Smoot and Davidson, 1993; Mather and Boslough, 1996).

Other experiments from balloons (Bernardis et al., 1994) and from the South Pole (Gundersen et al., 1995) have confirmed the result, some of them detecting anisotropy in the background radiation with angular scales of 1.5 to 1.8 degrees and comparable, or slightly larger, fluctuations in temperature (also see Scott, Silk and White, 1995 and Gawiser and Silk, 1998). Preliminary measurements of the cosmic background radiation with high angular resolution reveal finer details and may suggest a flat Universe; these balloon flights of the QMAD instrument are summarized by De Oliveira-Costa et al. (1998).

Although angular temperature fluctuations establish the template out of which the material Universe grew, the observations are not yet fine enough to directly connect to the structures we see today. That is, thanks to the expansion of the Universe and the resolution limit of COBE, the satellite could not directly probe the structures that eventually led to the galaxies, clusters of galaxies, walls and voids. With an angular resolution of 7 degrees, the smallest irregularities detected by COBE now spans a region larger than the Great Wall, and we need angular resolutions of minutes of arc to see those that led to clusters of galaxies.

In addition, there are several effects that can cause temperature fluctuations in the cosmic background radiation as a function of position on the sky (White, Scott and Silk, 1994). Both gravity and motion can produce changes, $\Delta \nu$, in the frequency, ν, that can result in observed temperature fluctuations of:

$$\Delta T/T_{\mathrm{o}} \approx \Delta \nu/\nu \ . \tag{5.918}$$

The radiation we receive has to move out of any gravitational potential, $\Delta\phi$, resulting from a mass fluctuation, ΔM, producing a gravitational redshift, $\Delta \nu$ in the frequency, ν, and the temperature fluctuation (Sachs and Wolfe, 1967):

$$\left(\frac{\Delta T}{T_{\mathrm{o}}}\right)_{\mathrm{G}} \approx \frac{\Delta \nu}{\nu} \approx \frac{\Delta\phi}{c^2} \approx \frac{G\Delta M}{Lc^2} \tag{5.919}$$

where $\Delta\phi$ is the change in Newtonian gravitational potential caused by a density fluctuation $\Delta\rho$ with mass change ΔM over a linear scale L. This Sachs–Wolfe effect dominates the temperature fluctuations for large linear scales

$$L > (ct)_{\mathrm{rec}}, \tag{5.920}$$

where $t_{\mathrm{rec}} \approx 4.4 \times 10^{12}(\Omega_{\mathrm{o}}h^2)^{-1/2}$, and

$$\left(\frac{\Delta T}{T_{\mathrm{o}}}\right)_{\mathrm{G}} \approx \left(\frac{\Delta\rho}{\rho_{\mathrm{o}}}\right)_{\mathrm{rec}} \left(\frac{L}{ct}\right)^2_{\mathrm{rec}} \tag{5.921}$$

for a density fluctuation $(\Delta\rho/\rho)_{\mathrm{rec}}$ at the recombination epoch.

Local motion produces the dipole anisotropy, discussed in Sect. 5.7.5, that must be removed from the observed data to measure temperature fluctuations at recombination. Perturbations of velocity, v, back then produce an additional temperature fluctuation:

$$\left(\frac{\Delta T}{T_{\mathrm{o}}}\right)_{v} \approx \frac{v}{c} \approx \frac{1}{c}\left(\frac{\Delta\rho}{\rho_{\mathrm{o}}}\right)_{\mathrm{rec}} \left(\frac{L}{ct_{\mathrm{rec}}}\right) \ . \tag{5.922}$$

Small-scale fluctuations of the relic radiation will arise from scattering of radiation on moving electrons. For an electron temperature, T_{e}, at the recombination time we have (Zeldovich and Sunyaev, 1969; Sunyaev and Zeldovich 1970, 1972);

$$\left(\frac{\Delta T}{T_{\mathrm{o}}}\right)_{\mathrm{rec}} = \frac{-2kT_e}{m_e c^2} \ , \tag{5.923}$$

where m_{e} is the electron mass and k is Boltzmann's constant. A similar angular temperature depression, called the Sunyaev–Zeldovich effect, is caused by the collisions of microwave photons with million-degree intergalactic electrons in clusters of galaxies (see Sect. 5.7.5).

Another source of temperature change is called isocurvature fluctuation in which particles perturb the specific entropy, ΔS, to produce (Peebles, 1987; Cen, Ostriker and Peebles, 1993; Peebles, 1993)

$$\left(\frac{\Delta T}{T_{\mathrm{o}}}\right) = -\frac{1}{3}\Delta S \propto \frac{\Delta\rho_B}{\rho_B} - \frac{3}{4}\frac{\Delta\rho_\gamma}{\rho_\gamma} \tag{5.924}$$

where ρ_{B} and ρ_γ respectively denote the baryon mass density and the radiation density.

While the detailed answers to galaxy formation probably lie on smaller angular scales then those observed so far, the existing data provides information on the spectrum of the primordial inhomogeneities (Wright et al., 1992) including an estimate of their index $n = 1.22 \pm 0.31$ (Gorski et al., 1994; also see Sect. 5.7.5). The observed temperature fluctuations are consistent, within the observational errors, with scale-invariant primordial density fluctuations that are independent of angular scale with an index n of unity. That is, COBE finds approximately the same number of irregularities of all sizes within the range that it measures (angular sizes larger than 7 degrees).

Still, due to the proliferation of theoretical parameters, we cannot arrive at other definitive conclusions about the growth of large-scale structures using the observed anisotropies in the cosmic microwave background radiation (Efstathiou, Bond and White, 1992; White, Scott and Silk, 1994; and Scott, Silk and White, 1995). Relatively recent proposals involve vast quantities of non-baryonic matter that might reduce the observed $\Delta T/T_0$ below that which would be obtained from purely baryonic matter (Gawiser and Silk, 1998). This therefore brings us to the more speculative aspects of the formation of structure in the Universe.

5.7.6.4 Formation of Structure in the Universe

We know that the expanding Universe was propelled into being from an explosive big bang, and that it was exceedingly hot, dense and uniform in its early life. Yet, cooled and thinned out by its expansion, the relatively cold and largely empty Universe now consists of stars and galaxies grouped into larger structures. How did this transformation occur? How did the Universe change from then to now, from smooth to lumpy? How did the galaxies come to be, and what arranged them into clusters and superclusters, separated by immense voids? Readable discussions of these fundamental questions are provided by Longair (1996) and Silk (1994).

There are two approaches to the formation problem. In one method, astronomers use powerful telescopes to peer ever farther into the shadow land of the dim and distant past, to watch galaxies evolve and perhaps catch a glimpse of their formation. They can now look back in time to before the Earth and Sun were born! In the other approach, one starts at the beginning and tries to reconstruct how the Universe evolved, matching the computations with the observed fluctuations in the background radiation and the large-scale structure of galaxies. In the simplest theories, an initial unevenness, caused by instabilities early in the big bang, could have been greatly magnified in the expansion of the Universe since then.

To see galaxies in their youth, astronomers use their telescopes as time machines, probing epochs when the Universe was only a fraction of its present age. Because of the time it takes light from distant galaxies to reach us, we see them as they were in relative youth. Traveling across the Universe, at the finite velocity of light for billions of years, the light from even the brightest of galaxies becomes so dimmed that only the most powerful telescopes can see it. By using the Hubble Space Telescope and the Keck Telescope, astronomers have reached out nearly to the edge of the visible Universe, pushing further and further into the no-man's land between smoothness and structure.

The time line of cosmic history is ordered by the redshift, z, that tells us about both distance and age. At any particular redshift, the Universe was a fraction $1/(1+z)$ of its present size. The relationship between age and redshift is more complex, depending on the mass density of the Universe, but we can use the redshift to establish an upper limit to age. The Universe at redshift, z, was at most $1/(1+z)$ times its present age, which is somewhere between 10 and 20 billion years old. Thus, at a redshift of 1, the Universe was only half as big as it is now and at most half as old as it is today.

The Universe is changing as it ages. Though seemingly immutable, the galaxies change and evolve over the eons of cosmic time. Like children, galaxies are more active in their youth. This is reflected by an apparent increase in star-formation rate in galaxies viewed at earlier times. Already in 1978, Harvey Butcher and Augustus Oemler showed that distant clusters of galaxies, at a redshift of $z \approx 0.4$, have a much larger percentage of blue galaxies than do their nearby present-day counterparts, indicating that the galaxies in clusters have changed dramatically in just the last 5 billion years or so (Butcher and Oemler, 1978, 1984). Subsequent research has shown that these blue spiral galaxies are undergoing vigorous star formation, often occurring in intense episodic bursts. A blue color is characteristic of massive, new-born stars that burn brightly for a short time of about 10 million years.

By taking a long, penetrating look at part of the sky, down to the 27 magnitude, Anthony Tyson (Tyson, 1988), and independently Richard Ellis, discovered a population of faint blue galaxies (Ellis, 1997). Moreover, stars in distant red galaxies were clearly mature when they emitted light we now observe, evidently having formed much earlier at some higher redshift, in some cases at $z > 4$ (Dunlop et al., 1996).

Deep imaging with the Hubble Space Telescope has opened a gateway to at least halfway back to the big bang, showing that galaxies formed early in the Universe and evolved dramatically since then (Macchetto and Dickinson, 1997). At redshifts of $z \approx 1$, elliptical galaxies are found resembling those seen nearby, suggesting that they took shape less than a few billion years after the big bang. Spirals formed early too, often with irregular distorted shapes, perhaps because the Universe was smaller and galaxies were closer together back then. In any event, the Universe of galaxies seems to have been fully formed out at a redshift of one.

Grown-up, well-developed galaxies are found at even larger redshifts, between 2.0 and 3.8, when the Universe was in its infancy and less than 20 percent of its present age. These galaxies have strong blue or ultraviolet light, but at great distances ultraviolet wavelengths are absorbed by hydrogen in the young galaxies and intervening space. So they are located by a cutoff in blue light, compared to red and green, and their distances then confirmed by spectroscopic analysis of their light using the Keck telescope (Steidel et al., 1995, 1996). They indicate that a substantial population of galaxies was already present when the Universe was less than a few billion years old.

controversy. This is the domain of high-energy particle physics where theories come and go, like the women talking of Michelangelo, with alarming speed.

One theory imagines topological defects or flaws in the structure of space-time itself, left over when a favored unification concept lost its symmetry and broke down. First proposed by Thomas Kibble, the cosmic flaws originated during a vacuum phase transition in the early Universe (Kibble, 1976). A familiar example of a phase transition is water freezing into ice; the white lines and planes in a ice cube are the localized defects. However, the topological defects in space-time are speculations not observations, and might conceivably be mental defects in the imagination of their originators.

One hypothetical defect is a thin line of concentrated energy, called a cosmic string. These strings have only one dimension and run through space without an end, so they are either infinitely long or loop into circles. Yakov Zeldovich applied string theory to galaxy formation, followed by Alexander Vilenkin (Zeldovich, 1980; Vilenkin, 1981, 1985). However, it was later shown that strings cannot produce the observed large-scale structures.

Another topological defect, called texture, sprang up to take the place of strings (Turok, 1989, 1991; Spergel and Turok, 1992). Texture models produce density fluctuations in the form of strong localized lumps. Supercomputers can be used to trace these irregularities through time, mimicking the structures that we observe today (Cen et al., 1991). Such numerical extrapolations also involve speculative, invisible particles that clump around the defects, and eventually gather perfectly ordinary matter around them. Such an idea was introduced a decade earlier (Peebles, 1982).

The second major problem for formation theories is related to the extraordinary smoothness of the cosmic microwave background radiation combined with the roughness of the present-day structures. Since gravitational instability growth is roughly linear with time, we expect the fractional density fluctuations to scale as $\Delta\rho/\rho_{\rm o} \propto 1/z$. Today with redshift $z = 0$, we have $\Delta\rho/\rho_{\rm o} = 1$, and back at the recombination time when the background fluctuations originated we had $z \approx 1,000$ and expect $\Delta\rho/\rho_{\rm o} \approx 10^{-3}$. The difficulty is that this is more than an order of magnitude larger than the observed value.

That is why some cosmologists propose the existence of dark matter that would have decoupled from the radiation much earlier than the gas did. Being unaffected by the radiation, the invisible matter could have begun clustering long before the ordinary kind. It helps save the phenomenon by extending the time available for formation and by making sure there was plenty of matter around when the Universe had cooled and thinned out enough for galaxies to form.

5.7.6.5 Using Invisible Matter to Force Formation

A crucial question for cosmologists is how the dramatic, large-scale features in the distribution and motions of galaxies could have evolved from an isotropic, homogeneous sea of radiation and matter. Small perturbations in density will grow in size as the Universe expands; but chance fluctuations, followed by gravitational attraction in an expanding Universe, may not conspire to create such large structures and empty places. There does not, for example, seem to be enough matter for gravity to create structures as large as the Great Wall out of density fluctuations that are consistent with the cosmic background temperature fluctuations. Some type of dark, unseen force seems to have molded the galaxies into their visible form.

The problem is relieved if dark matter rules the cosmos, providing the bulk of the mass of the Universe and creating invisible tracks along which the luminous galaxies form. The extra unseen mass could provide enough gravitational power to hold the galaxies, clusters of galaxies and superclusters together in a constantly expanding Universe.

In addition, because density fluctuations of dark matter might grow before their ordinary material counterparts, more time might be available to complete formation. If there was only ordinary baryonic matter, the growth of the primeval density perturbations could not have begun until matter was freed from the drag and ionizing power of the radiation, a few hundred thousand years after the big bang (Peebles, 1965). However, dark non-baryonic matter perturbations might have begun to grow at an earlier time.

With invisible dark matter, the process could begin as soon as the Universe became matter dominated, a few thousand years after the big bang. Because density inhomogeneities could have started growing sooner in the dark scenario, their initial amplitude could have been smaller, leading to smaller predicted variations in the cosmic microwave background radiation, in accordance with observations.

If the pundits are right, then perfectly ordinary matter was pulled into its present-day shape by a much greater amount of invisible matter. The glowing walls of galaxies seen through our telescopes could then be minor ingredients of the cosmos, and the voids might not then be empty but instead contain matter that has failed to become luminous. It is as if tiny, far-away ripples in the cosmic seas of long ago grew into a tidal wave, and all we can see is the white caps on a dark and stormy sea.

As long as you are imagining something that no one has ever seen, there might as well be plenty of it. Indeed, a number of cosmologists speculate that the Universe harbors up to a hundred times more dark matter than luminous material. The inward pull of all this invisible stuff would then be right on the edge of quelling the expansion of the Universe in the future, poised on the very brink of closure. In other words, they threw their hat into the inflationary ring, advocating a mass density, ρ_o, just equal to the critical value, ρ_c, so the density parameter is exactly unity, or $\Omega_o = 1.000$, where $\Omega_o = \rho_o/\rho_c = 8\pi G\rho_o/(3H_o^2)$ for a Hubble constant H_o.

To further shore up their speculations, the new cosmologists have used supercomputers to show how the gravity of their favorite kind of dark matter caused matter to condense in the aftermath of the big bang. The computer-generated portraits, or simulations, follow the shapes of ordinary matter as it grows and evolves over time, creating galaxies and larger structures under the dictates of gravity and expansion.

Where does the invisible substance required to form the galaxies come from? It is supposed to consist of swarms of exotic elementary particles left over from the big bang. All that dark stuff could have come out of the big bang moving very fast, and since higher temperatures usually go with larger velocities that kind is called hot dark matter. In contrast, the subatomic particles that make up cold dark matter come out of the big bang moving very slowly. Hypothetical examples are the neutralinos that do not move very fast because they are heavy, and axions that move slow because they were born with very low momentum.

We have long known from big-bang nucleosynthesis that there is not enough ordinary baryonic matter to fill out the Universe to the brink of closure (Reeves, Fowler and Hoyle, 1970, also see Sect. 5.7.4), so the hypothetical hot or cold particles must be non-baryonic if the Universe is truly at its critical density. (Protons and neutrons are baryons that make up ordinary matter, while non-baryonic matter has not yet been observed.) That is, these scenarios have also incorporated non-baryonic matter to fill out the Universe to a density parameter of $\Omega_0 = 1$. In their efforts to devise a grand unified theory of everything, particle physicists have invented a bevy of such hypothetical particles to mediate the interaction between forces. Both the hot and cold candidates for the hidden stuff are summarized in Sect. 5.4.6, so we will next focus on the two formation theories. The reader should nevertheless be forewarned that both models are speculative, with sufficient adjustable parameters to give an observer a permanent headache.

The leading dark matter candidate early on was the hot, fast-moving neutrino, a non-baryonic particle with no electric charge and little mass, traveling at or very near the velocity of light. These subatomic particles are so insubstantial, and interact so weakly with matter, that they streak through nearly everything in their path, like ghosts that move right through walls.

The elusive neutrino might have a tiny yet discernible mass. Because of the enormous numbers of neutrinos produced as the result of the big bang, they could control formation in an expanding Universe. Indeed, Ramanath Cowsik and his colleagues pointed out some time ago that neutrinos might have interesting cosmological implications (Cowsik and Mc Clelland, 1973; Cowsik and Ghosh, 1987). The relic neutrinos could eventually stop the expansion of the Universe if they have a mass of $m_\nu \approx \rho_c/n_\nu \approx 10^{-31}$ grams $=$ 100 eV, where the critical mass density needed to close the Universe is $\rho_c = 3H_0^2/(8\pi G)$ and n_ν is the number density of neutrinos. Since there are three types of neutrinos, this mass could be about 30 eV. The Super Kamiokande neutrino detector has obtained suggestions that muon neutrinos, generated in the Earth's atmosphere by cosmic rays, may undergo neutrino oscillations that would require a small neutrino rest mass (Fukuda et al., 1998, also see Volume I); but the mass (energy) is estimated at only about 0.05 eV, which would make little difference to cosmological closure or structure formation.

Moving at nearly the speed of light, neutrinos quickly cover a lot of territory, escaping from small regions of space and moving into larger ones. Any small-scale, primeval ripples are therefore wiped out and erased by the fast-moving neutrinos, as they stream out from regions of high density to lower density ones. Only the largest undulations are left to give ordinary matter its form, creating enormous structures early in the history of the Universe.

As argued by Lemaitre (1934), before the ghostlike neutrinos or dark matter were ever imagined, the largest mass concentrations might have been assembled at high redshift when the density of the expanding Universe was high, and smaller, less dense structures could have subsequently formed as the expansion continued and the Universe became more rarefied.

Yakov Zeldovich and his colleagues suggested one of the earliest hot dark matter models for the formation of structure in the Universe (Zeldovich, 1970, 1978; Bond and Szalay, 1983, 1984; Shandarin and Zeldovich, 1989). He called it the "pancake model"; others referred to it as the adiabatic theory. In this model, huge flat clouds of gas were the first material entities to emerge, spanning tens of millions of light years and containing about 10^{16} solar masses, comparable in shape, extent and mass to known superclusters of galaxies. In such a top-down cosmology, the largest pancake structures formed first, and smaller galaxies were created later as turbulence and shock waves fragmented the larger clouds into smaller pieces. Since this theory of galaxy formation predicts the formation of flattened, nonspherical superclusters and giant empty voids between them, it was initially heralded as the rosetta stone for creation (Zeldovich, Einasto and Shandarin, 1982; Silk, Szalay and Zeldovich, 1983).

The power spectrum of the density fluctuations in the hot dark matter model is given by Bond and Szalay (1983) and Kolb and Turner (1990):

$$\begin{aligned} |\delta_{\mathrm{k}}|^2 &= Ak^{1+6\alpha}10^{-2(k/k_\nu)^{1.5}} \\ &= Ak^{1+6\alpha}\exp\left[-4.61(k/k_\nu)^{1.5}\right] , \end{aligned} \tag{5.957}$$

where δk is the Fourier transform of the density contrast $\delta = \Delta\rho/\rho_0$ about the mean density ρ_0, the neutrino damping scale, or wavenumber, $k_\nu = 0.16(m_\nu/30\,\mathrm{eV})$ Mpc corresponding to a wavelength $\lambda_\nu = 40(m_\nu/30\,\mathrm{eV})^{-1}$ Mpc, the neutrino mass is m_ν, the normalization is A, and at the horizon $\delta = \Delta\rho/\rho_0 \propto M^{-\alpha}$.

Numerical simulations of the hot dark matter model can be adjusted to mimic reality, describing the shape and structure of bubbles destined to become voids and compressed regions that grow into thin intersecting sheets (Zeldovich, Einasto and Shandarin, 1982; Silk, Szalay and Zeldovich, 1983). Computer simulations based on the growth of adiabatic perturbations with neutrino masses corresponding to 10 eV to 100 eV produce superclusters with mass and dimensions comparable to those observed (Centrella and Melott, 1983).

Although the computed structures look similar to the observed large-scale ones, there are problems with the hot dark matter model. The simulations also indicate that the fast-moving neutrinos take too long to settle down into galaxies (White, Frenk and Davis, 1983), with an epoch of "pancaking" that happens relatively recently at low redshifts of $z \leq 1$. The formation of superclusters and their fragmentation down to the sizes of galaxies would take an appreciable fraction of the age of the Universe and galaxies would have only recently formed. This is difficult to reconcile with many galaxies of greater redshifts $z \geq 1$, as well as with quasars seen with $z \geq 4.5$, making it difficult to explain the existence of high-redshift galaxies that formed long ago. Moreover, our Galaxy is older than the local group of galaxies now coalescing around it, and the local sheet of galaxies is apparently forming from pre-existing ones (Peebles 1984, 1993), suggesting a bottom-up rather than a top-down scenario.

When hot dark matter became less popular, theorists shifted attention to the slow-moving cold variety. While the fast, relativistic neutrinos can only forge big structures first, cold dark matter, being more sluggish, initially constructs smaller ones. In this bottom-up process, structure is formed hierarchically, with tiny units blending into ever larger assemblies. Galaxies formed first, then gravitationally merged and consolidated into clusters, and later, superclusters as the Universe expanded and evolved. Unlike the hot case, galaxies form before their clusters, and there is no problem with the early formation of high-redshift galaxies and quasars.

The power spectrum of the scale-invariant fluctuations have been computed by Peebles (1982), Bond and Efstathiou (1984), and Davis et al. (1985). Numerical calculations indicate that the power spectral density, $P(k)$, at wavenumber, k, is given by (Efstathiou, Bond and White, 1992; Strauss and Willick, 1995):

$$P(k) = \frac{Ak^n}{\left\{1 + \left[\left(\frac{6.4k}{\Gamma}\right) + \left(\frac{3.0k}{\Gamma}\right)^{3/2} + \left(\frac{1.7k}{\Gamma}\right)^2\right]^{1.13}\right\}^{2/1.13}} , \tag{5.958}$$

where A is a normalizing constant, k is measured in units of h Mpc^{-1}, $\Gamma = \Omega_o h$, and n is the power-law index of the primordial power spectrum. The quantity h is the Hubble constant in units of $100\,\mathrm{km\,s^{-1}\,Mpc^{-1}}$. Standard cold dark matter sets $\Gamma = 0.5$ but a $\Gamma = 2$ provides a better comparison with observations and models with smaller Γ have more power on large scales. The quantity Γ is inversely proportional to the Hubble radius at matter-radiation equality, and thus sets the scale at which the power spectrum bends over. This equation (and the equality between Γ and $\Omega_o h$) holds only in a Universe with a negligible contribution of baryons to Ω_o. The normalization for the case of $n = 1$ and $\Omega_o = 1$ (standard case) can be written (Bunn, Scott, and White, 1995; White, Scott and Silk, 1994):

$$A = \frac{6\pi^2}{5} \frac{\langle Q \rangle^2}{T_o^2} \eta_o^4 = 1.45 \times 10^{16} \frac{\langle Q \rangle^2}{T_o^2} \left(h^{-1}\,\mathrm{Mpc}\right)^4 , \tag{5.959}$$

where η_o is the conformal time at the present, $\langle Q \rangle$ is the measured quadrupole anisotropy of the cosmic microwave background, and T_o is its measured temperature.

Such a model results in a plausible picture of galaxy formation (Blumenthal et al., 1984). Moreover, computer simulations based on this scenario seemed to reproduce observations of the distribution of galaxies very well (Davis, Efstathiou, Frenk and White, 1985; Frenk, White, Efstathiou and Davis, 1985; White, Frenk, Davis and Efstathiou, 1987; White, Davis, Efstathiou and Frenk, 1987; Park (1990), Weinberg and Gunn, 1990).

The cold dark matter model has some difficulties. It ran into trouble with the larger structures, like the Great Wall. Using cold matter near the critical density $\Omega_o = 1.0$, the model produces too many clumps of small masses, between 10^8 and 10^{10} solar mass, and not enough ones of larger mass. The computer-generated Universes made a close connection with reality with a lower density parameter $\Omega_o = 0.2$, in violation of the initial premise for inflation.

To get around some of these difficulties, the theoreticians added another adjustable parameter, a sort of fudge factor called biasing, that could enhance the large-scale structure and get rid of some of the smaller ones. To obtain a more favorable comparison with reality, they assumed that the luminous galaxies represent a biased, or inaccurate picture of the distribution of matter. To some, a biased view in this context is a narrow-minded one. Since dark matter rules, there might be a bias that effects the density fluctuation, and if so the invisible stuff is located outside places where the visible galaxies congregate. After all, the hypothetical dark matter will interact weakly with ordinary matter and radiation, or it would have been detected by now, and the apparent emptiness of the voids could be filled with dark stuff we cannot see.

In this adaptation, there is a bias that affects the density fluctuation. For a linear biasing model, the density fluctuations of ordinary matter, δ_{galaxies}, are defined to be a factor of b greater than that of dark matter, $\delta_{\text{dark matter}}$, so that

$$\delta_{\text{galaxies}} = b\delta_{\text{dark matter}} \quad . \tag{5.960}$$

A value for the biasing parameter, b, of $b = 2$ to 3 is, for example, inferred by comparing observed galaxy clusters with the amplitude of primordial fluctuations in the cold dark matter cosmogony (Frenk, White, Efstathiou and Davis, 1990).

To the extent that the linear biasing model holds, comparisons of observed data (peculiar velocities and gravity) are used to constrain the quantity β given by (Strauss and Willick, 1995)

$$\beta = \frac{f(\Omega_o, \Lambda)}{b} \approx \frac{\Omega^{0.6}}{b} \quad , \tag{5.961}$$

where (Lahav et al., 1991; Strauss and Willick, 1995)

$$f(\Omega_o, \Lambda) = \Omega_o^{0.6} + \frac{\Omega_\Lambda}{70}\left(1 + \frac{1}{2}\Omega_o\right) \tag{5.962}$$

$$\Omega_\Lambda = \Lambda/(3H_o)^2 \quad , \tag{5.963}$$

and the influence of the cosmological constant Λ on the dynamics at low redshift is minimal.

The current consensus is that biasing is relatively weak, with (Strauss and Willick, 1995)

$$\beta^{5/3} < \Omega_o < 2\beta^{5/3} \quad , \tag{5.964}$$

for $\beta = \Omega_o^{0.6}/b$. Observational constraints on both β and Ω_o are reviewed in table form by Strauss and Willick (1995), concluding that the range $0.3 \leq \Omega_o \leq 1$ is still possible, but with a preference the lower value of

References

Aaronson, M., et al. (1980): A distance scale from the infrared magnitude – H I velocity-width relation III. the expansion rate outside the local supercluster. Ap. J. **239**, 12.

Aaronson, M., et al. (1982): The velocity field in the Local Supercluster. Ap. J. **258**, 64.

Aaronson, M., et al. (1986): A distance scale from the infrared magnitude – H I velocity-width relation V. distance moduli to 10 galaxy clusters, and positive detection of bulk supercluster motion toward the microwave anisotropy. Ap. J. **302**, 536.

Aaronson, M., Mould, J. (1983): A distance scale from the infrared magnitude/HI velocity width relation IV. The morphological type dependence and scatter in the relation. The distances to nearby groups. Ap. J. **265**, 1.

Aaronson, M., Mould, J., Huchra, J. (1980): A distance scale from the infrared magnitude/HI velocity width relation I. The calibration. Ap. J. **237**, 655.

Abell, G. O. (1958): The distribution of rich clusters of galaxies. Ap. J. Supp. **3**, 211.

Abell, G. O. (1961): Evidence regarding second-order clustering of galaxies and interactions between clusters of galaxies. Astron. J. **66**, 607.

Abell, G. O. (1965): Clustering of galaxies. Ann. Rev. Astron. Astrophys. **3**, 1.

Abell, G. O., Margon, B. (1979): A kinematic model for SS 433. Nature **279**, 702.

Abramovici, A., et al. (1992): LIGO: The laser interferometer gravitational-wave observatory. Science **256**, 325.

Adams, J. C. (1853): On the secular variation of the moon's mean motion. Phil. Trans. **143**, 397.

Adams, J. C. (1860): Reply to various objections which have been brought against his theory of the secular acceleration of the moon's mean motion. M.N.R.A.S. **20**, 225.

Adams, W. S. (1914): An A-type star of very low luminosity. P.A.S.P. **26**, 198. Reproduced in: A source book in astronomy and astrophysics 1900–1975 (eds. K. R. Lang and O. Gingerich). Cambridge, Mass.: Harvard University Press 1979.

Adams, W. S. (1915): The spectrum of the companion of Sirius. P.A.S.P. **27**, 236. Reproduced in: A source book in astronomy and astrophysics 1900–1975 (eds. K. R. Lang and O. Gingerich). Cambridge, Mass.: Harvard University Press 1979.

Adams, W. S. (1925): The relativity displacement of the spectral lines in the companion of Sirius. Proc. Nat. Acad Sci. **11**, 382.

Adams, W. S. (1941): Some results with the Coude spectrograph of the mount wilson observatory. Ap. J. **93**, 11.

Adelberger, E. G., et al. (1990): Testing the equivalence principle in the field of the earth – particle physics at masses below one microelectronvolt. Phys. Rev. **D42**, 3267.

Albrecht, A., Steinhardt, P. J. (1982): Cosmology for grand unified theories with radiatively induced symmetry breaking. Phys. Rev. Lett. **48**, 1220.

Alcaino, G., Liller, W. (1988): The ages of globular clusters derived from BVRI CCD photometry. In: The Harlow Shapley symposium on globular cluster systems in galaxies (ed. J. E. Grindlay and A. G. Davis Philip). Boston: Kluwer Academic Pub..

Alcock, C. (1997): Searching for MACHOs with microlensing. In: Unsolved problems in astrophysics (eds. J. N. Bahcall and J. P. Ostriker). Princeton, New Jersey: Princeton University Press.

Alcock, C., et al. (1993): Possible gravitational microlensing of a star in the Large Magellanic Cloud. Nature **365**, 621.

Alcock, C., et al. (1995): Probable gravitational microlensing toward the galactic bulge. Ap. J. **445**, 133.

Alcock, C., et al. (1995): Experimental limits on the dark matter halo of the Galaxy from gravitational microlensing. Phys. Rev. Lett. **74**, 2867.

Alcock, C., et al. (1996): The MACHO project first-year Large Magellanic Cloud results: The microlensing rate and the nature of the galactic dark halo. Ap. J. **461**, 84.

Alcock, C., et al. (1996): The MACHO project: Limits on planetary mass dark matter in the galactic halo from gravitational microlensing. Ap. J. **471**, 774.

Alcock, C., et al. (1998): EROS and MACHO combined limits on planetary-mass dark matter in the galactic halo. Ap. J. (Letters) **499**, L9.

Aleph collaboration (1989): Determination of the number of light neutrino species. Physics Letters **B231**, 519.

Alfvén, H., Herlofson, N. (1950): Cosmic radiation and radio stars. Phys. Rev. **78**, 616. Reproduced in: A source book in astronomy and astrophysics 1900–1975 (eds. K. R. Lang and O. Gingerich). Cambridge, Mass.: Harvard University Press 1979.

Allen, C. W.: Astrophysical quantities. University of London, London: Athlone Press 1963.

Alpar, M. A., Cheng, A. F., Ruderman, M. A., Shaham, J. (1982): A new class of radio pulsars. Nature **300**, 728.

Alpher, R. A., Bethe, H., Gamow, G. (1948): The origin of the chemical elements. Phys. Rev. **73**, 803. Reproduced in: A source book in astronomy and astrophysics 1900-1975 (eds. K. R. Lang and O. Gingerich). Cambridge, Mass: Harvard University Press 1979.

Alpher, R. A., Follin, J. W., Herman, R. C. (1953): Physical conditions in the initial stages of the expanding universe. Phys. Rev. **92**, 1347.

Alpher, R. A., Herman, R. C. (1948): Evolution of the universe. Nature **162**, 774. Reproduced in: A source book in astronomy and astrophysics 1900–1975 (eds. K. R. Lang and O. Gingerich). Cambridge, Mass.: Harvard University Press 1979.

Alpher, R. A., Herman, R. C. (1949): Remarks on the evolution of the expanding universe. Phys. Rev. **75**, 1089.

Alpher, R. A., Herman, R. C. (1950): Theory of the origin and relative abundance distribution of the elements. Rev. Mod. Phys. **22**, 153.

Amaral, L. H., et al. (1996): The rotation curve of the Galaxy obtained from planetary nebulae and AGB stars. M.N.R.A.S. **281**, 339.

Ambartsumian, V. A. (1938): On the dynamics of open clusters. Ann. Leningrad State U. Astr. Series Issue **4**, No. 22. Translated in: Dynamics of star clusters – IAU symposium no. 113 (eds. J. Goodman and P. Hut). Dordrecht: Reidel 1985.

Ambartsumian, V. A.: On the evolution of galaxies. In: La structure et l'evolution de l'univers. Institut international de physique solvay (ed. R. Stoops). Brussels: Coudenberg 1958. Reproduced in: A source book in astronomy and astrophysics 1900-1975 (eds. K. R. Lang and O. Gingerich). Cambridge, Mass.: Harvard University Press 1979.

Anders, E. (1962): Meteorite ages. Rev. Mod. Phys. **34**, 287.

Anders, E. (1964): Origin, age and composition of meteorites. Space Sci. Rev. **3**, 583.

Anders, E., Grevesse, N. (1989): Abundances of the elements – meteoritic and solar. Geochim. cosmochim. acta **53**, 197. Partially reproduced in K. R. Lang, Astrophysical data. New York: Springer-Verlag 1992.

Anderson, J. (1991): Accurate masses and radii of normal stars. Astron. Astrophys. Rev. **3**, 91.

Anderson, J. D., et al. (1975): Experimental test of general relativity using time-delay data from Mariner 6 and Mariner 7. Ap. J. **200**, 221.

Anderson, J. D., et al. (1978): Tests of general relativity using astrometric and radio metric observations of the planets. Acta Astronautica **5**, 43.

Anderson, W. (1929): Über die Grenzdichte der Materie und der Energie (About the interface of matter and energy). Zeits. f. Phys. **56**, 851.

Angel, J. R. P. (1977): Magnetism in white dwarfs. Ap. J. **216**, 1.

Angel, J. R. P., Stockman, H. S. (1980): Optical and infrared polarization of active extragalactic objects. Ann. Rev. Astron. Astrophys. **18**, 165.

Anninos, P., et al. (1993): Collison of two black holes. Phys. Rev. Lett. **71**, 2851.
Anthony-Twarog, B. J., Twarog, B. A. (1985): Faint stellar photometry in clusters II. NGC 6791 and NGC 6535. Ap. J. **291**, 595.
Antonucci, R. (1993): Unified models for active galactic nuclei and quasars. Ann. Rev. Astron. Astrophys. **31**, 473.
Aoki, S., et al. (1982): The new definition of universal time. Astron. Astrophys. **105**, 359.
Aoki, S., Soma, M., Kinoshita, H., Inoue, K. (1983): Conversion matrix of epoch B1950.0 FK4-based positions of stars to epoch J2000.0 positions in accordance with new IAU resolutions. Astron. Astrophys. **128**, 263.
Araki, S., Lightman, A. P. (1983): Relativistic thermal plasmas: Effects of magnetic fields. Ap. J. **269**, 49.
Arnett, W. D. (1966): Gravitational collapse and weak interactions. Can. J. Phys. **44**, 2553.
Arnett, W. D. (1967): Mass dependence in gravitational collapse of stellar cores. Can. J. Phys. **45**, 1621.
Arnett, W. D. (1968): On supernova hydrodynamics. Ap. J. **153**, 341.
Arnett, W.D., Bahcall, J. N., Kirshner, R. P., Woosley, S. E. (1989): Supernova 1987A. Ann. Rev. Astron. Astrophys. **27**, 629.
Arnett, W. D., Branch, D., Wheeler, J. C. (1985): Hubble's constant and exploding carbon-oxygen white dwarf models for Type I supernovae. Nature **314**, 337.
Arnett, W. D., Cameron, A. G. W. (1967): Supernova hydrodynamics and nucleosynthesis. Can. J. Phys. **45**, 2953.
Arnould, M., Takahashi, K.: Inquietudes in nucleo-cosmochronology. In: Astrophyscial ages and dating methods (eds. Vangioni-Flam, E. Casse, M. Audouze J. and Tran Thanh Van J.). Paris: Editions Frontieres 1990.
Arp, H. (1962): The globular cluster M5. Ap. J. **135**, 311.
Ash, M. E., Shapiro, I. I., Smith, W. B. (1967): Astronomical constants and planetary ephemeris deduced from radar and optical observations. Astron. J. **72**, 338.
Ash, M. E., Shapiro, I. I., Smith, W. B. (1971): The system of planetary masses. Science **174**, 551.
Aston, F. W. (1929): The mass-spectrum of uranium lead and the atomic weight of protactinium. Nature **123**, 313.
Aubourg, E., et al. (1993): Evidence for gravitational microlensing by dark objects in the Galactic halo. Nature **365**, 623.
Aubourg, E., et al. (1995): Search for very low-mass objects in the Galactic halo. Astron. Astrophys. **301**, 1.
Audouze, J., Tinsley, B. M. (1974): Galactic evolution and the formation of the light elements. Ap. J. **192**, 487.
Audouze, J., Tinsley, B. M. (1976): Chemical evolution of galaxies. Ann. Rev. Astron. Astrophys. **14**, 43.
Ayres, D. S., et al. (1971): Measurements of the lifetimes of positive and negative pions. Phys. Rev. **D3**, 1051.
Baade, W. (1926): Über eine Möglichkeit, die Pulsationstheorie der delta Cephei-Veränderlichen zu prüfen (A possible proof of the pulsation theory of the variable delta-Cephei). Astron. Nach. **228**, 359.
Baade, W. (1944): The resolution of Messier 32, NGC 205, and the central region of the Andromeda nebula. Ap. J. **100**, 137. Reproduced in: A source book in astronomy and astrophysics 1900–1975 (eds. K. R. Lang and O. Gingerich). Cambridge, Mass.: Harvard University Press 1979.
Baade, W. (1952): A revision of the extra-galactic distance scale. Trans. I.A.U. **8**, 397. Reproduced in: A source book in astronomy and astrophysics 1900–1975 (eds. K. R. Lang and O. Gingerich). Cambridge, Mass.: Harvard University Press 1979.
Baade, W. (1956): Polarization in the jet of Messier 87. Ap. J. **123**, 550.
Baade, W., Minkowski, R. (1954): Identification of the radio sources in Cassiopeia, Cygnus A, and Puppis A. Ap. J. **119**, 206. Reproduced in: A source book in astronomy and

astrophysics 1900–1975 (eds. K. R. Lang and O. Gingerich). Cambridge, Mass.: Harvard University Press 1979.
Baade, W., Minkowski, R. (1954): On the identification of radio sources. Ap. J. **119**, 215.
Baade, W., Swope, H. H. (1955): The Palomar survey of variables in M 31. Astron. J. **60**, 151.
Baade, W., Swope, H. H. (1963): Variable star field 96' south preceding the nucleus of the Andromeda galaxy. Astron. J. **68**, 435.
Baade, W., Zwicky, F. (1934): On Super-Novae. Proc. Nat. Acad. Sci. **20**, 254. Reproduced in: A source book in astronomy and astrophysics 1900–1975 (eds. K. R. Lang and O. Gingerich). Cambridge, Mass.: Harvard University Press 1979.
Baade, W., Zwicky, F. (1934): Supernovae and cosmic rays. Phys. Rev. **45**, 138.
Babul, A., Katz, N. (1993): Does the baryon fraction in clusters imply an open universe? Ap. J. (Letters) **406**, L51.
Backer, D. C., et al. (1982): A millisecond pulsar. Nature **300**, 615.
Backer, D. C., Hellings, R. W. (1986): Pulsar timing and general relativity. Ann. Rev. Astron. Astrophys. **24**, 537.
Bahcall, J. N. (1978): Masses of neutron stars and black holes in X-ray binaries. Ann. Rev. Astron. Astrophys. **16**, 241.
Bahcall, J. N. (1984): K giants and the total amount of matter near the sun. Ap. J. **287**, 926.
Bahcall, J. N., Greenstein, J. L., Sargent, W. L. W. (1968): The absorption-line spectrum of the quasi-stellar radio source PKS 0237–23. Ap. J. **153**, 689.
Bahcall, J. N., Kirhakos, S., Schneider, D. P. (1994): HST images of nearby luminous quasars. Ap. J. (Letters) **435**, L11.
Bahcall, J. N., Salpeter, E. E. (1965): On the interaction of radiation from distant sources with the intervening medium. Ap. J. **142**, 1677.
Bahcall, J. N., Salpeter, E. E. (1970): Upper limits on the masses of quasi-stellar sources. Ap. J. (Letters) **159**, L135.
Bahcall, J. N., Schmidt, M., Soneira, R. M. (1983): The galactic spheroid. Ap. J. **265**, 730.
Bahcall, J. N., Tremaine, S. (1981): Methods for determining the masses of spherical systems I. test particles around a point mass. Ap. J. **244**, 805.
Bahcall, N. A. (1977): Clusters of galaxies. Ann. Rev. Astron. Astrophys. **15**, 505.
Bahcall, N. A.: Large-scale structure and motion traced by galaxy clusters. In: Large-scale motions in the universe (eds. V. C. Rubin and G. V. Coyne). Princeton, New Jersey: Princeton University Press 1988.
Bahcall, N. A. (1988): Large-scale structure in the universe indicated by galaxy clusters. Ann. Rev. Astron. Astrophys. **26**, 631.
Bahcall, N. A. (1993): Clusters, superclusters, and large-scale structure – a consistent picture. Proc. Nat. Acad. Sci. **90**, 4848.
Bahcall, N. A.: Large-scale structure in the universe. In: Unsolved problems in astrophysics (eds. J. N. Bahcall and J. P. Ostriker). Princeton, New Jersey: Princeton University Press 1997.
Bahcall, N. A., Cen, R. (1992): Galaxy clusters and cold dark matter: A low-density unbiased universe? Ap. J. (Letters) **398**, L81.
Bahcall, N. A., Lubin, L. M., Dorman, V. (1995): Where is the dark matter? Ap. J. (Letters) **447**, L81.
Bahcall, N. A., Soneira, R. M. (1983): The spatial correlation function of rich clusters of galaxies. Ap. J. **270**, 20.
Bahcall, N. A., West, M. J. (1992): The cluster correlation function: Consistent results from an automated survey. Ap. J. **392**, 419.
Baierlein, R. (1967): Testing general relativity with laser ranging to the moon. Phys. Rev. **162**, 1275.
Bailyn, C. D., et al. (1995): The optical counterpart of the superluminal source GRO J1655-40. Nature **374**, 701.
Bailyn, C. D., Orosz, J. A., Mc Clintock, J. E., Remillard, R. A. (1995): Dynamical evidence for a black hole in the eclipsing X-ray nova GRO J1655-40. Nature **378**, 157.

Baker, A. C., et al. (1994): Infrared spectroscopy of high-redshift quasars. M.N.R.A.S. **270**, 575.

Balbes, M. J., Boyd, R. N., Mathews, G. J. (1993): The primordial helium abundance as determined from chemical evolution of irregular galaxies. Ap. J. **418**, 229.

Balick, B., Heckman, T. M. (1982): Extranuclear clues to the origin and evolution of activity in galaxies. Ann. Rev. Astron. Astrophys. **20**, 431.

Balona, L. A. (1977): Application of the method of maximum likelihood to determination of cepheid radii. M.N.R.A.S. **178**, 231.

Bardas, D., et al., In: Cryogenic optical systems and instruments II (ed. R. K. Melugin). Bellingham, Washington: SPIE 1986.

Bardeen, J. M., Carter, B., Hawking, S. W. (1973): The four laws of black hole mechanics. Comm. Math. Phys. **31**, 161.

Bardeen, J. M., Petterson, J. A. (1975): The Lense-Thirring effect and accretion disks around Kerr black holes. Ap. J. (Letters) **195**, L65.

Bardeen, J. M., Press, W. H., Teukolsky, S. A. (1972): Rotating black holes: Locally nonrotating frames, energy extraction, and scalar synchrotron radiation. Ap. J. **178**, 347.

Bardeen, J. M., Steinhardt, P. J., Turner, M. S. (1983): Spontaneous creation of almost scale-free density perturbations in an inflationary universe. Phys. Rev. **D28**, 679.

Bardeen, J. M., Wagoner, R. V. (1969): Uniformly rotating disks in general relativity. Ap. J. (Letters) **158**, L65.

Bardeen, J. M., Wagoner, R. V. (1971): Relativistic disks I. Uniform rotation. Ap. J. **167**, 359.

Barnes, T. G., et al. (1977).: The distances of cepheid variables. M.N.R.A.S. **178**, 661

Barnes, T. G., Evans, D. S. (1976): Stellar angular diameters and visual surface brightness I. Late spectral types. M.N.R.A.S. **174**, 489.

Barnes, T. G., Evans, D. S., Moffet, T. J. (1978): Stellar angular diameters and visual surface brightness III. An improved definition of the relationship. M.N.R.A.S. **183**, 285.

Barnothy, J. M. (1965): Quasars and the gravitational image intensifier. Astron. J. **70**, 666.

Barnothy, J. M. (1970): On the masses of quasi-stellar sources. Ap. J. (Letters) **159**, L133.

Barnothy, J.: History of gravitational lenses and the phenomena they produce. In: Gravitational Lenses – Lecture Notes in Physics 300 (eds. J. M. Moran, J. N. Hewitt and K. Y. Lo). New York: Springer-Verlag 1989.

Barnothy, J., Barnothy, M. F. (1968): Galaxies as gravitational lenses. Science **162**, 348.

Bartlett, D. F., Van Buren, D. (1986): Equivalence of active and passive gravitational mass using the moon. Phys. Rev. Lett. **57**, 21.

Bartusiak, M.: Through a universe darkly. New York: Avon Books 1993.

Baum, W. A.: The diameter-red shift relation. In: External galaxies and quasi-stellar objects – I.A.U. Symp. No. 44 (ed. D. S. Evans). Dordrecht, Holland: D. Reidel 1972.

Baym, G., Pethick, C. (1979): Physics of neutron stars. Ann. Rev. Astron. Astrophys. **17**, 415.

Beall, E. F. (1970): Measuring the gravitational interaction of elementary particles. Phys. Rev. **D1**, 961.

Becker, W. (1940): Spektralphotometrische Untersuchungen an delta Cephei-Sternen (Photospectral research of delta Cephei variable stars). Zeits. f. Ap. **19**, 289.

Begelman, M. C. (1978): Accretion of $\gamma > 5/3$ gas by a Schwarzschild black hole. Astron. Astrophys. **70**, 583.

Begelman, M. C. (1978): Black holes in radiation-dominated gas: an analogue of the Bondi accretion problem. M.N.R.A.S. **184**, 53.

Begelman, M. C. (1979): Can a spherically accreting black hole radiate very near the Eddington limit? M.N.R.A.S. **187**, 237.

Begelman, M. C., Blandford, R. D., Rees, M. J. (1984): Theory of extragalactic radio sources. Rev. Mod. Phys. **56**, 255.

Begelman, M., Rees, M.: Gravity's fatal attraction – black holes in the universe. New York: W. H. Freeman 1996.

Bekenstein, J. D. (1972): Black holes and the second law. Lettere al Nuovo Cimento **4**, 737.

Bekenstein, J. D. (1973): Black holes and entropy. Phys. Rev. **D7**, 2333.

Bennett, C. L., et al. (1993): Scientific results from the Cosmic Background Explorer (COBE). Proc. Nat. Acad. Sci. **90**, 4766.

Bennett, C. L., et al. (1996): Four-year COBE DMR cosmic microwave background observations – maps and basic results. Ap. J. (Letters) **464**, L1.

Bennett, C. L., Turner, M. S., White, M. (1997): The cosmic Rosetta stone. Physics Today **50**, 32 – November.

Benson, J. M., et al. (1988): VLBI and MERLIN observations of 3C 120 at 1.7 GHz: Superluminal motions beyond 0.05". Ap. J. **334**, 560.

Bergeron, P., Liebert, J., Fulbright, M. S. (1995): Masses of DA white dwarfs with gravitational redshift determinations. Ap. J. **444**, 810.

Bernardis, P. De., et al. (1994): Detection of cosmic microwave background anisotropy at 1.8°: Theoretical implications on inflationary models. Ap. J. (Letters) **433**, L1.

Bernstein, J, Brown, L. S., Feinberg, G. (1989): Cosmological helium production simplified. Rev. Mod. Phys. **61**, 25.

Bertin, G., Lin, C.-C.: Spiral structure in galaxies: A density wave theory. Cambridge, Mass.: Massachusetts Institute of Technology Press 1996.

Bertotti, B., Ciufolini, I., Bender, P. L. (1987): New test of general relativity: Measurement of de Sitter geodetic precession rate for lunar perigee. Phys. Rev. Lett. **58**, 1062.

Bertsch, D. L., et al. (1992): Pulsed high-energy γ-radiation from Geminga (1E 0630 + 178). Nature **357**, 306.

Bertschinger, E. (1985): The self-similar evolution of holes in an Einstein-De Sitter universe. Ap. J. Supp. **58**, 1.

Bertschinger, E. (1985): Self-similar secondary infall and accretion in an Einstein-De Sitter universe. Ap. J. Supp. **58**, 39.

Bertschinger, E., et al. (1990): Potential, velocity, and density fields from redshift-distance samples: Application: Cosmography within 6000 kilometers per second. Ap. J. **364**, 370.

Bessell, F. W. (1839): Bestimmung der Entfernung des 61 Sterns des Schwans (A new determination of the distance of 61 Cygni). Astron. Nach **16**, No. 356.

Bethe, H. A. (1990): Supernova mechanisms. Rev. Mod. Phys. **62**, 801.

Bhattacharya, D., Van Den Heuvel, E. P. J. (1991): Formation and evolution of binary and millisecond radio pulsars. Physics Reports **203**, 1.

Bi, H., Davidsen, A. F. (1997): Evolution of structure in the intergalactic medium and the nature of the Lyα forest. Ap. J. **479**, 523.

Bignami, G. F., Caraveo, P. A., Lamb, R. C. (1983): An identification for "Geminga" (2CG 195 + 04) 1E 0630 + 178: A unique object in the error bars of the high-energy gamma-ray source. Ap. J. (Letters) **272**, L9.

Binggeli, B., Sandage, A., Tammann, G. A. (1988): The luminosity function of galaxies. Ann. Rev. Astron. Astrophys. **26**, 509.

Binney, J. (1982): Dynamics of elliptical galaxies and other spheroidal components. Ann. Rev. Astron. Astrophys. **20**, 399.

Binney, J., Tremaine, S.: Galactic Dynamics. Princeton: Princeton University Press 1987.

Birck, J. L.: Long-lived chronometers. In: Astrophysical ages and dating methods (eds. E. Vangioni-Flam, M. Casse, J. Audouze and J. Tran Thanh Van). Paris: Editions Frontieres 1990.

Biretta, J. A., Reid, M. J.: The nuclear jet in M87. In: Active galactic nuclei. Lecture notes in physics 307 (eds. H. R. Miller and P. J. Wiita). New York: Springer Verlag 1988.

Birkhoff, G. D.: Relativity and modern physics. Cambridge, Mass.: Harvard University Press 1923.

Birkhoff, G. D. (1943): Matter, electricity and gravitation in flat spacetime. Proc. Nat. Acad. Sci. **29**, 231.

Birkinshaw, M., Gull, S. F., Hardebeck, H. (1984): The Sunyaev-Zeldovich effect towards three clusters of galaxies. Nature **309**, 34.

Birkinshaw, M., Gull, S. F., Northover, K. J. E. (1978): Extent of hot intergalactic gas in the cluster Abell 2218. Nature **275**, 40.

Birkinshaw, M., Hughes, J. P. (1994): A measurement of the Hubble constant from the x-ray properties and the Sunyaev-Zeldovich effect of Abell 2218. Ap. J. **420**, 33.

Birkinshaw, M., Hughes, J. P., Arnaud, K. A. (1991): A measurement of the value of the Hubble constant from the x-ray properties and the Sunyaev-Zeldovich effect of Abell 665, Ap. J. **379**, 466.

Black, D. C. (1971): Some implications of a new value for the primordial solar deuterium-hydrogen ratio. Nature **234**, 148.

Black, D. C. (1972): On the origins of trapped helium, neon and argon isotopic variations in meteorites I. Gas-rich meteorites, lunar soil and breccia. Geochimica et Cosmochimica Acta **36**, 347.

Black, J. H. (1981): The physical state of primordial intergalactic clouds. M.N.R.A.S. **197**, 553.

Blades, J. C., Turnshek, D. A., Norman, C.A. (eds.): QSO absorption lines : Probing the universe. New York: Cambridge University Press 1988.

Blagg, M. A. (1913): On a suggested substitute for Bode's law. M.N.R.A.S. **73**, 414.

Blake, G. M. (1970): Observations of extragalactic radio sources having unusual spectra. Astrophys. Letters **6**, 201.

Blamont, J. E., Roddier, F. (1961): Precise observation of the profile of the Fraunhofer strontium resonance line, evidence for the gravitational red shift on the Sun. Phys. Rev. Lett. **7**, 437.

Bland-Hawthorn, J., Wilson, A. S., Tully, R. B. (1991): Ultramassive, about 10^{11} solar masses, dark core in the luminous infrared galaxy NGC 6240? Ap. J. (Letters) **371**, L19.

Blandford, R. D. (1976): Accretion disc electrodynamics – a model for double radio sources. M.N.R.A.S. **176**, 465.

Blandford, R. D. (1990): Gravitational lenses. Q.J.R.A.S. **31**, 305.

Blandford, R. D.: Unsolved problems in gravitational lensing. In: Unsolved problems in astrophysics (eds. J. N. Bahcall and J. P. Ostriker). Princeton, New Jersey: Princeton University Press 1997.

Blandford, R. D., Jaroszynski, M. (1981): Gravitational distortion of the images of distant radio sources in an inhomogeneous universe. Ap. J. **246**, 1.

Blandford, R. D., Kochanek, C. S., Kovner, I., Narayan, R. (1989): Gravitational lens optics. Science **245**, 824.

Blandford, R. D., Königl, A. (1979): Relativistic jets as compact radio sources. Ap. J. **232**, 34.

Blandford, R. D., Königl, A. (1979): A model for the knots in the M87 jet. Astrophys. Lett. **20**, 15.

Blandford, R. D., Kovner, I. (1988): Formation of arcs by nearly circular gravitational lenses. Phys. Rev. **38A**, 4028.

Blandford, R. D., Mc Kee, C. F. (1977): Radiation from relativistic blast waves in quasars and active galactic nuclei. M.N.R.A.S. **180**, 343.

Blandford, R. D., Mc Kee, C. F., Rees, M. J. (1977): Super-luminal expansion in extragalactic radio sources. Nature **267**, 211.

Blandford, R. D., Narayan, R. (1992): Cosmological applications of gravitational lensing. Ann. Rev. Astron. Astrophys. **30**, 311.

Blandford, R. D., Payne, D. G. (1982): Hydromagnetic flows from accretion discs and the production of radio jets. M.N.R.A.S. **199**, 883.

Blandford, R. D., Rees, M. J. (1974): A 'twin-exhaust' model for double radio sources. M.N.R.A.S. **169**, 395.

Blandford, R. D., Rees, M. J. (1978): Extended and compact extragalactic radio sources: Interpretation and theory. Physica Scripta **17**, 265.

Blandford, R. D., Teukolsky, S. A. (1975): On the measurement of the mass of PSR 1913+16. Ap. J. (Letters) **198**, L27.

Blandford, R. D., Teukolsky, S. A. (1976): Arrival-time analysis for a pulsar in a binary system. Ap. J. **205**, 580.

Blandford, R. D., Znajek, R. L. (1977): Electromagnetic extraction of energy from Kerr black holes. M.N.R.A.S. **179**, 433.
Blauuw, A., et al. (1960): The new IAU system of galactic coordinates. M.N.R.A.S. **121**, 123.
Blitz, L., et al. (1993): The centre of the Milky Way. Nature **361**, 417.
Blokhintsev, D. I. (1966): Basis for special relativity theory provided by experiments in high energy physics. Sov. Phys. Uspekhi **9**, 405.
Blumenthal, G. R., et al. (1984): Formation of galaxies and large-scale structure with cold dark matter. Nature **311**, 517.
Bode, J. E.: Anleitung zur Kenntnis des gestirnten Himmels (Guide to a knowledge of the heavenly bodies in the sky). Hamburg, 1772.
Bodmer, R., et al. (1995): Solar wind helium isotopic composition from SWICS/ULYSSES. Space Sci. Rev. **72**, 61.
Boesgaard, A. M., Steigman, G. (1985): Big bang nucleosynthesis – theories and observations. Ann. Rev. Astron. Astrophys. **23**, 319.
Böhringer, H.: In: Cosmological aspects of X-ray clusters of galaxies (ed. W. C. Seilter). Dordrecht: Kluwer 1994.
Bolte, M. (1989): The age of globular cluster NGC 288, the formation of the galactic halo, and the second parameter. Astron. J. **97**, 1688.
Bolte, M., Hogan, C. J. (1995): Conflict over the age of the Universe. Nature **376**, 399.
Bolton, C. T. (1972): Dimensions of the binary system HDE 226868: Cygnus X-1. Nature: Physical Science **240**, 124. Reproduced in: A source book in astronomy and astrophysics 1900–1975 (eds. K. R. Lang and O. Gingerich). Cambridge, Mass.: Harvard University Press 1979.
Bolton, J. G., Stanley, G. J., Slee, O. B. (1949): Positions of three discrete sources of galactic radio-frequency radiation. Nature **164**, 101. Reproduced in: A source book in astronomy and astrophysics 1900–1975 (eds. K. R. Lang and O. Gingerich). Cambridge, Mass.: Harvard University Press 1979.
Boltwood, B. B. (1907): The origin of radium. Nature **76**, 589.
Boltwood, B. B. (1907): On the ultimate disintegration products of the radioactive elements II. Am. J. Sci. **23**, 77.
Bond, J. R., Efstathiou, G. (1984): Cosmic background radiation anisotropies in universes dominated by nonbaryonic dark matter. Ap. J. (Letters) **285**, L45.
Bond, J. R., Efstathiou, G. (1987): The statistics of cosmic background radiation fluctuations. M.N.R.A.S. **226**, 655.
Bond, J. R., Szalay, A. S. (1983): The collisionless damping of density fluctuations in an expanding universe. Ap. J. **274**, 443.
Bond, J. R., Szalay, A. S. (1984): Adiabatic theories of galaxy formation and pancakes. Ann. New York Acad Sci. **422**, 82.
Bondi, H. (1952): On spherically symmetrical accretion. M.N.R.A.S. **112,** 195.
Bondi, H., Gold, T. (1948): The steady-state theory of the expanding universe. M.N.R.A.S. **108**, 252. Reproduced in: A source book in astronomy and astrophysics 1900–1975 (eds. K. R. Lang and O. Gingerich). Cambridge, Mass.: Harvard University Press 1979.
Bondi, H., Hoyle, F. (1944): On the mechanism of accretion by stars. M.N.R.A.S. **104**, 273.
Bondi, H., Van Der Burg, M. G. J., Metzner, A. W. K. (1962): Gravitational waves in general relativity VII. Waves from axisymmetric isolated systems. Proc. Roy. Soc. (London) **269**, 21.
Bonner, W. B. (1956): Boyle's law and gravitational instability. M.N.R.A.S. **116**, 351.
Bonnet-Bidaud, J.-M., Chardin, G. (1988): Cygnus X-3, a critical review. Physics Reports **170**, 325.
Bono, G., Caputo, F., Marconi, M. (1998): On the theoretical period-radius relation of classical Cepheids. Ap. J. (Letters) **497**, L43.
Borgeest, U. (1986): Determination of galaxy masses by the gravitational lens effect. Ap. J. **309**, 467.

Borkowski, K. M. (1987): Transformation of geocentric to geodetic coordinates without approximation. Astrophys. Space Sci. **139**, 1.

Borkowski, K. M. (1989): Accurate algorithms to transform geocentric to geodetic coordinates. Bulletin Geodesique **63**, 50.

Bosma, A. (1981): 21-cm line studies of spiral galaxies I. Observations of the galaxies NGC 5033, 3198, 5055, 2841, and 7331. Astron. J. **86**, 1791.

Bosma, A. (1981): 21-cm line studies of spiral galaxies II. The distribution and kinematics of neutral hydrogen in spiral galaxies of various morphological types. Astron. J. **86**, 1825.

Bottinelli, L., et al. (1980).: The 21 centimeter line width as an extragalactic distance indicator. Ap. J. (Lett.) **242**, L153

Bottinelli, L., et al. (1985): H I line studies of galaxies IV. distance moduli of 468 disk galaxies. Astron. Astrophys. Supp. **59**, 43.

Boughn, S. P., Cheng, E. S., Wilkinson, D. T. (1981): Dipole and quadrupole anisotropy of the 2.7 K radiation. Ap. J. Lett. **243**, L113.

Bourassa, R. R., Kantowski, R. (1975): The theory of transparent gravitational lenses. Ap. J. **195**, 13.

Bourassa, R. R., Kantowski, R. (1976): Multiple image probabilities for a spheroidal gravitational lens. Ap. J. **205**, 674.

Bourassa, R. R., Kantowski, R., Norton, T. D. (1973): The spheroidal gravitational lens. Ap. J. **185**, 747.

Bowyer, S. (1991): The cosmic far ultraviolet background. Ann. Rev. Astron. Astrophys. **29**, 59.

Boyer, R. H., Lindquist, R. W. (1967): Maximal analytic extension of the Kerr metric. J. Math. Phys. **8**, 265.

Bradley, J. (1728): A new apparent motion in the fixed stars discovered, its cause assigned, the velocity and equable motion of light deduced. Phil. Trans. **6**, 149.

Bradley, J. (1748): An apparent motion in some of the fixed stars. Phil. Trans. **10**, 32.

Bradt, H. V. D., Mc Clintock, J. E. (1983): The optical counterparts of compact galactic X-ray sources. Ann. Rev. Astron. Astrophys. **21**, 13.

Braginsky, V. B., Panov, V. I. (1972): Verification of the equivalence of inertial and gravitational mass. Sov. Phys. JETP **34**, 463.

Branch, D. (1988): Supernovae and the Hubble constant. In: The extragalactic distance scale (eds. S. Van Den Bergh and C. J. Pritchet). A.S.P. Conf. Ser. **4**, 146.

Branch, D., Tammann, G. A. (1992): Type Ia supernovae as standard candles. Ann. Rev. Astron. Astrophys. **30**, 359.

Brandt, J. C. (1960): On the distribution of mass in galaxies. I. The large scale structure of ordinary spirals with application to M 31. II. A discussion of the mass of the Galaxy. Ap. J. **131**, 293, 553.

Brandt, J. C., Belton, M. J. S. (1962): On the distribution of mass in galaxies III. Surface densities. Ap. J. **136**, 352.

Brans, C., Dicke, R. H. (1961): Mach's principle and a relativistic theory of gravitation. Phys. Rev. **124**, 925.

Brault, J. (1963): Gravitational redshift of solar lines. Bull. Am. Phys. Soc. **8**, 28.

Bray, I. (1984): Spheroidal gravitational lenses. M.N.R.A.S. **208**, 511.

Bridle, A. H., Davis, M. M., Formalont, E. B., Lequeux, J. (1972): Counts of intense extragalactic radio sources at 1,400 MHz. Nature-Phys. Sci. **235**, 135.

Bridle, A. H., Perley, R. A. (1984): Extragalactic radio jets. Ann. Rev. Astron. Astrophys. **22**, 319.

Briel, U. G., Henry, J. P., Böhringer, H. (1992): Observation of the Coma cluster of galaxies with ROSAT during the all-sky survey. Astron. Astrophys. **259**, L31.

Brillet, A., Hall, J. L. (1979): Improved laser test of the isotropy of space. Phys. Rev. Lett. **42**, 549.

Brout, R., Englert, F., Gunzig, E. (1978): The creation of the universe as a quantum phenomenon. Annals of Physics **115**, 78.

Brown, R. L., Liszt, H. S. (1984): Sagittarius A and its environment. Ann. Rev. Astron. Astrophys. **22**, 223.
Brown, T. M., Christensen-Dalsgaard, J. (1998): Accurate determination of the solar photospheric radius. Ap. J. (Letters) **500**, L195.
Brown, T. M., et al. (1989): Inferring the Sun's internal angular velocity from observed p mode frequency splittings. Ap. J. **343**, 526.
Brumberg, V. A., et al. (1975): Component masses and inclination of binary systems containing a pulsar, determined from relativistic effects. Sov. Astron. Lett. **1**, 2.
Bucher, M., Turok, N. (1995): Open inflation with an arbitrary false vacuum mass. Phys. Rev. **D52**, 5538.
Bunn, E. F., Scott, D., White, M. (1995): The COBE normalization for standard cold dark matter. Ap. J. (Letters) **441**, L9.
Burbidge, G. R. (1956): On synchrotron radiation from Messier 87. Ap. J. **124**, 416.
Burbidge, G. R. (1959): Estimates of the total energy in particles and magnetic field in the non-thermal radio sources. Ap. J. **129**, 849.
Burbidge, E. M., Burbidge, G. R.: The masses of galaxies. In: Galaxies and the universe (ed. A. Sandage, M. Sandage, and J. Kristian). Chicago: University of Chicago Press 1975.
Burbidge, E. M., et al. (1957): Synthesis of the elements in stars. Rev. Mod. Phys. **29**, 547. Reproduced in: A source book in astronomy and astrophysics 1900–1975 (eds. K. R. Lang and O. Gingerich). Cambridge, Mass.: Harvard University Press 1979.
Burbidge, G. R., Burbidge, E. M., Sandage, A. R. (1963): Evidence for the occurrence of violent events in the nuclei of galaxies. Rev. Mod. Phys. **35**, 947.
Burigana, C., De Zotti, G., Danese, L. (1991): Constraints on the thermal history of the universe from the cosmic microwave background spectrum. Ap. J. **379**, 1.
Burns, J. O. (1986): Very large structures in the universe. Scientific American **255**, 38 – July.
Burrows, A., Lattimer, J. M. (1986): The birth of neutron stars. Ap. J. **307**, 178.
Burrows, A., Liebert, J. (1993): The science of brown dwarfs. Rev. Mod. Phys. **65**, 301.
Burstein, D. (1990): Large-scale motions in the universe: a review. Rep. Prog. Phys. **53**, 421.
Burstein, D., Heiles, C. (1978): HI, galaxy counts and reddening – variation in the gas-to-dust ratio, the extinction at high galactic latitudes, and a new method for determining galactic reddening. Ap. J. **225**, 40.
Burstein, D., Heiles, C. (1982): Reddening derived from HI and galaxy counts – accuracy and maps. Astron. J. **87**, 1165.
Burton, W. B.: The structure of our Galaxy derived from observations of neutral hydrogen. In: Galactic and extragalactic radio astronomy (ed. G. L. Verschuur and K. I. Kellermann). New York: Springer-Verlag 1988.
Buta, R., Crocker, D. A., Elmegreen, B. G.: Barred galaxies IAU colloquium 157. San Francisco: Astronomical Society of the Pacific 1996.
Butcher, H., Oemler, A. (1978): The evolution of galaxies in clusters I. ISIT photometery of Cl 0024 + 1654 and 3C 295. Ap. J. **219**, 18.
Butcher, H., Oemler, A. (1978): The evolution of glaxies in clusters II. The galaxy content of nearby clusters. Ap. J. **226**, 559.
Butcher, H., Oemler, A. (1984): The evoluton of galaxies in clusters V. A study of populations since $z \approx 0.5$. Ap. J. **285**, 426.
Byrne, J., et al. (1990): Measurement of the neutron lifetime by counting trapped protons. Phys. Rev. Lett. **65**, 289.
Caldwell, C. O., et al. (1988): Laboratory limits on galactic cold dark matter. Phys. Rev. Lett. **61**, 510.
Caldwell, J. A. R., Coulson, I. M. (1986): The geometry and distance of the Magellanic clouds from cepheid variables. M.N.R.A.S. **218**, 223.
Caldwell, J. A. R., Ostriker, J. P. (1981): The mass distribution within our Galaxy: A three component model. Ap. J. **251**, 61.
Callahan, P. S. (1977): On the accretion disk model of QSOs. Astron. Astrophys. **59**, 127.
Cameron, A. G. W. (1957): Nuclear reactions in stars and nucleogenesis. P.A.S.P. **69**, 201.

Cameron, A. G. W. (1957): Stellar evolution, nuclear astrophysics and nucleogenesis. Atomic Energy of Canada, Chalk River Project AECL 454, CRL-41.

Camilo, F., Foster, R. S., Wolszczan, A. (1994): High-precision timing of PSR J1713 + 0747: Shapiro delay. Ap. J. (Letters) **437**, L39.

Campbell, A. (1992): On determining the primordial helium abundance from the spectra of H II galaxies. Ap. J. **401**, 157.

Canal, R., Isern, J., Labay, J. (1990): Theory of neutron stars in binary stars. Ann. Rev. Astron. Astrophys. **28**, 183.

Cannon, A. J., Pickering, E. C. (1918–1924): The Henry Draper catalogue. Ann. Harvard Coll. Obs. **91–99**.

Canuto, V. (1975): Equation of state at ultra-high densities II. Ann. Rev. Astron. Astrophys. **13**, 335.

Carney, B. W. (1980): The ages and distances of eight globular clusters. Ap. J. Supp. **42**, 481.

Carney, B. W.: The distances and ages of globular clusters. In: Calibration of stellar ages (ed. A. G. Davis Philip). Schnectady, N. Y.: L. Davis Press 1988.

Carney, B. W., et al. (1995): The distance to the galactic center obtained by infrared photometry of RR Lyrae variables. Astron. J. **110**, 1674.

Carr, B. (1990): Baryonic dark matter. Comments Astrophys. **14**, 257.

Carr, B. (1994): Baryonic dark matter. Ann. Rev. Astron. Astrophys. **32**, 531.

Carr, B.: The status of the dark matter problem and its baryonic solutions. In: The sun and beyond (eds. J. Tran Thanh Van, L. M. Celnikier, H. C. Trung and S. Vauclair). Paris: Editions Frontieres 1996.

Carroll, S. M., Press, W. H., Turner, E. L. (1992): The cosmological constant. Ann. Rev. Astron. Astrophys. **30**, 499.

Carswell, R. F., et al. (1994): Is there deuterium in the $z = 3.32$ complex in the spectrum of 0014 + 813? M.N.R.A.S. **268**, L1.

Carter, B. (1968): Global structure of the Kerr family of gravitational fields. Phys. Rev. **174**, 1559.

Carter, B. (1971): An axisymmetric black hole has only two degrees of freedom. Phys. Rev. Lett. **26**, 331.

Casares, J., Charles, P. A. (1994): Optical studies of V404 Cyg, the X-ray transient GS 2023+338-IV. The rotation speed of the companion star. M.N.R.A.S. **271**, L5.

Casares, J., Charles, P. A., Naylor, T. (1992): A 6.5-day periodicity in the recurrent nova V404 Cygni implying the presence of a black hole. Nature **355**, 614.

Cavaliere, A., Morrison, P., Sartori, L. (1971): Rapidly changing radio images. Science **173**, 525.

Cavendish, H. (1798): Experiments to determine the density of Earth. Phil. Trans. Roy. Soc. **88**, 469.

Cen, R. (1998): On the cluster Sunyaev-Zeldovich effect and Hubble constant. Ap. J. (Letters) **498**, L99.

Cen, R., et al. (1991): A hydrodynamic approach to cosmology: Texture-seedeed cdm and hdm cosmogonies. Ap. J. **383**, 1.

Cen, R., Miralda-Escoudé, J., Ostriker, J. P., Rauch, M. (1994): Gravitational collapse of small-scale structure as the origin of the Lyman-alpha forest. Ap. J. (Letters) **437**, L9.

Cen, R., Ostriker, J. P. (1994): A hydrodynamic approach to cosmology: The mixed dark matter cosmological scenario. Ap. J. **431**, 451.

Cen, R., Ostriker, J. P., Peebles, P. J. E. (1993): A hydrodynamic approach to cosmology: The primeval baryon isocurvature model. Ap. J. **415**, 423.

Centrella, J., Melott, A. L. (1983): Three-dimensional simulation of large-scale structure in the universe. Nature **305**, 196.

Chaboyer, B. (1995): Absolute ages of globular clusters and the age of the universe. Ap. J. (Letters) **444**, L9.

Chaboyer, B., et al. (1996): A lower limit on the age of the universe. Science **271**, 957.

Chaboyer, B., Sarajedini, A., Demarque, P. (1992): Ages of globular clusters and helium diffusion. Ap. J. **394**, 515.

Chadwick, J. (1932): The existence of a neutron. Proc. Roy. Soc. London **A136**, 692.

Chaffee, F. H. (1980): The discovery of a gravitational lens. Scientific American **243**, 70.

Chaffee, F. H. Jr., et al. (1985): High-resolution spectroscopy of selected absorption lines toward quasi-stellar objects II. The metal-to-hydrogen ratio in a "metal-free" cloud toward S5 0014+82. Ap. J. **292**, 362.

Chaffee, F. H. Jr., et al. (1986): On the abundance of metals and the ionization state in absorbing clouds toward QSOs. Ap. J. **301**, 116.

Chandler, S. C. (1891): On the variation of latitude. Astron. J. **11**, 59, 65, 75, 83; **12**, 17, 57, 65, 97 (1892); **13**, 159 (1893).

Chandler, S. C. (1898): Comparison of the observed and predicted motions of the pole, 1890–1898, and determination of revised elements. Astron. J. **19**, 105.

Chandrasekhar, S. (1931): The maximum mass of ideal white dwarfs. Ap. J. **74**, 81.

Chandrasekhar, S. (1934): Stellar configurations with degenerate cores. Observatory **57**, 373.

Chandrasekhar, S. (1935): The highly collapsed configurations of a stellar mass (Second Paper). M.N.R.A.S. **95**, 207.

Chandrasekhar, S. (1964): Dynamical instability of gaseous masses approaching the Schwarzschild limit in general relativity. Phys. Rev. Lett. **12**, 114 (1964). Ap. J. **140**, 417.

Chanmugam, G. (1992): Magnetic fields of degenerate stars. Ann. Rev. Astron. Astrophys. **30**, 143.

Chao, B. F. (1989): Length-of-day variations caused by el nino – southern oscillation and quasi-biennial oscillation. Science **243**, 923.

Chapman, G. A., Ingersoll, A. P. (1973): Photospheric faculae and the solar oblateness: A reply to "Faculae and the solar oblateness" by R. H. Dicke. Ap. J. **183**, 1005.

Chapront-Touze, M., Chapront, J. (1988): ELP 2000-85, a semi-analytical lunar ephemeris adequate for historical times. Astron. Astrophys. **190**, 342.

Charles, P. A., Wagner, R. M. (1996): Black holes in binary systems. Sky and Telescope **91**, 38.

Charlier, C. V. L. (1908): Wie eine unendliche Welt aufgebaut sein kann (How an infinite universe can be constructed). Ark. Mat. Astron. Phys. **4**, No. 24.

Charlier, C. V. L. (1922): How an infinite world may be built up. Arkiv för matematik, astronomi och fysik **16**, No. 22, 1.

Charlot, S., Silk, J. (1995): Signatures of white dwarf galaxy halos. Ap. J. **445**, 124.

Chen, K., Halpern, J. P., Filippenko, A. V. (1989): Kinematic evidence for a relativistic Keplerian disk: ARP 102B. Ap. J. **339**, 742.

Chen, K., Ruderman, M. (1993): Pulsar death lines and death valley. Ap. J. **402**, 264.

Cheng, E. S., et al. (1979): Large-scale anisotropy in the 2.7 K radiation. Ap. J. (Lett.) **232**, L139.

Chernoff, D. F., Finn, L. S. (1993): Gravitational radiation, inspiraling binaries, and cosmology. Ap. J. (Letters) **411**, L5.

Chincarini, G., Rood, H. J. (1971): Dynamics of the Perseus cluster of galaxies. Ap. J. **168**, 321.

Chincarini, G., Rood, H. J. (1975): The size of the Coma cluster. Nature **257**, 294.

Chiosi, C., Bertelli, G., Bressan, A. (1992): New developments in understanding the HR diagram. Ann. Rev. Astron. Astrophys. **30**, 235.

Chiosi, C., Maeder, A. (1986): The evolution of massive stars with mass loss. Ann. Rev. Astron. Astrophys. **24**, 329.

Chitre, S. M., Narlikar, J. V. (1979): On the apparent superluminal separation of radio source components. M.N.R.A.S. **187**, 655.

Christodoulou, D. (1970): Reversible and irreversible transformations in black-hole physics. Phys. Rev. Lett. **25**, 1596.

Chupp, T. E., et al. (1989): Results of a new test of local Lorentz invariance – a search for mass anisotropy in ^{21}Ne. Phys. Rev. Lett. **63**, 1541.
Chwolson, O. (1924): Über eine mögliche Form fiktiver Doppelsterne (Regarding a possible form of fictitious double stars). Astron. Nach. **221**, 329.
Ciardullo, R. B., Demarque, P. (1977): Tables of isochrones constructed from theoretical tracks of stellar evolution for ages between 200 and 25000 million years and chemical compositions in the ranges $0.1 \leq Y \leq 0.4$ and $0.00001 \leq Z \leq 0.1$. Trans. Astron. Obs. Yale University **33**, 1.
Ciardullo, R., et al. (1989): Planetary nebulae as standard candles II. The calibration in M31 and its companions. Ap. J. **339**, 53.
Ciardullo, R., Jacoby, G. H., Ford, H. C. (1989): Planetary nebulae as standard candles IV. A test in the Leo I group. Ap. J. **344**, 715.
Ciardullo, R., Jacoby, G. H., Harris, W. E. (1991): Planetary nebulae as standard candles VII. A test versus Hubble type in the NGC 1023 group. Ap. J. **383**, 487.
Clairaut, A. C.: Theory of the figure of the Earth, and the principles of hydrostatics. Paris: Du Mond 1743.
Clark, D. H.: The quest for SS 433. New York: Viking Penquin 1985.
Clark, D. H., Murdin, P. (1978): An unusual emission-line star/X-ray source/radio star, possibly associated with an SNR. Nature **276**, 44.
Clark, E. E. (1972): The uniform transparent gravitational lens. M.N.R.A.S. **158**, 233.
Clark, J. P. A., Eardley, D. M. (1977): Evolution of close neutron star binaries. Ap. J. **215**, 311.
Clark, J. P. A., Van Den Heuvel, E. P. J., Sutantyo, W. (1979): Formation of neutron star binaries and their importance for gravitational radiation. Astron. Astrophys. **72**, 120.
Clayton, D. D. (1964): Cosmoradiogenic chronologies of nucleosynthesis. Ap. J. **139**, 637.
Clayton, D. D. (1985): Astration of cosmological deuterium. Ap. J. **290**, 428.
Clayton, D. D. (1988): Nuclear cosmochronology within analytic models of the chemical evolution of the solar neighborhood. M.N.R.A.S. **234**, 1.
Clerke, A. M.: Problems in astrophysics. London: A. and C. Black 1903.
Cline, T. L., Desai, U. D., Klebesadel, R. W., Strong, I. B. (1973): Energy spectra of cosmic gamma-ray bursts. Ap. J. (Letters) **185**, L1.
Cohen, J. G. (1985): Nova shells II. calibration of the distance scale using novae. Ap. J. **292**, 90.
Cohen, J. G. (1988): Nova expansion parallaxes. In: The extragalactic distance scale (ed. S. Van Den Bergh and C. J. Pritchet). A.S.P. Conf. Ser. **4**, 114.
Cohen, J. G., et al. (1996): Redshift clustering in the Hubble deep field. Ap. J. (Letters) **471**, L5.
Cohen, M. (1995): The displacement of the Sun from the galactic plane using IRAS and FAUST source counts. Ap. J. **444**, 874.
Cohen, M. H., et al. (1971): The small-scale structure of radio galaxies and quasi-stellar sources at 3.8 centimeters. Ap. J. **170**, 207.
Cohen, M. H., et al. (1976): Rapid increase in the size of 3C 345. Ap. J. (Letters) **206**, L1.
Cohen, M. H., et al. (1979): Superluminal variations in 3C 120, 3C 273, and 3C 345. Ap. J. **231**, 293.
Cohen, M. H., Unwin, S. C.: Superluminal effects and bulk relativistic motion. In: VLBI and compact radio sources. IAU Symposium no. 110 (eds. R. Fanti, K. Kellermann and G. Setti). Boston: D. Reidel 1984.
Cohen, S. C., et al. (eds). (1985): Lageos scientific results. Journ. Geophys. Res. **90**, B11, 9215.
Coleman, S., Weinberg, E. (1973): Radiative corrections as the origin of spontaneous symmetry breaking. Phys. Rev. **D7**, 1888.
Coles, P., Ellis, G. (1994): The case for an open universe. Nature **370**, 609.
Colgate, S. A., White, R. H. (1966): The hydrodynamic behavior of supernovae explosions. Ap. J. **143**, 626.

Colley, W. N., Tyson, J. A., Turner, E. L. (1996): Unlensing multiple arcs in 0024+1654 – reconstruction of the source image. Ap. J. (Letters) **461**, L83.
Combes, F., Boissé, P., Mazure, A., Blanchard, A.: Galaxies and Cosmology. New York: Springer Verlag 1995.
Comella, J. M., et al. (1969): Crab nebula pulsar NP 0532. Nature **221**, 453.
Condon, J. J. (1984): Cosmological evolution of radio sources. Ap. J. **287**, 461.
Condon, J. J.: Radio sources and cosmology. In: Galactic and extragalactic radio astronomy. Second edition (eds. G. L. Verschuur and K. I. Kellermann). New York: Springer Verlag 1988.
Conklin, E. K. (1969): Velocity of the earth with respect to the cosmic background radiation. Nature **222**, 971.
Conklin, E. K., Bracewell, R. N. (1967): Isotropy of cosmic background radiation at 10.690 MHz. Phys. Rev. Lett. **18**, 614.
Conklin, E. K., Bracewell, R. N. (1967): Limits on small scale variations in the cosmic background radiation. Nature **216**, 777.
Conti, P. S., Leitherer, C., Vacca, W. D. (1996): Hubble space telescope ultraviolet spectroscopy of NGC 1741: A nearby template for distant energetic starbursts. Ap. J. (Letters) **461**, L87.
Copi, C. J., Schramm, D. N., Turner, M. S. (1995): Big bang nucleosynthesis and a new approach to galactic chemical evolution. Ap. J. (Letters) **455**, L95.
Copi, C. J., Schramm, D. N., Turner, M. S. (1995): Big-bang nucleosynthesis and the baryon density of the universe. Science **267**, 192.
Corey, B. E., Wilkinson, D. T. (1976): Bull. Amer. Astron. Soc. **8**, 351.
Cornell, J. (ed.): Bubbles, voids, and bumps in time: the new cosmology. New York: Cambridge University Press 1989.
Corwin, H. G. (1971): Notes on the Hercules cluster of galaxies. P.A.S.P. **83**, 320.
Counselman, C. C., et al. (1974): Solar gravitational deflection of radio waves measured by very-long-baseline interferometry. Phys. Rev. Lett. **33**, 1621.
Courteau, S., et al. (1993): Streaming motions in the local universe: Evidence for large-scale low-amplitude density fluctuations. Ap. J. (Letters) **412**, L51.
Cowan, J. J., Thielemann, F.-K., Truran, J. W. (1987): Nuclear chronometers from the r-process and the age of the Galaxy. Ap. J. **323**, 543.
Cowan, J. J., Thielemann, F.-K., Truran, J. W. (1991): Radioactive dating of the elements. Ann. Rev. Astron. Ap. **29**, 447.
Cowie, L. L., Songaila, A. (1993): Faint galaxy surveys. Proc. Nat. Acad. Sci. **90**, 4867.
Cowley, A. P. (1992): Evidence for black holes in stellar binary systems. Ann. Rev. Astron. Astrophys. **30**, 287.
Cowley, A. P., et al. (1983): Discovery of a massive unseen star in LMC X-3. Ap. J. **272**, 118.
Cowling, S. A. (1984): Gravitational light deflection in the solar system. M.N.R.A.S. **209**, 415.
Cowsik, R., Ghosh, P. (1987): Dark matter in the universe: massive neutrinos revisited. Ap. J. **317**, 26.
Cowsik, R., Mc Clelland, J. (1973): Gravity of neutrinos of nonzero mass in astrophysics. Ap. J. **180**, 7.
Cox, A. N. (1980): The masses of cepheids. Ann. Rev. Astron. Astrophys. **18**, 15.
Crawford, D. F., Jauncey, D. L., Murdoch, H. S. (1970): Maximum-likelihood estimation of the slope from number-flux-density counts of radio sources. Ap. J. **162**, 405.
Crézé, M., et al. (1998): The distribution of stars in phase space mapped by Hipparcos. Astron. Astrophys. **329**, 920.
Curtis, H. D. (1918): Descriptions of 762 nebulae and clusters photographed with the Crossley reflector. Pub. Lick Obs. **13**, 9.
Cutler, C., Flanagan, E. E. (1994): Gravitational waves from merging compact binaries: How accurately can one extract the binary's parameters from the inspiral waveform? Phys. Rev. **D49**, 2658.

Da Costa, L. N., et al. (1988): The southern sky redshift survey. Ap. J. **327**, 544.
Da Costa, L. N., et al. (1994): A complete southern sky redshift survey. Ap. J. (Letters) **424**, L1.
Da Costa, L. N., et al. (1996): The mass distribution in the nearby universe. Ap. J. (Letters) **468**, L5.
Dabbs, J. W. T., Harvey, J. A., Paya, D., Horstmann, H. (1965): Gravitational acceleration of free neutrons. Phys. Rev. **139**, B756.
Dabrowski, M., Stelmach, J. (1986): A redshift-magnitude formula for the universe with cosmological constant and radiation pressure. Astron. J. **92**, 1272.
Dabrowski, M. P., Stelmach, J. (1987): Astrophysical formulas for Friedman models with cosmological constant and radiation. Astron. J. **94**, 1373.
Dabrowski, M. P., Stelmach, J. (1989): Observable quantities in cosmological models with strings. Astron. J. **97**, 978.
Dahle, H., Maddox, S. J., Lilje, P. B. (1994): Deep imaging of the double quasar QSO 0957+561 – new constraints on H_o. Ap. J. (Letters) **435**, L79.
Damour, T.: The problem of motion in Newtonian and Einsteinian gravity. In: 300 years of gravitation (eds. S. W. Hawking and W. Israel). New York: Cambridge University Press 1987.
Damour, T., Gibbons, G. W., Taylor, J. H. (1988): Limits on the variability of G using binary-pulsar data. Phys. Rev. Lett. **61**, 1151.
Damour, T., Nordtvedt, K. (1993): General relativity as a cosmological attractor of tensor-scalar theories. Phys. Rev. Lett. **70**, 2217.
Damour, T., Taylor, J. H. (1991): On the orbital period change of the binary pulsar PSR 1913+16. Ap. J. **366**, 501.
Damour, T., Taylor, J. H. (1992): Strong-field tests of relativistic gravity and binary pulsars. Phys Rev. **D45**, 1840.
Damour, T., Vokrouhlicky, D. (1996): Equivalence principle and the moon. Phys. Rev. **D53**, 4177.
Danese, L, De ZOTTI, G. (1982): Double Compton process and the spectrum of the microwave background. Astron. Astrophys. **107**, 39.
Darwin, C. (1959): The gravity field of a particle. Proc. Roy. Soc. (London) **A249**, 180.
Dashevskii, V. M., Slysh, V. I. (1966): On the propagation of light in a nonhomogeneous universe. Sov. Astr. **9**, 671.
Dashevskii, V. M., Zeldovich, Y. B. (1965): The propagation of light in a nonhomogeneous nonflat universe II. Sov. Astr. **8**, 854.
David, M., Huchra, J. (1982): A survey of galaxy redshifts III. The density field and the induced gravity field. Ap. J. **254**, 437.
Davidsen, A. F. (1993): Far-ultraviolet astronomy on the astro-1 space shuttle mission. Science **259**, 327.
Davidsen, A. F., Kriss, G. A., Zheng, W. (1996): Measurement of the opacity of ionized helium in the intergalactic medium. Nature **380**, 47.
Davidson, K., Netzer, H. (1979): The emission lines of quasars and similar objects. Rev. Mod. Phys. **51**, 715.
Davis, M., et al. (1980): On the Virgo supercluster and the mean mass density of the universe. Ap. J. (Letters) **238**, L113.
Davis, M., et al. (1982): A survey of galaxy redshifts II. The large scale space distribution. Ap. J. **253**, 423.
Davis, M., et al. (1985): The evolution of large-scale structure in a universe dominated by cold dark matter. Ap. J. **292**, 371.
Davis, M., Geller, M. J., Huchra, J. (1978): The local mean mass density of the universe: New methods for studying galaxy clustering. Ap. J. **221**, 1.
Davis, M., Huchra, J. (1982): A survey of galaxy redshifts III. the density field and the induced gravity field. Ap. J. **254**, 437.

Davis, M., Peebles, P. J. E. (1983): Evidence for local anisotropy of the Hubble flow. Ann. Rev. Astron. Astrophys. **21**, 109.

Davis, M., Summers, F. J., Schlegel, D. (1992): Large-scale structure in a universe with mixed hot and cold dark matter. Nature **359**, 393.

Davis, R. L., et al. (1992): Cosmic microwave background probes models of inflation. Phys. Rev. Lett. **69**, 1856.

De Bernardis, P., et al. (1994): Detection of cosmic microwave background anisotropy at 1.8 degrees – theoretical implications on inflationary models. Ap. J. (Letters) **433**, L1.

De Chéseaux, J. P. De L.: Traité de la cométe qui a paru en Decembre 1743 et ene Janvier, Fevrier et Mars 1744 (Treatise of the comet that appeared in December 1743 and in January, February and March 1744). Lussanne et Geneve – Michel Bouguet et Compagnie, 1744. Relevant appendix reprod. in Dickson (1968) and Jaki (1969).

De Jager, C., Nieuwenhuijzen, H., Van Der Hucht, K. A. (1988): Mass loss rates in the Hertzsprung-Russell diagram. Astron. Astrophys. Supp. **72**, 259.

De Lapparent, V. – Lapparent, V. De

De Oliveira-Costa, A., et al. (1998): Mapping the CMB III: combined analysis of QMAP flights. Ap. J. (Letters) **509**, L77.

De Oliveira-Costa, A., Smoot, G. F. (1995): Constraints on the topology of the universe from the two-year COBE data. Ap. J. **448**, 477.

De Silva, L. N. K. – see Silva, L. N. K. De.

De Sitter, W. (1916): On Einstein's theory of gravitation and its astronomical consequences. M.N.R.A.S. **77**, 155.

De Sitter, W. (1916): On Einstein's theory of gravitation and its astronomical consequences II. M.N.R.A.S. **77**, 481.

De Sitter, W. (1917): On Einstein's theory of gravitation and its astronomical consequences III. M.N.R.A.S. **78**, 3.

De Sitter, W. (1917): On the relativity of inertia. Remarks concerning Einstein's latest hypothesis. Proc. Akad. Wetensch. Amst. **19**, 1217.

De Sitter, W. (1920): On Einstein's term in the motion of the lunar perigee and node. M.N.R.A.S. **81**, 102.

De Vaucouleurs, G. (1948): Recherches sur les nebuleuses extragalactiques 1. sur la technique de l'analyse microphotometrique des nebuleuses brillantes. Ann. d'Ap. **11**, 247.

De Vaucouleurs, G. (1953): Evidence for a local supergalaxy. Astron. J. **58**, 29.

De Vaucouleurs, G. (1953): On the distribution of mass and luminosity in elliptical galaxies. M.N.R.A.S. **113**, 134.

De Vaucouleurs, G. (1956): The distribution of bright galaxies and the local supergalaxy. Vistas in Astronomy **2**, 1584.

De Vaucouleurs, G. (1958): Further evidence for a local super-cluster of galaxies – rotation and expansion. Astron. J. **63**, 253.

De Vaucouleurs, G. (1959): The local super-cluster of galaxies. Astr. Z. **36**, 977. Sov. Astr. **3**, 897 (1960).

De Vaucouleurs, G. (1982): Five crucial tests of the cosmic distance scale using the Galaxy as a fundamental standard. Nature **299**, 303.

De Vaucouleurs, G. (1983): The Galaxy as fundamental calibration of the extragalactic distance scale I. The basic scale factors of the Galaxy and the two kinematic tests of the long and short distance scales. Ap. J. **268**, 451.

De Vaucouleurs, G. (1989): Who discovered the local supercluster of galaxies? Observatory **109**, 237.

De Vaucouleurs, G. (1993): The extragalactic distance scale. VIII. a comparison of distance scales. Ap. J. **415**, 10.

De Vaucouleurs, G. (1993): Tests of the long and short extragalactic distance scales. Proc. Nat. Acad. Sci. **90**, 4811.

De Vaucouleurs, G., Bollinger, G. (1979): The extragalactic distance scale. VII. The velocity-distance relations in different directions and the Hubble ratio within and without the local supercluster. Ap. J. **233**, 433.

De Vaucouleurs, G., De Vaucouleurs, A.: Reference catalogue of bright galaxies. Austin, Texas: Univ. of Texas Press 1964

De Young, D. S. (1976): Extended extragalactic radio sources. Ann. Rev. Astron. Astrophys. **14**, 447.

Dekel, A. (1994): Dynamics of cosmic flows. Ann. Rev. Astron. Astrophys. **32**, 371.

Deliyannis, C. P., et al. (1989): Primordial lithium and the standard model(s). Phys. Rev. Lett. **62**, 1583.

Dey, A., et al. (1998): A galaxy at $z = 5.34$. Ap. J. (Letters) **498**, L93.

Dicke, R. H. (1960): Eötvös experiment and the gravitational red shift. Am. J. Phys. **28**, 344.

Dicke, R. H. (1962): Mach's principle and invariance under transformation of units. Phys. Rev. **125**, 2163.

Dicke, R. H. (1964): The Sun's rotation and relativity. Nature **202**, 432.

Dicke, R. H. (1974): The oblateness of the Sun and relativity. Science **184**, 419.

Dicke, R. H., Goldenberg, H. M. (1967): Solar oblateness and general relativity. Phys. Rev. Lett. **18**, 313.

Dicke, R. H., Goldenberg, H. M. (1974): The oblateness of the sun. Ap. J. Supp. **27**, 131.

Dicke, R. H., Peebles, P. J. E.: The big bang cosmology – enigmas and nostrums. In: General Relativity – an Einstein centenary survey (eds. S. W. Hawking and W. Israel). Cambridge, England: Cambridge University Press 1979.

Dicke, R. H., Peebles, P. J. E., Roll, P. G., Wilkinson, D. T. (1965): Cosmic black-body radiation. Ap. J. **142**, 414.

Dickey, J. O., et al. (1994): Lunar laser ranging – a continuing legacy of the Apollo program. Science **265**, 482.

Dickson, F. P.: The bowl of the night. Cambridge, Mass.: M.I.T. Press 1968.

Dirac, P. A. M. (1937): The cosmological constants. Nature **139**, 323.

Dirac, P. A. M. (1938): A new basis for cosmology. Proc. Roy. Soc. London **A165**, 199.

Dirac, P. A. M. (1974): Cosmological models and the large numbers hypothesis. Proc. Roy. Soc. London **A338**, 439.

Disney, M.: The Hidden Universe. New York: Macmillan 1984.

Djorgovski, S., Davis, M. (1987): Fundamental properties of elliptical galaxies. Ap. J. **313**, 59.

Djorgovski, S., Spinrad, H. (1984): Discovery of a new gravitational lens. Ap. J. (Letters) **282**, L1.

Dodelson, S., Gates, E. I., Turner, M. S. (1996): Cold dark matter. Science **274**, 69.

Dodelson, S., Jubas, J. M. (1993): Microwave anisotropies in the light of the data from the COBE satellite. Phys. Rev. Lett. **70**, 2224.

Dolgov, A. D., Zeldovich, Y. B. (1981): Cosmology and elementary particles. Rev. Mod. Phys. **53**, 1.

Donnelly, R. H., Partridge, R. B., Windhorst, R. A. (1987): Six centimeter radio source counts and spectral index studies down to 0.1 millijansky. Ap. J. **321**, 94.

Doroshkevich, A. G., Novikov, I. D. (1964): Mean density of radiation in the metagalaxy and certain problems in relativistic cosmology. Sov. Phys. Doklady **9**, 111.

Doroshkevich, A. D., Novikov, I. D. (1978): Space-time and physical fields in black holes. Zhur. Eks. Teor. Fiziki **74**, 3. English translation in Sov. Phys. JETP **47**, 1 (1978).

Doroshkevich, A. D., Zeldovich, Y. B., Novikov, I. D. (1965): Gravitational collapse of nonsymmetric and rotating masses. Zhurn. Eks. Teor. Fiziki **49**, 170. English translation in Sov. Phys JETP **22**, 122 (1966).

Dressler, A. (1987): The large-scale streaming of galaxies. Scientific American **257**, 38 – September.

Dressler, A. (1988): The supergalactic plane redshift survey – a candidate for the great attractor. Ap. J. **329**, 519.

Dressler, A. (1991): The great attractor: do galaxies trace the large-scale mass distribution? Nature **350**, 391.

Dressler, A., et al. (1987): Spectroscopy and photometry of elliptical galaxies I. A new distance estimator. Ap. J. **313**, 42.

Dressler, A., Faber, S. M. (1990): New measurements of distances to spirals in the great attractor: Further confirmation of the large-scale flow. Ap. J. (Letters) **354**, L45.

Dressler, A., Faber, S. M., Burstein, D., Davies, R. L., Lynden-Bell, D., Terlevich, R. J., Wegner, G. (1987): Spectroscopy and photometry of elliptical galaxies: A large-scale streaming motion in the local universe. Ap. J. (Letters) **313**, L37.

Dressler, A., Richstone, D. O. (1988): Stellar dynamics in the nuclei of M31 and M32: Evidence for massive black holes. Ap. J. **324**, 701.

Drever, R. W. P. (1961): A search for anisotropy of inertial mass using a free precession technique. Phil. Mag. **6**, 683.

Dreyer, J. L. E. (1888): New general catalogue of nebulae and clusters of stars. Mem. R.A.S. **49**, 1. Republished with the index catalogues by the Royal Astronomical Society. (1953)

Dreyer, J. L. E. (1895): Index catalogue of nebulae found in the years 1888–1894, with notes and corrections to NGC. Mem. R.A.S. **51**, 185.

Dreyer, J. L. E. (1908): Second index catalogue of nebulae and clusters of stars, continuing objects found in the years 1895–1907, with notes and corrections to NGC and to first IC. Mem. R.A.S. **59**, 105.

Droste, J. (1916): The field of a single centre in Einstein's theory of gravitation, and the motion of a particle in that field. Proc. Acad. Sci. (Amsterdam) **19**, 197.

Dunlop, J., et al. (1996): A 3.5-gyr-old galaxy at redshift 1.55. Nature **381**, 581.

Dupree, A. K. (1986): Mass loss from cool stars. Ann. Rev. Astron. Astrophys. **24**, 377.

Duvall, T. L., et al. (1984): Internal rotation of the sun. Nature **310**, 22.

Dyer, C. C. (1976): The gravitational perturbation of the cosmic background radiation by density concentrations. M.N.R.A.S. **175**, 429.

Dyer, C. C., Roeder, R. C. (1972): The distance-redshift relation for universes with no intergalactic medium. Ap. J. (Letters) **174**, L115.

Dyer, C. C., Roeder, R. C. (1973): Distance-redshift relations for universes with some intergalactic medium. Ap. J. (Letters) **180**, L31.

Dyer, C. C., Roeder, R. C. (1974): Observations in locally inhomogeneous cosmological models. Ap. J. **189**, 167.

Dyer, C. C., Roeder, R. C. (1982): A method for estimating the masses of some quasars. Ap. J. **256**, 386.

Dyson, F. J.: Gravitational machines. In: Interstellar communication (ed. A.G.W Cameron). New York: W. A. Benjamin 1963.

Dyson, F. W., Eddington, A. S., Davidson, C. (1920): A determination of the deflection of light by the sun's gravitational field, from observations made at the total eclipse of May 29, 1919. Phil. Trans. Roy. Soc. (London) **220**, 291. Reproduced in: A source book in astronomy and astrophysics 1900–1975 (eds. K. R. Lang and O. Gingerich). Cambridge, Mass.: Harvard University Press 1979.

Eardley, D. M. (1975): Observable effects of a scalar gravitational field in a binary pulsar. Ap. J. (Letters) **196**, L59.

Eardley, D. M.: In: Gravitational radiation (eds. N. Deruelle and T. Piran). Amsterdam: North Holland 1983.

Eardley, D. M., et al. (1973): Gravitational-wave observations as a tool for testing relativistic gravity. Phys. Rev. Lett. **30**, 884.

Eardley, D. M., Lee, D. L., Lightman, A. P. (1973): Gravitational-wave observations as a tool for testing relativistic gravity. Phys. Rev. **D8**, 3308.

Eardley, D. M., Press, W. H. (1975): Astrophysical processes near black holes. Ann. Rev. Astron. Astrophys. **13**, 381.

Eckart, A., Genzel, R. (1996): Observations of stellar proper motions near the Galactic centre. Nature **383**, 415.

Eddington, A. S.: Report on the relativity theory of gravitation. London: Physical Society of London 1918.
Eddington, A. S. (1919): The total eclipse of 1919 May 29 and the influence of gravitation on light. Observatory **42**, 119.
Eddington, A. S. (1920): The internal constitution of the stars. Nature **106**, 14, Observatory **43**, 341 (1920). Reproduced in: A source book in astronomy and astrophysics 1900–1975 (eds. K. R. Lang and O. Gingerich). Cambridge, Mass.: Harvard University Press 1979.
Eddington, A. S.: Space, time and gravitation. Cambridge, England: Cambridge University Press 1920.
Eddington, A. S.: The mathematical theory of relativity. Cambridge, England: Cambridge University Press 1923.
Eddington, A. S. (1924): A comparison of Whitehead's and Einstein's formulae. Nature **113**, 192.
Eddington, A. S. (1924): On the relation between the masses and luminosities of the stars. M.N.R.A.S. **84**, 308. Reproduced in: A source book in astronomy and astrophysics 1900–1975 (eds. K. R. Lang and O. Gingerich). Cambridge, Mass.: Harvard University Press 1979.
Eddington, A. S.: The internal constitution of the stars. Cambridge, England: Cambridge University Press 1926. Republished New York: Dover Publ. 1959.
Efstathiou, G. (1991): Large-scale structure in the universe. Physica Scripta **T36**, 88.
Efstathiou, G. (1993): Galaxy clustering on large scales. Proc. Nat. Acad. Sci. **90**, 4859.
Efstathiou, G., Bond, J. R., White, S. D. M. (1992): COBE background radiation anisotropies and large-scale structure in the universe. M.N.R.A.S. **258**, 1p.
Efstathiou, G., Ellis, R. S., Peterson, B. A. (1988): Analysis of a complete galaxy redshift survey – II. the field-galaxy luminosity function. M.N.R.A.S. **232**, 431.
Efstathiou, G., Rees, M. J. (1988): High-redshift quasars in the cold dark matter cosmogony. M.N.R.A.S. **230**, 5p.
Efstathiou, G., Sutherland, W. J., Maddox, S. J. (1990): The cosmological constant and cold dark matter. Nature **348**, 705.
Eggen, O. J., Lynden-Bell, D., Sandage, A. R. (1962): Evidence from the motions of old stars that the Galaxy collapsed. Ap. J. **136**, 748.
Ehlers, J., et al. (1976): Comments on gravitational radiation damping and energy loss in binary systems. Ap. J. (Letters) **208**, L77.
Eichler, D., et al. (1989): Nucleosynthesis, neutrino bursts and γ-rays from coalescing neutron stars. Nature **340**, 126.
Einasto, J., et al. (1997): A 120-Mpc periodicity in the three-dimensional distribution of galaxy superclusters. Nature **385**, 139.
Einasto, J., Jôeveer, M., Saar, E. (1980): Structure of superclusters and supercluster formation. M.N.R.A.S. **193**, 353.
Einasto, J., Kaasik, A., Saar, E. (1974): Dynamic evidence on massive coronas of galaxies. Nature **250**, 309.
Einstein, A. (1905): Über einen die Erzeugung und Verwandlung des Lichtes betreffenden heuristischen Gesichtspunkt (On a heuristic point of view concerning the generation and transformation of light). Ann. Phys. **17**, 132.
Einstein, A. (1905): Zur Elektrodynamic bewegter Körper (On the electrodynamics of moving bodies). Ann. Physik **17**, 891. Eng. trans. in: The principle of relativity. New York: Dover 1952.
Einstein, A. (1905): Ist die Trägheit eines Körpers von seinem Energieinhalt abhängig? (Does the inertia of a body depend on its energy?). Ann Phys. **18**, 639. English trans. in: A source book in astronomy and astrophysics 1900–1975 (eds. K. R. Lang and O. Gingerich). Cambridge, Mass.: Harvard University Press 1979.
Einstein, A. (1906): Das Prinzip von der Erhaltung der Schwerpunktsbewegung und die Trägheit der Energie (The principle of the conservation of the motion of the center of gravity and the inertia of energy). Ann. Phys. **20**, 627. English translation in: A source

book in astronomy and astrophysics 1900–1975 (eds. K. R. Lang and O. Gingerich). Cambridge, Mass.: Harvard University Press 1979.

Einstein, A. (1907): Über die vom Relativitätsprinzip geforderte Trägheit der Energie (Variation of intertia with energy on the principle of relativity). Ann. Phys. **23**, 371.

Einstein, A. (1907): Über das Relativitätsprinzip und die aus demselben gezogenen Fogerungen (About the principle of relativity and the conclusions drawn therefrom). Jahrb. Radioact. Elekt. **4**, 411. English translation in: The principle of relativity (ed. A. Sommerfeld). New York: Dover 1952.

Einstein, A. (1909): Zum gegenwärtigen Stand des Strahlungsproblems (Additional new opinions on radiation problems). Phys. Z. **10**, 185.

Einstein, A. (1911): Über den Einfluß der Schwerkraft auf die Ausbreitung des Lichtes (On the influence of gravitation on the propagation of light). Ann. Physik **35**, 898. English translation in: The principle of relativity (ed. A. Sommerfeld). New York: Dover 1952.

Einstein, A. (1915): Explanation of the perihelion motion of Mercury by means of the general theory of relativity. Sitz. Preuss. Akad. Wiss. **11**, 831. English translation in: A source book in astronomy and astrophysics 1900–1975 (eds. K. R. Lang and O. Gingerich). Cambridge, Mass.: Harvard University Press 1979.

Einstein, A. (1916): Die Grundlage der allgemeinen Relativitätstheorie (The foundation of the general theory of relativity). Ann. Physik **49**, 769. English trans. in: The principle of relativity (ed. A. Sommerfeld). New York: Dover 1952.

Einstein, A. (1917): Kosmologische Betrachtungen zur allgemeinen Relativitätstheorie (Cosmological considerations of the general theory of relativity). Acad. Wiss. **1**, 142. English translation in: The principle of relativity. New York: Dover 1952.

Einstein, A. (1917): Zur Quantentheorie der Strahlung (On the quantum theory of radiation). Phys. Z. **18**, 121.

Einstein, A. (1919): Spielen Gravitationsfelder im Aufbau der materiellen Elementarteilchen eine wesentliche Rolle? (Do gravitational fields play an essential part in the structure of the elementary particles of matter?). Acad. Wiss. **1**, 349. English trans. in: The principle of relativity. New York: Dover 1952.

Einstein, A. (1924): Quantentheorie des einatomigen idealen Gases (The quantum theory of the monatomic perfect gas). Preuss. Acad. Wiss. Berl. Berlin Sitz. **22**, 261; **1**, 3, 18 (1925).

Einstein, A. (1936): Lens-like action of a star by the deviation of light in the gravitational field. Science **84**, 506.

Einstein, A. (1939): On a stationary system with spherical symmetry consisting of many gravitating masses. Ann. Math. **40**, 922.

Einstein, A: The meaning of relativity. Princeton: Princeton University Press, 1945.

Einstein, A. (1950): On the generalized theory of gravitation. Scientific American **182**, 13.

Einstein, A., De Sitter, W. (1932): On the relation between the expansion and the mean density of the universe. Proc. Nat. Acad. Sci. (Wash.) **18**, 213. Reproduced in: A source book in astronomy and astrophysics 1900–1975 (ed. K. R. Lang and O. Gingerich). Cambridge, Mass.: Harvard University Press 1979.

Ellis, G. F. R. (1984): Alternatives to the big bang. Ann. Rev. Astron. Astrophys. **22**, 157.

Ellis, R. A. (1997): Faint blue galaxies. Ann. Rev. Astron. Astrophys. **35**, 389.

Elmegreen, B. G., Elmegreen, D. M. (1985): Properties of barred spiral galaxies. Ap. J. **288**, 438.

Elmegreen, B. G., Elmegreen, D. M. (1990): Optical tracers of spiral wave resonances in galaxies: Applications to NGC 1566. Ap. J. **355**, 52.

Elmegreen, B. G., Elmegreen, D. M., Montenegro, L. (1992): Optical tracers of spiral wave resonances in galaxies II. Hidden three-arm sprials in a sample of 18 galaxies. Ap. J. Supp. **79**, 37.

Elmegreen, D. M., Elmegreen, B. G. (1984): Blue and near-infrared surface photometry of spiral structure in 34 nonbarred grand design and flocculent galaxies. Ap. J. Supp. **54**, 127.

Elsworth, Y., et al. (1995): Slow rotation of the Sun's interior. Nature **376**, 669.

Encrenaz, T., Combes, M. (1982): On the C/H and D/H ratios in the atmospheres of Jupiter and Saturn. Icarus **52**, 54.

Eötvös, R. (1889): Über die Anziehung der Erde auf verschiedene Substanzen (On the attraction by the earth on different materials). Math. Nat. Ber. Ungarn **8**, 65.

Eötvös, R. V., Pekar, V., Fekete, E. (1922): Beitrage zum Gesetze der Proportionalität von Trägheit und Gravität (Contributions to the law of proportionality of inertia and gravity). Ann. Phys. **68**, 11.

Epstein, E. E. (1967): On the small-scale distribution at 3.4-mm wavelength of the reported 3 degree background radiation. Ap. J. (Letters) **148**, L157.

Epstein, R. (1977): The binary pulsar, post-Newtonian timing effects. Ap. J. **216**, 92.

Epstein, R., Wagoner, R. V. (1975): Post-Newtonian generation of gravitational waves. Ap. J. **197**, 717.

Epstein, R. I., Lattimer, J. M., Schramm, D. N. (1976): The origin of deuterium. Nature **263**, 198.

Esposito, L. W., Harrison, E. R. (1975): Properties of the Hulse-Taylor binary pulsar system. Ap. J. (Letters) **196**, L1.

Essen. L. (1969): The measurement of time. Vistas in astronomy **2**, 45.

Essen, L., Parry, J. V. L. (1957): The caesium resonator as a standard of frequency and time. Phil. Trans Roy. Soc. (London) **A250**, 46.

Estermann, I., Simpson, O. C., Stern, O. (1938): The free fall of molecules. Phys. Rev. **53**, 947.

Etherington, I. M. H. (1933): On the definition of distance in general relativity. Phil. Mag. **15**, 761.

Everitt, C. W. F.: The gyroscope experiment I. General description and analysis of gyroscope performance. In: Experimental gravitation (ed. B. Bertotti). New York: Academic 1974.

Fabbiano, G. (1989): X-rays from normal galaxies. Ann. Rev. Astron.Astrophys. **27**, 87.

Faber, S. M., Burstein, D.: Motions of galaxies in the neighborhood of the local group. In: Large-scale motions in the universe (eds.V. C. Rubin and G. V. Coyne). Princeton, New Jersey: Princeton University Press 1988.

Faber, S. M., Gallagher, J. S. (1979): Masses and mass-to-light ratios of galaxies. Ann. Rev. Astron. Astrophys. **17**, 135.

Faber, S. M., Jackson, R. E. (1976): Velocity dispersions and mass-to-light ratios for elliptical galaxies. Ap. J. **204**, 668.

Fabian, A. (1979): Theories of the nuclei of active galaxies. Proc. Roy. Soc. (London) **A366**, 449.

Fabian, A. C., Barcons, X. (1992): The origin of the X-ray background. Ann. Rev. Astron. Astrophys. **30**, 429.

Fabian, A. C., Rees, M. J. (1979): SS 433: a double jet in action? M.N.R.A.S. **187**, 13p.

Fabricant, D., Gorenstein, P. (1983): Further evidence for M87's massive dark halo. Ap. J. **267**, 535.

Feast, M. W., Catchpole, R. M. (1997): The Cepheid PL zero-point from Hipparcos trigonometrical parallaxes. M.N.R.A.S. **286**, L1.

Feast, M. W., Walker, A. R. (1987): Cepheids as distance indicators. Ann. Rev. Astron. Astrophys. **25**, 345.

Feast, M., Whitelock, P. (1997): Galactic kinematics of Cepheids from Hipparcos proper motions. M.N.R.A.S. **291**, 683.

Fedorov, E. P., et al.: The motion of the pole of the Earth from 1890.0 to 1969.0. Naukova Dumka, Kiev: Ukrainian Academy of Sciences 1972.

Feibelman, W. A. (1966): Gravitational lens effect – an observational test. Science **151**, 73.

Felten, J. E. (1985): Galaxy luminosity functions, M/L relations, and closure of the universe: Numbers and problems. Comments Astrophys. **11**, 53.

Ferguson, H. C., Williams, R. E., Cowie, L. L. (1997): Probing the faintest galaxies. Physics Today **50**, 24 – April.

Fernie, J. D. (1969): The period-luminosity relation – a historical review. P.A.S.P. **81**, 707.

Garnavich, P. M., et al. (1998): Constraints on cosmological models from Hubble space telescope observations of high-z supernovae. Ap. J. (Letters) **493**, L53.

Gates, E. I., et al. (1998): No need for Machos in the halo. Ap. J. (Letters) **500**, L145.

Gates, E. I., Gyuk, G., Turner, M. S. (1995): Microlensing and halo cold dark matter. Phys. Rev. Lett. **74**, 3724.

Gauss, C. F.: Theoria Motus ... (Theory of the motion of heavenly bodies moving about the Sun in conic sections). Trans. by C. H. Davis, Boston: Little Brown 1857.

Gawiser, E., Silk, J. (1998): Extracting primordial density fluctuations. Science **280**, 1405.

Geikie, A. (1871): On modern denudation. Trans. Geol. Soc. Glasgow **3**, 153.

Geiss, J.: Primordial abundances of hydrogen and helium isotopes. In: Origin and evolution of the elements (ed. N. Prantzos, E. Vangioni-Flam and M. Cassé). New York: Cambridge University Press 1993.

Geiss, J., Reeves, H. (1972): Cosmic and solar system abundances of deuterium and helium-3. Astron. Astrophys. **18**, 126.

Geller, M.: Mapping the universe: slices and bubbles. In: Bubbles, voids, and bumps in time: the new cosmology. New York: Cambridge University Press 1989.

Geller, M. J., Huchra, J. P. (1989): Mapping the universe. Science **246**, 857.

Genzel, R., Townes, C. H. (1987): Physical conditions, dynamics and mass distribution in the center of the Galaxy. Ann. Rev. Astron. Astrophys. **25**, 377.

Georgelin, Y. M., Georgelin, Y. P. (1976): The spiral structure of our Galaxy determined from H II regions. Astron. Astrophys. **49**, 57.

Gershtein, S. S., Zeldovich, Y. B. (1966): Rest mass of muonic neutrino and cosmology. ZhETF Pis'ma **4**, No. 5, 174 (1966). English translation in JETP Letters **4**, 120.

Giacconi, R., et al. (1962): Evidence for X-rays from sources outside the solar system. Phys. Rev. Lett. **9**, 439. Reproduced in: A source book in astronomy and astrophysics 1900–1975 (eds. K. R. Lang and O. Gingerich). Cambridge, Mass.: Harvard University Press 1979.

Giacconi, R., et al. (1972): The UHURU catalogue of X-ray sources. Ap. J. **178**, 281.

Gibbons, G. W., Hawking, S. W. (1977): Action integrals and partition functions in quantum gravity. Phys. Rev. **D15**, 2752.

Gieren, W. P. (1988): The galactic cepheid period-luminosity relation from the visual surface brightness method. Ap. J. **329**, 790.

Gieren, W. P., Barnes, T. G., Moffett, T. J. (1989): The period-radius relation for classical cepheids from the visual surface brightness technique. Ap. J. **342**, 467.

Gieren, W. P., Barnes, T. G., Moffett, T. J. (1993): The Cepheid period-luminosity relation from independent distances of 100 galactic variables. Ap. J. **418**, 135.

Gies, D. R., Bolton, C. T. (1986): The optical spectrum of HDE 226868 = Cygnus X-1. II. Spectrophotometry and mass estimates. Ap. J. **304**, 371.

Gilmore, G., Wyse, R. F. G., Kuijken, K. (1989): Kinematics, chemistry, and structure of the Galaxy. Ann. Rev. Astron. Astrophys. **27**, 555.

Ginzburg, V. L. (1956): The nature of cosmic radio emission and the origin of cosmic rays. Nuovo Cimento Supplement **3**, 38. Reproduced in: A source book in astronomy and astrophysics 1900–1975 (eds. K. R. Lang and O. Gingerich). Cambridge, Mass.: Harvard University Press 1979.

Ginzburg, V. L. (1964): The magnetic fields of collapsing masses and the nature of superstars.: Dokl. Akad. Nauk SSSR **156**, 43. English translation in Sov. Phys. Doklady **9**, 329 (1964).

Giovanelli, R., et al. (1997): The Tully-Fisher relation and H_o. Ap. J. (Lettters) **477**, L1.

Giovanelli, R., Haynes, M. P. (1991): Redshift surveys of galaxies. Ann. Rev. Astron. Astrophys. **29**, 499.

Giovanelli, R., Haynes, M. P., Chincarini, G. L. (1986): Morphological segregation in the Pisces-Perseus supercluster. Ap. J. **300**, 77.

Giroux, M. L., Fardal, M. A., Shull, J. M. (1995): Intergalactic helium absorption toward high-redshift quasars. Ap. J. **451**, 477.

Glanz, J. (1997): Clouds gather over deuterium sighting. Science **275**, 158.

Glauert, H. (1915): The rotation of the Earth. M.N.R.A.S. **75**, 489.

Godwin, J. G., Metcalfe, N., Peach, J. V. (1983): The Coma cluster – I. a catalogue of magnitudes, colours, ellipticities and position angles for 6724 galaxies in the field of the Coma cluster. M.N.R.A.S. **202**, 113.

Gold, T. (1968): Rotating neutron stars as the origin of the pulsating radio sources. Nature **218**, 731. Reproduced in: A source book in astronomy and astrophysics 1900–1975 (eds. K. R. Lang and O. Gingerich). Cambridge, Mass.: Harvard University Press 1979.

Goldsmith, D.: Einstein's greatest blunder? The cosmological constant and other fudge factors in the physics of the universe. Cambridge, Mass.: Harvard University Press 1995.

Goldstein, H.: Classical mechanics. Reading, Mass.: Addison-Wesley 1950.

Gooding, A. K., Spergel, D. N., Turok, N. (1991): The formation of galaxies and quasars in a texture-seeded cold dark matter cosmogony. Ap. J. (Letters) **372**, L5.

Gorenstein, M. V., et al. (1984): The milli-arcsecond images of Q 0957+561. Ap. J. **287**, 538.

Gorenstein, M. V., et al. (1988): VLBI observations of the gravitational lens system 0957+561 - structure and relative magnification of the a and b images. Ap. J. **334**, 42.

Gorenstein, M. V., Smoot, G. F. (1981): Large-angular-scale anisotropy in the cosmic background radiation. Ap. J. **244**, 361.

Gorski, K. M., et al. (1994): On determining the spectrum of primordial inhomogeneity from the COBE DMR sky maps: Results of two-year data analysis. Ap. J. (Letters) **430**, L89.

Gorski, K. M., et al. (1996): Power spectrum of primordial inhomogenieity determined from the four-year COBE DMR sky maps. Ap. J. (Letters) **464**, L11.

Gorski, K. M., Silk, J. Vittorio, N. (1992): Cold dark matter confronts the cosmic microwave backgound – large-angular-scale anisotropies in cosmological density parameter plus cosmological constant equal one models. Phys. Rev. Lett. **68**, 733.

Gott, J. R. (1983): Gravitational lenses. Am. Scientist **71**, 150.

Gott, J. R., Gunn, J. E. (1974): The double quasar 1548+115 a,b as a gravitational lens. Ap. J. (Letters) **190**, L105.

Gott, J. R., Gunn, J. E., Schramm, D. N., Tinsley, B. M. (1974): An unbound universe? Ap. J. **194**, 543.

Gould, A. (1993): An estimate of the COBE quadrupole. Ap. J. (Letters) **403**, L51.

Gould, R. J. (1982): Processes in relativistic plasmas. Ap. J. **254**, 755.

Gould, R. J. (1982): The temperature of thermal X-ray and γ-ray sources. Ap. J. **258**, 131.

Gould, R. J. (1982): Effects of nuclear forces on ion thermalization in high-temperature plasmas. Ap. J. **263**, 879.

Gratton, R. G., et al. (1997): Ages of globular clusters from Hipparcos parallaxes of local subdwarfs. Ap. J. **491**, 749.

Green, E. M., Demarque, P., King, C. R.: The revised Yale isochrones and luminosity functions. New Haven: Yale University Observatory, 1987.

Greenfield, P. E., Burke, B. F., Roberts, D. H. (1980): The double quasar 0957+561 – examination of the gravitational lens hypothesis using the Very Large Array. Science **208**, 495.

Greenfield, P. E., Roberts, D. H., Burke, B. F. (1980): The double quasar 0957+561 as a gravitational lens – further VLA observations. Nature **286**, 865.

Greenfield, P. E., Roberts, D. H., Burke, B. F. (1985): The gravitationally lensed quasar 0957+561 – VLA observations and mass models. Ap. J. **293**, 370.

Greenhill, L. J., et al. (1996): VLBI imaging of water maser emission from the nuclear torus of NGC 1068. Ap. J. (Letters) **472**, L21.

Greenstein, J. L., Matthews, T. A. (1963): Red-shift of the unusual radio source: 3C 48. Nature **197**, 1041. Reproduced in: A source book in astronomy and astrophysics 1900–1975 (eds. K. R. Lang and O. Gingerich). Cambridge, Mass.: Harvard University Press 1979.

Greenstein, J. L., Oke, J. B., Shipman, H. L. (1971): Effective temperature, radius, and gravitational redshift of Sirius B. Ap. J. **169**, 563.

Greenstein, J. L., Trimble, V. (1967): The Einstein redshift in white dwarfs. Ap. J. **149**, 283.

Greenstein, J. L., Trimble, V. (1972): The gravitational redshift of 40 Eridani B. Ap. J. (Letters) **175**, L1.

Gregory, S. A., Thompson, L. A. (1978): Low mass portion of the galaxy clustering spectrum. Nature **274**, 450.

Gregory, S. A., Thompson, L. A. (1978): The Coma/A1367 supercluster and its environs. Ap. J. **222**, 784.

Gregory, S. A., Thompson, L. A. (1982): Superclusters and voids in the distribution of galaxies. Scientific American **246**, 106 – March.

Gregory, S. A., Thompson, L. A., Tifft, W. G. (1981): The Perseus supercluster. Ap. J. **243**, 411.

Grenon, M.: The age of the galactic disk. In: Astrophysical ages and dating methods (eds. E. Vangioni-Flam, M. Cassé, J. Audouze and J. Tran Thanh Van). Paris: Editions Frontieres 1990.

Gribbin, J.: The omega point. The search for the missing mass and the ultimate fate of the universe. London: William Heinemann 1987.

Gribbin, J.: In the beginning: After cobe and before the big bang. Boston: Little, Brown and Co. 1993.

Grossman, S. A., Narayan, R. (1988): Arcs from gravitational lensing. Ap. J. (Letters) **324**, L37.

Grossman, S. A., Narayan, R. (1989): Gravitationally lensed images in Abell 370. Ap. J. **344**, 637.

Groth, E. J., Peebles, P. J. E. (1977): Statistical analysis of catalogs of extragalactic objects VII. Two- and three-point correlation functions for the high-resolution Shane-Wirtanen catalog of galaxies. Ap. J. **217**, 385.

Gum, C. S., De Vaucouleurs, G. (1952): Large ring-like H II regions as extra-galactic distance indicators. Observatory **73**, 152.

Gundersen, J. O., et al. (1995): Degree-scale anisotropy in the cosmic microwave background: SP94 results. Ap. J. (Letters) **443**, L57.

Guerra, E. J., Daly, R. A. (1998): Central engines of active galactic nuclei: Properties of collimated outflows and applications for cosmology. Ap. J. **493**, 536.

Gunn, J. E. (1967): A fundamental limitation on the accuracy of angular measurements in observational cosmology. Ap. J. **147**, 61.

Gunn, J. E. (1974): A worm's-eye view of the mass density in the universe. Comm. Astrophys. Space Sci. **6**, 7.

Gunn, J. E. (1977): Massive galactic halos I. Formation and evolution. Ap. J. **219**, 592.

Gunn, J. E.: In: Observatonal Cosmology (eds. A. Maeder, L. Martinet and G. Tammann). Sauverny: Geneva Observatory 1978.

Gunn, J. E., Oke, J. B. (1975): Spectrophotometry of faint cluster galaxies and the Hubble diagram: An approach to cosmology. Ap. J. **195**, 255.

Gunn, J. E., Peterson, B. E. (1965): On the density of neutral hydrogen in intergalactic space. Ap. J. **142**, 1633.

Gursky, H., Schwartz, D. A. (1977): Extragalactic X-ray sources. Ann. Rev. Astron. Astrophys. **15**, 541.

Gurzadyan, V. G., Savvidy, G. K. (1986): Collective relaxation of stellar systems. Astron. Astrophys. **160**, 203.

Gush, H. P., Halpern, M., Wishnow, E. H. (1990): Rocket measurement of the cosmic-background-radiation mm-wave spectrum. Phys. Rev. Lett. **65**, 537.

Guth, A. H. (1981): Inflationary universe: a possible solution to the horizon and flatness problems. Phys. Rev. **D23**, 347.

Guth, A. H. (1993): Inflation. Proc. Nat. Acad. Sci. **90**, 4871.

Guth, A. H.: The inflationary universe. The quest for a new theory of cosmic origins. Reading, Mass.: Addison-Wesley 1997.

Guth, A. H., Pi, S.-Y. (1982): Fluctuations in the new inflationary universe. Phys. Rev. Lett. **49**, 1110.

Guth, A. H., Steinhardt, P. J. (1984): The inflationary universe. Scientific American **250**, 116 – May.

Guth, A. H., Steinhardt, P.: The infationary universe. In: the new physics (ed. P. Davies). Cambridge, England: Cambridge University Press, 1989.

Guth, A. H., Weinberg, E. J. (1981): Cosmological consequences of a first-order phase transition in the SU_3 grand unified model. Phys. Rev. **D23**, 876.

Hafele, J. C., Keating, R. E. (1972): Around-the-world atomic clocks – predicted relativistic time gains. Science **177**, 166.

Hafele, J. C., Keating, R. E. (1972): Around-the-world atomic clocks – observed relativistic time gains. Science **177**, 168.

Hainebach, K. L., Schramm, D. N. (1976): Galactic evolution models and the rhenium-187/osmium-187 chronometer – a greater age for the Galaxy. Ap. J. Lett. **207**, L87.

Hale, G. E. (1928): The possibilities of large telescopes. Harper's magazine **156**, 639. Reproduced in: A source book in astronomy and astrophysics 1900-1975 (eds. K. R. Lang and O. Gingerich). Cambridge, Mass.: Harvard University Press 1979.

Hall, A. (1894): A suggestion in the theory of Mercury. Astron. J. **14**, 49.

Hall, D. N. B. (1975): Spectroscopic detection of solar ^{3}He. Ap. J. **197**, 509.

Halley, E. (1718): Considerations on the change in the latitudes of some of the prinicipal fixed stars. Phil. Trans. **30**, 756.

Halley, E. (1720): On the infinity of the sphere of fix'd stars. Phil. Trans. **31**, 22.

Halley, E. (1720): Of the number, order and light of the fix'd stars. Phil. Trans. **31**, 24. Reprod. in Jaki (1969).

Halpern, J. P., Holt, S. S. (1992): Discovery of soft X-ray pulsations from the γ-ray source Geminga. Nature **357**, 222.

Hammersley, P. L., Garzon, F., Mahoney, T., Calbet, X. (1995): The tilted old galactic disc and the position of the Sun. M.N.R.A.S. **273**, 206.

Hamuy, M., et al. (1995): A Hubble diagram of distant type Ia supernovae. Astron. J. **109**, 1.

Han, M., Mould, J. R. (1992): Peculiar velocities of clusters in the Perseus-Pisces supercluster. Ap. J. **396**, 453.

Hancock, S., et al. (1994): Direct observation of structure in the cosmic microwave background. Nature **367**, 333.

Hanes, D. A. (1977): Globular clusters and the Virgo cluster distance modulus. M.N.R.A.S. **180**, 309.

Hanes, D. A. (1979): A new determination of the Hubble constant. M.N.R.A.S. **188**, 901.

Hanes, P. A. (1982): A re-examination of the Sandage-Tammann extragalactic distance scale. M.N.R.A.S. **201**, 145.

Hanson, R. B.: The Hyades cluster distance. In: star clusters IAU symposium 85 (ed. J. E. Hesser). Dordrecht: Riedel 1980.

Hardee, P. E., Owen, F. N., Cornwell, T. J.: VLA observations of the M87 jet and small scale jet structures. In: Active galactic nuclei. Lecture notes in physics 307 (eds. H. R. Miller and P. J. Wiita). New York: Springer Verlag 1988.

Hargreaves, R. (1908): Integral forms and their connexion with physical equations. Trans. Camb. Phil. Soc. **21**, 107.

Harmon, B. A., et al. (1995): Correlations between X-ray outbursts and relativistic ejections in the X-ray transient GRO J1655–40. Nature **374**, 703.

Harms, R. J., et al. (1994): HST FOS spectroscopy of M87: Evidence for a disk of ionized gas around a massive black hole. Ap. J. (Letters) **435**, L35.

Harris, D. E., Grindlay, J. E. (1979): The prospects for X-ray detection of inverse-Compton emission from radio source electrons and photons of the microwave background. M.N.R.A.S. **188**, 25.

Harris, D. L., Strand, K. A., Worley, C. E.: Empirical data on stellar masses, luminosities and radii. In: Basic astronomical data – Stars and stellar systems III. (ed. K. Strand). Chicago: University of Chicago Press 1963.

Harris, W. E. (1991): Globular cluster systems beyond the local group. Ann. Rev. Astron. Astrophys. **29**, 543.

Harrison, B. K., Wakano, M., Wheeler, J. A.: Matter-energy at high density: End point of thermonuclear evolution. In: La structure et l'evolution de l'univers. Institut intrenational de physique Solvay (ed. R. Stoops). Brussels: Coudenberg 1958.

Harrison, E. R. (1970): Fluctuations at the threshold of classical cosmology. Phys. Rev. **D1**, 2726.

Harrison, E. (1988): Black holes in history. Q.J.R.A.S. **29**, 87.

Hartle, J. B. (1978): Bounds on the mass and moment of inertia of non-rotating neutron stars. Physics Reports **46**, 201.

Hartwick, F. D. A., Schade, D. (1990): The space distribution of quasars. Ann. Rev. Astron. Astrophys. **28**, 437.

Harwit, M.: Astrophysical concepts, second edition. New York: Springer-Verlag 1988.

Haselgrove, B., Hoyle, F. (1959): Main-sequence stars. M.N.R.A.S. **119**, 112.

Hata, N., et al. (1996): Predicting big bang deuterium. Ap. J. **458**, 637.

Hatcher, D. A. (1984): Simple formulae for Julian day numbers and calendar dates. Q.J.R.A.S. **25**, 53.

Hatcher, D. A. (1985): Generalized equations of Julian day numbers and calendar dates. Q.J.R.A.S. **26**, 151.

Haugan, M. P., Will, C. M. (1987): Modern tests of special relativity. Phys. Today **40**, 69 – May.

Hauser, M. G., Peebles, P. J. E. (1973): Statistical analysis of catalogs of extragalactic objects II. The Abell catalog of rich clusters. Ap. J. **185**, 757.

Hawking, S. W. (1966): The occurrences of singularities in cosmology. Proc. Roy. Soc. (London) **A294**, 511.

Hawking, S. W. (1971): Gravitational radiation from colliding black holes. Phys. Rev. Lett. **26**, 1344.

Hawking, S. W. (1972): Black holes in general relativity. Comm. Math. Phys. **25**, 152.

Hawking, S. W. (1974): Black hole explosions? Nature **248**, 30.

Hawking, S. W. (1976): Black holes and thermodynamics. Phys. Rev. **D13**, 191.

Hawking, S. W. (1982): The development of irregularities in a single bubble inflationary universe. Phys. Lett. **115B**, 295.

Hawking, S. W., Penrose, R. (1970): The singularities of gravitational collapse and cosmology. Proc. Roy. Soc. (London) **A314**, 529.

Hawking, S. W., Penrose, R.: The nature of space and time. Princeton, New Jersey: Princeton University Press 1996. Also see Scientific American **275**, 60 (1996) – July.

Hawking, S. W., Turok, N. (1998): Open inflation without false vacua. Physics Letters **B425**, 25.

Hayakawa, S., Matsuoka, M. (1964): Origin of cosmic X-rays. Prog. Theor. Phys. Supp. (Japan) **30**, 217.

Hayashi, C. (1950): Proton-neutron concentration ratio in the expanding universe at the stages preceding the formation of the elements. Prog. Theor. Phys. (Japan) **5**, 224.

Haynes, M. P., Giovanelli, R. (1986): The connection between Pisces-Perseus and the local supercluster. Ap. J. (Letters) **306**, L55.

Hazard, C., Mackey, M. B., Shimmins, A. J. (1963): Investigation of the radio source 3C 273 by the method of lunar occultations. Nature **197**, 1037. Reproduced in: A source book in astronomy and astrophysics 1900-1975 (eds. K. R. Lang and O. Gingerich). Cambridge, Mass.: Harvard University Press 1979.

Heckel, B. R., et al. (1989): Experimental bounds on interactions mediated by ultralow-mass bosons. Phys. Rev. Lett. **63**, 2705.

Hege, E. K., et al. (1980): Morphology of the triple QSO PG1115+08. Nature **287**, 416.

Hege, E. K., et al. (1981): Speckle interferometry observations of the triple QSO PG 1115+08. Ap. J. (Letters) **248**, L1.

Hegyi, D. J., Olive, K. A. (1983): Can galactic halos be made of baryons? Phys. Lett. **126B**, 28.

Hellings, R. W., et al. (1983): Experimental test of the variability of G using Viking lander ranging data. Phys. Rev. Lett. **51**, 1609.

Helmholtz, H. (1856): On the interaction of natural forces. Popular lecture on 7 February 1854. Phil. Mag. Ser 4. **11**, 489.

Henderson, T. (1839): Parallax of α Centauri, communicated to the astronomical society january 9, 1839. See Clerk, A.: A popular history of astronomy during the nineteenth century. Edinburgh: Adam and Charles Black 1885.

Henry, P. S. (1971): Isotropy of the 3 K background. Nature **231**, 516.

Henry, R. C. (1991): Ultraviolet background radiation. Ann. Rev. Astron. Astrophys. **29**, 89.

Herbig, T., et al. (1995): A measurement of the Sunyaev-Zeldovich effect in the Coma cluster of galaxies. Ap. J. (Letters) **449**, L5.

Hernanz, M., et al.: The age of the Galaxy obtained from white dwarfs. In: Astrophysical ages and dating methods (eds. E. Vangioni-Flam, M. Cassé, J. Audouze and J. Tran Thanh Van). Paris: Editions Frontieres 1990.

Hernquist, L., et al. (1996): The lyman-alpha forest in the cold dark matter model. Ap. J. (Letters) **457**, L51.

Herschel, J. (1864): A general catalogue of nebulae and star clusters. Philosphical Transactions **114**, 27.

Herschel, Sir F. W. (1753): On the proper motion of the sun and solar system, with an account of several changes that have happened among the fixed stars since the time of Mr. Flamsteed. Phil. Trans. **73**, 247.

Hertzsprung, E. (1905): On the radiation of stars. Zeits. f. Wiss. Photo. **3**, 429. English translation in: A source book in astronomy and astrophysics 1900–1975 (eds. K. R. Lang and O. Gingerich). Cambridge, Mass.: Harvard University Press 1979.

Hertzsprung, E. (1913): Über die räumliche Verteilung der Veränderlichen vom delta Cephei-Typus. (About the space distribution of the variables of the delta Cephei type). Aston. Nach. **196**, 201.

Hesser, J. E., et al. (1987): A ccd color-magnitude study of 47 Tucanae. P.A.S.P. **99**, 739.

Hewish, A., et al. (1968): Observation of a rapidly pulsating radio source. Nature **217**, 709. Reproduced in: A source book in astronomy and astrophysics 1900–1975 (eds. K. R. Lang and O. Gingerich). Cambridge, Mass.: Harvard University Press 1979.

Hewitt, J. N., et al. (1987): The triple radio source 0023 + 171 – a candidate for a dark gravitational lens. Ap. J. **321**, 706.

Hewitt, J. N., et al. (1988): Unusual radio source MG 1131 + 0456 – a possible Einstein ring. Nature **333**, 537.

Heyl, P. R. (1930): A redetermination of the constant of gravitation. J. Res. Nat. Bur. Standards **5**, 1243.

Higbie, J. (1981): Gravitational lens. Am. J. Phys. **49**, 652.

Higdon, J. C., Lingenfelter, R. E. (1990): Gamma-ray bursts. Ann. Rev. Astron. Astrophys. **28**, 401.

Hill, E. R. (1960): The component of the galactic gravitational field perpendicular to the galactic plane. B.A.N. **15**, 1.

Hill, H. A., et al. (1974): Solar oblateness, excess brightness and relativity. Phys. Rev. Lett. **33**, 1497.

Hill, H. A., Stebbins, R. T. (1975): The intrinsic visual oblateness of the Sun. Ap. J. **200**, 471.

Hill, J. M. (1971): A measurement of the gravitational deflection of radio waves by the Sun. M.N.R.A.S. **153**, 7p.

Hillas, A. M. (1984): The origin of ultra-high-energy cosmic rays. Ann. Rev. Astron. Astrophys. **22**, 425.

Hils, D., Hall, J. L. (1990): Improved Kennedy-Thorndike experiment to test special relativity. Phys. Rev. Lett. **64**, 1697.

Hinshaw, G., et al. (1996): Band power spectra in the COBE DMR four-year anisotropy maps. Ap. J. (Letters) **464**, L17.

Hipparchus, 125 B.C. in Ptolemy's Almagest. English translation by G. J. Toomer (London: Duckworth 1984).
Hjellming, R. M. (1973): An astronomical puzzle called Cygnus X-3. Science **182**, 1089.
Hjellming, R. M.: Radio stars: In: Galactic and extragalactic radio astronomy. Second edition (eds. G. L. Verschuur and K. I. Kellermann). New York: Springer Verlag 1988.
Hjellming, R. M., Johnston, K. J. (1981): An analysis of the proper motions of SS 433 radio jets. Ap. J. (Letters) **246**, L141.
Hjellming, R. M., Johnston, K. J. (1981): Structure, strength, and polarization changes in radio source SS 433. Nature **290**, 100.
Hjellming, R. M., Rupen, M. P. (1995): Episodic ejection of relativistic jets by the X-ray transient GRO J1655–40. Nature **375**, 464.
Hobbs, L. M., Pilachowski, C. (1988): Lithium in an extreme halo star. Ap. J. (Letters) **326**, L23.
Hodge, P. W. (1981): The extragalactic distance scale. Ann. Rev. Astron. Astrophys. **19**, 357.
Hoffman, G. L., Salpeter, E. E. (1982): Dynamical models and our virgocentric deviation from Hubble flow. Ap. J. **263**, 485.
Höflich, P., Khokhlov, A. (1996): Explosion models for Type Ia supernovae – a comparison with observed light curves, distances, H_o, and q_o. Ap. J. **457**, 500.
Hogan, C. J. (1996): Primordial deuterium and the big bang. Scientific American **275**, 68 – December.
Hogg, D. E., et al. (1969): Synthesis of brightness distribution in radio sources. Astron. J. **74**, 1206.
Hogg, D. E., MacDonald, G. H., Conway, R. G., Wade, C. M. (1969): Synthesis of brightness distribution in radio sources. Astron. J. **74**, 1206.
Holden, D. J. (1966): An investigation of the clustering of radio sources. M.N.R.A.S. **133**, 225.
Hollywood, J. M., et al. (1995): General relativistic flux modulations in the galactic center black hole candidate Sagittarius A*. Ap. J. (Letters) **448**, L21.
Holmes, A., Lawson, R. W. (1926): Radioactivity of potassium and its geological significance. Phil. Mag. **2**, 1218.
Hooley, A., Longair, M. S., Riley, J. M. (1978): The angular diameter–redshift test for quasi-stellar radio sources with large redshifts. M.N.R.A.S. **182**, 127.
Howell, T. F., Shakeshaft, J. R. (1966): Measurement of the minimum cosmic background radiation at 20.7-cm wavelength. Nature **210**, 1318.
Howell, T. F., Shakeshaft, J. R. (1967): Spectrum of the 3° K cosmic microwave radiation. Nature **216**, 753.
Hoyle, F. (1948): A new model for the expanding universe. M.N.R.A.S. **108**, 372.
Hoyle, F.: The relation of radio astronomy to cosmology. In Radio astronomy, IAU symposium No. 9 (ed. R. N. Bracewell). Stanford, Calif.: Stanford University Press 1959.
Hoyle, F. (1968): Review of recent developments in cosmology. Proc. Roy. Soc. (London) **A308**, 1.
Hoyle, F., Burbidge, G. R., Sargent, W. L. W. (1966): On the nature of quasi-stellar sources. Nature **209**, 751.
Hoyle, F., Fowler, W. A. (1960): Nucleosynthesis in supernovae. Ap. J. **132**, 565.
Hoyle, F., Fowler, W. A. (1963): Nature of strong radio sources. Nature **197**, 533.
Hoyle, F., Fowler, W. A. (1963): On the nature of strong radio sources. M.N.R.A.S. **125**, 169.
Hoyle, F., Lyttleton, R. A. (1939): The evolution of the stars. Proc. Camb. Phil. Soc. **35**, 592.
Hoyle, F., Narlikar, J. V. (1966): A conformal theory of gravitation. Proc. Roy. Soc. (London) **A294**, 138.
Hoyle, F., Sandage, A. R. (1956): The second-order term in the redshift-magnitude relation. P.A.S.P. **68**, 301.

Hoyle, F., Tayler, R. J. (1964): The mystery of the cosmic helium abundance. Nature **203**, 1108.

Hubble, E. P. (1925): Cepheids in spiral nebulae. P.A.S.P. **5**, 261, Observatory **48**, 139 (1925). Reproduced in: A source book in astronomy and astrophysics 1900–1975 (eds. K. R. Lang and O. Gingerich). Cambridge, Mass.: Harvard University Press 1979.

Hubble, E. P. (1926): Extra-galactic nebulae. Ap. J. **64**, 321. Reproduced in: A source book in astronomy and astrophysics 1900–1975 (eds. K. R. Lang and O. Gingerich). Cambridge, Mass.: Harvard University Press 1979.

Hubble, E. P. (1929): A relation between distance and radial velocity among extra-galactic nebulae. Proc. Nat. Acad. Sci. (Wash.) **15**, 168. Reproduced in: A source book in astronomy and astrophysics 1900–1975 (eds. K. R. Lang and O. Gingerich). Cambridge, Mass.: Harvard University Press 1979.

Hubble, E. P. (1934): The distribution of extra-galactic nebulae. Ap. J. **79**, 8.

Hubble, E. P.: The realm of the nebulae. New Haven, Conn.: Yale University Pess 1936. Republ. New York: Dover 1958.

Hubble, E. P. (1936): The luminosity function of nebulae. Ap. J. **84**, 158.

Hubble, E. P. (1939): The motion of the galactic system among the nebulae. J. Franklin Institute **228**, 131.

Hubble, E. P., Humason, M. L. (1931): The velocity-distance relation among extra-galactic nebulae. Ap. J. **74**, 43.

Hubble, E. P., Humason, M. L. (1934): The velocity-distance relation for isolated extra-galactic nebulae. Proc. Nat. Acad. Sci. **20**, 264.

Huchra, J. P.: The cosmological distance scale. In: 13th Texas Symposium on Relativistic Astrophysics (ed. M. P. Ulmer). Singapore: World Scientific Pub. 1987.

Huchra, J. P. (1988): On infall into the Virgo cluster. In: The extragalactic distance scale (ed. S. Van Den Bergh and C. J. Pritchet). Proc. A.S.P. 100th Ann. Symp. **4**, 257.

Huchra, J., et al. (1985): 2237+0305 – a new and unusual gravitational lens. Astron. J. **90**, 691.

Huchra, J. P., Geller, M. J. (1982): Groups of galaxies I. Nearby groups. Ap. J. **257**, 423.

Hughes, R. B., Longair, M. S. (1967): Evidence on the isotropy of faint radio sources. M.N.R.A.S. **135**, 131.

Hughes, V. W., Robinson, H. G., Beltran-Lopez, V. (1960): Upper limit for the anisotropy of inertial mass from nuclear resonance experiments. Phys. Rev. Lett. **4**, 342.

Hulse, R. A. (1994): The discovery of the binary pulsar. Rev. Mod. Phys. **66**, 699.

Hulse, R. A., Taylor, J. H. (1975): Discovery of a pulsar in a binary system. Ap. J. (Letters) **195**, L51.

Humason, M. L., Mayall, N. U., Sandage, A. R. (1956): Redshifts and magnitudes of extra-galactic nebulae. Astron. J. **61**, 97. Reproduced in: A source book in astronomy and astrophysics 1900-1975 (eds. K. R. Lang and O. Gingerich). Cambridge, Mass.: Harvard University Press 1979.

Humphreys, R. M., Larsen, J. A. (1995): The sun's distance above the galactic plane. Astron. J. **110**, 2183.

Hutchings, J. B., et al. (1987): Optical and uv spectroscopy of the black hole candidate LMC X-1. Astron. J. **94**, 340.

Iben, I. Jr., Renzini, A. (1983): Asymptotic giant branch evolution and beyond. Ann. Rev. Astron. Astrophys. **21**, 271.

Iben, I. Jr., Renzini, A. (1984): Single star evolution I. Massive stars and early evolution of low and intermediate mass stars. Phys. Rpt. **105**, 329.

Iben, I. Jr., Rood, R. T. (1970): Metal-poor stars I. evolution from the main sequence to the giant branch. Ap. J. **159**, 605.

Iben, I. Jr., Rood, R. T. (1970): Metal-poor stars III. on the evolution of horizontal-branch stars. Ap. J. **161**, 587.

Iben, I., Jr., Tuggle, R. S. (1972): Comments on a PLC relationship for Cepheids and on the comparison between pulsation and evolution masses for Cepheids. Ap. J. **178**, 441.

Kaufman, S. E. (1971): A complete redshift-magnitude formula. Astron. J. **76**, 751.

Kaufman, S. E., Schucking, E. L. (1971): A generalized redshift-magnitude formula. Astron. J. **76**, 583.

Kaula, W. M.: Theory of satellite geodesy. Waltham, Mass.: Blaisdell 1966.

Kayser, R., et al. (1990): New observations and gravitational lens models of the cloverleaf quasar H1413+117. Ap. J. **364**, 15.

Kayser, R., Refsdal, S. (1983): The difference in light travel time between gravitational lens images. Astron. Astrophys. **128**, 156.

Kayser, R., Refsdal, S. (1989): Detectability of gravitational microlensing in the quasar QSO 2237+0305. Nature **338**, 745.

Kazanas, D. (1980): Dynamics of the universe and spontaneous symmetry breaking. Ap. J. (Letters) **241**, L59.

Kellermann, K. I. (1985): Superluminal radio sources. Comments. Astrophys. **11**, No. 2, 69.

Kellermann, K. I., Davis, M. M., Pauliny-Toth, I. I. K. (1970): Counts of radio sources at 6-centimeter wavelength. Ap. J. (Letters) **170**, L1.

Kellermann, K. I., Owen, F. N.: Radio galaxies and quasars. In: Galactic and extragalactic radio astronomy. Second edition. New York: Springer-Verlag 1988.

Kellermann, K. I., Pauliny-Toth, I. I. K. (1968): Variable Radio Sources. Ann. Rev. Astron. Astrophys. **6**, 417.

Kellermann, K. I., Pauliny-Toth, I. I. K. (1981): Compact radio sources. Ann. Rev. Astron. Astrophys. **19**, 373.

Kelvin, Lord (W. Thomson) (1862): On the age of the sun's heat. Macmillan's magazine **5**, 288; Popular Lectures **I**, 349 (1862).

Kelvin, Lord (W. Thomson) (1862): On the rigidity of the Earth. Proc. Roy. Soc. (London) **12**, 103; Phil. Trans. Roy. Soc. (London) **153**, 573 (1863).

Kelvin, Lord (W. Thomson) (1863): On the secular cooling of the Earth. Phil. Mag. Ser. 4, **25**, 1.

Kelvin, Lord (W. Thomson) (1899): The age of the Earth as an abode fitted for life. J. Victoria Inst. **31**, 11.

Kemp, J. C., et al. (1970): Discovery of circularly polarized light from a white dwarf star. Ap. J. (Letters) **161**, L77. Reproduced in: A source book in astronomy and astrophysics 1900–1975 (eds. K. R. Lang and O. Gingerich). Cambridge, Mass.: Harvard University Press 1979.

Kennedy, R. J., Thorndike, E. M. (1932): Experimental establishment of the relativity of time. Phys. Rev. **42**, 400.

Kennicutt, R. C., Jr. (1979): HII regions as extragalactic distance indicators. II. Applications of isophotal diameters. Ap. J. **228**, 696.

Kennicutt, R. C., Jr. (1979): HII regions as extragalactic distance indicators III. Applications of the HII region fluxes and galaxy diameters. Ap. J. **228**, 704.

Kennicutt, R. C., Jr. (1981): HII regions as distance indicators IV. The Virgo cluster. Ap. J. **247**, 9.

Kent, S. M. (1986): Dark matter in spiral galaxies I. galaxies with optical rotation curves. Astron. J. **91**, 1301.

Kent, S. M. (1987): Dark matter in spiral galaxies II. galaxies with HI rotation curves. Astron. J. **93**, 816.

Kepler, J.: De harmonice monde (The harmony of the world). Lintz, Austria: Johann Blancken 1619.

Kernan, P. J., Krauss, L. M. (1994): Refined big bang nucleosynthesis constraints on Ω_B and N_ν. Phys. Rev. Lett. **72**, 3309.

Kerr, F. J. (1962): Galactic velocity models and the interpretation of 21-cm surveys. M.N.R.A.S. **123**, 327.

Kerr, F. J. (1969): The large-scale distribution of hydrogen in the Galaxy. Ann. Rev. Astron. Astrophys. **7**, 39.

Kerr, F. J., Lynden-Bell, D. (1986): Review of galactic constants. M.N.R.A.S. **221**, 1023.

Kerr, R. P. (1963): Gravitational field of a spinning mass as an example of algebraically special metrics. Phys. Rev. Lett. **11**, 237.
Kiang, T. (1961): The galaxian luminosity function. M.N.R.A.S. **122**, 263.
Kibble, T. W. B. (1976): Topology of cosmic domains and strings. J. Phys **A9**, 1387.
Kiepenheuer, K. O. (1950): Cosmic rays as the source of general galactic radio emission. Phys. Rev. **79**, 738. Reproduced in: A source book in astronomy and astrophysics 1900–1975 (eds. K. R. Lang and O. Gingerich). Cambridge, Mass.: Harvard University Press 1979.
Kim, D.-W., Fabbiano, G. (1995): ROSAT pspc observations of the early-type galaxies NGC 507 and NGC 499 – central cooling and mass determination. Ap. J. **441**, 182.
King, C. R., Demarque, P., Green, E. M.: Isochrone fitting of color magnitude diagrams of old star clusters – implications for stellar ages. In: Calibration of stellar ages (ed. A. G. Davis Philip). Schnectady, N. Y.: L. Davis Press 1988.
King, I. R., Minkowski, R.: In IAU Symposium 44. External galaxies and quasi-stellar objects (ed. D. J. Evans). Dordrecht: Reidel 1972.
King-Hele, D. G. (1961): The earth's gravitational potential, deduced from the orbits of artificial satellites. Geophys. J. **4**, 20.
Kinoshita, H. (1977): Theory of the rotation of the rigid Earth. Celes. Mech. **15**, 277.
Kirshner, R. P. et al. (1981): A million cubic megaparsec void in Bootes? Ap. J. (Letters) **248**, L57.
Kirshner, R. P., et al. (1983): A deep survey of galaxies. Astron. J. **88**, 1285.
Kirshner, R. P., et al. (1987): A survey of the Bootes void. Ap. J. **314**, 493.
Kirshner, R. P., Oemler, A., Schechter, P. L. (1979): A study of field galaxies II. The luminosity function and space distribution of galaxies. Astron. J. **84**, 951.
Klebesadel, R. W., Strong, I. B., Olson, R. A. (1973): Observations of cosmic rays of cosmic origin. Ap. J. (Letters) **182**, L85.
Klimov, Y. G. (1963): The deflection of light rays in the gravitational fields of galaxies. Sov. Phys. Doklady **8**, 119.
Klypin, A. A., Kopylov, A. I. (1983): The spatial covariance function for rich clusters of galaxies. Sov. Astron. Lett. **9**, 41.
Knapen, J. H., Beckman, J. E. (1996): Global morphology and physical relations between the stars, gas and dust in the disc and arms of M100. M.N.R.A.S. **283**, 251.
Kochanek, C. S. (1996): The mass of the Milky Way. Ap. J. **457**, 228.
Kochanek, C. S., Piran, T. (1993): Gravitational waves and γ-ray bursts. Ap. J. (Letters) **417**, L17.
Koester, L. (1976): Verification of the equivalence of gravitational and inertial mass for the neutron. Phys. Rev. **D14**, 907.
Kogut, A., et al. (1996): Microwave emission at high galactic latitudes in the four-year DMR sky maps. Ap. J. (Letters) **464**, L5.
Kolb, E. W., Turner, M. S.: The early universe. Reading, Mass.: Addison-Wesley 1990.
Kolb, E. W., Turner, M. S.: The early universe – reprints. Reading, Mass.: Addison-Wesley 1990.
Komesaroff, M. M. (1960): Ionospheric refraction in radio astronomy. Austr. J. Phys. **13**, 153.
Königl, A. (1989): Self-similar models of magnetized accretion discs. Ap. J. **342**, 208.
Kormendy, J. (1988): Evidence for a supermassive black hole in the nucleus of M31. Ap. J. **325**, 128.
Kormendy, J. (1988): Evidence for a central dark mass in NGC 4594 (the sombrero galaxy). Ap. J. **335**, 40.
Kormendy, J., Djorgovski, S. (1989): Surface photometry and the structure of elliptical galaxies. Ann. Rev. Astron. Astrophys. **27**, 235.
Kormendy, J., Knapp, G. R. (eds.): Dark matter in the universe. Proc. IAU Symp. 117. Boston: Reidel 1987.
Kormendy, J., Richstone, D. (1992): Evidence for a supermassive black hole in NGC 3115. Ap. J. **393**, 559.

Kormendy, J., Richstone, D. (1995): Inward bound – the search for supermassive black holes in galactic nuclei. Ann. Rev. Astron. Astrophys. **33**, 581.
Kovner, I. (1987): Giant luminous arcs from gravitational lensing. Nature **327**, 193.
Kovner, I. (1987): The quadrupole gravitational lens. Ap. J. **312**, 22.
Kowal, C. T. (1968): Absolute magnitudes of supernovae. Astron. J. **73**, 1021.
Kozai, Y. (1959): The motion of a close earth satellite. Astron. J. **64**, 367.
Kozai, Y. (1961): The gravitational field of the earth derived from motions of three satellites. Astron. J. **66**, 8.
Kozai, Y.: New determination of zonal harmonic coefficients of the earth's gravitational potential. In: Satellite geodesy 1958–1964. Nat. Aeronaut. and Space Admin. 1966.
Kraan-Korteweg, R. C., Cameron, L. M., Tammann, G. A. (1988): 21 centimeter line width distances of cluster galaxies and the value of H_o. Ap. J. **331**, 620.
Kraan-Korteweg, R. C., et al. (1996): A nearby massive galaxy cluster behind the milky way. Nature **379**, 519.
Krabbe, A., et al. (1995): The nuclear cluster of the Milky Way: Star formation and velocity dispersion in the central 0.5 parsec. Ap. J. (Letters) **447**, L95.
Kraft, R. P. (1964): Binary stars among cataclysmic variables III. Ten old novae. Ap. J. **139**, 457. Reproduced in: A source book in astronomy and astrophysics 1900–1975 (eds. K. R. Lang and O. Gingerich). Cambridge, Mass.: Harvard University Press 1979.
Kraus, L. M. (1986): Dark matter in the universe. Scientific American **255**, 58 – December.
Krauss, L. M.: The fifth essence. The search for dark mater in the universe. New York: Basic Books 1989.
Krauss, L. M., Kernan, P. J. (1994): Recent deuterium observations and big bang nucleosynthesis constraints. Ap. J. (Letters) **432**, L79.
Krauss, L. M., Kernan, P. J. (1995): Big bang nucleosynthesis constraints and light element abundance estimates. Physics Letters **B347**, 347.
Krauss, L. M., Romanelli, P. (1990): Big bang nucleosynthesis: predictions and uncertainties. Ap. J. **358**, 47.
Krauss, L. M., Tremaine, S. (1988): Test of the weak equivalence principle for neutrinos and photons. Phys. Rev. Lett. **60**, 176.
Krauss, L. M., White, M. (1992): Grand unification, gravitational waves, and the cosmic microwave background anisotropy. Phys. Rev. Lett. **69**, 869.
Kreuzer, L. B. (1968): Experimental measurement of the equivalence of active and passive gravitational mass. Phys. Rev. **169**, 1007.
Krisciunas, K. (1993): Look-back time, the age of the universe, and the case for a positive cosmological constant. J.R.A.S. Canada **87**, 223.
Krisciunas, K.: Fundamental cosmological parameters. In: Encyclopedia of cosmology (ed. N. S. Hetherington). New York: Garland Pub. 1993.
Krisher, T. P., Anderson, J. D., Campbell, J. K. (1990): Test of the gravitational redshift effect at Saturn. Phys. Rev. Lett. **64**, 1322.
Krisher, T. P., Anderson, J. D., Taylor, A. H. (1991): Voyager 2 test of the radar time-delay effect. Ap. J. **373**, 665.
Krisher, T. P., et al. (1990): Test of the isotropy of the one-way speed of light using hydrogen-maser frequency standards. Phys. Rev. **D42**, 731.
Kristian, J., Sandage, A., Westphal, J. A. (1978): The extension of the Hubble diagram II. New redshifts and photometry of very distant galaxy clusters. First indication of a deviation of the Hubble diagram from a straight line. Ap. J. **221**, 383.
Kruskal, M. D. (1960): Maximal extension of the Schwarzschild metric. Phys. Rev. **119**, 1743.
Kuiper, G. P. (1941): On the interpretation of β Lyrae and other close binaries. Ap. J. **93**, 133.
Kulessa, A. S., Lynden-Bell, D. (1992): The mass of the milky way Galaxy. M.N.R.A.S. **255**, 205.

Kulkarni, S. R., et al. (1998): Identification of a host galaxy at redshift $z = 3.42$ for the γ-ray burst of 14 December 1997. Nature **393**, 35.

Kulkarni, S. R., Hester, J. J. (1988): Discovery of a nebula around PSR 1957+20. Nature **335**, 801.

Kunth, D., Sargent, W. L. W. (1983): Spectrophotometry of 12 metal-poor galaxies: implications for the primordial helium abundance. Ap. J. **273**, 81.

Kuroda, K. (1995): Does the time-of-swing method give a correct value of the Newtonian gravitational constant? Phys. Rev. Lett. **75**, 2796.

Kwee, K. K., Muller, C. A., Westerhout, G. (1954): The rotation of the inner parts of the galactic system. B.A.N. **12**, 211.

L3 Collaboration(1992): A direct determination of the number of light neutrino families from $e^+e^- \rightarrow \nu\nu\gamma$ at LEP. Physics Letters **B275**, 209.

Lahav, O. (1987): Optical dipole anisotropy. M.N.R.A.S. **225**, 213.

Lahav, O., Lilje, P. B., Primack, J. R., Rees, M. J. (1991): Dynamical effects of the cosmological constant. M.N.R.A.S. **251**, 128.

Lambeck, K. (1978): Tidal dissipation in the oceans – astronomical, geophysical and oceanographic consequences. Phil. Trans. Roy. Soc. (London) **A287**, 545.

Lambeck, K.: The earth's variable rotation. Cambridge, England: Cambridge University Press 1980.

Lamoreaux, S. K., et al. (1986): New limits on spatial anisotropy from optically pumped ^{201}Hg and ^{199}Hg. Phys. Rev. Lett. **57**, 3125.

Lanczos, K. (1922): Bemerkung zur de Sitterschen Welt (Comments on the de Sitter universe). Phys. Z. **23**, 539.

Lanczos, K. (1923): Über die Rotverschiebung in der de Sitterschen Welt (On the redshift in the de Sitter universe). Z. Physik **17**, 168.

Landau, L. D. (1932): On the theory of stars. Phys. Zeits Sowjetunion **1**, 285. Reproduced in: A source book in astronomy and astrophysics 1900–1975 (eds. K. R. Lang and O. Gingerich). Cambridge, Ma.: Harvard University Press 1979.

Landau, L. D. (1938): Origin of stellar energy. Nature **141**, 333.

Landau, L. D., Lifshitz, E. M.: Fluid mechanics. Reading, Mass.: Addison-Wesley 1959.

Landau, L. D., Lifshitz, E. M.: The classical theory of fields. Reading, Mass.: Addison-Wesley 1962.

Landstreet, J. D. (1992): Magnetic fields at the surfaces of stars. Astron. Astrophys. Rev. **4**, 35.

Landy, S. D., et al. (1996): The two-dimensional power spectrum of the las campanas redshift survey: Detection of excess power on 100 h^{-1} scales. Ap. J. (Letters) **456**, L1.

Laney, C. D., Stobie, R. S. (1986): Infrared photometry of Magellanic cloud cepheids. Intrinsic properties of cepheids and the spatial structure of the clouds. M.N.R.A.S. **222**, 449.

Lang, K. R. (1985): The uncertainties of space and time. Vistas in Astronomy **28**, 277.

Lang, K. R.: Astrophysical data: planets and stars. New York: Springer-Verlag 1992.

Lang, K. R.: Sun, earth and sky. New York: Springer-Verlag 1995.

Lang, K. R., et al. (1975): The composite Hubble diagram. Ap. J. **202**, 583.

Lang, K. R., Whitney, C. A.: Wanderers in space – exploration and discovery in the solar system. New York: Cambridge University Press 1991.

Lange, B. (1995): Proposed high-accuracy gyroscope test of general relativity and the search for a massless scalar field. Phys. Rev. Lett. **74**, 1904.

Langston, G. I., et al. (1989): MG1654+1346 – an Einstein ring image of a quasar radio lobe. Astron. J. **97**, 1283.

Lanzetta, K. M., et al. (1991): A new spectroscopic survey for damped Ly α absorption lines from high-redshift galaxies. Ap. J. Supp. **77**, 1.

Laplace, P. S. Marquis De (1782): Théorie des attractions des sphéroides et de la figure de planétes (Theory of the attraction of spheres and the figure of planets). Mém. de l'Acad. 113.

Lindblad, B. (1925): Star-streaming and the structure of the stellar system. Arkiv för matematik, astronomi och fysik **19A**, no. 21, 1. Reproduced in: A source book in astronomy and astrophysics 1900–1975 (eds. K. R. Lang and O. Gingerich). Cambridge, Mass.: Harvard U. Press 1979.

Lindblad, B. (1927): The small oscillations of a rotating stellar system and the development of spiral arms. Medd. Astron. Obs. Uppsala **19**, 1.

Linde, A. D. (1982): A new inflationary universe scenario – a possible solution of the horizon, flatness, homogeneity, isotropy and primordial monopole problems. Phys. Lett. **108B**, 389.

Linde, A. D. (1983): Chaotic inflating universe. JETP Lett. **38**, 177.

Linde, A. D. (1983): Chaotic inflation. Phys. Lett. **129B**, 177.

Lineweaver, C. H., et al. (1996): The dipole observed in the COBE DMR 4 year data. Ap. J. **470**, 38.

Linsky, J. L. et al. (1993): Goddard high-resolution spectrograph observations of the local interstellar medium and the deuterium/hydrogen ratio along the line of sight toward Capella. Ap. J. **402**, 694.

Lipa, J. A., Everitt, C. W. F. (1978): The role of cryogenics in the gyroscope experiment. Acta Astronautica **5**, 119.

Lipa, J. A., Fairbank, W. M., Everitt, C. W. F.: The gyroscope experiment II. Development of the London-moment gyroscope and of cryogenic technology for space. In: Experimental gravitation (ed. B. Bertotti). New York: Academic 1974.

Liszt, H. S.: The galactic center. In: Galactic and extragalactic astronomy (eds. G. L. Verschuur, K. I. Kellermann). New York: Springer-Verlag 1988.

Lo, K. Y. (1986): The galactic center; Is it a massive black hole? Science **233**, 1394.

Lo Presto, J. C., Schrader, C., Pierce, A. K. (1991): Solar gravitational redshift from the infrared oxygen triplet. Ap. J. **376**, 757.

Lodge, O. J. (1919): Gravitation and light. Nature **104**, 354.

Long, D. R. (1976): Experimental examination of the gravitational inverse square law. Nature **260**, 417.

Longair, M. S. (1966): On the interpretation of radio source counts. M.N.R.A.S. **133**, 421.

Longair, M. S. (1971): Observational cosmology. Rep. Prog. Phys. **34**, 1125.

Longair, M. S.: Radio astronomy and cosmology. In: Observational cosmology (eds. A. Maeder, L. Martinet, and G. Tammann). Sauverny, Switzerland: Geneva Observatory 1978.

Longair, M. S.: Our evolving universe. Cambridge, England: Cambridge University Press 1996.

Longair, M. S., Ryle, M., Scheuer, P. A. G. (1973): Models of extended radio sources. M.N.R.A.S. **164**, 243.

Longo, M. J. (1988): New precision tests of the Einstein equivalence principle from SN1987A. Phys. Rev. Lett. **60**, 173.

Lorentz, H. A. (1904): Electromagnetic phenomena in a system moving with any velocity smaller than that of light. Proc. Am. Acad. Sci **6**, 809. Reproduced in: The Principle of Relativity. New York: Dover 1952.

Loveday, J., et al. (1992): The stromlo-apm redshift survey I. the luminosity function and space density of galaxies. Ap. J. **390**, 338.

Lovelace, R. V., Wang, J. C., Sulkanen, M. E. (1987): Self-collimated electromagnetic jets from magnetised accretion disks. Ap. J. **315**, 504.

Lubin, P., et al. (1985): A map of the cosmic background radiation at 3 millimeters. Ap. J. (Letters) **298**, L1.

Lubkin, G. B. (1980): Multiple quasar may indicate another gravitational lens. Phys. Today **33**, 17.

Lund Observatory (1961): Tables for the conversion of equatorial and galactic coordinates. Ann. Obs. Lund **15**, **16**, **17**.

Lundmark, K. (1924): The determination of the curvature of space-time in de Sitter's world. M.N.R.A.S. **84**, 747.

Luppino, G. A., et al. (1993): The complex gravitationally lensed arc system in the X-ray-selected cluster of galaxies MS 0440+0204. Ap. J. **416**, 444.

Luther, G. G., Towler, W. R. (1982): Redetermination of the Newtonian gravitational constant G. Phys Rev. Lett. **48**, 121.

Lynden-Bell, D. (1967): Statistical mechanics of violent relaxation in stellar systems. M.N.R.A.S. **136**, 101.

Lynden-Bell, D. (1969): Galactic nuclei as collapsed old quasars. Nature **223**, 690.

Lynden-Bell, D. (1977): Hubble's constant determined from super-luminal radio sources. Nature **270**, 396.

Lynden-Bell, D. (1978): Gravity power. Physica Scripta **17**, 185.

Lynden-Bell, D. (ed.) (1982): The big bang and element creation. Phil. Trans. Roy. Society (London) **A307**, 1–148.

Lynden-Bell, D., et al. (1988): Spectroscopy and photometry of elliptical galaxies V. Galaxy streaming toward the new supergalactic center. Ap. J. **326**, 19.

Lynden-Bell, D., Lahav, O.: Whence arises the local flow of galaxies? In: Large-scale motions in the universe. Princeton, New Jersey: Princeton University Press 1988.

Lynden-Bell, D., Lahav, O., Burstein, D. (1989): Cosmological deductions from the alignment of local gravity and motion. M.N.R.A.S. **241**, 325.

Lynden-Bell, D., Pringle, J. E. (1974): The evolution of viscous discs and the origin of the nebular variables. M.N.R.A.S. **168**, 603.

Lynden-Bell, D., Rees, M. J. (1971): On quasars, dust and the galactic centre. M.N.R.A.S. **152**, 461.

Lynds, R. (1971): The absorption-line spectrum of 4C 05.34. Ap. J. (Letters) **164**, L73.

Lynds, R., Petrosian, V. (1986): Giant luminous arcs in galaxy clusters. Bull. A.A.S. **18**, 1014.

Lynds, R., Petrosian, V. (1989): Luminous arcs in clusters of galaxies. Ap. J. **336**, 1.

Lyne, A. G., Bailes, M. (1990): The mass of the PSR 2303+46 system. M.N.R.A.S. **246**, 15p.

Lyne, A. G., Graham-Smith, F.: Pulsar Astronomy. Cambridge, England: Cambridge University Press 1990.

Lyttleton, R. A., Fitch, J. P. (1978): Effect of a changing G on the moment of inertia of the earth. Ap. J. **221**, 412.

Macchetto, F. D., Dickinson, M. (1997): Galaxies in the young universe. Scientific American **276**, 92 – May.

Mac Donald, D., Thorne, K. S. (1982): Black-hole electrodynamics: an absolute-space/universal-time formulation. M.N.R.A.S. **198**, 345.

Maddox, J. (1984): Continuing doubt on gravitation. Nature **310**, 723.

Maddox, S. J., et al. (1990): Galaxy correlations on large scales. M.N.R.A.S. **242**, 43p.

Madore, B. F., Freedman, W. L. (1991): The cepheid distance scale. P.A.S.P. **103**, 933.

Madsen, J., Epstein, R. (1985): Improved astronomical limits on the neutrino mass. Phys. Rev. Lett. **54**, 2720.

Magain, P., et al. (1988): Discovery of a quadruply lensed quasar – the 'clover leaf' H1413+117. Nature **334**, 325.

Malmquist, K. G.: A study of the stars of spectral type A. Lund: Scientia 1920.

Mampe, W., et al. (1989): Neutron lifetime measured with stored ultracold neutrons. Phys. Rev. Lett. **63**, 593.

Mampe, W., et al. (1993): Measuring neutron lifetime by storing ultracold neutrons and detecting inelastically scattered neutrons. JETP Lett. **57**, 82.

Mandolesi, N., Vittorio, N.: The cosmic microwave background 25 years later. Dordrecht: Kluwer Academic Pub. 1990.

Maraschi, L., Maccacaro, T., Ulrich, M.-H. (eds.): BL Lac objects. Lecture notes in physics 334. New York: Springer Verlag 1989.

Narlikar, J. V., Rana, N. C. (1985): Newtonian n-body calculations of the advance of Mercury's perihelion. M.N.R.A.S. **213**, 657.

Nemiroff, R. J. (1991): Limits on compact dark matter from null results of searches for lensing of quasistellar objects. Phys. Rev. Lett. **66**, 538.

Newcomb, S. (1892): On the dynamics of the earth's rotation, with respect to the variations of latitude. M.N.R.A.S. **52**, 336.

Newcomb, S. (1892): Remarks on Mr. Chandler's law of variation of terrestrial latitudes. Astron. J. **12**, 49.

Newcomb, S.: The elements of the four inner planets and the fundamental constants. Washington, D.C.: U.S. Govt. Office, 1895.

Newcomb, S. (1895–1898): Am. Pap. Am. Ephem. **6**, parts 2, 3 and 4.

Newcomb, S. (1898): Tables of the motion of the earth on its axis around the Sun. A. P. Am. Ephem. **6**, 7.

Newcomb, S. (1909): Fluctuations of the moon's mean motion. M.N.R.A.S. **69**, 164.

Newkirk, G. Jr. (1983): Variations in solar luminosity. Ann. Rev. Astron. Astrophys. **21**, 429.

Newman, D., et al. (1978): Precision experimental verification of special relativity. Phys. Rev. Lett. **40**, 1355.

Newman, E. T., et al. (1965): Metric of a rotating, charged mass. J. Math. Phys. **6**, 918.

Newton, H. W., Nunn, M. L. (1951): The Sun's rotation derived from sunspots 1934–1944. M.N.R.A.S. **111**, 413.

Newton, I. (1687): Philsophiae naturalis principia mathematica (Natural philosophy and the principles of mathematics). English translation by A. Motte, revised and annotated by F. Cajori. University of California Press 1966.

Neyman, J., Scott, E. L., Shane, C. D. (1953): On the spatial distribution of galaxies a specific model. Ap. J. **117**, 92.

Niebauer, T. M., Mc Hugh, M. P., Faller, J. E. (1987): Galilean test for the fifth force. Phys. Rev. Lett. **59**, 609.

Nieto, M. M.: The Titius-Bode law of planetary distances – its history and theory. Oxford: Pergamon Press 1972.

Nobili, A. M., Will, C. M. (1986): The real value of Mercury's perihelion advance. Nature **320**, 39.

Noonan, T. W. (1971): The effect of the Einstein light deflection on observed properties of clusters of galaxies. Astron. J. **76**, 765.

Noonan, T. W. (1983): Image distortion by gravitational lensing. Ap. J. **270**, 245.

Nordsieck, K. H. (1973): The angular momentum of spiral galaxies II. detailed models and correlations for 17 galaxies. Ap. J. **184**, 735,.

Nordstrom, G. (1918): On the energy of the gravitation field in Einstein's theory. K. Ned. Akad. Wet. Amst. Proc. Sec. Sci. **20**, 1238.

Nordtvedt, K. (1968): Equivalence principle for massive bodies I. Phenomenology, II. Theory. Phys. Rev. **169**, 1014, 1017.

Nordtvedt, K. (1968): Testing relativity with laser ranging to the Moon. Phys. Rev. **170**, 1186.

Nordtvedt, K. (1970): Solar-system Eötvös experiments. Icarus **12**, 91.

Nordtvedt, K. (1970): Post-Newtonian metric for a general class of scalar-tensor gravitational theories and observational consequences. Ap. J. **161**, 1059.

Nordtvedt, K. (1971): Equivalence principle for massive bodies, IV. Planetary orbits and modified Eötvös-type experiments. Phys. Rev. **D3**, 1683.

Nordtvedt, K. (1972): Gravitation theory – empirical status from solar-system experiments. Science **178**, 1157.

Nordtvedt, K. (1973): Post-Newtonian gravitational effects in lunar laser ranging. Phys. Rev. **D7**, 2347.

Nordtvedt, K. (1985): A post-newtonian gravitational lagrangian formalism for celestial body dynamics in metric gravity. Ap. J. **297**, 390.

Nordtvedt, K. (1987): Probing gravity to the second post-newtonian order and to one part in 10^7 using the spin axis of the sun. Ap. J. **320**, 871.
Nordtvedt, K. (1988): Gravitomagnetic interaction and laser ranging to earth satellites. Phys. Rev. Letters **61**, 2647.
Nordtvedt, K. (1988): Lunar laser ranging and laboratory Eötvös-type experiments. Phys. Rev. **D37**, 1070.
Nordtvedt, K. (1990): (dG/dt)/G and a cosmological acceleration of gravitationally compact bodies. Phys. Rev. Letters **65**, 953.
Nordtvedt, K. (1991): Lunar laser ranging reexamined: The non-null relativistic contribution. Phys. Rev. **D43**, 3131.
Nordtvedt, K. (1995): The relativistic orbit observables in lunar laser ranging. Icarus **114**, 51.
Nordtvedt, K. (1996): From Newton's moon to Einstein's moon. Physics Today **49**, 26 – May.
Nordtvedt, K, Will, C. M. (1972): Conservation laws and preferred frames in relativistic gravity II. Experimental evidence to rule out preferred-frame theories of gravity. Ap. J. **177**, 775.
Norris, J. P., et al. (1995): Duration distributions of bright and dim BATSE gamma-ray bursts. Ap. J. **439**, 542.
Novikov, I. D., Thorne, K. S.: Astrophysics of black holes. In: Black holes (eds. C. De Witt and B. S. De Witt). New York: Gordon and Breach 1972.
Nugent, D., et al. (1995): Low Hubble constant from the physics of type Ia supernovae. Phys. Rev. Lett. **75**, 394.
O'Dell, C. R. (1963): Photoelectric spectrophotometry of planetary nebulae. Ap. J. **138**, 1018.
O'Keefe, J. A., Eckels, A., Squires, R. K. (1959): The gravitational field of the earth. Astron. J. **64**, 245.
Oegerle, W. R., Hoessel, J. G. (1991): Fundamental parameters of brightest cluster galaxies. Ap. J. **375**, 15.
Oke, J. B. (1971): Redshifts and absolute spectral energy distributions of galaxies in distant clusters. Ap. J. **170**, 193.
Oke, J. B., Sandage, A. R. (1968): Energy distributions, K corrections and the Stebbins-Whitford effect for giant elliptical galaxies. Ap. J. **154**, 21.
Olbers, W. (1826): Über die Durchsichtigkeit des Weltraums (About the transparency of the universe). Bode Jb. 15. Reproduced in Dickson (1968) and Jaki (1969).
Olive, K. A., et al. (1981): Big-bang nucleosynthesis as a probe of cosmology and particle physics. Ap. J. **246**, 557.
Olive, K. A., et al. (1990): Big-bang nucleosynthesis revisited. Physics Letters **B236**, 454.
Olive, K. A., Steigman, G. (1995): On the abundance of primordial helium. Ap. J. Supp. **97**, 49.
Olive, K. A., Steigman, G., Walker, T. P. (1991): The upper bound to the primordial abundance of helium and the consistency of the hot big bang model. Ap. J. (Letters) **380**, L1.
Oliveira-Costa, A. De, et al. (1998): Mapping the CMB III: combined analysis of QMAP flights. Ap. J. (Letters) **509**, L77.
Oliveira-Costa, A. De, Smoot, G. F. (1995): Constraints on the topology of the universe from the 2 year COBE data. Ap. J. **448**, 477.
Olson, B. I. (1975): On the ratio of total to selective absorption. P.A.S.P. **87**, 349.
Oort, J. H. (1927): Observational evidence confirming Lindblad's hypothesis of a rotation of the galactic system. B.A.N. **3**, 275. Reproduced in: A source book in astronomy and astrophysics 1900–1975 (eds. K. R. Lang and O. Gingerich). Cambridge, Mass.: Harvard University Press 1979.
Oort, J. H. (1928): Dynamics of the galactic system in the vicinity of the Sun. B.A.N. **4**, 269.
Oort, J. H. (1932): The force exerted by the stellar system in the direction perpendicular to the galactic plane and some related problems. B.A.N. **6**, 249.

Oort, J. H.: Distribution of galaxies and the density of the universe. In: La structure et l'evolution de l'univers – Institut International de Physique Solvay (ed. R. Stoop). Bruxelles: Coodenberg 1958.

Oort, J. H. (1960): Note on the determination of K_z and on the mass density near the Sun. B.A.N. **15**, 45.

Oort, J. H. (1981): Superclusters and Lyman α absorption lines in quasars. Astron. Astrophys. **94**, 359.

Oort, J. H. (1983): Superclusters. Ann. Rev. Astron. Astrophys. **21**, 373.

Oort, J. H., Kerr, F. J., Westerhout, G. (1958): The galactic system as a spiral nebula. M.N.R.A.S. **118**, 379. Reproduced in: A source book in astronomy and astrophysics 1900–1975 (eds. K. R. Lang and O. Gingerich). Cambridge, Mass.: Harvard University Press 1979.

Öpik, E. (1922): An estimate of the distance of the Andromeda nebula. Ap. J. **55**, 406.

Öpik, E. (1938): Stellar structure, source of energy and evolution. Pub. Obs. Astron. Univ. Tartu **30**, No. 3, 1. Reproduced in: A source book in astronomy and astrophysics 1900–1975 (eds. K. R. Lang and O. Gingerich). Cambridge, Mass.: Harvard University Press 1979.

Öpik, E. J. (1951): Stellar models with variable composition II. Sequences of models with energy generation proportional to the fifteenth power of temperature. Proc. Royal Irish Academy **54A**, 49.

Oppenheimer, J. R., Serber, R. (1938): On the stability of stellar neutron cores. Phys. Rev. **54**, 608.

Oppenheimer, J. R., Snyder, H. (1939): On continued gravitational contraction. Phys. Rev. **56**, 455. Reproduced in: A source book in astronomy and astrophysics 1900–1975 (eds. K. R. Lang and O. Gingerich). Cambridge, Mass.: Harvard University Press 1979. Also reproduced in: Neutron stars, black holes and binary X-ray sources (eds. H. Gursky and R. Ruffini). Boston: Reidel 1975.

Oppenheimer, J. R., Volkoff, G. M. (1939): On massive neutron cores. Phys. Rev. **55**, 374. Reproduced in: A source book in astronomy and astrophysics 1900–1975 (eds. K. R. Lang and O. Gingerich). Cambridge, Mass.: Harvard University Press 1979.

Orr, M. J. L., Browne, I. W. A. (1980): Relativistic beaming and quasar statistics. M.N.R.A.S. **200**, 1067.

Ortolani, S. et al. (1995): Near-coeval formation of the galactic bulge and halo inferred from globular cluster ages. Nature **377**, 701.

Osterbrock, D. E. (1984): Active galactic nuclei. Q.J.R.A.S. **25**, 1.

Osterbrock, D. E. (1991): Active galactic nuclei. Rep. Prog. Phys. **54**, 579.

Osterbrock, D. E., Mathews, W. G. (1986): Emission-line regions of active galaxies and QSOs. Ann. Rev. Astron. Astrophys. **24**, 171.

Osterbrock, D. E., Miller, J. S. (eds.): Active galactic nuclei. Proc. IAU symposium 134. Boston: Kluwer Academic 1989.

Ostic, R. G., Russell, R. D., Reynolds, P. H. (1963): A new calculation for the age of the earth from abundances of lead isotopes. Nature **199**, 1150.

Ostriker, J. P.: Mass determinations and dark matter at intermediate scales. In: Dark matter in the universe (eds. J. Kormendy and G. R. Knapp). Boston: Reidel 1987.

Ostriker, J. P. (1993): Astronomical tests of the cold dark matter scenario. Ann. Rev. Astron. Astrophys. **31**, 689.

Ostriker, J. P., Cowie, L. L. (1981): Galaxy formation in an intergalactic medium dominated by explosions. Ap. J. (Letters) **243**, L127.

Ostriker, J. P., Gunn, J. E. (1969): On the nature of pulsars I. Theory. Ap. J. **157**, 1395.

Ostriker, J. P., Ikeuchi, S. (1983): Physical properties of the intergalactic medium and the Lyman-alpha absorbing clouds. Ap. J. (Letters) **268**, L63.

Ostriker, J. P., Mc Kee, C. F. (1988): Astrophysical blastwaves. Rev. Mod. Phys. **60**, 1.

Ostriker, J. P., Peebles, P. J. E., Yahil, A. (1974): The size and mass of galaxies, and the mass of the universe. Ap. J. (Letters) **193**, L1.

Ostriker, J. P., Steinhardt, P. J. (1995): The observational case for a low-density universe with a non-zero cosmological constant. Nature **377**, 600.

Ostriker, J. P., Suto, Y. (1990): The mach number of the cosmic flow: A critical test for current theories. Ap. J. **348**, 378.

Owen, F. N., Burns, J. O., Rudnick, L. (1978): VLA observations of NGC 1265 at 4886 MHz. Ap. J. (Letters) **226**, L119.

Ozernoy, L. M., Chernin, A. D. (1968): The fragmentation of matter in a turbulent matagalactic medium. Sov. Astoron. A. J. **11**, 907.

Ozernoy, L. M., Chibisov, G. V. (1971): Galactic parameters as a consequence of cosmological turbulence. Astrophys. Lett. **7**, 201.

Ozernoy, L. M., Sazonov, V. N. (1969): The spectrum and polarization of a source of synchrotron emission with components flying apart at relativistic velocities. Astrophys. and Space Sci. **3**, 395.

Pacholczyk, A. G.: Radio astrophysics – nonthermal processes in galactic and extragalactic sources. San Francisco: W. H. Freeman 1970.

Pacini, F. (1967): Energy emission from a neutron star. Nature **216**, 567. Reproduced in: A source book in astronomy and astrophysics 1900–1975 (eds. K. R. Lang and O. Gingerich). Cambridge, Mass.: Harvard University Press 1979.

Paczynski, B. (1983): Mass of Large Magellanic Cloud X-3. Ap. J. (Letters) **273**, L81.

Paczynski, B. (1986): Gravitational microlensing by the galactic halo. Ap. J. **304**, 1.

Paczynski, B. (1986): Gamma-ray bursters at cosmological distances. Ap. J. (Letters) **308**, L43.

Paczynski, B. (1987): Giant luminous arcs discovered in two clusters of galaxies. Nature **325**, 572.

Paczynski, B. (1996): Gravitational microlensing in the local group. Ann. Rev. Astron. Astrophys. **34**, 419.

Page, T. (1960): Average masses and mass-luminosity ratios of the double galaxies. Ap. J. **132**, 910.

Page, T. (1961): Average masses of the double galaxies. In: Fourth Berkeley symposium on mathematical statistics and probability **3**, 277.

Page, T. (1965): Statistical evidence of the masses and evolution of galaxies. Smithsonian Astrophys. Obs. Rpt. No. 195.

Pagel, B. E. J. (1977): On the limits to past variability of the proton-electron mass ratio set by quasar absorption lines. M.N.R.A.S. **179**, 81P.

Pagel, B. E. J. (1982): Abundances of elements of cosmological interest. Phil. Trans. Roy. Soc. (London) **A307**, 19.

Pagel, B. E. J. (1993): Abundances of light elements. Proc. Nat. Acad. Sci. **90**, 4789.

Pagel, B. E. J., Simonson, E. A., Terlevich, R. J., Edmunds, M. G. (1992): The primordial helium abundance from observations of extragalactic HII regions. M.N.R.A.S. **255**, 325.

Pais, A.: Inward Bound. New York: Oxford University Press 1986.

Panagia, N., et al. (1991): Properties of the SN 1987A circumstellar ring and the distance to the Large Magellanic Cloud. Ap. J. (Letters) **380**, L23.

Panagia, N., Felli, M. (1975): The spectrum of the free-free radiation from extended envelopes. Astron. Astrophys. **39**, 1.

Panov, V. I., Forntov, V. N. (1979): The Cavendish experiment at large distances. Sov. Phys. JETP **50**, 852.

Papaloizou, J. C. B., Lin, D. N. C. (1995): Theory of accretion disks I.: Angular momentum transport processes. Ann. Rev. Astron. Astrophys. **33**, 505.

Paradijs, J. Van et al. (1997): Transient optical emission from the error box of the γ-ray burst of 28 February 1997. Nature **386**, 686.

Park, C. (1990): Large n-body simulations of a universe dominated by cold dark matter. M.N.R.A.S. **242**, 59p.

Pierce, M. J., et al. (1994): The Hubble constant and Virgo cluster distance from observations of Cepheid variables. Nature **371**, 385.
Pierce, M. J., Tully, R. B. (1988): Distances to the Virgo and Ursa Major clusters and a determination of H_o. Ap. J. **330**, 579.
Pierce, M. J., Tully, R. B. (1992): Luminosity-line width relations and the extragalactic distance scale I. Absolute calibration. Ap. J. **387**, 47.
Piran, T. (1995): Binary neutron stars. Scientific American **272**, 52 – May.
Piran, T., Stark, R. F.: In: Dynamical spacetimes and numerical relativity (ed. J. M. Centrella). New York: Cambridge University Press 1986.
Planck, M. (1901): Über das Gesetz der Energieverteilung im Normalspectrum (On the law of energy distriubtion in a normal spectrum). Ann. Physik **4**, 553.
Planck, M.: The theory of heat radiation. 1913. Reprod. New York: Dover 1969.
Pogson, N. (1856): Magnitude of 36 of the minor planets. M.N.R.A.S. **17**, 12.
Poincaré, H. (1904): The present and future state of mathematical physics. Bull. Sci. Math. **2**, 28.
Ponman, T. J., Bertram, D. (1993): Hot gas and dark matter in a compact galaxy group. Nature **363**, 51.
Pooley, G. G., et al. (1979): Radio studies of the double QSO 0957 + 561 a, b: Nature **280**, 461.
Popper, D. M. (1954): Redshift in the spectrum of 40 Eridani B. Ap. J. **120**, 316.
Popper, D. M. (1980): Stellar masses. Ann. Rev. Astron. Astrophys. **18**, 115.
Porcas, R. W.: Summary of known superluminal sources. In: Superluminal radio sources (eds. J. A. Zensus and T. J. Pearson). New York: Cambridge University Press 1987.
Porcas, R. W., et al. (1979): VLBI observations of the double QSO 0957 + 561 a, b. Nature **282**, 385.
Postman, M.: Using galaxy clusters to trace the large scale velocity field. In: Dark matter. AIP conference proceedings 336. New York: AIP Press 1995.
Potter, H. H. (1923): Some experiments on the proportionality of mass and weight. Proc. Roy. Soc. (London) **104**, 588.
Pound, R. V., Rebka, G. A. (1959): Gravitational redshift in nuclear resonance. Phys. Rev. Lett. **3**, 439.
Pound, R. V., Rebka, G. A. (1960): Apparent weight of photons. Phys. Rev. Lett. **4**, 337.
Pound, R. V., Snider, J. L. (1964): Effect of gravity on nuclear resonance. Phys. Rev. Lett. **13**, 539.
Pound, R. V., Snider, J. L. (1965): Effect of gravity on gamma radiation. Phys. Rev. **140**, B788.
Prendergast, K. H., Burbidge, G. R. (1968): On the nature of some galactic X-ray sources. Ap. J. (Letters) **151**, L83.
Press, W. H., Gunn, J. E. (1973): Method for detecting a cosmological density of condensed objects. Ap. J. **185**, 397.
Press, W. H., Spergel, D. N. (1989): Cosmic strings: Topological fossils of the hot big bang. Physics Today **42**, 29 – March.
Press, W. H., Thorne, K. S. (1972): Gravitational wave astronomy. Ann. Rev. Astron. Astrophys. **10**, 335.
Press, W. H., Vishniac, E. T. (1980): Tenacious myths about cosmological perturbations larger than the horizon size. Ap. J. **239**, 1.
Prestage, J. D., et al. (1985): Limits for spatial anisotropy by use of nuclear-spin-polarized $^9Be^+$ ions. Phys. Rev. Lett. **54**, 2387.
Primack, J. R., Seckel, D., Sadoulet, B. (1988): Detection of cosmic dark matter. Ann. Rev. Nuclear Particle Sci. **38**, 751.
Pringle, J. E. (1981): Accretion discs in astrophysics. Ann. Rev. Astron. Astrophys. **19**, 137.
Pringle, J. E., Rees, M. J. (1972): Accretion disc models for compact X-ray sources. Astron. Astrophys. **21**, 1.
Proffitt, C. R., Vandenberg, D. A. (1991): Implications of helium diffusion for globular cluster isochrones and luminosity functions. Ap. J. Supp. **77**, 473.

Protheroe, R. J., Szabo, A. P. (1992): High energy cosmic rays from active galactic nuclei. Phys. Rev. Lett. **69**, 2885.

Quashnock, J. M., Vanden Berk, D. E., York, D. G. (1996): High-redshift superclustering of quasi-stellar object absorption-line systems on 100 h^{-1} scales. Ap. J. (Letters) **472**, L69.

Ramaty, R., Kozlovsky, B., Lingenfelter, R. (1998): Cosmic rays, nuclear gamma rays and the origin of the light elements. Physics Today **51**, 30 – April.

Ramsey, N. F. (1968): The atomic hydrogen maser. Am. J. Sci. **56**(4), 420.

Ramsey, N. F. (1972): History of atomic and molecular standards of frequency and time. I.E.E.E. Trans. Instrum. and Meas. **21** (2), 90.

Rana, N. C. (1991): Chemical evolution of the Galaxy. Ann. Rev. Astron. Astrophys. **29**, 129.

Rastall, P. (1976): A theory of gravity. Can. J. Phys. **54**, 66.

Rastall, P. (1977): The maximum mass of a neutron star. Ap. J. **213**, 234.

Rastall, P. (1979): The Newtonian theory of gravitation and its generalization. Can. J. Phys. **57**, 944.

Ratnatunga, K. U., Ostrander, E. J., Griffiths, R. E., Im, M. (1995): New "Einstein cross" gravitational lens candidates in Hubble Space Telescope WFDPC2 survey images. Ap. J. (Letters) **453**, L5.

Ratra, B., Peebles, P. J. E. (1994): Cold dark matter cosmogony in an open universe. Ap. J. (Letters) **432**, L5.

Rawlings, S., et al. (1995): A radio galaxy at redshift 4.41. Nature **383**, 502.

Raychaudhuri, A. (1955): Relativistic cosmology I. Phys. Rev. **98**, 1123.

Raychaudhuri, A. (1957): Relativistic and Newtonian cosmology. Z. Astrophys. **43**, 161.

Raychaudhuri, A. K., Banerji, S., Banerjee, A.: General relativity, astrophysics and cosmology. New York: Springer Verlag 1992.

Rayleigh, Lord (Robert Strutt) (1900): Remarks upon the law of complete radiation. Phil. Mag. **49**, 539.

Rayleigh, Lord (Robert Strutt) (1905): The rate of formation of radium. Nature **72**, 365.

Rayleigh, Lord (Robert Strutt) (1905): The dynamical theory of gases and of radiation. Nature **72**, 54.

Readhead, A. C. S., et al. (1978): Bent beams and the overall size of extragalactic radio sources. Nature **276**, 768.

Readhead, A. C. S., et al. (1989): A limit on the anisotropy of the microwave background radiation on arc minute scales. Ap. J. **346**, 566.

Readhead, A. C. S., Lawrence, C. R. (1992): Observations of the isotropy of the cosmic microwave background radiation. Ann. Rev. Astron. Astrophys. **30**, 653.

Readhead, A. C. S., Pearson, T. J., Barthel, P. D.: Relativistic beaming models and VLBI observations of a complete sample of radio sources. In: The impact of VLBI on astrophysics and geophysics. Proc. IAU symposium 129 (eds. M. J. Reid and J. M. Moran). Boston: Kluwer Academic 1988.

Reasenberg, R. D., et al. (1979): Viking relativity experiment: Verification of signal retardation by solar gravity. Ap. J. (Letters) **234**, L219.

Rebolo, R., et al. (1996): Brown dwarfs in the Pleiades cluster confirmed by the lithium test. Ap. J. (Letters) **469**, L53.

Rebolo, R., Molaro, P., Beckman, J. E. (1988): Lithium abundances in metal-deficient dwarfs. Astron. Astrophys. **192**, 192.

Rees, M. J. (1966): Appearance of relativistically expanding radio sources. Nature **211**, 468.

Rees, M. J. (1967): Studies in radio source structure I. A relativistically expanding model for variable quasi-stellar radio sources. M.N.R.A.S. **135**, 345.

Rees, M. J. (1967): Studies in radio source structure III. Inverse Compton radiation from radio sources. M.N.R.A.S. **137**, 429.

Rees, M. J. (1971): New interpretation of extragalactic radio sources. Nature **229**, 312, 510.

Rees, M. J. (1976): Opacity-limited hierarchical fragmentation and the masses of protostars. M.N.R.A.S. **176**, 483.

Rees, M. J. (1976): Present status of double radio sources. Comments on Astrophys. **6**, 113.
Rees, M. J. (1978): The M87 jet: internal shocks in a plasma beam? M.N.R.A.S. **184**, 61p.
Rees, M. J. (1978): Relativistic jets and beams in radio galaxies. Nature **275**, 516.
Rees, M. J. (1978): Accretion and the quasar phenomenon. Physica Scripta **17**, 193.
Rees, M. J. (1984): Black hole models for active galactic nuclei. Ann. Rev. Astron. Astrophys. **22**, 471.
Rees, M. J.: The emergence of structure in the universe: Galaxy formation and 'dark matter'. In: Three hundred years of gravitation (eds. S. W. Hawking and W. Israel). New York: Cambridge University Press 1987.
Rees, M. J. (1989): Galaxy formation and dark matter. Highlights of Astronomy **8**, 45.
Rees, M. J. (1990): Black holes in galactic centers. Scientific American **263**, 56 – November.
Rees, M. J. (1993): Causes and effects of the first quasars. Proc. Nat. Acad. Sci. **90**, 4840.
Rees, M. J., Mészáros, P. (1992): Relativistic fireballs: energy conversion and time-scales. M.N.R.A.S. **258**, 41P.
Rees, M. J., Sciama, D. W. (1967): Possible large-scale clustering of quasars. Nature **213**, 374.
Rees, M. J., Silk, J. (1970): The origin of galaxies. Scientific American **222**, 26 – June.
Rees, M. J., Simon, M. (1968): Evidence for relativistic expansion in variable radio sources. Ap. J. (Letters) **152**, L145.
Reeves, H. (1994): On the origin of the light elements ($Z < 6$). Rev. Mod. Phys. **66**, 193.
Reeves, H., Audouze, J., Fowler, W. A., Schramm, D. N. (1973): On the origin of light elements. Ap. J. **179**, 909.
Reeves, H., Fowler, W. A., Hoyle, F. (1970): Galactic cosmic ray origin of Li, Be and B in stars. Nature **226**, 727.
Refsdal, S. (1964): The gravitational lens effect. M.N.R.A.S. **128**, 295.
Refsdal, S. (1964): On the possibility of determining Hubble's parameter and the masses of galaxies from the gravitational lens effect. M.N.R.A.S. **128**, 307.
Refsdal, S. (1966): On the possibility of testing cosmological theories from the gravitational lens effect. M.N.R.A.S. **132**, 101.
Refsdal, S. (1966): On the possibility of determining the distances and masses of stars from the gravitational lens effect. M.N.R.A.S. **134**, 315.
Refsdal, S. (1970): On the propagation of light in universes with inhomogeneous mass distribution. Ap. J. **159**, 357.
Regos, E., Geller, M. J. (1989): Infall patterns around rich clusters of galaxies. Astron. J. **98**, 755.
Reid, M. J.: The distance to the galactic center, R_0. In: The center of our Galaxy – proc. 136th symp. IAU (ed. M. Morris). Boston: Kluwer 1989.
Reid, M. J. (1993): The distance to the center of the Galaxy. Ann. Rev. Astron. Astrophys. **31**, 345.
Reid, M. J., et al. (1989): Subluminal motion and limb brightening in the nuclear jet of M87. Ap. J. **336**, 112.
Reinhardt, V. S., Lavanceau, J.: A comparison of the cesium and hydrogen hyperfine frequencies by means of Loran-C and portable clocks. Proceedings of the twenty-eighth annual symposium on frequency control, Ft. Monmouth, New Jersey 1974.
Reissner, H. (1916): Über die Eigengravitation des elektrischen Feldes nach der Einsteinschen Theorie (On the special gravitation of the electric field given by Einstein's theory). Ann. Phys. **50**, 106.
Remillard, R. A., Mc Clintock, J. E., Bailyn, C. D. (1992): Evidence for a black hole in the X-ray binary nova muscae 1991. Ap. J. (Letters) **399**, L145.
Renn, J., Sauer, T., Stachel, J. (1997): The origin of gravitational lensing: A postcript to Einstein's 1936 Science paper. Science **275**, 184.
Renzini, A., Pecci, F. F. (1988): Tests of evolutionary sequences using color-magnitude diagrams of globular clusters. Ann. Rev. Astron. Astrophys. **26**, 199.
Rephaeli, Y. (1995): Comptonization of the cosmic microwave background: The Sunyaev-Zeldovich effect. Ann. Rev. Astron. Astrophys. **33**, 541.

Review of Particle Properties: Phys. Rev. **D50**, 1218 (1994).
Rhoades, C. E. Jr., Ruffini, R. (1974): Maximum mass of a neutron star. Phys. Rev. Lett. **32**, 324.
Richards, D. W., Comella, J. M. (1969): The period of the pulsar NP 0532. Nature **222**, 551.
Richardson, D. E. (1945): Distances of the planets from the Sun and of satellites from their primaries in the satellites of Jupiter, Saturn and Uranus. Pop. Astron. **53**, 14.
Richter, O.-G., Huchtmeier, W. K. (1984): Is there a unique relation between absolute (blue) luminosity and total 21 cm linewidth of disk galaxies? Astron. Astrophys. **132**, 253.
Riegler, G. R., Blandford, R. D. (eds.): The galactic center. AIP Conference Proceedings No. 83. New York: American Institute of Physics 1982.
Riess, A. G., Press, W. H., Kirshner, R. P. (1995): Using Type Ia supernova light curve shapes to measure the Hubble constant. Ap. J. (Letters) **438**, L17.
Riess, A. G., Press, W. H., Kirshner, R. P. (1996): A precise distance indictor: Type Ia supernova multicolor light-curve shapes. Ap. J. **473**, 88.
Riis, E., et al. (1988): Test of the isotropy of the speed of light using fast-beam laser spectroscopy. Phys. Rev. Lett. **60**, 81.
Riley, J. M. (1973): A measurement of the gravitational deflection of radio waves by the Sun during 1972 October. M.N.R.A.S. **161**, 11P.
Rindler, W. (1956): Visual horizons in world-models. M.N.R.A.S. **116**, 662.
Riordan, M., Schramm, D. N.: The shadows of creation. Dark matter and the structure of the universe. New York: W. H. Freeman and Company 1991.
Roberts, D. H., et al. (1985): The multiple images of quasar 0957+561. Ap. J. **293**, 356.
Roberts, D. H., et al. (1991): The Hubble constant from VLA measurement of the time delay in the double quasar 0957+561. Nature **352**, 43.
Roberts, D. H., Greenfield, P. E., Burke, B. F. (1979): The double quasar 0957+561 – a radio study at 6-centimeters wavelength. Science **205**, 894.
Roberts, M. S. (1969): Integral properties of spiral and irregular galaxies. Astron. J. **74**, 859.
Roberts, M. S.: Radio observations of neutral hydrogen in galaxies. In: Galaxies and the universe (eds. A. Sandage, M. Sandage, and J. Kristian). Chicago, Ill.: University of Chicago Press 1975.
Roberts, M. S., Rots, A. H. (1973): Comparison of rotation curves of different galaxy types. Astron. Astrophys. **26**, 483.
Robertson, D. S. (1991): Geophysical applications of very-long-baseline interferometry. Rev. Mod. Phys. **63**, 899.
Robertson, D. S., Carter, W. E., Dillinger, W. H. (1991): New measurement of solar gravitational deflection of radio signals using VLBI. Nature **349**, 768.
Robertson, H. P. (1928): On relativistic cosmology. Phil. Mag. **5**, 835.
Robertson, H. P. (1933): Relativistic cosmology. Rev. Mod. Phys. **5**, 62.
Robertson, H. P. (1935): Kinematics and world-structure. Ap. J. **82**, 284.
Robertson, H. P. (1936): Kinematics and world-structure II, III. Ap. J. **83**, 187, 257.
Robertson, H. P. (1938): The two-body problem in general relativity. Ann. Math. **39**, 101.
Robertson, H. P. (1949): Postulate versus observation in the special theory of relativity. Rev. Mod. Phys. **21**, 378.
Robertson, H. P. (1955): The theoretical aspects of the nebular redshift. P.A.S.P. **67**, 82.
Robertson, H. P.: Relativity and cosmology. In: Space age astronomy (ed. A. J. Deutsch and W. B. Klemperer). New York: Academic Press 1962.
Robertson, H. P., Noonan, T. W.: Relativity and cosmology. Philadelphia: W. B. Saunders 1968.
Roeder, R. C. (1975): Apparent magnitudes, redshifts and inhomogeneities in the universe. Ap. J. **196**, 671.
Rogerson, J. B., York, D. G. (1973): Interstellar deuterium abundance in the direction of Beta Centauri. Ap. J. (Letters) **186**, L95. Reproduced in: A source book in astronomy and astrophysics 1900–1975 (eds. K. R. Lang and O. Gingerich). Cambridge, Mass.: Harvard University Press 1979.

Rogstad, D. H., Shostak, G. S. (1972): Gross properties of five Scd galaxies as determined from 21-centimeter observations. Ap. J. **176**, 315.

Roll, P. G., Krotov, R., Dicke, R. H. (1964): The equivalence of inertial and passive gravitational mass. Ann. Phys. (N. Y.) **26**, 442.

Roll, P. G., Wilkinson, D. T. (1966): Cosmic background radiation at 3.2 cm – support for cosmic black-body radiation. Phys. Rev. Lett. **16**, 405.

Rood, H. J. (1981): Clusters of galaxies. Rep. Prog. Phys. **44**, 1077.

Rood, H. J. (1988): Voids. Ann. Rev. Astron. Astrophys. **26**, 245.

Rood, H. J., Page, T. L., Kintner, E. C., King, I. R. (1972): The structure of the Coma cluster of galaxies. Ap. J. **175**, 627.

Rood, H. J., Rothman, V. C.A., Turnrose, B. E. (1970): Empirical properties of the mass discrepancy in groups and clusters of galaxies. Ap. J. **162**, 411.

Rose, R. D., Parker, H. M., Lowry, R. A., Kuhlthau, A. R., Beams, J. W. (1969): Determination of the gravitational constant G. Phys. Rev. Lett. **23**, 655.

Rosen, N. (1974): A theory of gravitation. Ann. Phys. (N.Y.) **84**, 455.

Roseveare, N. T.: Mercury's perihelion from Le Verrier to Einstein. Oxford: Clarendon Press 1982.

Ross, D. K., Schiff, L. I. (1966): Analysis of the proposed planetary radar reflection experiment. Phys. Rev. **141**, 1215.

Roth, J., Primack, J. R. (1996): Cosmology. Sky and Telescope **91**, 20.

Rowan-Robinson, M.: The cosmological distance ladder. New York: W. H. Freeman 1985.

Rowan-Robinson, M. (1988): The extragalactic distance scale. Space Sci. Rev. **48**, 1.

Rowan-Robinson, M. (1993): Derivation of the cosmological density paramter Ω_o from large-scale flows. Proc. Nat. Acad. Sci. **90**, 4822.

Rowan-Robinson, M.: Ripples in the cosmos: A view behind the scenes of the new cosmology. New York: W. H. Freeman 1993.

Rowan-Robinson, M., et al. (1990): A sparse-sampled redshift survey of IRAS galaxies – I. The convergence of the IRAS dipole and the origin of our motion with respect to the microwave background. M.N.R.A.S. **247**, 1.

Roy, A. E. (1953): Miss Blagg's formula. J. Brit. Astron. Assoc. **63**, 212.

Rubin, V. C. (1951): Differential rotation of the inner metagalaxy. Astron. J. **56**, 47.

Rubin, V. C. (1979): Extended optical-rotation curves of spiral galaxies. Comm. Astrophys. **8**, 79.

Rubin, V. C. (1983): The rotation of spiral galaxies. Science **220**, 1339.

Rubin, V. C. (1983): Dark matter in spiral galaxies. Scientific American **248**, 96 – June.

Rubin, V. C.: Constraints on the dark matter from optical rotation curves. In: Dark matter in the universe (eds. J. Kormendy and G. R. Knapp). Boston: Reidel 1987.

Rubin, V. C. (1993): Galaxy dynamics and the mass density of the universe. Proc. Nat. Acad. Sci. **90**, 4814.

Rubin, V. C., Coyne, G. V. (eds.): Large-scale motions in the universe. Princeton, New Jersey: Princeton University Press 1988.

Rubin, V. C., et al. (1976): Motion of the Galaxy and the local group determined from the velocity anisotropy of distant Sc I galaxies I. The data. II. The analysis for the motion. Astron. J. **81**, 687, 719.

Rubin, V. C., Ford, W. K. (1970): Rotation of the andromeda nebula from a spectroscopic survey of emission regions. Ap. J. **150**, 379.

Rubin, V. C., Ford, W. K., Thonnard, N. (1980): Rotational properties of 21 Sc galaxies with a large range of luminosities and radii. From NGC 4605 (R = 4 kpc) to UCG 2885 (R = 122 kpc). Ap. J. **238**, 471.

Ruderman, M., Shaham, J., Tavani, M. (1989): Accretion turnoff and rapid evaporation of very light secondaries in low-mass X-ray binaries. Ap. J. **336**, 507.

Rugers, M, Hogan, C. J. (1996): Confirmation of high deuterium abundance in quasar absorbers. Ap. J. (Letters) **459**, L1.

Russell, H. N. (1914): Relations between the spectra and other characteristics of stars. Pop. Astron. **22**, 275. Reproduced in: A source book in astronomy and astrophysics 1900–1975 (eds. K. R. Lang and O. Gingerich). Cambridge, Mass.: Harvard University Press 1979.

Russell, H. N., Dugan, R. S., Stewart, J.: Astronomy. Boston, Mass.: Gunn 1926.

Rutherford, E. (1905): Radium – the cause of the earth's heat. Harper's Magazine Feb., 390.

Rutherford, E. (1929): Origin of actinium and age of the Earth. Nature **123**, 313.

Rutherfurd, L. M. (1863): Astronomical observations with the spectroscope. Sill. Amer. J. **35**, 71.

Ryba, M. F., Taylor, J. H. (1991): High-precision timing of millisecond pulsars I. Astrometry and masses of the PSR 1855 + 09 system. Ap. J. **371**, 739.

Ryba, M. F., Taylor, J. H. (1991): High-precision timing of millisecond pulsars II. Astrometry, orbital evolution, and eclipses of PSR 1957 + 20. Ap. J. **380**, 557.

Ryle, M. (1958): The nature of the cosmic radio sources. Proc. Roy. Soc. London **248**, 289.

Ryle, M. (1968): The counts of radio sources. Ann. Rev. Astron. Astrophys. **6**, 249.

Ryle, M., Caswell, J. L., Hine, G., Shakeshaft, J. (1978): A new class of radio star. Nature **276**, 571.

Ryle, M., Clarke, R. W. (1961): An examination of the steady-state model in the light of some recent observations of radio sources. M.N.R.A.S. **122**, 349.

Ryle, M., Longair, M. S. (1967): A possible method for investigating the evolution of radio galaxies. M.N.R.A.S. **136**, 123.

Sachs, R. K., Wolfe, A. M. (1967): Perturbations of a cosmological model and angular variations of the microwave background. Ap. J. **147**, 73.

Sackmann, I.-J., Boothroyd, A. I., Kraemer, K. E. (1993): Our Sun. III. present and future. Ap. J. **418**, 457.

Sadoulet, B.: Nonbaryonic dark matter: In: The sun and beyond (eds. J. Tran Thanh Van, L. M. Celnikier, H. C. Trung and S. Vauclair). Paris: Editions Frontieres 1996.

Saikia, D. J., Kulkarni, V. K. (1979): On the interpretation of the angular size–redshift diagram for radio sources. M.N.R.A.S. **189**, 393.

Saikia, D. J., Salter, C. J. (1988): Polarization properties of extragalactic radio sources. Ann. Rev. Astron. Astrophys. **26**, 93.

Salopek, D. S. (1992): Consequences of the COBE satellite results for the inflationary scenario. Phys. Rev. Lett. **69**, 3602.

Salpeter, E. (1952): Nuclear reactions in stars without hydrogen. Ap. J. **115**, 336. Reproduced in: Souce book in astronomy and astrophysics 1900–1975 (eds. K. R. Lang and O. Gingergich). Cambridge, Mass.: Harvard University Press 1979.

Salpeter, E. E. (1964): Accretion of interstellar matter by massive objects. Ap. J. **140**, 796.

Sancisi, R., Van Albada, T. S.: HI rotation curves of galaxies. In: Dark matter in the universe (eds. J. Kormendy and G. R. Knapp). Boston: Reidel 1987.

Sandage, A. (1953): The color-magnitude diagram for the globular cluster M3. Astron. J. **58**, 61.

Sandage, A. (1957): Observational approach to evolution I. luminosity functions. Ap. J. **125**, 422.

Sandage, A. (1957): Observational approach to evolution III. semiempirical evolution tracks for M67 and M3. Ap. J. **126**, 326.

Sandage, A. (1958): Current problems in the extragalactic distance scale. Ap. J. **127**, 513.

Sandage, A. (1961): The ability of the 200-inch telescope to discriminate between selected world models. Ap. J. **133**, 355.

Sandage, A. (1961): The light travel time and the evolutionary correction to magnitudes of distant galaxies. Ap. J. **134**, 916.

Sandage, A.: The distance scale. In: Problems in extragalactic research (ed. G. Mc Vittie). New York: Macmillan 1962.

Sandage, A. (1970): Main-sequence photometry, color-magnitude diagrams, and ages for the globular clusters M3, M13, M15 and M92. Ap. J. **162**, 841.

Sandage, A.: The age of the galaxies and globular clusters, problems in finding the Hubble constant and deceleration parameter. In: Nuclei of galaxies, pontifica academia scientiavm (ed. D. K. O'Connell). Amsterdam, Holland: North Holland 1971.

Sandage, A. (1972): Classical cepheids – cornerstone to extragalactic distances. Q.J.R.A.S. **13**, 202.

Sandage, A. (1975): On the ratio of extinction to reddening for interstellar matter using galaxies. I. A limit on the neutral extinction from photometry of the 3C 129 group. P.A.S.P. **87**, 853.

Sandage, A. (1982): The Oosterhoff period groups and the age of globular clusters III. the age of the globular cluster system. Ap. J. **252**, 553.

Sandage, A. (1983): On the age of M92 and M15. Astron. J. **88**, 1159.

Sandage, A. (1986): The population concept, globular clusters, subdwarfs, ages, and the collapse of the Galaxy. Ann. Rev. Astron. Astrophys. **24**, 421.

Sandage, A. (1986): The redshift-distance relation IX. perturbation of the very nearby velocity field by the mass of the local group. Ap. J. **307**, 1.

Sandage, A. (1988): Observational tests of world models. Ann. Rev. Astron. Astrophys. **26**, 561.

Sandage, A. (1988): The case for $H_o = 55$ from the 21 centimeter linewidth absolute magnitude relation for field galaxies. Ap. J. **331**, 605.

Sandage, A. (1994): Bias properties of the extragalactic distance scale I. The Hubble constant does not increase outward. Ap. J. **430**, 1.

Sandage, A. (1994): Bias properties of extragalactic distance indicators II. bias corrections to Tully-Fisher distances for field galaxies. Ap. J. **430**, 13.

Sandage, A., et al. (1994): The cepheid distance to NGC 5253 – calibration of M(max) for the Type Ia supernovae SN 1972E and SN 1895B. Ap. J. (Letters) **423**, L13.

Sandage, A., et al. (1996): Cepheid calibration of the peak brightness of type Ia supernovae: Calibration of SN 1990N in NGC 4639 averaged with six earlier type Ia supernova calibrations to give H_o directly. Ap. J. (Letters) **460**, L15.

Sandage, A., Kristian, J., Westphal, J. A. (1976): The extension of the Hubble diagram I. New redshifts and bvr photometry of remote cluster galaxies, and an improved richness correction. Ap. J. **205**, 688.

Sandage, A., Schwarzschild, M. (1952): Inhomogeneous stellar models II. models with exhausted cores in gravitational contraction. Ap. J. **116**, 463. Reproduced in: A source book in astronomy and astrophysics 1900–1975 (eds. K. R. Lang and O. Gingerich). Cambridge, Mass.: Harvard University Press 1979.

Sandage, A., Tammann, G. A. (1974): Steps toward the Hubble constant I. calibration of the linear sizes of extragalactic H II regions. Ap. J. **190**, 525.

Sandage, A., Tammann, G. A. (1974): Steps toward the Hubble constant IV. distances to 39 galaxies in the general field leading to a calibration of the galaxy luminosity classes and a first hint of the value of H_o. Ap. J. **194**, 559.

Sandage, A., Tammann, G. A. (1975): Steps toward the Hubble constant V. The Hubble constant from nearby galaxies and the regularity of the local velocity field. Ap. J. **196**, 313.

Sandage, A., Tammann, G. A. (1976): Steps toward the Hubble constant. VII. distances to NGC 2403, M101 and the Virgo cluster using 21 centimeter line widths compared with optical methods – the global value of H_o. Ap. J. **210**, 7.

Sandage, A., Tammann, G. A.: A revised Shapley-Ames catalog of bright galaxies. Washington, D.C.: Carnegie Institution of Washington 1981.

Sandage, A., Tammann, G. A. (1982): Steps toward the Hubble constant VIII. the global value. Ap. J. **256**, 339.

Sandage, A., Tammann, G. A. (1984): The Hubble constant as derived from 21 cm line widths. Nature **307**, 326.

Sandage, A., Tammann, G. A. (1990): Steps toward the Hubble constant IX. The cosmic value of H_O freed from all local velocity anomalies. Ap. J. **365**, 1.

Sandage, A., Tammann, G. A. (1993): The Hubble diagram in V for supernovae of Type Ia and the value of H_0 therefrom. Ap. J. **415**, 1.

Sandage, A., Tammann, G. A. (1995): Steps toward the Hubble constant X. the distance of the Virgo cluster core using globular clusters. Ap. J. **446**, 1.

Sandage, A., Tammann, G. A. (1998): Confirmation of previous ground-based Cepheid P–L zero-points using Hipparcos trigonometric parallaxes. M.N.R.A.S. **293**, L23.

Sandage, A., Tammann, G. A., Federspiel, M. (1995): Bias properties of extragalactic distance indicators IV. Demonstration of the population incompleteness bias inherent in the Tully-Fisher method applied to clusters. Ap. J. **452**, 1.

Sanders, D. B. (1989): Continuum energy distribution of quasars – shapes and origins. Ap. J. **347**, 29.

Sanitt, N. (1971): Quasi-stellar objects and gravitational lenses. Nature **234**, 199.

Santiago, B. X., et al. (1995): The optical redshift survey – sample selection and the galaxy distribution. Ap. J. **446**, 457.

Sarajedini, A., King, C. R. (1989): Evidence for an age spread among the galactic globular clusters. Astron. J. **98**, 1624.

Sarazin, C. L. (1986): X-ray emission from clusters of galaxies. Rev. Mod. Phys. **58**, 1.

Sarazin, C. L.: X-ray emissions from clusters of galaxies. New York: Cambridge University Press 1988.

Sargent, W. L. W., et al. (1978): Dynamical evidence for a central mass concentration in the galaxy M87. Ap. J. **221**, 731.

Sargent, W. L. W., et al. (1980): The distribution of Lyman-alpha absorption lines in the spectra of six QSOs: Evidence for an intergalactic origin. Ap. J. Supp. **42**, 41.

Sasaki, M., Tanaka, T., Yamamoto, K. (1995): Euclidean vacuum mode functions for a scalar field on open de Sitter space. Phys. Rev. **D51**, 2979.

Saslaw, W. C., Narasimha, D., Chitre, S. M. (1985): The gravitational lens as an astronomical diagnostic. Ap. J. **292**, 348.

Sasselov, D., Goldwirth, D. (1995): A new estimate of the uncertainties in the primordial helium abundance – new bounds on Ω baryons. Ap. J. (Letters) **444**, L5.

Sato, K. (1981): First-order phase transition of a vacuum and the expansion of the universe. M.N.R.A.S. **195**, 467.

Saunders, W., et al. (1991): The density field of the local universe. Nature **349**, 32.

Savage, B. D., Mathis, J. S. (1979): Observed properties of interstellar dust. Ann. Rev. Astron. Astrophys. **17**, 73.

Scaramella, R., et al. (1989): A marked concentration of galaxy clusters: Is this the origin of large-scale motions? Nature **338**, 562.

Schaffer, S. (1979): John Michell and black holes. J. Hist. Astron. **10**, 42.

Schaller, G., et al. (1992): New grids of stellar models from 0.8 to 120 solar masses at $z=0.020$ and $z=0.001$. Astron. Astrophys. Supp. **96**, 269.

Scharlemann, E. T., Wagoner, R. V. (1972): Electromagnetic fields produced by relativistic rotating disks. Ap. J. **171**, 107.

Schatzman, E. (1949): Remarques sur le phenomene de novae II. Annales d'Astrophysique **12**, 281.

Schechter, P. (1976): An analytic expression for the luminosity function for galaxies. Ap. J. **203**, 297.

Scheuer, P. A. G. (1965): A sensitive test for the presence of atomic hydrogen in intergalactic space. Nature **207**, 963.

Scheuer, P. A. G. (1974): Models of extragalactic radio sources with a continuous energy supply from a central object. M.N.R.A.S. **166**, 513.

Scheuer, P. A. G.: Radio astronomy and cosmology. In: Stars and stellar systems, vol IX – Galaxies and the universe (ed. A. R. Sandage). Chicago, Ill.: University of Chicago Press 1974.

Scheuer, P. A. G. (1976): AO 0235+16.4 – Another source exceeding the speed limit. M.N.R.A.S. **177**, 1p.

Scheuer, P. A. G.: Explanations of superluminal motion. In: VLBI and compact radio sources. IAU symposium no. 110 (eds. R. Fanti, K. Kellermann and G. Setti). Boston: Reidel 1984.

Scheuer, P. A. G., Readhead, A. C. S. (1979): Superluminally expanding radio sources and the radio-quiet QSOs. Nature **277**, 182.

Schiff, L. I. (1960): Motion of a gyroscope according to Einstein's theory of gravitation. Proc. Nat. Acad. Sci. **46**, 871.

Schiff, L. I. (1960): On experimental tests of the general theory of relativity. Am. J. Phys. **28**, 340.

Schiff, L. I. (1960): Possible new test of general relativity theory. Phys. Rev. Lett. **4**, 215.

Schild, R., Oke, J. B. (1971): Energy distributions and K-corrections for the total light from giant elliptical galaxies. Ap. J. **169**, 209.

Schmidt, M. (1956): A model of the distribution of mass in the galactic system. B.A.N. **13**, 15.

Schmidt, M. (1957): Spiral structure in the inner parts of the galactic system derived from the hydrogen emission at 21 cm wavelength. B.A.N. **13**, 247.

Schmidt, M. (1963): 3C 273: A star-like object with large red-shift. Nature **197**, 1040. Reproduced in: A source book in astronomy and astrophysics 1900–1975 (eds. K. R. Lang and O. Gingerich). Cambridge, Mass.: Harvard University Press 1979.

Schmidt, M.: Rotation parameters and distribution of mass in the Galaxy. In: Galactic structure – stars and stellar systems V (ed. A. Blauuw and M. Schmidt). Chicago: Univ. of Chicago Press 1965.

Schmidt, T. (1957): Die lichtkurven-leuchtkraft-beziehung neuer sterne (The relationship of the curvature of starlight and its luminosity). Zeits. f. Ap. **41**, 182.

Schmitt, J. L. (1968): BL Lac identified as a radio source. Nature **218**, 663.

Schneider, D. P., et al. (1988): High-resolution CCD imaging and derived gravitational lens models of 2237+0305. Astron. J. **95**, 1619.

Schneider, D. P., Schmidt, M., Gunn, J. E. (1991): PC 1247 + 3406: An optically selected quasar with a redshift of 4.897. Astron. J. **102**, 837.

Schneider, P. (1984): The amplification caused by gravitational bending of light. Astron. Astrophys. **140**, 119.

Schneider, P., Ehlers, J., Falco, E. E.: Gravitational lenses. New York: Springer Verlag 1992.

Schönberg, M., Chandrasekhar, S. (1942): On the evolution of main-sequence stars. Ap. J. **96**, 161.

Schramm, D. N.: The age of the universe – concordance. In: Astrophysical ages and dating methods (ed. E. Vangioni-Flam, M. Cassé, J. Audouze, and J. Tran Thanh Van). Paris: Editions Frontieres 1990.

Schramm, D. N. (1993): Cosmological implications of light element abundances – theory. Proc. Nat. Acad. Sci. **90**, 4782.

Schramm, D. N., Wagoner, R. V. (1974): What can deuterium tell us? Physics Today **27**, 41 – December.

Schramm, D. N., Wagoner, R. V. (1977): Element production in the early universe. Ann. Rev. Nuc. Sci. **27**, 37.

Schramm, D. N., Wasserburg, G. J. (1970): Nucleochronologies and the mean age of the elements. Ap. J. **162**, 57.

Schreier, E., et al. (1972): Evidence for the binary nature of Centaurus X-3 from UHURU X-ray observations. Ap. J. (Letters) **172**, L79.

Schutz, B. F. (1986): Determining the Hubble constant from gravitational wave observations. Nature **323**, 310.

Schwarzschild, K. (1916): Über das Gravitationsfeld eines Massenpunktes nach der Einsteinschen Theorie (The gravitational field of a point mass according to Einstein's theory). Sitz. Acad. Wiss., Physik-Math Kl. **1**, 189. English translation in: A source book in astronomy and astrophysics 1900–1975 (Eds. K. R. Lang and O. Gingerich). Cambridge, Mass.: Harvard University Press 1979.

Schwarzschild, K. (1916): Über das gravitationsfeld einer Kugel aus inkompressibler Flussigkeit nach der Einsteinschen Theorie (On the gravitational field of a sphere of non-compressible liquid according to Einstein's theory).. Sitz. Akad. Wiss., Physik-Math Kl. **1**, 424.

Schwarzschild, M, Schwarzschild, B. (1950): A spectroscopic comparison between high- and low-velocity F dwarfs. Ap. J. **112**, 248.

Sciama, D. W. (1976): Black holes and their thermodynamics. Vistas in Astronomy **19**, 385.

Sciama, D. W.: Modern cosmology and the dark matter problem. Cambridge, England: Cambridge University Press 1993.

Scott, D., Silk, J., White, M. (1995): From microwave anisotropies to cosmology. Science **268**, 829.

Scott, P. F., et al. (1996): Measurements of structure in the cosmic background radiation with the cambridge cosmic anisotropy telescope. Ap. J. (Letters) **461**, L1.

Searle, L., Zinn, R. (1978): Compositions of halo clusters and the formation of the galactic halo. Ap. J. **225**, 357.

Secchi, P. A.: Spettri prismatici delle stelle fisse (Spectra of the fixed stars). Memoira Roma, 1868, Firenze, 1869.

Secker, J. (1992): A statistical investigation into the shape of the globular cluster luminosity distribution. Astron. J. **104**, 1472.

Segal, I. E. (1993): The redshift-distance relation. Proc. Nat. Acad. Sci. **90**, 4798.

Seidelmann, K. P. (ed.): Explanatory supplement to the astronomical almanac. Mill Valley, Ca.: University Science Books 1992.

Seielstad, G. A., Sramek, R. A., Weiler, K. W. (1970): Measurement of the deflection of 9.602 GHz radiation from 3C 279 in the solar gravitational field. Phys. Rev. Lett. **24**, 1373.

Seldner, M., et al. (1977): New reduction of the Lick catalog of galaxies. Astron. J. **82**, 249.

Sérsic, J. L. (1959): The H II regions as distance indicators. Observatory **79**, 54.

Sérsic, J. L. (1960): The H II regions in galaxies. Zeits. f. Ap. **50**, 168.

Seyfert, C. K. (1943): Nuclear emission in spiral nebulae. Ap. J. **97**, 28. Reproduced in: A source book in astronomy and astrophysics 1900–1975 (eds. K. R. Lang and O. Gingerich). Cambridge, Mass.: Harvard University Press 1979.

Shafi, Q., Stecker, F. W. (1984): Implications of a class of grand-unified theories for large-scale structure in the universe. Phys. Rev. Lett. **53**, 1292.

Shahbaz, T., et al. (1994): The mass of the black hole in V404 Cygni. M.N.R.A.S. **271**, L10.

Shakeshaft, J. R., Ryle, M., Baldwin, J. E., Elsmore, B. Thompson, J. H. (1955): A survey of radio sources between declinations $-38°$ and $+83°$. Mem. R. A. S. **67**, 106.

Shaklan, S. B., Hege, E. K. (1986): Detection of the lensing galaxy in PG 1115+08. Ap. J. **303**, 605.

Shakura, N. I., Sunyaev, R. A. (1973): Black holes in binary systems. Observational appearance. Astron. Astrophys. **24**, 337.

Shakura, N. I., Sunyaev, R. A. (1976): A theory of the instability of disk accretion on to black holes and the variability of binary X-ray sources, galactic nuclei and quasars. M.N.R.A.S. **175**, 613.

Shandarin, S. F., Zeldovich, Y. B. (1989): The large-scale structure of the universe: Turbulence, intermittency, structures in a self-gravitating medium. Rev. Mod. Phys. **61**, 185.

Shane, C. D., Wirtanen, C. A. (1967): The distribution of galaxies. Pub. Lick Observatory **22**, 1.

Shane, W. W., Bieger-Smith, G. P. (1966): The galactic rotation curve derived from observations of neutral hydrogen. B.A.N. **18**, 263.

Shapiro, I. I. (1964): Fourth test of general relativity. Phys. Rev. Lett. **13**, 789.

Shapiro, I. I. (1966): Testing general relativity with radar. Phys. Rev. **141**, 1219.

Shapiro, I. I. (1967): New method for detection of light deflection by solar gravity. Science **157**, 806.

Shapiro, I. I. (1968): Fourth test of general relativity – preliminary results. Phys. Rev. Lett. **20**, 1265.
Shapiro, I. I.: Spin and orbital motions of the planets. In: Radar astronomy (ed. J. V. Evans and T. Hagfors). New York: McGraw-Hill 1968.
Shapiro, I. I., Counselman, C. C., King, R. W. (1976): Verification of the principle of equivalence for massive bodies. Phys. Rev. Lett. **36**, 555.
Shapiro, I. I., et al. (1971): Fourth test of general relativity - new radar result. Phys. Rev. Lett. **26**, 1132. Reproduced in: A source book in astronomy and astrophysics 1900–1975 (eds. K. R. Lang and O. Gingerich). Cambridge, Mass.: Harvard University Press 1979.
Shapiro, I. I., et al. (1971): General relativity and the orbit of Icarus. Astron. J. **76**, 588.
Shapiro, I. I., et al. (1971): Gravitational constant – experimental bound on its time variation. Phys. Rev. Lett. **26**, 27.
Shapiro, I. I., et al. (1972): Mercury's perihelion advance – determination by radar. Phys. Rev. Lett. **28**, 1594.
Shapiro, I. I., et al. (1977): The Viking relativity experiment. J. Geophys. Res. **82**, 4329.
Shapiro, I. I., et al. (1988): Measurement of the de Sitter precession of the moon: A relativistic three-body effect. Phys. Rev. Lett. **61**, 2643.
Shapiro, S. L. (1971): The density of matter in the form of galaxies. Astron. J. **76**, 291.
Shapiro, S. L., Teukolsky, S. A.: Black holes, white dwarfs, and neutron stars. New York: John Wiley & Sons 1983.
Shapley, H. (1918): On the determination of the distances of globular clusters. Ap. J. **48**, 89.
Shapley, H. (1934): On some structural features of the metagalaxy. M.N.R.A.S. **94**, 791.
Shapley, H., Ames, A. (1932): A survey of the external galaxies brighter than the thirteenth magnitude. Ann. Astron. Obs. Harvard College. **88**, No. 2, 1.
Sharp, N. A. (1986): The whole-sky distribution of galaxies. P.A.S.P. **98**, 740.
Shaver, P. A. (1991): Radio surveys and large scale structure. Aust. J. Phys. **44**, 759.
Shaver, P. A., et al. (1996): Decrease in the space density of quasars at high redshift. Nature **384**, 439.
Shi, X. (1995): The uncertainties in the age of globular clusters from their helium abundance and mass loss. Ap. J. **446**, 637.
Shipman, H. L., et al. (1997): The mass and radius of 40 Eridani B from Hipparcos: An accurate test of stellar interior theory. Ap. J. (Letters) **488**, L43.
Shklovskii, I. S. (1953): On the nature of the luminescence of the Crab Nebula. Doklady Akad. Nauk SSSR **90**, 983. English translation in: A source book in astronomy and astrophysics 1900–1975 (eds. K. R. Lang and O. Gingerich). Cambridge, Mass.: Harvard University Press 1979.
Shklovskii, I. S. (1962): On the nature of radio galaxies. Astron. Zhurnal **39**, No. 4, 591 – July, August. English translation in Sov. Astron. AJ **6**, 465 (1963).
Shklovskii, I. S. (1964): Physical conditions in the gaseous envelope of 3C 173. Astron. Zh. **41**, 408. Sov. Astron. **8**, 638 (1965).
Shklovskii, I. S. (1967): On the nature of the source of X-ray emission of Sco XR-1. Ap. J. (Letters) **148**, L1.
Shklovskii, I. S. (1977): The nature of the jet in M87. Astron. Zhurnal **54**, 713 (1977) – July, August. English translation in Sov. Astron, **21**, No. 4, 401.
Shu, F. H.: The physical universe. Mill Valley, Ca: University Science Books 1982.
Shu, F. H.: The physics of astrophysics, volume 2, gas dynamics. Mill Valley, California: University Science Books 1992.
Silk, J. (1967): Fluctuations in the primordial fireball. Nature **215**, 1155.
Silk, J. (1968): Cosmic black-body radiation and galaxy formation. Ap. J. **151**, 459.
Silk, J. (1974): Large-scale inhomogeneity of the universe: Implications for the deceleration parameter. Ap. J. **193**, 525.
Silk, J. (1977): Large-scale inhomogeneity of the universe: Spherically symmetric models. Astron. Astrophys. **59**, 53.
Silk, J. (1987): The formation of galaxies. Physics Today **40**, 28 – April.

Silk, J. (1989): Is cosmic drift a cosmic myth? Ap. J. (Letters) **345**, L1.
Silk, J. (1993): Dissipative processes in galaxy formation. Proc. Nat. Acad. Sci. **90**, 4835.
Silk, J.: A short history of the universe. New York: W. H. Freeman 1994.
Silk, J., Szalay, A. S., Zeldovich, Y. B. (1983): The large-scale structure of the universe. Scientific American **249**, 72 – October.
Silk, J., White, S. D. M. (1978): The determination of the deceleration parameter using X-ray and microwave observations of galaxy clusters. Ap. J. (Letters) **226**, L103.
Silva De, L. N. K. (1970): Quasi-stellar objects and gravitational lenses. Nature **228**, 1180.
Silva De, L. N. K. (1974): On gravitational-lens quasars. Ap. J. **189**, 177.
Singh, K. P., Westergaard, N. J., Schnopper, H. W. (1988): EXOSAT observations of the hot gas in the A1060 cluster of galaxies. Ap. J. **330**, 620.
Skillman, E. D., et al. (1994): Spatially resolved optical and near-infrared spectroscopy of the low-metallicity galaxy UGC 4483. Ap. J. **431**, 172.
Skillman, E. D., Kennicutt, R. C. Jr. (1993): Spatially resolved optical and near-infrared spectroscopy of I Zw 18. Ap. J. **411**, 655.
Slipher, V. M. (1917): A spectrographic investigation of spiral nebulae. Proc. Amer. Phil. Soc. **56**, 403. Reproduced in: A source book in astronomy and astrophysics 1900–1975 (eds. K. R. Lang and O. Gingerich). Cambridge, Mass.: Harvard University Press 1979.
Smarr, L. L., Blandford, R. (1976): The binary pulsar: Physical processes, possible companions, and evolutionary histories. Ap. J. **207**, 574.
Smith, C. A., et al. (1989): Mean and apparent place computations in the new IAU system I. the transformation of astrometric catalog systems to the equinox J2000.0. Astron. J. **97**, 265.
Smith, F. G. (1979): A gravitational lens. Nature **279**, 374.
Smith, H. J., Hoffleit, D. (1963): Light variations in the superluminous radio galaxy 3C 273. Nature **198**, 650.
Smith, M. S., Kawano, L. H., Malaney, R. A. (1993): Experimental, computational, and observational analysis of primordial nucleosynthesis. Ap. J. Supp. **85**, 219.
Smith, R. W. (1979): The origins of the velocity distance relation. J.H.A. **10**, 133.
Smith, S. (1936): The mass of the Virgo cluster. Ap. J. **83**, 23.
Smith, W. H., Schempp, W. V., Baines, K. H. (1989): The D/H ratio for Jupiter. Ap. J. **336**, 967.
Smoot, G. F., Davidson, K.: Wrinkles in time: The imprint of creation. London: Little, Brown and Co. 1993.
Smoot, G. F., et al. (1991): Preliminary results from the COBE differential microwave radiometers – large angular scale isotropy of the cosmic microwave background. Ap. J. (Letters) **371**, L1.
Smoot, G. F., et al. (1992): Structure in the COBE differential microwave radiometer first-year maps. Ap. J. (Letters) **396**, L1.
Smoot, G. F., Gorenstein, M. V., Muller, R. A. (1977): Detection of anisotropy in the cosmic blackbody radiation. Phys. Rev. Lett. **39**, 898.
Smoot, G. F., Lubin, P. M. (1979): Southern hemisphere measurements of the anisotropy in the cosmic microwave background radiation. Ap. J. (Letters) **234**, L83.
Snider, J. L. (1972): New measurement of the solar gravitational redshift. Phys. Rev. Lett. **28**, 853.
Snider, J. L. (1974): Comments on two recent measurements of the solar gravitational redshift. Solar Phys. **36**, 233.
Soldner, J.: Über die Ablenkung eines Lichtstrahls von seiner geradlinigen Bewegung durch die Attraktion eines Weltkörpers, an welchem er nahe vorbeigeht (Concerning the deflection of a light ray from its straight path due to the attraction of a heavenly body which it passes closely). Berliner Astronomische Jahrbuch. Berlin: Späthen 1801.
Solheim, J. E. (1966): Relativistic world models and redshift-magnitude observations. M.N.R.A.S. **133**, 321.
Soma, M., Aoki, S. (1990): Transformation from FK4 system to FK5 system. Astron. Astrophys. **240**, 150.

Sonett, C. P., et al. (1996): Late proterozoic and paleozoic tides, retreat of the moon, and rotation of the earth. Science **273**, 100.

Songaila, A., et al. (1994): Deuterium abundance and background radiation temperature in high-redshift primordial clouds. Nature **368**, 599.

Songaila, A., Hu, E. M., Cowie, L. L. (1995): A population of very diffuse Lyman–α clouds as the origin of the He^+ absorption signal in the intergalactic medium. Nature **375**, 124.

Songaila, A., Wampler, E. J., Cowie, L. L. (1997): A high deuterium abundance in the early universe. Nature **385**, 137.

Soucail, G., et al. (1987): A blue ring-like structure in the center of the A 370 cluster of galaxies. Astron. Astrophys. **172**, L14.

Soucail, G., et al. (1987): Further data on the blue ring-like structure in A 370. Astron. Astrophys. **184**, L7.

Soucail, G., et al. (1988): The giant arc in A 370 – spectroscopic evidence for gravitational lensing from a source at $z = 0.754$. Astron. Astrophys. **191**, L19.

Spencer Jones, H. (1939): The rotation of the earth and the secular accelerations of the sun, moon and planets. M.N.R.A.S. **99**, 541.

Spergel, D.: Particle dark matter. In: Unsolved problems in astrophysics (eds. J. N. Bahcall and J. P. Ostriker). Princeton, New Jersey: Princeton University Press 1997.

Spergel, D. N., Turok, N. G. (1992): Textures and cosmic structure. Scientific American **266**, 52 – March.

Spinrad, H. (1986): Faint galaxies and cosmology. P.A.S.P. **98**, 269.

Spinrad, H., Dey, A., Graham, J. R. (1995): Keck observations of the most distant galaxy 8C 14355 + 63 at $z = 4.25$. Ap. J. (Letters) **438**, L51.

Spite, F., Spite, M. (1982): Abundance of lithium in unevolved halo stars and old disk stars – interpretation and consequences. Astron. Astrophys. **115**, 357.

Spite, M., Maillard, J. P., Spite, F. (1984): Abundance of lithium in another sample of halo dwarfs, and in the spectroscopic binary BD-0°4234. Astron. Astrophys. **141**, 56.

Spitzer, L. (1940): The stability of isolated clusters. M.N.R.A.S. **100**, 396.

Spitzer, L., Härm, R. (1958): Evaporation of stars from isolated clusters. Ap. J. **127**, 544.

Spitzer, L., Hart, M. H. (1971): Random gravitational encounters and the evolution of spherical systems I. Method. Ap. J. **164**, 399.

Spruch, L. (1991): Padogogic notes on Thomas-Fermi theory (and on some improvements) – atoms, stars, and the stability of bulk matter. Rev. Mod. Phys. **63**, 151.

Sramek, R. A. (1971): A measurement of the gravitational deflection of microwave radiation near the Sun 1970 October. Ap. J. (Letters) **167**, L55.

St. John, C. E. (1917): The principle of generalized relativity and the displacement of Fraunhofer lines toward the red. Ap. J. **46**, 249.

St. John, C. E. (1928): Evidence for the gravitational displacement of lines in the solar spectrum predicted by Einstein's theory. Ap. J. **67**, 195.

Staelin, D. H., Reifenstein, E. C. (1968): Pulsating radio sources near the Crab Nebula. Science **162**, 1481.

Standish, E. M. Jr. (1982): Conversion of positions and proper motions from B1950.0 to IAU system at J2000.0. Astron. Astrophys. **115**, 20.

Stanek, K. Z., Garnavich, P. M. (1998): Distance to M31 with the Hubble Space Telescope and Hipparcos red clump stars. Ap. J. (Letters) **503**, L31.

Stanek, K. Z., Zaritsky, D., Harris, J. (1998): A "short" distance to the Large Magellanic Cloud with the Hipparcos calibrated red clump stars. Ap. J. (Letters) **500**, L141.

Stark, R. F., Piran, T. (1985): Gravitational-wave emission from rotating gravitational collapse. Phys. Rev. Lett. **55**, 891.

Starobinsky, A. A. (1980): A new type of isotropic cosmological models without singularity. Phys. Lett. **91B**, 99.

Starobinsky, A. A. (1982): Dynamics of phase transition in the new inflationary universe scenario and generation of perturbations. Phys. Lett. **117B**, 175.

Stauffer, J. R. (1987): Dynamical mass determinations for the white dwarf components of HZ 9 and Case 1. Astron. J. **94**, 996.
Stebbins, J., Whitford, A. E. (1943): Six-color photometry of stars: I. The law of space reddening from the colors of O and B stars. Ap. J. **98**, 20.
Stebbins, J., Whitford, A. E. (1948): Six-color photometry of stars.VI. The colors of extragalactic nebulae. Ap. J. **108**, 413.
Steidel, C. C., et al. (1996): Spectroscopic confirmation of a population of normal star-forming galaxies at redshifts $z > 3$. Ap. J. (Letters) **462**, L17.
Steidel, C. C., Pettini, M., Hamilton, D. (1995): Lyman limit imaging of high-redshift galaxies III. New observations of four QSO fields. Astron. J. **110**, 2519.
Steidel, C. C., Sargent, W. L. W., Boksenberg, A. (1988): QSO absorption lines and the time scale for initial heavy element enrichment in galaxies. Ap. J. (Letters) **333**, L5.
Steigman, G. (1993): Challenges to the standard model of big bang nucleosynthesis. Proc. Nat. Acad. Sci. **90**, 4779.
Steigman, G. (1996): Cosmic lithium – going up or coming down? Ap. J. **457**, 737.
Steigman, G., Schramm, D. N., Gunn, J. E. (1977): Cosmological limits to the number of massive leptons. Physics Letters **66B**, 202.
Steigman, G., Tosi, M. (1992): Galactic evolution of D and ^{3}He. Ap. J. **401**, 150.
Steigman, G., Tosi, M. (1995): Generic evolution of deuterium and ^{3}He. Ap. J. **453**, 173.
Steinhardt, P. J. (1995): Cosmology confronts the cosmic microwave background. Int. J. Mod. Phys **A10**, 1091.
Stephenson, C. B., Sanduleak, N. (1977): New h-alpha emission stars in the milky way. Ap. J. Supp. **33**, 459.
Stephenson, F. R., Lieske, J. H. (1988): Changes in the earth's rate of rotation between A.D. 1672 and 1806 deduced from solar eclipse timings. Astron. Astrophys. **200**, 218.
Stephenson, F. R., Morrison, L. V. (1984): Long-term changes in the rotation of the Earth – 700 B. C. to A. D. 1980. Phil. Trans. Roy. Soc. (London) **A313**, 47.
Stetson, P. B., et al. (1989): CCD photometry of the anomalous globular cluster Palomar 12. Astron. J. **97**, 1360.
Stetson, P. B., Harris, W.E. (1988): CCD photometry of the globular cluster M92. Astron. J. **96**, 909.
Stinebring, D. R., et al. (1990): Cosmic gravitational-wave background – limits from millisecond pulsar timing. Phys. Rev. Lett. **65**, 285.
Stockton, A. (1980): The lens galaxy of the twin QSO 0957+561. Ap. J. (Letters) **242**, L141.
Stoner, E. C. (1930): The equilibrium of dense stars. Phil. Mag. **9**, 944.
Strauss, M. A., Cen, R., Ostriker, J. (1993): The cosmic mach number: Direct comparisons of observations and models. Ap. J. **408**, 389.
Strauss, M. A., Davis, M.: The peculiar velocity field predicted by the distribution of IRAS galaxies. In: Large-scale motions in the universe (eds. V. C. Rubin and G. V. Coyne). Princeton, New Jersey: Princeton University Press 1988.
Strauss, M. A., et al. (1992): A redshift survey of IRAS galaxies V. the acceleration of the local group. Ap. J. **397**, 395.
Strauss, M. A., Willick, J. A. (1995): The density and peculiar velocity fields of nearby galaxies. Physics Reports **261**, 271.
Strom, K. M., Strom, S. E. (1982): Galactic evolution: A survey of recent progress. Science **216**, 571 – May 7.
Struble, M. F., Rood, H. J. (1991): A compilation of redshifts and velocity dispersions for Abell clusters (epoch 1991.2). Ap. J. Supp. **77**, 363.
Su, Y., et al. (1994): New tests of the universality of free fall. Phys. Rev. **D50**, 3614.
Subramanian, K., Rees, M. J., Chitre, S. M. (1987): Gravitational lensing by dark galactic haloes. M.N.R.A.S. **224**, 283.
Suess, H. E., Urey, H. C. (1956): Abundances of the elements. Rev. Mod. Phys. **28**, 53.
Sulentic, J. W., Tifft, W. G.: The revised new general catalogue of nonstellar astronomical objects. Tucson, Arizona: University of Arizona Press 1973.

Sunyaev, R. A., Zeldovich, Y. B. (1970): Small-scale fluctuations of relic radiation. Astrophys. Space Sci. **7**, 3.

Sunyaev, R. A., Zeldovich, Y. B. (1972): The observation of relic radiation as a test of the nature of X-ray radiation from the clusters of galaxies. Comm. Astrophys. Space Phys. **4**, 173.

Sunyaev, R. A., Zeldovich, Y. B. (1980): Microwave background radiation as a probe of the contemporary structure and history of the universe. Ann. Rev. Astron. Astrophys. **18**, 537.

Surdej, J., et al. (1987): A new case of gravitational lensing. Nature **329**, 695.

Surdej, J. et al. (1988): Observations of a the new gravitational lens system UM 673 = Q0142–100. Astron. Astrophys. **198**, 49.

Susskind, L. (1997): Black holes and the information paradox. Scientific American **276**, 52 – April.

Suto, Y., Cen, R., Ostriker, J. P. (1992): Statistics of the cosmic mach number from numerical simulations of a cold dark matter universe. Ap. J. **395**, 1.

Svensson, R. (1982): The pair annihilation process in relativistic plasmas. Ap. J. **258**, 321.

Svensson, R. (1982): Electron-positron pair equilibria in relativistic plasmas. Ap. J. **258**, 335.

Svensson, R. (1984): Steady mildly relativistic thermal plasmas: processes and properties. M.N.R.A.S. **209**, 175.

Swenson, F. J., et al. (1994): The Hyades lithium problem revisited. Ap. J. **425**, 286.

Symbalisty, E. M. D., Schramm, D. N. (1981): Nucleocosmochronology. Rep. Prog. Phys. **44**, 293.

Szalay, A. S. (1981): Nucleosynthesis with nonzero lepton numbers: Is there a limit on the neutrino flavors? Physics Lettters **101B**, 453.

Szalay, A. S., et al. (1993): Redshift survey with multiple pencil beams at the galactic poles. Proc. Nat. Acad. Sci. **90**, 4853.

Szekeres, G. (1960): On the singularities of a Riemannian manifold. Pub. Math. Debrecen **7**, 285.

Tammann, , G. A. (1972): Remarks on the radial velocities of galaxies in the Virgo cluster. Astron. Astrophys. **21**, 355.

Tammann, G. A., Leibundgut, B. (1990): Supernova studies IV. The global value of H_o from supernovae Ia and the peculiar motion of field galaxies. Astron. Astrophys. **236**, 9.

Tammann, G. A., Sandage, A. (1985): The infall velocity toward Virgo, the Hubble constant, and a search for motion toward the microwave background. Ap. J. **294**, 81.

Tammann, G. A., Sandage, A. (1995): The Hubble diagram for supernovae of type Ia. II. The effect on the Hubble constant of a correlation between absolute magnitude and light decay rate. Ap. J. **452**, 16.

Tanaka, Y., et al. (1995): Gravitationally redshifted emission implying an accretion disk and massive black hole in the active galaxy MCG-6-30-15. Nature **375**, 659.

Tananbaum, H., Gursky, H., Kellogg, E. M., Levinson, R., Schreier, E., Giacconi, R. (1972): Discovery of a periodic pulsating binary X-ray source from UHURU. Ap. J. (Letters) **174**, L143.

Tanvir, N. R., et al. (1995): Determination of the Hubble constant from observations of Cepheid variables in the galaxy M96. Nature **377**, 27.

Tayler, R. J. (1982): Introduction to big bang nucleosynthesis: open and closed models, anisotropies. Phil. Trans Roy. Soc. (London) **A307**, 3.

Taylor, J. H. (1975): Discovery of a pulsar in a binary system. Ann. N. Y. Acad. Sci. **262**, 490.

Taylor, J. H. (1994): Binary pulsars and relativistic gravity. Rev. Mod. Phys. **66**, 711.

Taylor, J. H., et al. (1976): Further observations of the binary pulsar PSR 1913+16. Ap. J. (Letters) **206**, L53.

Taylor, J. H., Fowler, L. A., Mc Culloch, P. M. (1979): Measurements of general relativistic effects in the binary pulsar PSR 1913+16. Nature **277**, 437.

Taylor, J. H., Mc Culloch, P. M. (1980): Evidence for the existence of gravitational radiation from measurements of the binary pulsar PSR 1913+16. Ann. N. Y. Acad. Sci. **336**, 442.

Taylor, J. H., Stinebring, D. R. (1986): Recent progress in the understanding of pulsars. Ann. Rev. Astron. Astrophys. **24**, 285.
Taylor, J. H., Weisberg, J. M. (1989): Further experimental tests of relativistic gravity using the binary pulsar PSR 1913+16. Ap. J. **345**, 434.
Taylor, S. R.: Planetary science – a lunar perspective. Huston: Lunar and planetary science institute 1982.
Teerikorpi, P. (1997): Observational selection bias affecting the determination of the extragalactic distance scale. Ann. Rev. Astron. Astrophys. **35**, 101.
Tenorio-Tagle, G., Bodenheimer, P. (1988): Large-scale expanding superstructures in galaxies. Ann. Rev. Astron. Astrophys. **26**, 145.
Terrell, J. (1966): Quasi-stellar objects: Possible local origin. Science **154**, 1281.
Terrell, J.: (1977): Size limits on fluctuating astronomical sources. Ap. J. (Letters) **213**, L93.
Terrell, J. (1977): The luminosity distance equation in Friedmann cosmology. Am. J. Phys. **45**, 869.
Thackrah, A., Jones, H., Hawkins, M. (1997): Lithium detection in a field brown dwarf candidate. M.N.R.A.S. **284**, 507.
Thaddeus, P., Clauser, J. F. (1966): Cosmic microwave radiation at 2.63 mm from observations of interstellar CN. Phys. Rev. Lett. **16**, 819.
The Astronomical Almanac. Washington, D. C.: U.S. Government Printing Office 1996 and any other year.
Thomson, W.: see Kelvin, Lord.
Thorburn, J. A. (1994): The primordial lithium abundance from extreme subdwarfs: New observations. Ap. J. **421**, 318.
Thorne, K. S. (1980): Multipole expansions of gravitational radiation. Rev. Mod. Phys. **52**, 299.
Thorne, K. S.: Gravitational radiation. In: Three hundred years of gravitation (eds. S. W. Hawking and W. Israel). New York: Cambridge University Press 1987.
Thorne, K. S.: Black holes and time warps. New York: W. W. Norton 1994.
Thorne, K. S., Blandford, R. D.: Black holes and the origin of radio sources. In: Extragalactic radio sources (eds. D. S. Heeschen and C. M. Wade) Boston: D. Reidel 1982.
Thorne, K. S., Lee, D. L., Lightman, A. P. (1973): Foundations for a theory of gravitation theories. Phys. Rev. **D7**, 3563.
Thorne, K. S., Will, C. M. (1971): Theoretical frameworks for testing relativistic gravity I. Foundations. Ap. J. **163**, 595.
Thorsett, S. E., et al. (1993): The masses of two binary neutron star systems. Ap. J. (Letters) **405**, L29.
Thuan, T. X., Izotov, Y. I., Lipovetsky, V. A. (1996): Hubble space telescope observations of the unusual blue compact dwarf galaxy Makarian 996. Ap. J. **463**, 120.
Tingay, S. J., et al. (1995): Relativistic motion in a nearby bright X-ray source. Nature **374**, 141.
Tinsley, B. M. (1975): Nucleochronology and chemical evolution. Ap. J. **198**, 145.
Tinsley, B. M. (1977): Chemical evolution in the solar neighborhood III. time scales and nucleochronology. Ap. J. **216**, 548.
Tipler, F. J. (1988): Johann Mädler's resolution of Olbers' paradox. Q.J.R.A.S. **29**, 313.
Titius, J. D.: Betrachtung über die Natur, vom Herrn Karl Bonnet (An overview of nature, by Mr. Karl Bonnet), Leipzig: Johann Friedrich Junius 1766.
Tolman, R. C. (1930): On the estimation of distance in a curved universe with a non-static line element. Proc. Nat. Acad. Sci. **16**, 511.
Tolman, R. C.: Relativity, thermodynamics and cosmology. Oxford: At the Clarendon Press 1934.
Tolman, R. C. (1939): Static solutions of Einstein's field equations for spheres of fluid. Phys. Rev. **55**, 364.
Tomczyk, S., Schou, J., Thompson, M. J. (1995): Measurement of the rotation rate in the deep solar interior. Ap. J. (Letters) **448**, L57.

Tonry, J. L. (1987): A central black hole in M32. Ap. J. **322**, 632.
Tonry, J. L. (1991): Surface brightness fluctuations: A bridge from M31 to the Hubble constant. Ap. J. (Letters) **373**, L1.
Tonry, J. L., Ajhar, E. A., Luppino, G. A. (1989): Surface brightness fluctuations and the distance to the Virgo cluster. Ap. J. (Letters) **346**, L57.
Tonry, J. L., Davis, M. (1981): Velocity dispersions of elliptical and S0 galaxies II. Infall of the Local Group to Virgo. Ap. J. **246**, 680.
Tonry, J. L., Schneider, D. P. (1988): A new technique for measuring extragalactic distances. Astron. J. **96**, 807.
Townes, C. H.: The galactic center. In: The center of the galaxy. Proc IAU symposium 136 (ed. M. Morris). Boston: Kluwer Academic 1989.
Trefil, J.: The dark side of the universe. A scientist explores the mysteries of the cosmos. New York: Charles Scribner's Sons 1988.
Tremaine, S.: The centers of elliptical galaxies. In: Unsolved problems in astrophysics (eds. J. N. Bahcall and J. P. Ostriker). Princeton, New Jersey: Princeton University Press 1997.
Trimble, V. (1982): Supernovae. part 1 – the events. Rev. Mod. Phys. **54**, 1183.
Trimble, V. (1983): Supernovae. part 2 – the aftermath. Rev. Mod. Phys. **55**, 511.
Trimble, V. (1987): Existence and nature of dark matter in the universe. Ann. Rev. Astron. Astrophys. **25**, 425.
Trimble, V. (1988): 1987A – the greatest supernova since Kepler. Rev. Mod. Phys **60**, 859.
Trimble, V. (1991): The origin and abundances of the chemical elements revisited. Astron. Astrophys. Rev. **3**, 1.
Trimble, V., Greenstein, J. L. (1972): The Einstein redshift in white dwarfs III. Ap. J. **177**, 441.
Trinchieri, G., et al. (1994): Rosat pspc observations of NGC 4636 – interaction with Virgo gas? Ap. J. **428**, 555.
Truran, J. W., Cameron, A. G. W. (1971): Evolutionary models of nucleosynthesis in the Galaxy. Astrophys. Space Sci. **14**, 179.
Tsujimoto, T., Miyamoto, M., Yoshii, Y. (1998): The absolute magnitude of RR Lyrae stars derived from the Hipparcos catalogue. Ap. J. (Letters) **492**, L79.
Tucker, W., Tucker, K.: The dark matter – contemporary science's quest for the mass hidden in our universe. New York: William Morrow and Company 1988.
Tully, R. B. (1986): Alignment of clusters and galaxies on scales up to 0.1 c. Ap. J. **303**, 25.
Tully, R. B. (1987): More about clustering on a scale of 0.1 c. Ap. J. **323**, 1.
Tully, R. B.: Nearby galaxies catalog. New York: Cambridge University Press 1988.
Tully, R. B.: The Hubble constant. In: Astrophysical ages and dating methods (ed. E. Vangioni-Flam, M. Cassé, J. Audouze and J. Tran Thanh Van). Paris: Editions Frontieres 1990.
Tully, R. B. (1993): The Hubble constant. Proc. Nat. Acad. Sci. **90**, 4806.
Tully, R. B., Fisher, J. R. (1977): A new method of determining distances to galaxies. Astr. Ap. **54**, 661.
Turneaure, J. P., et al. (1983): Test of the principle of equivalence by a null gravitational redshift experiment. Phys. Rev. **D27**, 1705.
Turneaure, P., et al.: In: Proceedings of the fourth Marcel Grossman meeting on general relativity (ed. R. Ruffini). Amsterdam: Elsevier 1986.
Turner, E. L. (1976): Binary galaxies I. a well-defined statistical sample. Ap. J. **208**, 20.
Turner, E. L. (1976): Binary galaxies II. dynamics and mass-to-light ratios. Ap. J. **208**, 304.
Turner, M. S. (1991): Dark matter in the universe. Physica Scripta **T36**, 167.
Turner, M. S. (1993): Dark matter: Theoretical perspectives. Proc. Nat. Acad. Sci. **90**, 4827.
Turner, M. S. (1993): Why is the temperature of the universe 2.726 Kelvin? Science **262**, 861.
Turner, T. J., Pounds, K. A. (1989): The EXOSAT spectral survey of AGN. M.N.R.A.S. **240**, 833.
Turok, N. (1989): Global texture as the origin of cosmic structure. Phys. Rev. Lett. **63**, 2625.
Turok, N. (1991): Global texture as the origin of cosmic structure. Physica Scripta **T36**, 135.

Turok, N., Hawking, S. W. (1998): Open inflation, the four form and the cosmological constant. Physics Letters **B432**, 271.

Twarog, B. A., Anthony-Twarog, B. J. (1989): NGC 188, the age of the galactic disk and the evolution of the Li abundance. Astron. J. **97**, 759.

Tyson, J. A. (1988): Deep CCD survey: Galaxy luminosity and color evolution. Astron. J. **96**, 1.

Tyson, J. A., Giffard, R. P. (1978): Gravitational-wave astronomy. Ann. Rev. Astron. Astrophys. **16**, 521.

Tyson, J. A., Seitzer, P., Weymann, R. J., Foltz, C. (1986): Deep CCD images of 2345 + 007: Lensing by dark matter. Astron. J. **91**, 1274.

Tyson, J. A., Valdes, F., Wenk, R. A. (1990): Detection of systematic gravitaional lens galaxy image alignments: Mapping dark matter in galaxy clusters. Ap. J. (Letters) **349**, L1.

Tytler, D., Fan, X.-M., Buries, S. (1996): Cosmological baryon density derived from the deuterium abundance at redshift $z = 3.57$. Nature **381**, 207.

Udalski, A., et al. (1993): The optical gravitational lensing experiment. Discovery of the first candidate microlensing event in the direction of the galactic bulge. Acta Astronomica **43**, 289.

Ulrich, M.-H., Maraschi, L., Urry, C. M. (1997): Variability of active galactic nuclei. Ann. Rev. Astron. Astrophys. **35**, 445.

Unsöld, A., Baschek, B.: The new cosmos – fourth completely revised edition. New York: Springer-Verlag 1991.

Unwin, S. C., et al. (1983): Superluminal motion in the quasar 3C 345. Ap. J. **271**, 536.

Urry, C. M.: X-ray timing of active galactic nuclei. In: Active galactic nuclei. Lecture notes in physics 307 (eds. H. R. Miller and P. J. Wiita). New York: Springer Verlag 1988.

Uson, J. M., Bagri, D. S., Cornwell, T. J. (1991): Radio detections of neutral hydrogen at redshift $z = 3.4$. Phys. Rev. Lett. **67**, 3328.

Van Albada, T. S., et al. (1985): Distribution of dark matter in the spiral galaxy NGC 3198. Ap. J. **295**, 305.

Van Dalen, A., Schaefer, R. K. (1992): Structure formation in a universe with cold plus hot dark matter. Ap. J. **398**, 33.

Van Den Bergh, S. (1961): A preliminary luminosity classification for galaxies of type Sb. Ap. J. **131**, 558.

Van Den Bergh, S. (1988): The galactic luminosity, rotational velocity, nova and supernova rates. Comm. Astrophys. **12**, 131.

Van Den Bergh, S. (1989): The cosmic distance scale. Astron. Ap. Rev. **1**, 111.

Van Den Bergh, S. (1992): The age and size of the universe. Science **258**, 421.

Van Den Bergh, S. (1992): The Hubble parameter. P.A.S.P. **104**, 861.

Van Den Bergh, S. (1994): The Hubble parameter revisited. P.A.S.P. **106**, 1113.

Van Den Bergh, S. (1995): A new method for the determination of the Hubble parameter. Ap. J. (Letters) **453**, L55.

Van Den Bergh, S., Mc Clure, R. D., Evans, R. (1987): The supernova rate in Shapley-Ames galaxies. Ap. J. **323**, 44.

Van Den Bergh, S., Pritchet, C. J. (1986): Novae as distance indicators. P.A.S.P. **98**, 110.

Van Den Bergh, S., Pritchet, C. J. (eds.): The extragalactic distance scale. San Francisco: Astronomical Society of the Pacific 1988.

Van Der Kruit, P. C. (1986): Surface photometry of edge-on spiral galaxies V. The distribution of luminosity in the disk of the Galaxy derived from Pioneer 10 background experiment. Astron. Astrophys. **157**, 230.

Van Der Kruit, P. C., Allen, R. J. (1978): The kinematics of spiral and irregular galaxies. Ann. Rev. Astron. Astrophys. **16**, 103.

Van Der Laan, H., Perola, G. C. (1969): Aspects of radio galaxy evolution. Astron. Astrophys. **3**, 468.

Van Der Marel, R. P. (1991): The velocity dispersion anisotropy and mass-to-light ratio of elliptical galaxies. M.N.R.A.S. **253**, 710.

Subject Index

D

E

S